THE NEW NATURA

A SURVEY OF BRITISH NATURAL HISTORY

GULLS

*

The aim of this series is to interest the general reader in the wildlife of Britain by recapturing the enquiring spirit of the old naturalists. The editors believe that the natural pride of the British public in the native flora and fauna, to which must be added concern for their conservation, is best fostered by maintaining a high standard of accuracy combined with clarity of exposition in presenting the results of modern scientific research.

THE NEW NATURALIST LIBRARY

GULLS

JOHN C. COULSON

This edition published in 2019 by William Collins,
an imprint of HarperCollins Publishers

HarperCollins Publishers
1 London Bridge Street
London SE1 9GF

WilliamCollinsBooks.com

First published 2019

A CIP catalogue record for this book is available
from the British Library.

Set in FF Nexus

Edited and designed by
D & N Publishing
Baydon, Wiltshire

Printed in China by RR Donnelley APS

Hardback
ISBN 978-0-00-820142-5

Paperback
ISBN 978-0-00-820143-2

Contents

Editors' Preface vii
Author's Foreword and Acknowledgements ix

1 An Overview of Gulls 1

2 Black-headed Gull 43

3 Mediterranean Gull 91

4 Common Gull (Mew Gull) 103

5 European Herring Gull 121

6 Lesser Black-backed Gull 179

7 Great Black-backed Gull 209

8 Black-legged Kittiwake 231

9 Yellow-legged Gull 295

10 Little Gull 301

11 Rare Gulls in Britain and Ireland 309

12 Methods Used to Study Gulls 355

13 Urban Gulls 373

14 Conservation, Management and Exploitation of Gulls 415

Appendices 439
Select Bibliography and Further Reading 447
Species Index 459
General Index 466

Editors' Preface

Gulls are a family of distinctive and easily recognised birds that are familiar to us all, whether we live close to the sea or many miles inland, and whether we live in urban or rural surroundings. Long gone are the days when the sight of flocks of gulls overhead or in nearby fields was taken as a sure and ominous sign of storms at sea. For most of us today, birdwatchers or not, gulls are genuinely everyday birds, as common a sight as Woodpigeons (*Columba palumbus*) or the various members of the crow family. Other than in immature plumages (which may cause problems), the handful of commoner species are readily identifiable to those with any interest.

Seeking an expert author for a new book on gulls was a relatively easy task for the editors. Dr John Coulson was, until his retirement, Reader in Animal Ecology at the University of Durham, and has a lifetime's experience of research into many aspects of gull (and other seabird) biology and ecology, with a particular focus on the Black-legged Kittiwake (*Rissa tridactyla*). He is an expert of world renown, and recipient in 1992 of the British Ornithologists' Union's highest accolade, the Godman-Salvin Medal, and in 1993 of the Waterbird Society's Robert Cushman Murphy Prize.

A fulsome initial chapter introduces the gull family as a whole, including a succinct assessment of the rather complex current (worldwide) status of gull taxonomy. Things have changed dramatically since the eighteenth century, when Carl Linnaeus first established the single genus *Larus*, containing all the then known species! After this, the reader is treated to nine chapters – each in effect a treatise – on the regular British and Irish gulls, including the latest newcomer, the rapidly increasing and very elegant Mediterranean Gull (*Ichthyaetus melanocephalus*). These are followed by chapters on the rarer species; methods used in studying gulls; urban problems; and conservation, management and exploitation.

In common with many components of our fauna and flora, times have changed for gulls, and not always for the better – as the author describes. They may not universally be seen as slender-winged, elegant seabirds, and to some –

especially those living in or visiting towns or cities with rooftop gull colonies – they are raucously noisy neighbours. Similarly, many seaside promenades are blighted by the menacing presence of gulls with unblinking eyes seeking scraps, often boldly. Since the late twentieth century, even conservation bodies seeking to encourage nesting terns or other seabirds have resorted to culling in an attempt to limit predatory gull activity. In the last few decades, however, Herring Gull numbers, boosted in the second half of the twentieth century by the ready feeding opportunities offered by poorly covered landfill sites, have fallen to the extent that they themselves have become of conservation concern.

Such problems, as well as the many fascinating but less nefarious aspects of our British and Irish gull populations, are dealt with in substantial detail in John Coulson's admirably comprehensive text. This is a most noteworthy addition to the New Naturalist Library.

Author's Foreword and Acknowledgements

My first recollection of gulls was when I was about eight years old, when a maritime pilot on the Tyne told me that the gulls that frequently followed commercial ships entering the busy river were the reincarnations of past pilots, and were keeping a supportive eye on the steering of current pilots. An individual gull the pilots recognised because it called frequently was said to be the reincarnation of pilot 'Clagger' Purvis, who apparently had had as much to say for himself in life. Even at that age I found this story hard to believe, but the pilot was right about the frequency of gulls following both large and small vessels into the river.

My interest in birds, and particularly in gulls, developed during my school days. As a member of a small group mentored by Fred Grey, a master at South Shields High School for Boys and a person with a lifelong passion for birds, I learned much about the identification of birds during mid-morning breaks and on field trips. Through his guidance, my questioning eventually expanded from 'What is it?' to 'What is it doing?' I was also very much influenced by the Reverend Edward Armstrong's book *Bird Display and Behaviour* (1942), David Lack's *The Life of the Robin* (1943) and, in particular, Sir Arthur Landsborough Thomson's *Bird Migration* (1936), which highlighted some of the information that could be gained by ringing birds.

While still in the sixth form at school, I wrote to Elsie Leach, who ran the national bird-ringing scheme in an honorary capacity for the British Trust for Ornithology (BTO) from a small office and a cupboard in the Natural History Museum in London, asking her how to become a ringer. I received a prompt response requesting a letter of support from an acknowledged ornithologist that I could identify birds. I obtained this from a local doctor; he had an excellent knowledge of the birds of northern Norway and I had previously been able to draw his attention to uncommon birds visiting the vicinity, which he would stop off to see while making slight detours en route to his house calls.

Having sent the recommendation to Miss Leach, I received a reply within a few days simply asking what numbers and sizes of rings I wanted. At that time, 1948, there was no training whatsoever for potential ringers, nor an age limit. Every ring used had to be recorded on a thin cardboard sheet, which took only six ring entries and was guillotined into separate strips if the ringed bird was recovered. Ringing was on a very much smaller scale in those days! I remained an active ringer for 62 years, and have seen many changes to the British ringing scheme, including the intensive training now required for those who wish to ring. I also received one of the first mist nests in the country, and had to work out for myself how to use it – all I was told was to hang it between two poles!

One day in early June 1950, Edward White (a school and university friend) and I walked along the coast to where Black-legged Kittiwakes (*Rissa tridactyla*, here referred to simply as Kittiwakes) were nesting on a stack called Marsden Rock and on the nearby mainland cliffs. As we walked on the beach below the colonies, we noticed that about a third of the nests on the stack contained chicks, while only one out of a hundred nests on the mainland cliffs had chicks. Two weeks later, we excluded predation as the cause of this difference, as almost all nests now contained chicks, even though those in nests on the mainland cliff tended to be smaller than those on the stack.

This brief observation started a chain of interest and speculation. Why was there this difference, which was obviously not due to predation? Our attention was drawn to Frank Fraser Darling's book *Bird Flocks and the Breeding Cycle* (1938). The local public library managed to borrow a copy from another library and we were able to read about the ecologist's idea of social stimulation, which he believed affected the timing and success of breeding in colonial birds. Some authors suggested that the differences Fraser Darling had observed in small colonies of large gulls were due to young birds breeding later than old individuals, and that more young birds were breeding in some colonies than in others. However, we failed to find any information on the effect of age on the breeding biology of gulls and very little on birds of any species.

Edward and I were both undergraduates at Durham University, and we planned to make twice-weekly visits the following year to the Kittiwake colonies at Marsden to count the numbers of nests containing chicks (counting when eggs were laid was not possible, as most nests could not be viewed from above). The difference we had observed the previous year was repeated, and we presented our first results at one of the delightfully informal annual student bird conferences held at St Hugh's College in Oxford each January, organised by David Lack. This resulted in the opportunity to meet and engage with many knowledgeable researchers, including Niko Tinbergen, Charles Elton, Arthur Cain, Mike and

Esther Cullen, 'Mick' Southern, Reg Moreau, David Snow, Lance Richdale and, of course, David Lack and his students.

The problem of whether the differences we found between colonies were an effect of age or social conditions required knowledge of the breeding biology of Kittiwakes of different ages and, in turn, this needed marked individuals whose ages were known. The nests on the limestone cliffs at Marsden were inaccessible, but by chance, while crossing the river Tyne on a ferry two years previously, I had noticed a few pairs of Kittiwakes nesting on the window ledges of a riverside warehouse at North Shields. At Durham University, the head of zoology, Professor J. B. Cragg, was kind enough to take an interest in our study and in 1953 he obtained access for us to the building from the owners, Smith's Dock Co. The company actively protected the Kittiwakes and the birds were the delight of the two employees working there. We were able to reach the nests from inside the warehouse, allowing us to capture and individually mark the breeding adults and chicks. Fate was very kind to us, because the windows on the ledges used by the Kittiwakes opened inwards, and we were able to snare adults without pushing off or even disturbing the nests and their contents. Our main tool was a 60 cm length of wire, which had been used to bind a crate of oranges. We shaped one end into a crook that could be passed out of a narrow gap when the window was ajar and slipped around the leg of the adult. The Kittiwakes were not alarmed by the wire, and indeed were often curious and pecked at it. Pairs on neighbouring windowsills did not react when an individual suddenly disappeared into the 'cliff' to be colour-ringed. Within a few years we were able to measure the timing of laying in individuals of different ages.

The study eventually spanned more than 35 years, although there was never any guarantee that we would be able to continue it beyond a year or two at a time. At that time, long-term studies on birds were virtually unknown, the exception being that undertaken by Huijbert Kluijver on Great Tits (*Parus major*) in the Netherlands (which began in 1912), and it was many years before the research councils in the UK appreciated their considerable value and supported some of them.

The opportunity arose for both Edward and me to study for a PhD, but independently we came to the same opinion that a doctorate in ornithology was unlikely to lead to employment. There were only three people at that time professionally employed as ornithologists in Britain, and one of these held a post of reader in entomology! Only two universities employed ornithologists (Oxford and Aberdeen), and there was one ornithologist who worked for the Ministry of Agriculture and Fisheries. An eminent Scottish professor of zoology at that time openly stated that the study of birds and butterflies was not part of zoology, and ecology was only just starting to develop in Oxford, under the influence of Charles Elton, David Lack and Alister Hardy.

This situation was in marked contrast to the present day, when many of the now much more numerous universities have one or usually more staff researching in ornithology and ecology, while currently Natural England, the BTO, the Game and Wildlife Conservation Trust, Scottish Natural Heritage and the Royal Society for the Protection of Birds (RSPB) all employ professional ornithologists. Because of the situation in 1953, Edward and I both decided to work on entomological topics for our higher degrees, and studying Kittiwakes remained a happy hobby.

Our cooperative studies of Kittiwakes had to cease after we both completed our PhD studies. Edward went to work at a college in Freetown, Sierra Leone, and then moved to New Zealand. I went into school teaching but continued the Kittiwake studies in my spare time. Eventually, I obtained a post-doctoral fellowship from the Royal Commission for the Exhibition of 1851, and this allowed me to develop a full-time research programme of my own choice – the study of both Kittiwakes and moorland insects.

Two years later, I was offered a lectureship in zoology at Durham University, and in the following years I supervised a series of students working on seabirds and also on moorland invertebrates. Those studying seabirds included Jeff Brazendale, Brian Springett, Dick Potts, Andy Hodges, Ron Wooler, Jean Horobin, Callum Thomas, Jennifer Butterfield, Julie Porter, John Chardine, Fiona Dixon, David Jackson, Nicholas Aebischer and Jackie Fairweather, many of whom worked with me using the individually marked Kittiwakes nesting on the warehouse at North Shields from 1961 to 1990. Étienne Danchin visited from France and contributed much to Kittiwake studies during his post-doctoral year at Durham. In addition, Tom Pearson made a comprehensive study of the feeding biology of seabirds, including Kittiwakes, European Herring Gulls (*Larus argentatus*, here referred to simply as Herring Gulls) and Lesser Black-backed Gulls (*L. fuscus*) breeding on the Farne Islands, Northumberland.

Dick Potts started a PhD study on Shags (*Phalocrocorax aristotelis*) in 1962, and this took us to the Isle of May in Scotland's Firth of Forth. There, we met those managing the Isle of May Bird Observatory, who were concerned about the lack of use their accommodation and facilities received during the summer. As a result, they encouraged me to start studies on the tens of thousands of Herring Gulls nesting on the island. First, Jasper Parsons, and then Neil Duncan and George Chabrzyk, developed studies on the large gull colony, while the contribution started by Margaret Emmerson was sadly curtailed by her untimely death. We continued studies on Herring Gulls for 11 years, until the major and 'experimental' reduction of the large gulls on the island that was carried out by the Nature Conservancy Council. Elsewhere in Britain, Niko Tinbergen had researchers working on gulls at Ravenglass and the Isle of Walney in Cumbria, and more recently several others have studied large gulls nesting on islands in south Wales.

Herring Gulls started to invade and breed in coastal towns in north-east England in the 1960s, and Pat Monaghan was the first to study these and others feeding at landfill sites in detail. We mapped the distribution and rapidly increasing numbers of large gull species nesting in urban areas throughout Britain and Ireland in 1977, building on Stanley Cramp's initial study. Twenty years later, Susan Raven and I repeated the national census and found both a dramatic spread and increases in numbers.

Investigations on Lesser Black-backed Gulls breeding on Tarnbrook Fell were started by Neil Duncan, and later developed further by Mark O'Connor and Nick Royle, with the aim of establishing non-lethal methods to reduce the extent of the large gull colony there. From 1999 to date, my wife, Becky, and I have studied and counted Herring Gulls nesting in South Shields, Sunderland, Newcastle upon Tyne and Berwick-upon-Tweed in north-east England, and Lesser Black-backed Gulls in Dumfries in south-west Scotland.

In 1992, a new Kittiwake colony was established on the low cliffs of Coquet Island, Northumberland. With the goodwill of the RSPB, Becky and I were able to ring every chick reared there for the first 15 years of the colony's existence, and so were able to identify the huge immigration of adults reared in other colonies as the Coquet colony rapidly grew in size.

The Natural Environmental Research Council supported my research on gulls in collaboration with their Unit of Virology, which was concerned with the possibility of viruses being carried and distributed by the birds. This involved post-doctoral studies by Neil Duncan, Callum Thomas and Jennifer Butterfield, aided by Nicholas Aebischer, David Jackson and Sarah Wanless. The range of the study was expanded from the north of England to include Scotland, with Pat Monaghan working from Glasgow University, where she had recently been appointed as a lecturer. This work subsequently led to a joint study by Susan Greig, Pat Monaghan and myself on gull feeding methods at landfill sites, which relied on video recordings and then frame-by-frame play-backs of the feeding activity of individually marked gulls of known sex and age during their mass feeding frenzies. Collaborative work was developed with the Public Health Laboratory Service over several years on the microorganisms carried by gulls (particularly *Salmonella*, *Campylobacter* and *Cryptosporidium*). Later, Gabriella MacKinnon and David Baines carried out studies on wintering Black-headed Gulls (*Chroicocephalus ridibundus*) and Common Gulls (*Larus canus*).

I have also had considerable exchange of ideas with gull researchers from other countries, in particular Rudi Drent and Arie Spaans in the Netherlands, and Ian Nisbet, Michael Gochfeld and Joanna Burger in the United States. In addition, I have had the pleasure of meeting very many others at international ornithological congresses.

My studies on gulls developed rapidly at Durham with the cooperation of a large series of dedicated and able researchers, well supported by a number of technicians, particularly Eric Henderson, Michael Bone and John Richardson. In addition, we received welcome financial support at different times from the Department of Scientific and Industrial Research, the Science Research Council and the Natural Environment Research Council, supporting facilities supplied by Durham University itself.

Ornithology, including the study of seabirds, has expanded considerably since 1950, and the worldwide output of publications on aspects of the biology of gulls has consequently increased at a remarkable rate. Many, but by no means all, ornithological publications are listed annually in Zoological Record, and it is remarkable how the number of scientific papers included there has increased 15-fold since 1950. Journals such as Ibis, Bird Study, Ardea, Auk and British Birds, along with several journals published in North America, have greatly increased the annual number of pages dedicated to ornithology, while many new journals have also been started. Having been an editor of three ornithological journals over a period of 40 years, I have been particularly aware of this change and, incidentally, the increasing number of papers with multiple authors. In part, this results from more cooperative studies, but perhaps also from the increasing demands (and needs) of students to publish with tenured members of staff so their work can be included in research assessments. In the 1950s and 1960s, most scientific papers in ornithology had single authorship, with a minority having two authors. Recently, however, there has been a change – for example, in a recently published issue of 24 papers in one symposium, the contributions averaged more than four authors per paper, two having 12 authors and one 13, while only two were written by a single author.

The numbers of ornithological papers published in ecological and behavioural journals has also increased dramatically. These, together with information on gulls published elsewhere, such as annual county reports and those of conservation bodies, theses, records, online reports and books, have increased almost beyond belief, to such an extent that a complete bibliography for this book would have been taken up more than fifty pages and the main text would have been littered with multiple references to data sources. In 2011, my monograph on the Kittiwake required 16 pages of bibliography for this one species alone. In dealing currently with the Herring Gull, the references consulted exceeded a thousand items, and since this book covers the information on all gulls on the British list, I found that including a complete bibliography was a major problem. Providing references in books is not new problem, however, and was one faced by both James Fisher when writing *The Fulmar* (New Naturalist Monograph 6) in 1952 and Niko Tinbergen in his *Herring Gull's World* in 1953.

This book has had the main objective of bringing together and digesting information on the gull species that occur in Britain and Ireland, and is primarily aimed at readers who are interested in gulls but not necessarily in researching their biology in depth. In discussing this problem with the editors and others, I have been encouraged to reduce the number of references reported in the text appreciably, and instead present a Select Bibliography and Further Reading section at the end of the book (p. 447), and listing select key studies mentioned in the text, along with other informative publications. Readers wishing for more information should be able to obtain further details from the extensive sources available on the Internet. The extent of published (and unpublished) information on gulls is now so extensive that I have had to be selective about the information presented for several species, and I have been able to include information and analyses not previously published or available to most readers.

This book is not meant as an identification guide, although some of the salient features of each species and their geographical distribution are briefly mentioned. Those wishing to identify gulls should use one of the excellent field guides available, as listed in the Select Bibliography (p. 447). The world distributions of all gull species considered here have been fully described in standard identification texts, including the gull sections in volume 3 of the *Birds of the Western Palaearctic* and volume 3 of the *Birds of the World*.

It has not been possible to write this book without encountering areas of controversy, which include taxonomy, conservation, and the ability to identify some closely related species and subspecies in the field. Ornithology has thrived on controversy in the past and it will continue to do so in the future. In many cases, disagreement has led to the development of new methods of study and more in-depth investigations. In writing this book, debate has reared its head about where a species is placed in red or yellow categories of conservation concern based on a series of possible criteria, the existence of any one of which is enough to indicate that it is threatened. Some ornithologists feel that the system should be improved, and that a more critical scientific approach is needed to interpret data used to estimate the current risk of extinction of individual gull species.

For the same reason, there is no comment on climate change in this book. While there have been several claims that this has affected gulls, sound evidence in favour of such effects is currently poor and is not supported by evidence that adequately allows other possible causes to be excluded. Increasing studies have been made to estimate the potential risks to gulls from offshore wind farms and the rotating arms of the turbines. At present, most of these are based on informed speculation and such factors as the flight height of individual species and their numbers in the key areas. In the future, this problem will be investigated in more detail and the level of the perceived risk to gulls and other seabirds will be based on actual information, not just models of the situation, but the research is not yet at this stage.

In recent years, the common names of some gulls have lengthened, allegedly to avoid international confusion. Hence, the Black-legged Kittiwake, European Herring Gull, American Herring Gull and Yellow-legged Gull join the Lesser Black-backed Gull, Great Black-backed Gull, Black-headed Gull and Slender-billed Gull, which already have long names. Proposals have been made to change the Great Black-headed Gull to the shorter Pallas's Gull, while it has been suggested that the Common Gull is changed to the Mew Gull. In this book, as mentioned earlier, Herring Gull is used for the European Herring Gull and the American Herring Gull is referred to in full. Similarly, and as mentioned above, the Black-legged Kittiwake is referred to as the Kittiwake, while its sibling species, the Red-legged Kittiwake (which is infrequently mentioned) is written out in full. In most cases this follows the vernacular English names listed in the ninth edition of 'The Simple British List Based on a Checklist of Birds of Britain' (2018). In general, this edition retained names already familiar to most readers.

Statistical tests are important in evaluating differences in quantitative data, but to many the presentation of these and their outcomes are but an irritation. In general, I have commented on quantitative differences only when they have been shown to be statistically significant and so are likely to be real and meaningful, although I have not given details of the tests used.

ACKNOWLEDGEMENTS

Much of the information in this book derives from two sources. Some comes from professional research, but much is the result of the activities of amateur birdwatchers who spend their spare time visiting sites where they are likely to encounter unusual bird species or birds in exceptionally large numbers, and who then send the details to local recorders or contribute to national surveys and national censuses. Over the years, the number of observers has increased markedly and systems of notifying others of the presence of unusual birds have developed. Both categories of people studying birds have swelled progressively over the past 60 years, and the information and numbers of records have increased to a remarkable extent. In addition, the methods of identifying and confirming rare species have been increasingly supported by good-quality photographs. Accompanying these trends has been a dramatic increase in the numbers of gulls ringed both in Britain and Ireland, and also elsewhere in Europe. Capturing and ringing adult gulls has increased dramatically in recent years with the development of cannon nets and more frequent visits to landfill sites by teams of ringers. All of these additional efforts are appreciated, for without them, much of the information in this book would not have been available.

I would like to acknowledge the detailed contribution of Alan Dean, both for his detailed analysis of the records of gulls in the West Midlands in England over many years and for the photographs he (and others) have so willingly contributed.

The data used in writing this book has been collected over many years and it is possible here to identify only a small number of the hundreds of contributors. The many research students I supervised and advised have all made appreciable contributions over the years, and many have since made further contributions in ornithology and science in general while holding permanent posts both in the UK and North America. I was fortunate in having such an able set of students who contributed wholeheartedly and consistently to studies often made under difficult conditions. They all willingly volunteered to take part in teamwork as required, often at the most unsocial hours of the day and in adverse weather conditions, all while advancing their own studies.

I owe a great deal to the BTO, not only for the ringing scheme they manage so efficiently, but also for the extensive data they have collected. When I first had contact with the BTO in the early 1950s, it had a small office at the top of a set of outside stairs in a side street in Oxford, with a single salaried member of staff, Dr Bruce Campbell. Here, I acknowledge the trust's willingness to allow me to reproduce maps of ringing recoveries of gulls and to use the most recently available maps of the breeding season distribution of gulls in Britain and Ireland in 2007–11.

Many people have offered photographs for inclusion in the book, and these have all been credited in the captions. In particular, I thank Nicholas Aebischer, Pep Arcos, Rob Barrett, Colin Carter, Becky Coulson, Anthony Davison, Alan Dean, Andrew Easton, Phil Jones, John Kemp, Mark Leitch, Fred van Olphen, Daniel Oro, Mike Osborne, Viola Ross-Smith, Steven Seal, Charles Sharp, Michael Southcott, Brett Spencer, Norman Deans van Swelm, Thermos (fi.wikipedia), Dan

Turner and an anonymous photographer for their photographic help and willing permission to allow their excellent images to be included in this book.

The extensive data set on British seabirds now archived and maintained by the Joint Nature Conservancy Council and readily accessible online is a valuable asset and source of information. My thanks also go to Natural England for granting a freedom of information request concerning culls and licences issued to collect gull eggs.

Help and information have been supplied by many, in particular Nicholas Aebischer, David Baines, Robert Barrett, Peter Bell, Richard Bevan, Tim Birkhead, Bill Bourne, Joanna Burger, David Cabot, Kees Camphuysen, Colin Carter, Keith Clarkson, Ian Court, J. B. Cragg, Francis Daunt, Ian Deans, Greg Douglas, Andy Douse, Steve Dudley, Tim Dunn, George Dunnet, Andrew Easton, Julie Ellis, Mike Erwin, Sheila Frazer, Bob Furness, Michael Gochfeld, Thalassa Hamilton, Gill Hartley, Scott Hatch, Martin Heubeck, Grace Hickling, Keith Houston, Jon Jonsson, Heather Kyle, Susan Lindsay, Roddy Mavor, Jim Mills, Ian Nisbet, Daniel Oro, Ian Patterson, Ray Pierotti, Jean-Marc Pons, Julie Porter, Dick Potts, Richard Procter, Chris Redfern, Jim Reid, Sam Rider, Peter Rock, Robin Sellars, Peter Shield, Robert Swann, Mike Swindells, Martin Taylor, Mike Toms, Andrew Tongue, Mike Trees, Daniel Turner, Sarah Wanless, Matt Wood, Vero Wynne-Edwards, Bernard Zonfrillo and Jan Zorgdrager. Over the years, many others have helped in studies and investigations, and I appreciate the assistance of them all.

I owe a major debt of gratitude to my wife, Becky, for the many ways she has supported and helped my gull studies, solved many computing problems, and given help in preparing and checking the text of this book.

CHAPTER 1

An Overview of Gulls

Gulls are conspicuous web-footed, long-winged, medium or large seabirds that are readily recognised by the public. Adults are mainly white with shades of grey or black on the mantle and wings. Most species have black wing-tips, some have white 'mirrors' within the black areas, but a few species – mainly those restricted to an Arctic breeding distribution – have entirely white wing-tips. In the breeding season, adults of different species either have entirely pure white or very dark (black or brown) heads, and all revert to white heads in the autumn and winter, often with small grey marks behind the eye or grey streaking on the neck.

Gulls are widely known to the public because of their size and the habit of many species to frequent harbours, follow ships, visit landfill sites and visit outdoor areas also frequented by humans, such as seaside resorts, sports fields, beaches, rivers, picnic areas and large car parks at shopping complexes. In recent years, they have become even more familiar in parts of Europe and North America because their numbers have increased and several species have taken to nesting on buildings in urban areas. This habit of urban breeding has developed independently several times in different countries during the twentieth century and has spread rapidly. Urban nesting is now occurring in several species, and has almost certainly arisen through the marked increases in the size of gull populations, coupled with the increased protection given to them over the last century. The presence of gulls in urban areas has been given considerable adverse publicity, including reported cases of adults protecting their unfledged young by diving close to people's heads, or of gulls snatching food from unsuspecting members of the public. Such reports have resulted in gulls, and particularly the large species, acquiring pest status in certain areas.

EVOLUTION OF GULLS

The Charadriiformes constitutes a single large and distinctive lineage of modern birds, and includes waders, skuas, auks, terns, gulls and a few apparently aberrant species such as jacanas. Although the skuas appear to be similar to gulls, current evidence – including DNA studies – suggests that their ancestry may be nearer to the auks than to gulls.

The lightly built bones of birds associated with flight are fragile and therefore do not often produce good fossil remains. As a result, the evolution of present-day birds is poorly known, and much less so than that of reptiles, fish or mammals. Fossil remains attributable to gulls are particularly scarce. Many of those that have been found have not or cannot be attributed to the presently recognised genera, and certainly not to present-day species. More recently, several fossil bones initially attributed to gulls have been found to belong to other avian groups.

Fossil bones attributed to gulls and possibly members of the genus *Larus* have been reported both in Europe and the USA from deposits from the Middle Miocene, 20–15 million years ago. The relationship of these fossils to modern-day gulls is unclear; fossilised bird bones are often, and perhaps uncritically, given different specific names to those of currently existing species, ignoring the fact that the bones of a present-day species vary considerably in size according to sex and locality.

TAXONOMY OF GULLS

Initially, gull species were separated and identified on the basis of plumage, skeletal structure and size. In many species, specimens from different geographical areas held in museums and private collections were often named and given subspecies status based on minor differences in size and plumage, but all too frequently this relied on small numbers collected from only a few localities. Many of these named subspecies are still used today and, while the majority are probably justified, others that were described and named many years ago should be re-evaluated using modern techniques and larger samples. Some subspecies have already been rejected on this basis, and it is likely that others will not stand critical re-examination and will also be rejected.

In other cases, some existing subspecies have been promoted to the status of a full species, as has occurred recently within the Herring Gull complex in Europe and Asia. Still others may show only gradual changes in size, structure or plumage shades over their geographical range, a concept not recognised by early taxonomists until Julian Huxley applied the term clines to these groups in 1942. Such clines have already been demonstrated for the Black-legged Kittiwake (*Rissa tridactyla*), the Puffin (*Fratercula arctica*) and the Common Guillemot (*Uria aalge*) breeding in the North Atlantic. Questionable subspecies of gulls still exist, and some are discussed in more detail later in this chapter.

Initially, the eighteenth-century taxonomist Carl Linnaeus placed all gulls in the genus *Larus*, and most species remained there in what became a very large taxon. Eventually, the two species of kittiwake were removed from *Larus* and placed in the genus *Rissa*, while the Ivory Gull was moved to the genus *Pagophila* (*P. eburnea*), Sabine's Gull was transferred to the genus *Xema* (*X. sabini*), and Ross's Gull was placed in the genus *Rhodostethia* and then, more recently, to *Hydrocoloeus* (*H. rosea*), alongside the Little Gull (*H. minutus*). The Swallow-tailed Gull became the sole species in the genus *Creagrus* (*C. furcatus*). These separations were not unreservedly accepted, however, and as late as 1998, Philip Chu proposed returning all gulls to a single genus, *Larus*. Seven years later, the intensive study of the mitochondrial DNA of many gull species made by Jean-Marc Pons, Alexandre Hassanin and Pierre-Andre Crochet (2005) moved in the opposite direction and separated gulls into nine genera, and in so doing created the new genera *Chroicocephalus* (with 10 species worldwide), *Hydrocoloeus* (with two species) and *Saundersilarus* (comprising only Saunders' Gull, *S. saundersi*, in China). Worldwide, at least 24 gull species, especially those with white heads in the breeding season, are still retained in the large genus *Larus*. Aside from *Saundersilarus*, three other genera are composed of only one species: *Creagrus*, containing the Galapagos Islands' Swallow-tailed Gull; *Xema*, containing the High Arctic Sabine's Gull; and *Pagophila*, including the Ivory Gull, also breeding in the High Arctic. Like *Hydrocoloeus*, the genus *Rissa* also contains two species.

One of the major findings made during the in-depth investigation by Pons *et al.* (2005) was that the gulls that had dark heads as adults did not form a single taxonomic group, as had been suggested by studies made in the second half of the twentieth century, but were composed of three distinct groups of species. These groups were called the 'black-headed gulls' and placed in the genus *Ichthyaetus*, while 'hooded gulls' were separated into another new genus, *Leucophaeus*. The third group, including the Black-headed Gull, were called 'masked gulls' and were placed in the genus *Chroicocephalus*. To an extent, this separation of gulls with dark heads in the breeding season is supported by similar courtship behaviour within each group, as originally suggested by Niko Tinbergen and his co-workers in the 1950s and supported by more extensive recent studies.

GULL SPECIES WORLDWIDE

Currently, there are about 50 species of gulls in the world. This total has increased in recent years and will probably be increased further as improved molecular techniques are used to revise their status; even the definition of a species may be modified or revised. The uncertainty about the precise number of species reflects the fact that the gulls as a group present a taxonomic nightmare,

Speciation concepts

The decision as to whether and under what circumstances two populations of animals that occur in different geographical areas can be considered distinct species remains a taxonomic problem, because the level of genetic difference between the two that justifies specific status is often an arbitrary decision and one that is not always universally accepted.

One major taxonomic problem relates to the Herring Gull and Lesser Black-backed Gull complex of subspecies. In 1942, the evolutionary biologist Ernst Mayr suggested that the subspecies formed a chain around the northern hemisphere, starting with the Lesser Black-backed Gull in Europe, and then further subspecies occurring eastwards through Asia, each having progressively lighter-coloured wings and leading to the American Herring Gull in North America, and finally completing the ring with the Herring Gull of western Europe. The theory is that, by the time this chain of subspecies has spread eastwards around the northern hemisphere and the ends meet up again in Europe, the Lesser Black-backed Gull and the Herring Gull are obviously separate species and interbreed only very rarely. This beautiful explanation of a series of subspecies first spreading, then part of each becoming isolated, allowing the formation of further subspecies around the northern hemisphere and eventually producing two distinct species, was widely acclaimed and has been frequently quoted in books, scientific papers and lectures on genetics and speciation.

However, the recent development and application of mitochondrial DNA techniques has shown that the American Herring Gull is not the closest relative of the European Herring Gull as was previously thought, nor did it spread eastwards historically from North America to Europe to evolve into the European Herring Gull. That said, while the fascinating concept of a ring of gull subspecies spreading around the northern hemisphere and ending with two distinct species has been discredited, it may soon, with some minor modifications, become viable again. This is because the Lesser Black-backed Gull is currently spreading from Europe to North America via Iceland and Greenland, and is beginning to breed in Canada. As such, it is establishing new end points of the chain of subspecies, this time in North America and involving the American Herring Gull.

and this has resulted in years of confusion and disagreement. For example, the American Ornithologists' Union (AOU) considers the Herring Gull breeding in North America to be a subspecies of the European Herring Gull (*Larus argentatus smithsonianus*), while the British Ornithologists' Union (BOU) regards it as a separate species, *Larus smithsonianus*. There is still much confusion within the extensive Herring Gull and Lesser Black-backed Gull complex of species and subspecies, particularly those occurring in Asia (see box opposite).

The Iceland Gull (*Larus glaucoides*), which breeds in Greenland and parts of arctic Canada, has entirely white primaries and is a well-established species, but there is conflict over the status of two similar gulls, Thayer's and Kumlien's gulls, both of which show some black on the tips of the primaries. The AOU recognises Thayer's Gull as a distinct species (*L. thayeri*), but regards Kumlien's Gull as a subspecies of Thayer's Gull (*L. thayeri kumlieni*). Within Europe, there is considerable disagreement about the status of the three forms, with some national bodies (such as those in Ireland) agreeing with the AOU classification, and others (including the BOU) considering both Thayer's Gull and Kumlien's gull as subspecies of the Iceland Gull (*L. glaucoides thayeri* and *L. g. kumlieni*). Yet other bodies believe that they represent three distinct species. In this book and without strong conviction on the matter, I have treated Thayer's Gull as a distinct species but have followed the BOU and regarded Kumlien's Gull as a subspecies of the Iceland Gull. Fortunately, most individuals that visit Britain are typical Iceland Gulls and lack any black or brown on the wing-tips.

Elsewhere in the world, birds in the Kelp Gull and Dominican Gull complex (currently all known as *Larus dominicanus*) are similar and obviously related to the Great Black-backed Gull (*L. marinus*) of the North Atlantic. They are also a taxonomic problem and currently are regarded as consisting of five geographically separated subspecies. Just as some of the former subspecies of the Herring Gull have been recognised as distinct species, some of the *L. dominicanus* subspecies may also be elevated to species status when more intensive genetic and ecological investigations have been completed.

CURRENT GEOGRAPHICAL RANGES

Gulls breed on all continents, with Kelp Gulls extending their southern range into Antarctica and several gull species breeding in the High Arctic. The number of gull species breeding in each 10-degree zone of latitude varies considerably, with two peaks of abundance, one in each hemisphere (Fig. 1). In the northern hemisphere, the number of species peaks between 40°N and 60°N, and in the southern hemisphere, a smaller peak occurs between 20°S and 40°S. Few gull species breed in the tropics, Antarctica or the High Arctic regions. This variation in species abundance, particularly between the two hemispheres, correlates reasonably closely with the amount of land within each latitude zone. This may offer a partial explanation as to why appreciably fewer species of gulls are found and breed in the southern hemisphere.

Few individual gull species breed over a wide range, with about 80 per cent spread over less than 20 degrees of latitude, and very few breed in both hemispheres. These patterns differ markedly from the terns, where many species breed in both the northern and southern hemispheres. A comparison of the ratio of gull and tern species breeding in different latitude zones throughout the world is shown in Fig. 2. There are more gull than tern species in only the zones north of 40°N, which begs the question as to why fewer gull species occur in the other zones. Could this be the result of competition between gulls on the one hand, and with petrels and shearwaters in the southern hemisphere?

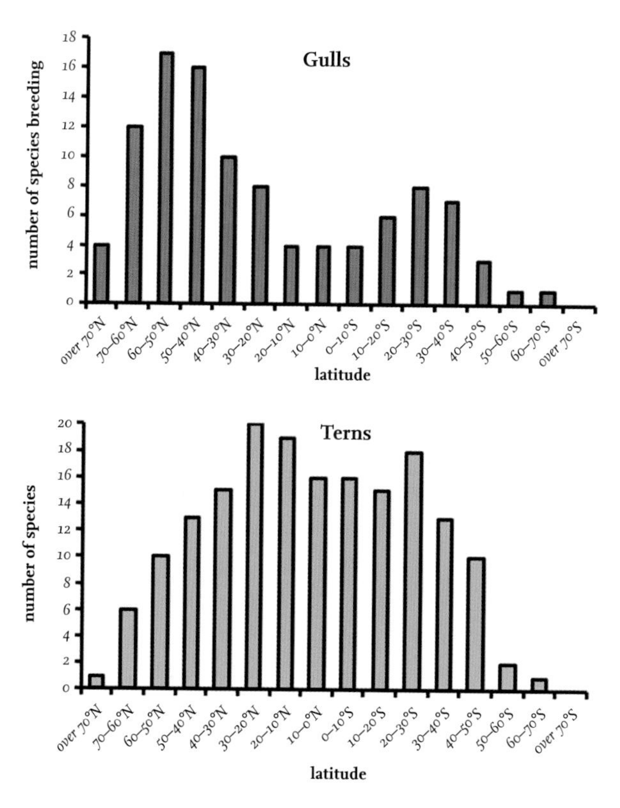

FIG 1. The number of gull species (upper graph) and tern species (lower graph) breeding throughout the world in each zone of 10-degrees latitude. The distribution of gull species is clearly bimodal, while that for tern species peaks in the tropical zone between 30°N and 30°S. Tern data from Cabot & Nisbet (2013).

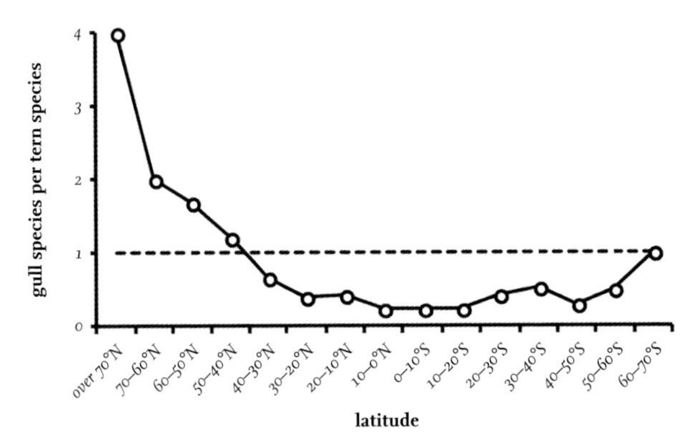

FIG 2. The ratio of the number of breeding gull species to the number of tern species in relation to zones of latitude. Gull species are more numerous than tern species only north of 40°N. The dashed line indicates equality.

The dominance of gulls over terns in temperate and arctic regions of the northern hemisphere is even greater when numbers of individuals are considered. For example, the average numbers of gulls per species breeding in Britain are very much greater than for terns. Using the figures from the national census in Britain and Ireland in 2000, there was a total of 1,810,000 breeding gulls (of seven species), but only 176,000 breeding terns (of five species), indicating a ratio of 10 gulls for every tern. Each gull species was represented by an average of seven times the average numbers for each tern species. Similar large differences are evident elsewhere in Europe and in North America.

The numerical dominance of gull species in the northern hemisphere suggests that much of their speciation occurred there. However, species of the genera *Larus, Leucophaeus* and *Chroicocephalus* breed in both the northern and southern hemispheres, indicating that in the past at least one species belonging to each genus must have spread, as breeding birds, across the equator on at least one occasion.

GULL HABITATS

The majority of gull species frequent coastal areas, marshes, rivers, estuaries and large inland lakes. Many occupy the same habitat zones used by marsh and sea terns, and in this respect they contrast markedly with shearwaters and petrels, which are pelagic. The smaller species often feed and breed inland, while the larger gulls breed mainly at coastal sites. Within the last hundred years, several species of large gulls have bred inland more frequently, a change in behaviour that has coincided with their overall increase in abundance.

Gulls breeding on the coast move only moderate distances from the shore. They are tied by relatively short incubation shifts and the need to feed their young frequently and regularly. In general, the density of gulls at sea tends to decline rapidly as the distance from shore increases, although the Kittiwake does not show this tendency in winter. Outside the breeding season, most gulls remain within daily flying distance of the shore, preferring to roost overnight on land or on sheltered coastal waters. The exception to this is when they are migrating. Only the two kittiwake species, Sabine's Gull and Ross's Gull, occur regularly in oceanic waters far from land throughout the long non-breeding season.

GULL SPECIES RECORDED IN BRITAIN AND IRELAND

The box (opposite) is the current list of 26 species recorded in Britain and Ireland as breeding species, regular visitors or occasional vagrants. The list represents about half of all gull species in the world. Kumlien's Gull is listed, but is retained as a subspecies of the Iceland Gull.

An approximate phylogenetic tree of the evolution of gull species recorded in Britain and Ireland (mainly based on the research by Pons and colleagues) is shown in Fig. 3 and involves eight genera. Such a representation can be only approximate, as their evolution has most likely been multi-dimensional and so cannot be presented accurately in two dimensions. There is still considerable uncertainty about the relationships between the species in the genus *Larus*, and no attempt is made in the order shown in Fig. 3 to indicate these, other than to suggest that the species with totally white wing-tips probably represent a distinct group.

Audouin's Gull, which lacks a black head at any time, is placed in the same genus (*Ichthyaetus*) as two black-headed species on the British list (Mediterranean Gull and Great Black-headed Gull), along with three other black-headed species that occur elsewhere in the world, so its inclusion is surprising. Similarly, the Slender-billed Gull, which has a white head in all seasons, is included with the dark-headed Black-headed and Bonaparte's gulls. However, Jean-Marc Pons in response to my query believes that 'the dark hood is not a good character to construct evolutionary relationships because it has repeatedly been lost during the evolution of gulls'. In addition, he confirms that there is additional evidence indicating that the Black-headed and Bonaparte's gulls should be included in the genera *Ichthyaetus* and *Chroicocephalus*, respectively.

The national censuses of gulls and other seabirds have been incredibly important and informative, and at last we have a sound knowledge of both the distribution and numbers of adults. However, we do not have a census value for

Gulls recorded in Britain and Ireland

Species	Status
Genus *Hydrocoloeus*	
Little Gull (*H. minutus*)	Regular visitor, very occasional breeder
Ross's Gull (*H. roseus*)	Vagrant
Genus *Xema*	
Sabine's Gull (*X. sabini*)	Regular visitor, usually in small numbers
Genus *Pagophila*	
Ivory Gull (*P. eburnea*)	Vagrant
Genus *Chroicocephalus*	
Slender-billed Gull (*C. genei*)	Rare vagrant
Bonaparte's Gull (*C. philadelphia*)	Vagrant
Black-headed Gull (*C. ridibundus*)	Abundant breeder and winter visitor
Genus *Larus*	
Common Gull (*L. canus*)	Common breeder and winter visitor
Ring-billed Gull (*L. delawarensis*)	Vagrant
Great Black-backed Gull (*L. marinus*)	Common breeder and winter visitor
Glaucous-winged Gull (*L. glaucescens*)	Vagrant
Glaucous Gull (*L. hyperboreus*)	Regular winter visitor
Iceland Gull (*L. glaucoides*)	Regular winter visitor
Kumlien's Gull (*L. glaucoides kumlieni*)	Vagrant subspecies
Thayer's Gull (*L. thayeri*)	Vagrant
European Herring Gull (*L. argentatus*)	Abundant breeder and winter visitor
American Herring Gull (*L. smithsonianus*)	Vagrant
Caspian Gull (*L. cachinnans*)	Currently vagrant but increasingly recorded
Yellow-legged Gull (*L. michahellis*)	Visitor and now an occasional breeder
Lesser Black-backed Gull (*L. fuscus*)	Abundant breeder
Slaty-backed Gull (*L. schistisagus*)	Vagrant
Genus *Ichthyaetus*	
Great Black-headed Gull or Pallas's Gull (*I. ichthyaetus*)	Vagrant
Mediterranean Gull (*I. melanocephalus*)	Rapidly increasing breeder
Audouin's Gull (*I. audouinii*)	Vagrant
Genus *Leucophaeus*	
Laughing Gull (*L. atricilla*)	Vagrant
Franklin's Gull (*L. pipixcan*)	Vagrant
Genus *Rissa*	
Black-legged Kittiwake (*R. tridactyla*)	Abundant breeder

numbers of immature individuals that have never bred for any of the gull species. Since immature birds may include up to five year classes (varying according to species), the numbers involved are appreciable and can be estimated only

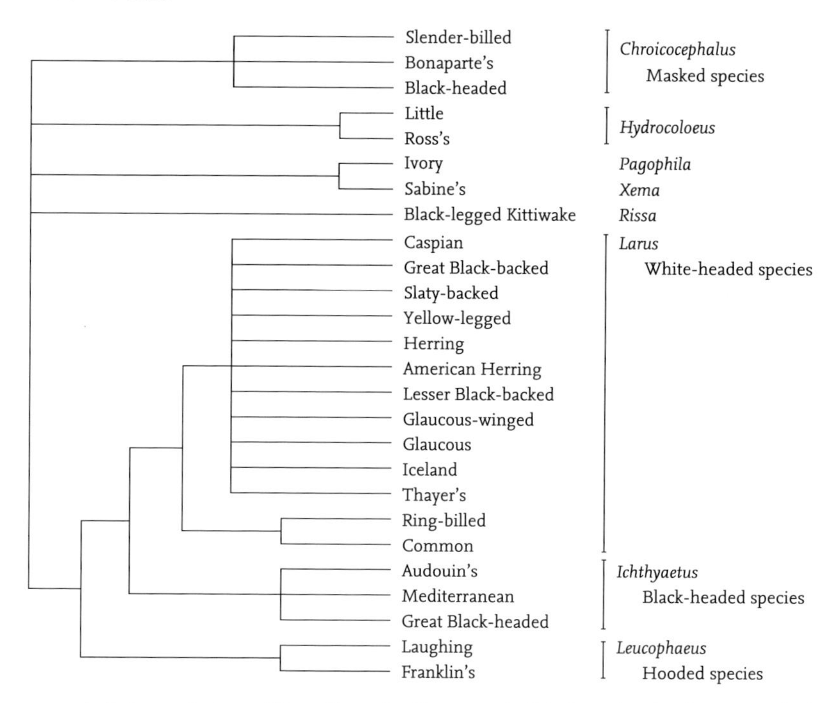

FIG 3. A phylogenetic tree of the gull species that occur in Britain and Ireland. In general, the shorter the line leading from each species, the more recently that species is presumed to have arisen. This does not apply to the large genus *Larus*, where more work is necessary to elucidate the affinities of the different species. This diagram includes all of the genera in the world, with the exception of the genus *Saundersilarus* (where Saunders's Gull, *S. saundersi*, is the only species) and the genus *Creagrus* (where the Swallow-tailed Gull, *C. furcatus*, is the only species), neither of which occur in Europe. The Vega Gull (*L. vegae*) has not been included, and its presence in Britain has yet to be confirmed, while Kumlein's Gull is regarded as a subspecies of the Iceland Gull. Based in part on the work of Pons *et al.* (2005).

from a life table formed from survival rates obtained from detailed marking studies. Table 1 gives rough estimates of the proportion of immature gulls of the six commonest species occurring in Britain and Ireland. The figures are approximations but indicate that, by autumn, there is a large proportion (probably about 40 per cent) of individuals of every species listed that have not yet matured and bred. The proportions of immature individuals will have decreased by early spring because the mortality rate of young birds in their first year of life is usually markedly higher than that of adults, but they will still form an appreciable minority of the numbers of each species.

TABLE 1. The approximate numbers of adult and immature individuals in early autumn and the total wintering of the six commonest gull species breeding in Britain and Ireland in 2000, and compared with the estimated world numbers.

Gull species	*Number of adults in early autumn in Britain and Ireland*	*Estimated number of immatures in early autumn in Britain and Ireland*	*Total of all ages in early autumn in Britain and Ireland*	*Percentage of immatures in early autumn in Britain and Ireland*	*World total of adults estimated by Bird Life International*	*British and Irish breeding adults as a percentage of worldwide adult population*	*Total winter numbers in Britain and Ireland*
Kittiwake	800,000	600,000	1.4 million	43%	10–12.5 million	6%	100
Herring Gull	300,000	250,000	550,000	45%	2.0–2.2 million**	15%	620,000
Black-headed Gull	280,000	220,000	500,000	44%	4.7–6.5 million	4–6%	1.64 million
Lesser Black-backed Gull	244,000	220,000	464,000	47%	700,000–830,000*	29–35%	200,000
Common Gull	100,000	75,000	175,000	43%	1.0–1.5 million	7–10%	700,000
Great Black-backed Gull	40,000	40,000	80,000	50%	440,000–470,000	9%	100,000
Total	1.7 million	1.4 million	3.2 million		18.8–24 million		About 3.3 million

* Depends upon species definitions. ** Restricted to European Herring Gull *sensu stricto*, and excluding the American Herring Gull (*Larus smithsonianus*).

Table 1 also shows estimates of numbers of the six most abundant gulls in Britain and Ireland in about 2000. Again, these figures are estimates, but they indicate that for all six species combined there are about 3.2 million gulls of all ages in Britain and Ireland in the early autumn. This total is probably about 3.3 million in winter, when the numbers of departing Kittiwakes are replaced by immigrant Black-headed, Herring and Common gulls. By late spring, the winter visitors have departed and Kittiwakes have returned to their colonies, maintaining numbers at about 2.6 million just before breeding begins.

SUBSPECIES IN WESTERN EUROPE

There are only a few western European gull species that are represented by more than one subspecies. As mentioned earlier in the chapter, the Herring Gull was previously split into several named subspecies, some of which have now been elevated to species (Yellow-legged Gull, Caspian Gull). Two subspecies occur in Britain and Ireland: *Larus argentatus argenteus*, which breeds here; and the larger, darker *L. a. argentatus*, which breeds in northern Scandinavia and north-west Russia, and winters in Britain and the North Sea region. The nominate Common Gull subspecies, *L. canus canus*, occurs widely in western Europe, while the larger, darker subspecies *L. c. heinei*, which breeds in Russia, probably (based on ringing data) occurs occasionally in Britain, but is difficult to identify. In the Atlantic, the Kittiwake is represented by a single subspecies (*Rissa tridactyla tridactyla*); individuals are progressively larger towards the north of its range, but this gradual change does not justify separate subspecies status and is described as a cline. There are probably several other clines among gulls that are yet to be recognised, including the Black-headed Gull, Glaucous Gull and Lesser Black-backed Gull.

SURVIVAL AND LONGEVITY

Gulls are medium- to long-lived birds, with an average expectation of adult life and number of breeding years of different species ranging between four and 12 or more years. Annual adult survival rates for different species vary between 80 per cent and 92 per cent, and often vary appreciably from year to year and over longer periods of time. The annual survival rate tends to be higher in the larger gulls, which also have a longer period of immaturity. This delay in reaching breeding age in gulls appears to be associated with the time that is necessary to acquire competence in obtaining food, but why this should be longer in the large species is not evident.

Ringed Herring Gulls that are nearly 35 years old have been recorded, but these represent a few extreme individuals comprising less than 1 per cent of those that reached maturity. In several gull species, the peak of mortality occurs during and just after the breeding season, when the adults are at their lowest weights during the year, suggesting that breeding is a significant stress. Data suggest that survival of gulls is usually high in winter, but the Kittiwake may be an exception, with most mortality occurring while the species is in its pelagic wintering range.

Less is known of the survival of immature gulls, but there is usually a lower survival rate in the 12 months following fledging, after which the survival rate approaches that of the adults.

The longevity records based on birds ringed as nestlings and living under natural conditions are given below, although several individuals are known to have lived longer in captivity.

Mediterranean Gull	22 years 1 month
Little Gull	20 years 11 months
Black-headed Gull	30 years 7 months
Common Gull	33 years 8 months
Lesser Black-backed Gull	34 years 10 months
European Herring Gull	34 years 9 months
Great Black-backed Gull	29 years 2 months
Black-legged Kittiwake	28 years 6 months
Ivory Gull	23 years 11 months
Laughing Gull	22 years 1 month
Ring-billed Gull	27 years 6 months
Glaucous-winged Gull	23 years 10 months

The species with the longest recorded lifespans are mainly those that have been ringed in large numbers, and therefore have more chance of an exceptional record. It should be kept in mind that these lifespans are reached by exceptional individuals – perhaps one in several thousand – and so it is very likely that the maximum known age of many gull species in the wild will increase in future years as more recoveries of marked birds accumulate.

SIZE DIFFERENCES BETWEEN SPECIES

Gull species vary considerably in size. The Little Gull is the smallest and weighs about 100 g (the weight of an Arctic Tern, *Sterna paradisaea*), while the largest is the Great

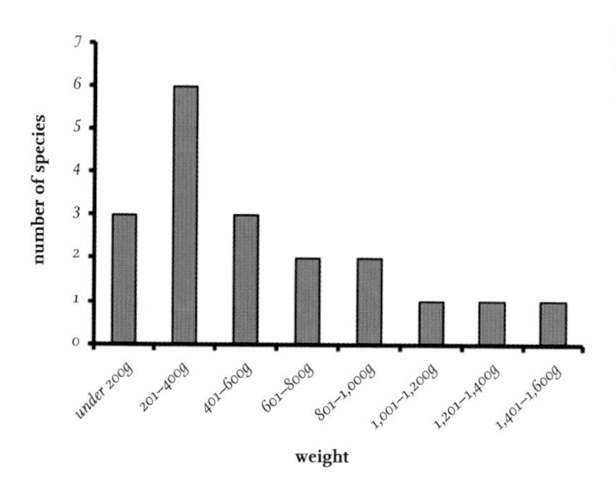

FIG 4. The average weight of the females of 19 gull species on the British list.

Black-backed Gull, with males averaging 1,800 g and some individuals exceeding 2,000 g. Fig. 4 shows the average weights of adult females of 19 gull species on the British list. The weights of females of nine of these species are less than 400 g on average and overlap with terns, of which the adult females of all except one species on the British breeding list weigh under 400 g. The distribution chart for male weights is similar, but is shifted to the right because of their slightly greater size.

INDIVIDUAL VARIATION

Like all animals, individuals of each gull species show variation in many characters, including size, colour and age at first breeding. Males are larger than females and tend to have a more substantial bill, and size within a species can also vary geographically.

Variation in the immature plumage is widespread in gulls of the same age and this is frequently overlooked in the field identification of species, particularly within the genus *Larus*. The occurrence of hybrid individuals further adds to plumage variation and typical examples of hybrids can often be identified in the field by experienced observers; the characteristics used often overlap with those of other species. Consequently a proportion of immature and even adult birds that are infrequently recorded in Britain and Ireland may fail to be identified because of potential confusion with other species.

Immature plumages

The plumage, leg and bill colours of recently fledged chicks are very different from those of their parents, to such an extent that, many years ago, a first-year

Kittiwake was claimed as a species new to science, despite the adult having already been described and named some years earlier. The first plumage of the young of most gull species is made up of feathers of varying shades of brown and grey, producing a cryptic pattern that helps to conceal them in vegetation in a colony and also appears to reduce aggression from adults. In the smaller gull species, the plumage is replaced by one that resembles the adult at the first annual moult; that is, when the bird is 13 months or so old. In the larger species, all feathers are replaced each year, but only some of the new ones resemble those of the adults and the full adult plumage pattern is not achieved until four years after hatching. These progressive changes in plumage at successive annual moults can vary between individuals and produce a series of different plumage patterns that make it a challenge to identify both the species and the age of immature individuals.

The slow and progressive acquisition of the adult plumage through successive moults contrasts with the rapid growth of bones and wing feathers, which reach full size within a few weeks after hatching because they are necessary before flight can be achieved. Why the acquisition of the adult plumage takes longer in the larger species of gull than in the smaller ones is not clear. The mechanism determining plumage patterns is obviously controlled by hormones and is linked with the greater length of immaturity in the larger gull species, but it is not evident why the large species delay reaching maturity for so long.

Differences between the sexes

The sexes of gulls have identical plumage and differ only in that females are usually smaller and tend to have slightly less substantial bills. Fig. 5 shows the relationship between the differences in weight of male and female gulls of seven species, using data from different parts of their geographical ranges where available. The extent of the difference between sexes is not constant from species to species, but increases with the weight of the species, ranging from a 5 per cent increase in the Black-headed Gull to more than 20–25 per cent in large species.

Standard measurements of wing length also tend to be longer in males than in females, but the magnitude of the difference is much smaller, ranging from 1 per cent in some small species to 6 per cent in the largest species (Fig. 6). Even when this difference is converted to wing area, it still results in the wing loading being higher in the large gulls, which explains why these species typically have a more laboured flight, with a slower, more powerful wing-beat. The small gull species, which are of similar weight to many tern species, have a characteristic buoyant flight similar to that of terns.

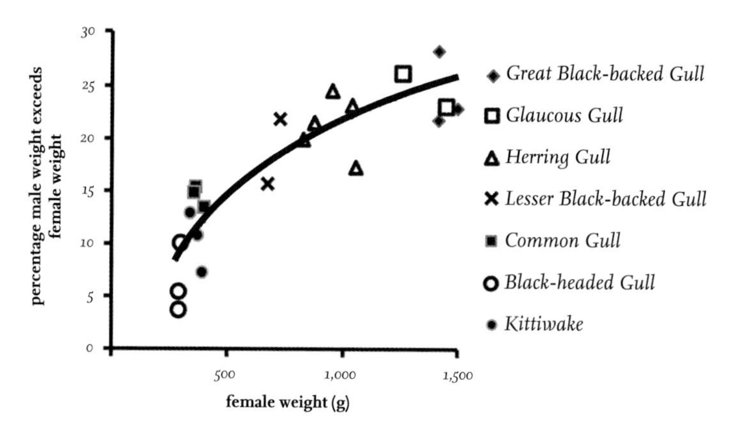

FIG 5. The percentage by which male gulls of several species are heavier than females, based on data for seven well-studied species. It is evident that there is a much greater difference between the size of males and females in the larger species of gulls.

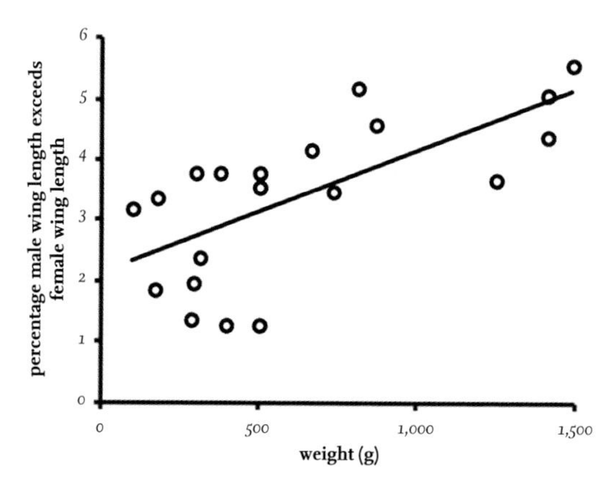

FIG 6. The relationship between adult weight and the extent to which the male has a longer wing than females. The percentage difference in wing length between the sexes increases in heavier species, but is much less than the difference of body weight shown in Fig. 5.

The reason why there is a greater size difference between the sexes in large gull species is not known, and currently it is possible only to speculate. Perhaps there is a greater need in the large species to reduce competition for food between the sexes, or perhaps the dimorphism is related to the greater need for males of large species to defend nesting territories. The reader might speculate further, bearing in mind that in the skuas, females are invariably larger than males, while male terns are only 1–3 per cent heavier than females.

Despite the average size differences, there is an overlap in the range of sizes of male and female gulls. Niko Tinbergen claimed that, despite the overlap in size between the sexes of Herring Gulls, invariably the male is larger in every pair.

Sexing gulls

As male and female gulls have identical plumage features, distinguishing them in the field can be very difficult. The most reliable way to determine the sex of an individual bird – without killing it and then dissecting it to examine the gonads – is by carrying out a DNA analysis on samples obtained from feathers or blood. While this method is highly efficient, it is time consuming and it is expensive when large numbers of birds are being studied. In the field, biometric measurements taken while a bird is temporarily captured for ringing can also be used for sexing the individual. I found that the best measure was the head and bill length (from the back of the head to the bill tip), which also had the advantage of showing the highest degree of consistency when measured by different people. Further, the proportionate difference in head and bill length between the sexes is almost twice that for wing length (for example, 9.6 per cent compared to 5 per cent in the Great Black-backed Gull). The only disadvantage of this measure is that in some museum specimens part of the back of the skull was removed during preparation, which prevents it being used in these cases. As shown in Table 2, the head and bill measurement is satisfactory in sexing 92–98 per cent of individuals of several gull species. Including two other body measurements (wing length and bill depth) in a discriminant analysis increased the accuracy of sexing only by less than 2 per cent points.

TABLE 2. The head and bill lengths of adults of six species of gulls in Britain, the measure separating the sexes and the proportions sexed correctly by this single measurement. The data are based on samples of at least 80 individuals of each sex breeding in Britain, except for Common Gulls (*Larus canus*), which were captured in winter and so were from unknown breeding areas. Based on Coulson *et al.* (1983a) and additional data.

	Average female length (mm)	*Average male length (mm)*	*Separation value; males greater than (mm)*	*Percentage of females sexed correctly*	*Percentage of males sexed correctly*	*Percentage difference between sexes*
Black-headed Gull	79	84	81	96%	94%	6.1%
Common Gull	85	92	89	95%	94%	7.9%
Lesser Black-backed Gull	109	119	113	98%	95%	8.8%
Herring Gull	112	123	118	98%	95%	9.3%
Great Black-backed Gull	135	149	142	97%	98%	9.9%
Black-legged Kittiwake	86	91	88	92%	94%	5.6%

continued

Sexing gulls *continued*

When a group of breeding gulls is being studied, the behaviour of marked individuals can be used as a reliable method of sexing. Copulation is totally reliable in this respect, as is courtship feeding of the female by the male and intensive food begging by the female. More details on the methodology used to sex gulls are given in Chapter 12.

Because of the average difference in size between the sexes, by chance the male will be larger than the female in many pairs, and more so in the larger gull species, but I have not found evidence that the male is invariably larger than the female. Size, and particularly the size of the bill, may play a part in individual birds recognising the sex of other gulls, but it is more likely that behaviour – particularly during courtship – plays the major role in sex recognition in gulls, especially in smaller species.

Adult plumages

There is considerable variation in the shade of grey on the wing and mantle in adult gulls of the same species, which is evident in birds nesting in the same colony. This is illustrated in Fig. 7, which shows the extent of such variation in Herring Gulls breeding on the Isle of May in Scotland (subspecies *Larus argentatus argenteus*) and in northern Norway (subspecies *L. a. argentatus*). Because of the variation, there is overlap in wing shades between the two subspecies of Herring Gulls and most, but not all, individuals can be identified on this basis alone (see also box on p. 17). Even using more measurements of body size does not completely separate all *argenteus* males from *argentatus* females.

In Lesser Black-backed Gulls breeding in the Netherlands (Fig. 7), there is also considerable variation in wing shade, with the palest approaching the darkest shade of Herring Gulls breeding in northern Norway. The darkest shade reported in Lesser Black-backs in the Netherlands is said to fall within the shade range of the subspecies *Larus fuscus fuscus*, which breeds in eastern Scandinavia and typically has a black mantle and wings very similar to those of the Great Black-backed Gull. The majority of Lesser Black-backed Gulls breeding in the Netherlands have a range of shades found in both the subspecies *L. f. intermedius* (breeding in north-west Europe) and *L. f. graellsii* (breeding in Britain).

Identification of the Lesser Black-backed Gull subspecies *intermedius* and *graellsii* in the field is further complicated by whether the individual is seen in bright sunlight or under dull conditions, and also by the direction of the light, all of which affect the apparent shade of grey of the same individual recorded in photographs or observed in the field. Reliable records of shade need to be measured with the bird

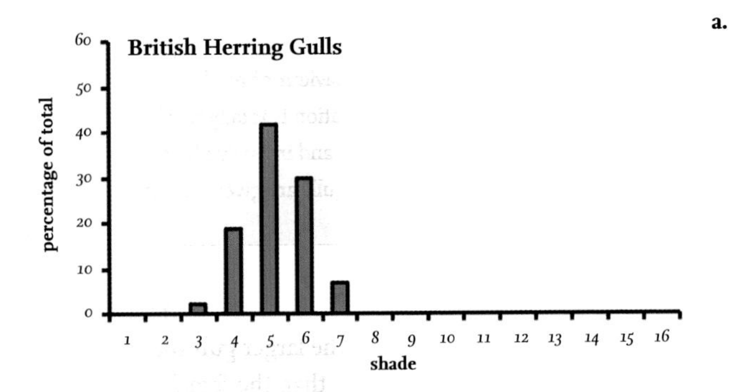

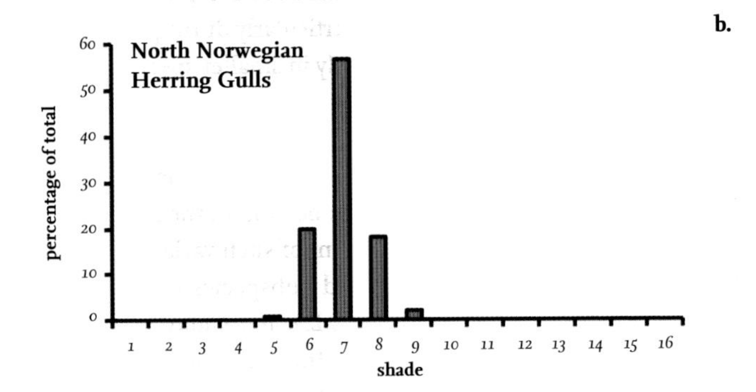

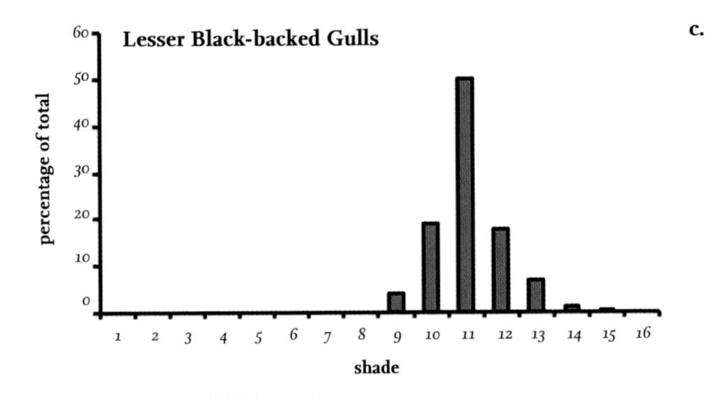

FIG 7. The shade of the wings of adult (a) Herring Gulls (*Larus argentatus*) breeding in Britain (n = 1,591) and (b) northern Norway (n = 140), and (c) Lesser Black-backed Gulls (*L. fuscus*) breeding in the Netherlands (n = 899). The shades increase in darkness from left to right and correspond approximately to shades of grey, which range from 1 (white) to 20 (black). The data for Lesser Black-backed Gulls are taken from Muusse *et al.* (2011).

in the hand, using standard lighting and comparing it against a reliable shade chart, but even this would not identify the subspecies of all individuals.

Differences in wing pigmentation

A major source of plumage variation within gull species is the pattern of black or brown pigmentation on the wings of immature individuals. These show progressive changes at each annual moult, until adult plumage is eventually achieved. In addition, this patterning varies appreciably between individuals of the same age, even within a single colony. For example, fully adult Herring Gulls in the same colony showed a range in the number of primaries that are tipped with black pigment (Table 3), and they also showed variation in the extent of white on the tip of the longest (10th) primary. Part of this variation is linked to the age of the birds (Table 4), with change continuing for several years after individuals reach breeding age, but the patterning is not related to gender.

TABLE 3. The percentage of fully adult Herring Gulls (*Larus argentatus*) examined in breeding colonies with different numbers of black-tipped outer primaries. Dutch data from Muusse *et al.* (2011), Norwegian data mainly from Barth (1975).

Location	*Number of birds*	*Percentage with four black tips*	*Percentage with five black tips*	*Percentage with six black tips*	*Percentage with seven black tips*
Isle of May	461	2%	15%	78%	5%
Scarborough	140	5%	20%	70%	5%
Netherlands	100	0%	17%	73%	9%
South Norway	91	1%	41%	56%	1%
North Norway	269	3%	66%	30%	1%

TABLE 4. The wing-tip pattern in Herring Gulls (*Larus argentatus*) of known age breeding in colonies in Britain. The extent of white on the tip of the 10th primary tends to increase with age.

Age (years)	*Number of birds*	*Percentage for which 10th primary has white tip at least 30 mm long*	*Percentage for which six or more primaries have black markings*
3	17	0%	100%
4	40	15%	87%
5–7	54	30%	83%
8–15	58	36%	84%
Adults of unknown ages	416	25%	86%

Part of the variation is also geographical, as seen in adult Herring Gulls breeding on the east and west sides of England and Scotland (Table 5). This difference between east and west in the black-and-white patterning on the primaries is maintained in the winter, presumably because the north–south dispersive movements of British Herring Gulls mainly follow either the eastern or the western coastlines, with relatively few individuals crossing between the two coasts. Figs 8 and 9 illustrate the variation in two adult Herring Gulls captured in north-east England, with one showing the *thayeri*-type pattern on the ninth primary (second from left), where the black does not spread across the whole of the width of the feather.

TABLE 5. Comparison of wing-tip patterns of adult British Herring Gulls (*Larus argentatus*) breeding on the east and west coasts of England and Scotland. The differences in both characters are significant.

Location	*Number of birds*	*Percentage for which 10th primary has white tip at least 30 mm long*	*Percentage for which six or more primaries have black markings*
West coast	502	10%	89%
East coast	717	28%	73%

FIG 8. The wing-tip pattern of an adult Herring Gull (*Larus argentatus*) captured in north-east England, showing what is known as the *thayeri*-type pattern on the ninth primary. The bird was subsequently found breeding in northern Norway. (John Coulson)

FIG 9. A typical wing-tip pattern of a Herring Gull (*Larus argentatus*) breeding in Scotland, with the 10th primary still growing. (John Coulson)

Differences caused by hybrids

One definition of a biological species is that the individuals can form a group of interbreeding or potentially interbreeding organisms that produce viable offspring. If hybrids occur between two species, such as in the classical case of a horse and a donkey, the hybrid offspring of the two (a mule or hinny) are usually sterile, probably because the two parent species have different numbers of chromosomes. In the case of gulls, hybrids are not uncommon and are reported far more frequently than in major taxa, such as terns. Studies have revealed that *Larus* gulls have the same number of chromosomes (72) and as a result, hybrids are usually fertile and they have been reported breeding successfully. There are now numerous records of gulls of different species and even different genera pairing and rearing hybrid offspring, as listed below:

Mediterranean Gull × Black-headed Gull
Herring Gull × Lesser Black-backed Gull
Herring Gull × Yellow-legged Gull
Lesser Black-backed Gull × Yellow-legged Gull
Western Gull (*Larus occidentalis*) × Glaucous-winged Gull
Great Black-backed Gull × American Herring Gull
Herring Gull × Glaucous Gull
Glaucous-winged Gull × Glaucous Gull
American Herring Gull × Kelp Gull
Iceland Gull × Thayer's Gull
Common Gull × Ring-billed Gull
Mediterranean Gull × Common Gull
Laughing Gull × Black-headed Gull
Laughing Gull × Ring-billed Gull
Herring Gull × Caspian Gull

In most cases, adults that are believed to be hybrids have been recognised by the intermediate nature of their plumage and the colouring of their legs and bill, but in only a very few instances has the plumage been described for adults that are known to be hybrids and were ringed as such before they fledged. It is usually believed that hybrid gulls show intermediate characters of their parents in terms of plumage, bill colour and leg colour, but this is not always the case, and in several instances they display minor characteristics not evident in either parent.

There is little doubt that some hybrids can share similarities with, and resemble, other gull species. As a result, it is sometimes difficult to accept a new sight record of a species from a geographical area where it has not been previously or convincingly been recorded before, and to confirm that it is not a hybrid between species that breed nearby. Rarity committees have a particularly difficult job with gulls, and ideally need DNA samples obtained from feathers of the presumed rarity to be certain of the record.

Some hybrid gulls, when adult, have been known to pair and mate with an individual of one of their parent species, producing offspring known as back-crosses. Even less is known about the plumage of these offspring, but it is likely that they differ both from the original species and from the hybrid parent. Breeding between pairs of hybrid gulls has not been recorded. However, hybridisation and subsequent breeding is likely to produce at least three different types of individuals, all of which vary in some respect from the original parent species as well as from each other. The immature plumages of hybrid gulls are poorly known and in many cases their origin has been assumed only because of their intermediate characteristics.

Eventually, after several generations of breeding, a particular gene can be transferred via the offspring of a hybrid from one of the parent species to the other. This has been recorded in the American Herring Gull, which appears to have acquired a gene from the Great Black-backed Gull in North America, presumably through hybrids between the two species. To date, this gene has not been recorded in the European Herring Gull.

In Belgium and the Netherlands, mixed pairs of Yellow-legged Gulls and either Lesser Black-backed Gulls or Herring Gulls have occurred particularly frequently. For example, more than 15 mixed pairs were reported in Rotterdam annually from 1986 to 1998 (van Swelm, 1998) and more in more recent years, and others have been frequently identified in at least five other colonies in the Netherlands and Belgium. Hybrid individuals that have reached adulthood and that are presumed to be crosses between Yellow-legged and Lesser Black-backed gulls have also been recorded in Belgium breeding with Lesser Black-backed Gulls, producing back-crosses.

Inter-species breeding is more frequent when one of the gull species is rare and spreading into the main range of the other. For example, when Lesser Black-backed Gulls first started to breed in the Netherlands in the 1930s, a few individuals joined large colonies of Herring Gulls and several mixed breeding pairs were recorded. Despite the fact that both species are now numerous and breed in the same colonies, hybrid pairs still occur, although they are infrequent. Very few pairings between these two species have been reported in Britain, except when experimentally induced (see below).

When Herring Gulls spread to Iceland in the 1920s, individuals formed mixed pairs with Glaucous Gulls, and by 1966 about half of the adults there were considered to be hybrids. These were distinctive in showing small but variable amounts of dark pigment on the tips of the primaries (Ingolfsson, 1970).

When Mediterranean Gulls first started to breed in Britain, early pioneers frequently paired with Black-headed Gulls (as they have done so elsewhere). In fact, this is ongoing, as a few individuals continue to spread north from the south coast of England. The recent arrival of a few adult Yellow-legged Gulls in Britain has seen them join both Herring Gull and Lesser Black-backed Gull colonies. Again, they have formed mixed pairs that on some occasions have managed to fledge hybrid chicks. Perhaps this inter-species breeding occurs because individuals arriving in new areas are mainly of the same sex and fail to find a mate of the same species.

In a study carried out on the island of Skokholm in south-west Wales, Mike Harris (1970) switched large numbers of eggs between Lesser Black-backed Gull and Herring Gull nests. The chicks that subsequently hatched imprinted on their foster parents and apparently considered that they were the same species, so that when they matured they chose a mate of that species, forming a series of mixed-species pairs. The young produced and reared by these mixed pairs were hybrids between the parent species and usually (but not always) showed plumage and leg colour intermediate between the two. At least 40 of these hybrids later returned to breed on Skokholm and on nearby Skomer, and most paired with adult Herring or Lesser Black-backed gulls. The chicks they produced were back-crosses and, when adult, were more similar to one of the parent species than the first generation of hybrids. While some of these hybrids reared chicks, it is not known whether they and their offspring were less viable. However, as each generation was produced, presumably both parent species incorporated small amounts of the genetic material belonging to the other species into their make-up despite appearing to be 'pure' Herring or Lesser Black-backed gulls (as discussed above for the American Herring and Great Black-backed gulls in North America).

BREEDING

Gulls are monogamous, although a few cases of male Kittiwakes breeding simultaneously with two females at different nest sites have been recorded. Pairs of gulls produce only one brood each breeding season, but if their eggs are lost, many will lay a replacement clutch. While most gulls breed annually during a well-defined breeding season, some individuals skip breeding for a year. The exception is the Swallow-tailed Gull on the Galapagos Islands, which does not have a clear-cut breeding season and nests throughout the year, with individuals breeding at nine- to 10-month intervals.

Breeding sites

Gulls typically favour bare ground and areas with short vegetation for nesting, or floating vegetation on lakes or marshes. The main exceptions are Bonaparte's Gull, which regularly nests in trees; Common, Black-headed and American Herring Gulls, which occasionally nest in low trees at a small number of localities; Kittiwakes, which favour narrow ledges on steep sea cliffs; and Herring, Glaucous and Ivory gulls, which sometimes use larger cliff ledges.

Ground nesting makes gulls particularly susceptible to mammalian predators, and most species nest only at sites where these predators are usually unable to reach the colonies, such as small islands or isolated peninsulas. Gulls vary in their ability to deter avian predators. Adults will attack birds of prey and corvids, but in parts of northern Scandinavia White-tailed Eagles (*Haliaeetus albicilla*) are having an increasing impact on breeding gulls – this is a future risk for Britain, since the species has been reintroduced here and its numbers are increasing. In addition, adult gulls at breeding sites suffer occasional predation from Peregrine Falcons (*Falco peregrinus*). Ravens and crows are a problem for some smaller gulls, but in general they are attacked and prevented from entering dense gull colonies. Individual Herring, Lesser Black-backed and Great Black-backed gulls, as well as Great Skuas (*Stercorarius skua*), have developed the ability to reach and prey on eggs and young gulls at otherwise well-protected nesting sites (sometimes even attacking their own species).

Mammalian predators such as Red Foxes (*Vulpes vulpes*) and Badgers (*Meles meles*) have now reached some gull colonies in Britain after being absent for many years, and these and the spread of American Mink (*Neovison vison*) has often resulted in the sudden desertion of sites used by breeding gulls. This desertion may be immediate, while in other cases the decreases in numbers of adults are spread over several years, apparently because new recruits to the colony are deterred by the presence and activity of predators. In particular, Foxes have become very much more abundant in Britain in recent years, and have captured and killed many incubating gulls at night. Six gull species now also nest on buildings in urban areas (p. 373); these have the same characteristics as natural sites, in that mammalian predators cannot normally reach them and they are generally given public protection.

Humans entering gull colonies are not usually attacked by the smaller gulls, which instead tend to fly overhead giving alarm calls. Large gulls do frequently dive at human intruders, however, usually from behind. While they pass closely overhead, they seldom actually strike. I have been struck only once by a large gull, although one of my students was knocked to the ground by a particularly aggressive Lesser Black-backed Gull defending its unfledged chicks.

Colony and nest site fidelity

Adults gulls are highly site faithful, provided that the nesting site remains safe and is not subject to high levels of predation on eggs, young or adults. Males – particularly those that were successful in rearing young in the previous breeding season – tend to return to the same nesting sites. In contrast, young birds are much less likely to return to the place where they were reared as chicks. A proportion – usually dominated by males – does so, and these birds often return to the same part of a large colony in which they were reared. This behaviour is called philopatry, the extent of which is influenced by many factors, including competition for nesting sites and food availability. In the past, the proportion of birds moving elsewhere has often been underestimated because of the much greater difficulty in locating those that have moved. In some species, the majority of the young that survive to breed move to other colonies, with some moving 100 km or more away. In several species, young individuals have been recorded visiting a number of colonies while approaching maturity, including their natal colony, only then to move and breed elsewhere. Such movements are, of course, necessary to form new colonies.

Colonial breeding

Colonial breeding is widespread among gulls, and only a few species regularly breed both as isolated pairs and in colonies. Some gull colonies are composed of mixed species, and the smaller species frequently nest in or alongside colonies of terns. In many gull species, it appears that single pairs cannot breed in isolation and the presence of a group of gulls of the same or even different species is necessary before egg-laying is possible, thus making colonial breeding essential. The main exceptions are Common and Great Black-backed gulls, where a proportion of breeding pairs nest in isolation from others. It is obvious that colonial nesting is not forced upon gull species as a result of shortage of suitable nesting sites, and it is usually regarded that there is an advantage to breeding close together. An obvious reason is that it improves defence of eggs and unfledged young against predation, but while this is evident in deterring Ravens (*Corvus corax*) and crows, it is less evident that colonial nesting prevents mammalian predators from raiding colonies and consuming eggs, young and even adults.

The reliance upon a group to ensure breeding suggests that stimulation from other individuals is necessary. This was first suggested in 1938 by Frank Fraser Darling, who noted that the display from neighbours within the colony stimulated pairs to breed and in larger groups or colonies to lay eggs earlier, and that it resulted in greater synchrony with neighbouring pairs. He suggested that the effect of this synchrony was that eggs or chicks were available over a shorter time period, and therefore fewer were predated and breeding success was enhanced. At the time, Fraser Darling's idea appealed to some, but others were critical of the concept. Unfortunately, his original data on Herring and Lesser Black-backed gulls only hinted at this effect, and others later showed that the variations he found between colonies of different sizes could have been produced by chance and were not statistically significant.

In the 1950s, Edward White and I collected information on the timing of breeding in a series of colonies of Kittiwakes. We found that the laying season in large colonies was more spread out than in small ones, the opposite of Fraser Darling's contention. While some pairs in the larger colonies started laying earlier than those in the smaller ones, pairs of late-breeding birds (which we now know were mainly young, first-time breeders) were recorded in all the colonies studied.

However, we did discover that breeding in smaller groups of 20–50 Kittiwake pairs within a single colony was more synchronous than in the colony as a whole, and that the average breeding date of each of these groups was earlier as the density of pairs increased. This finding suggested that social stimulation was occurring at the more local level, among groups of breeding pairs, and not in the colony as a whole.

I know of no single pair of Kittiwakes that has been recorded breeding in isolation. As a result, it is difficult for members of the species to form new colonies because this requires a group of individuals to be attracted to, and display at, a new potential breeding site before egg-laying can occur. Up to 20 pairs will collect at the site of a potential new colony for a year or more, and before a few of the pairs succeed in building a nest and laying eggs. However, in subsequent years the numbers breeding increase rapidly. In contrast, Herring Gulls sometimes nest in isolation, but these are birds that have bred before. Young Herring Gulls attempting to breed for the first time do not seem to be able to breed without joining a group already breeding together.

Colonial nesting allows gull species to benefit from the stimulation of their neighbours engaging in courtship activity, and probably encourages earlier nesting in some individuals. This was clearly evident in Kittiwakes when observing a pair reuniting at the colony after one member had been away on a fishing trip (Coulson & Dixon, 1979; Coulson, 2001). The birds engaged in a vocal display of calling, which immediately stimulated similar mutual calling in neighbouring pairs. Where Kittiwake nests were denser, the frequency of this stimulation from neighbours was more frequent. Anyone who has visited a Kittiwake colony will recognise the loud, periodic outburst of calling by groups of pairs as a major feature of the pre-laying period.

New colonies of Kittiwakes and Herring Gulls are highly attractive to potential breeders, and as a result they increase rapidly in size. Part of this effect is a result of the low density of breeding birds in a new colony, with relatively large areas available where recruits can settle without aggression from neighbours. As colonies grow, the area available for new recruits within, and at the immediate edge, decreases relative to their size, and their rate of growth decreases over time. Whether this eventually determines the maximum size for colonies of some species is unclear, but remarkably large colonies are known, numbering tens of thousands of pairs. In contrast, Great Black-backed Gull colonies rarely exceed 200 pairs in Britain (although much larger colonies are known in North America), while large colonies of Common Gulls do occur but are exceptional.

Mixed colonies of Herring Gulls and Lesser Black-backed Gulls are common, but the two species show a tendency to select slightly different areas within the colony. On natural sites, they often nest side by side, but the Herring Gull more frequently nests on open ground with little or short vegetation, while the Lesser Black-backed Gull shows a partial preference for areas with taller vegetation. At urban sites, both species frequently nest side by side and there is no obvious or consistent difference in the sites they select. Great Black-backed Gulls often nest in isolation or in small numbers of pairs in colonies of Herring Gulls or Lesser Black-backed Gulls, where they frequently select sites on raised ground.

Terns often nest in association with Black-headed Gulls, with the latter arriving and laying earlier and taking the first choice of sites. However, this is not always the case – the first Black-headed Gulls to nest on Coquet Island in Northumberland did so alongside Sandwich Terns (*Thalasseus sandvicensis*), and only arrived after the terns had laid.

Age at first breeding and mate fidelity

Within a species, the age at first breeding is variable and can span three or more years. Smaller gulls tend to breed when they are two or three years old, while many of the large gulls do not normally start breeding until they are four or five years old. This long period of immaturity often allows individuals to visit colonies for one or more years before they breed themselves.

The claim that gulls pair for life is not justified. While the same individuals frequently re-form as a pair in successive years, 'divorce' often occurs, with both individuals taking new partners. Change of mate also occurs when one partner dies.

Nests

In general, gulls' nests are often substantial structures composed of plant material collected by the adults nearby or, in some cases, picked up either from the sea surface, on the tideline or from cliff tops. Gulls that breed on marshes and ponds probably start nest-building earlier than those on drier sites, and they produce a platform above the water level on which most of the displaying and courtship takes place. In situations where the water level rises after rain, there is often an attempt by the pair of gulls to build up the height of the nest to keep the eggs dry. However, this is effective only when relatively small increases in the water level occur.

On several occasions I have seen Kittiwakes flying along the coast carrying nest material as far as 15 km from the nearest colony. This contrasts with Herring Gulls, which usually collect plant material for the nest from within their territory, although if it is not available, such as on roofs of buildings, material is collected from further away. Having collected material, adults shape a central hollow in the centre of the nest by compressing the material with their breasts. This hollow is large enough to hold the usual two or three eggs of the clutch together and prevents the eggs from rolling away by accident.

While nests of most gulls are formed by an accumulation of plant material, the exception is the Kittiwake. The adults collect marine algae and mud to form a base for the nest, and as this dries, it becomes firmly cemented to the rock surface, forming a base on which the rest of the nest is then built. The nest is completed by bringing in grass to form the top. This species is exceptional in that the adults at the nest defecate onto and over the edge of the structure, leaving obvious white streaks on both the nest and the cliff face below, and thus clearly make no effort to disguise the nest. Other gull species move away from the nest to defecate. The Kittiwake also differs from other gulls in that nest-building is often highly social, with a hundred or more individuals repeatedly flying back and forth between a particular source of mud and vegetation and the nest sites. On one occasion when in spring a local farmer spread straw and dung onto a cliff-top field, within a week the nearby Kittiwake colony was entirely draped with long strands of golden straw.

Recently fledged young and immature one- and two-year-old Kittiwakes often pick up potential nest material from the tideline and fly about with it in their beaks, but eventually drop it without taking it into a colony. I have not seen immature individuals of other gull species doing this. Non-breeding three-year-old Kittiwakes often occupy sites in a colony and bring in small quantities of nesting material very late in the breeding season, when most pairs of Kittiwakes have half- or full-grown chicks. In some cases, this forms a base for a nest in the following year. The chicks of most gulls leave the nest when a few days old, seeking cover and protection, but the Kittiwake is again exceptional in that the chicks remain on the nest until they fledge five or more weeks after hatching.

Depending on the position of the colony, some Kittiwake nests are washed off in winter storms. However, those in more sheltered positions remain over winter and until the following breeding season when they are used as the base of new nests. One nest on the Farne Islands was supported in a vertical cleft in the rock face, and as material accumulated year after year, it reached a height of more than 100 cm.

Eggs

All gull eggs have a brown or grey background with numerous spots or streaks of darker pigment, offering a degree of camouflage if they are left uncovered in the nest. The extent to which camouflage prevents or reduces predation of gull eggs has not been investigated, but it can only be one of several factors affecting successful breeding.

The size of egg laid by gulls varies with the average weight of each species, but not in a simple linear manner (Fig. 10). At one extreme, the Little Gull lays an egg that, on average, weighs 18.6 per cent of the normal body weight of females, while the egg of the largest species, the Great Black-backed Gull, weighs only 7.3 per cent of the female's body weight. The change in egg size as a proportion of the adult weight is a common effect in many bird groups and not just a peculiarity of gulls. Large eggs, irrespective of the species laying them, usually require a longer

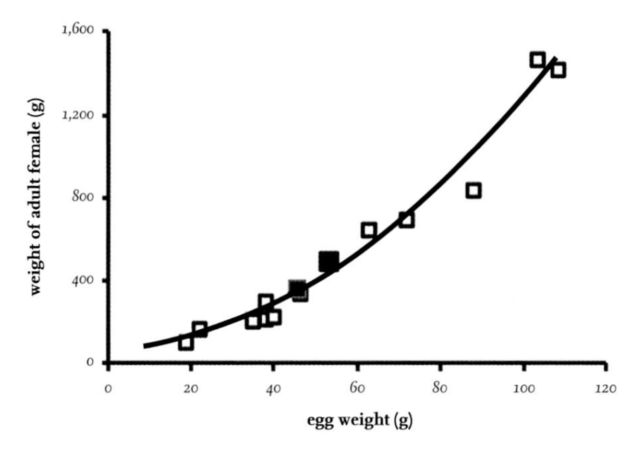

FIG 10. The relationship between the average weight of adult females of 13 gull species and the average weight of the eggs they lay. The gull species that normally lay two rather than three eggs (Kittiwake, *Rissa tridactyla*, indicated by the lower of the two filled squares; and the Ivory Gull, *Pagophila eburnea*, indicated by the upper filled square) sit exactly on the curve calculated for the other species, which normally lay three-egg clutches, indicating that despite having smaller clutches they do not lay bigger eggs.

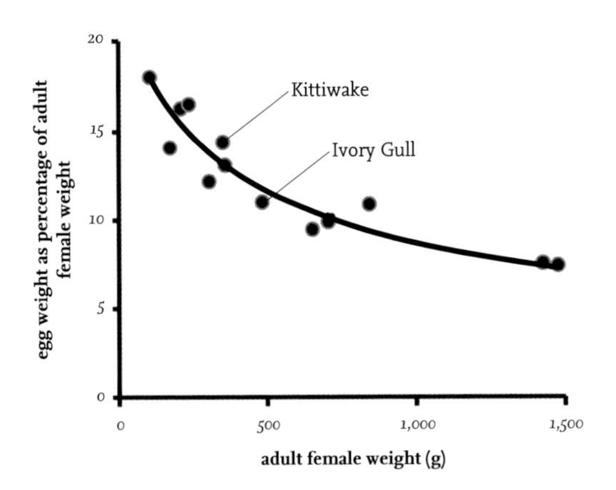

FIG 11. The relationship between the average weight of adult females of 13 gull species on the British list and the average weight of one of their eggs, expressed as a percentage of the female's weight.

incubation period, although the reason for this is not immediately obvious. Presumably the body of a chick of a large gull species has more cells and requires more cell divisions before it is ready to hatch. However, that is not the only factor involved, since the eggs of birds in some other taxa, such as the Procellariiformes (petrels and shearwaters), require a much longer incubation period even when the eggs are of the same size.

While the most frequent clutch size in gulls is three eggs, two species that occur in Britain typically lay two-egg clutches, namely the Kittiwake and Ivory Gull. This raises the question as to whether they respond by laying larger eggs. At first inspection, Kittiwake eggs look large, but this is an apparent effect of their paler background in comparison to the eggs of most other gulls (see above). Fig. 11 shows the relationship between female weight and egg weight for a series of species. The relationship is curvilinear, but the Ivory Gull and Kittiwake sit almost exactly on the curve, fitting the data based on the other gull species.

Each egg is laid at about two-day intervals. Usually, the last egg in the clutch is the smallest and tends to differ in shape, being longer and narrower than the rest. The infrequent cases of four eggs in a clutch are probably the result of a second female contributing one or more eggs. This can occur when female–female pairs are formed, caused by a shortage of males. Incubation lasts for more than three weeks and is longer in the larger species. Both members of the pair contribute to the incubation and feeding of the young.

Incubation period

Ideally, the precise measure of the incubation period is the length of time that the eggs are maintained at a temperature just below the adult's body temperature of about 40 °C. There is a lower temperature (which may be as high as 25 °C) below which the development of the embryo ceases; in temperate regions, this is usually above the environmental temperature. As a result, development soon stops when the eggs are not covered by an adult and the temperature falls below the critical temperature for development. If the incubation of the eggs is frequently interrupted, the time taken for the eggs to complete development to hatching is lengthened.

In practical terms, the incubation period is usually measured as the number of days taken from the start of incubation to the time the egg hatches, but hatching takes several hours and confusion has existed as to whether the end of incubation is when the shell is first pierced by the chick or when it becomes entirely free of the shell. Covering the eggs by an adult gull often takes place before the final egg of the clutch is laid, but in some cases the vascularisation of the brood patches on the adult may not have become fully developed with supplementary blood vessels and the optimal temperature for embryo development may not be reached. Consequently, there may be a delay of a day or so until the eggs reach the critical temperature for development. As a result, the recorded incubation period for a species will vary. The incubation period of birds, and in gulls in particular, is slightly longer in the larger species, but the differences are small and the variation recorded between individual pairs is considerable (Table 6).

In gulls, the incubation of the eggs is often initiated by the female, but overall the sexes tend to share incubation equally in a shift system. Either sex may incubate overnight. In most cases, the incubating individual remains on the eggs

TABLE 6. Estimates of the incubation period in eight well-studied gull species, presented in order of adult size (smallest first). Data mainly from Cramp & Simmons (1983), Fisher & Lockley (1954) and other sources where appropriate, including personal data for the Black-legged Kittiwake.

Species in order of adult size	*Estimated average incubation period*	*Range of recorded incubation periods*
Little Gull	24 days	23–25 days
Black-headed Gull	24 days	23–26 days
Mediterranean Gull	24 days	23–25 days
Black-legged Kittiwake	27 days	25–30 days
Common Gull	26 days	24–29 days
Lesser Black-backed Gull	27 days	24–27 days
Herring Gull	28 days	26–32 days
Great Black-backed Gull	28 days	27–28 days

until relieved by its partner, so that once incubation starts, the eggs are covered by an adult for about 90–95 per cent of the time.

Periodically – perhaps five to eight times a day – the incubating adult will stand up and roll the eggs. Egg-turning is believed to facilitate the absorption of the albumen by the embryo and prevents the embryo from adhering to the shell membrane. In domestic hens, the failure to roll eggs in an incubator does not necessarily prevent chicks from hatching, although the success rate is reduced and the chicks often have a lower weight. Egg-turning is unique to birds and the reason why reptile eggs do not need to be rolled is as yet unclear.

In gulls that lay three-egg clutches, it is common for the third egg – which is laid two or three days later than the second egg – to hatch one or sometimes two days after the first two. This asynchronous hatching suggests that effective incubation in these cases starts during the normal two-day period between the laying of the second and third eggs.

Chicks

At hatching, chicks of most gull species have a patterned down that offers some cryptic protection; this is replaced by a pattern of dark feathers that helps to conceal the young birds when they attempt to hide under vegetation. The dark patterning at hatching is missing in Kittiwake chicks, which hatch with a pale grey down and then acquire white, grey and black feathering when older, all of which could also be regarded as aiding camouflage among the whitewashed nests and rocks on the cliffs where they nest.

An effect of the relationship between egg size and adult weight is that, at hatching, the chicks of larger gulls are smaller in proportion to the adults than those of small gull species, and they therefore must achieve more growth in order to reach their adult size and weight. This is presumably one of several reasons why the period of chick growth (and the fledging period) tends to be longer in large gull species than in small ones.

The time between the chick hatching and first flying is known as the fledging period and also shows much individual variation (Table 7). There is a weak trend for it to be longer in the larger species of gulls. The fledging period of the Kittiwake tends to be longer than in other gulls of a similar size. In the case of the Kittiwake, the first flight has to be totally successful from the cliff-nesting site and so is an all-or-nothing event, with the young bird either flying well enough to reach and return to the nest, or to achieve sustained flight immediately, leave the colony and start an independent life.

Fledging is often a gradual process and tends to take longer in the larger species. It is longer in Kittiwakes, where the young cannot fly until they are at least

TABLE 7. Estimates of the fledging period in eight well-studied gull species, presented in order of adult size (smallest first). Data mainly taken from Cramp & Simmons (1983), Fisher & Lockley (1954) and modified from other sources where appropriate. Note that the considerable variation in several species is genuine, and depends on the rate of growth and development of the chicks and whether there had been a stimulus to fly, for example, by the presence of a predator or the observer.

Species in order of adult size	*Estimated average fledging period*	*Range of fledging periods*
Little Gull	22 days	21–24 days
Black-headed Gull	34 days	Unknown
Mediterranean Gull	37 days	35–40 days
Black-legged Kittiwake	42 days	35–54 days
Common Gull	35 days	Unknown
Lesser Black-backed Gull	35 days	30–40 days
Herring Gull	37 days	35–40 days
Great Black-backed Gull	52 days	49–56 days

five weeks old and many are at least a week older before their first flight. Unlike most terns, parental feeding of young gulls often ceases at the time of fledging, but sometimes family parties of large gulls stay together for some weeks. During this time they visit feeding sites together and make loud contact calls while moving between areas, and they may even return to the colony in the evening. Exceptionally, a Herring Gull has been observed feeding chicks three months after they fledged.

Breeding success

Under favourable conditions, breeding success of gulls is high, with about half the eggs laid producing young that fledged. However, lower success rates occur when there are food shortages or predation occurs. No gulls are known to breed as one-year-olds and they do so only when they have acquired almost adult plumage.

WALKING AND FLYING

Gulls have relatively short legs, and this is particularly true of Kittiwakes and Ivory Gulls. Gulls typically walk or run only short distances, and rapid movement beyond a few metres is usually achieved by flying. The exception to this occurs when gulls are searching for food on grassland and mudflats, when walking – with stops to pick up food – is the norm. Of all gull species, Kittiwakes walk less than any other. During an appreciable part of the year they are oceanic, either flying over or floating on the sea. Immature Kittiwakes that visit land during moult typically remain where they alight, and the adults fly directly onto their nests on precipitous sea cliffs without needing to take more than a step or two as they land. This behaviour is associated with the species having short legs (an adaptation to produce a low centre of gravity to enable the birds to withstand strong gusts of wind while nesting on sea cliffs), while the Kittiwake's hind toe, which in other birds – including gulls – assists with walking, is reduced to a small protrusion.

The long, narrow wings of gulls, petrels, albatrosses and shearwaters are an adaptation for flight over water, particularly where the sea surface has peaks and troughs that affect wind movement. Many gulls fly with steady wing-beats, while at other times they introduce periods of gliding, between wing-beats. They utilise rising air currents, particularly those over the sea produced by waves or swell, although it is rare for them to fly more than a few metres above the surface of the open sea. In addition, they glide in thermals over land and cliffs, and those produced by buildings in urban areas. Several species of gulls use thermals to rise to considerable heights over land without the need for flapping flight, although the smaller species more rarely soar. Favourable conditions for soaring occur on warm, sunny days. Such apparent effortless flight can be seen in Herring Gulls that breed on the coast but visit the highest ground on Snowdon in north Wales, where they search for food discarded by people walking and climbing here.

MOVEMENTS

Gulls breeding in extreme northern and southern areas of the world are migratory, but this is usually less pronounced or even absent among those living in temperate or equatorial regions. In general, the distances gulls migrate are much shorter than those made by terns, with only Sabine's and Franklin's gulls making regular large trans-equatorial migrations. In other gull species, a few individuals may reach or cross the equator, but most remain in the hemisphere in which they breed or were reared. Whether gulls migrate mainly depends on whether food is readily available in winter in their breeding areas.

The seasonal movements of many gulls are better described as dispersal rather than migration, with individuals from the same area or even a single colony moving in different directions and for variable yet often short distances. Immature individuals tend to move further than adults and often do not visit their natal or other colonies where they will ultimately breed for several years.

MOULTING

In all gulls, the 11 primary feathers on the wing are moulted in sequence, with the inner primary being replaced first. The remaining primaries are then progressively replaced outward towards the longest, the 10th, and then finally the small 11th primary on the bastard wing (Fig. 13). With the exception of one species, the flight and tail feathers are moulted and replaced once a year, but some of the body feathers are moulted twice each year.

The exception to this is Franklin's Gull, which moults all its feathers – including the primaries – twice a year, and as a result adults are missing

FIG 12. Herring Gull (*Larus argentatus*) presumably starting its third year of life, still lacking white mirrors and with the new outer primaries and the outer tail feathers still growing. (Mike Osborne)

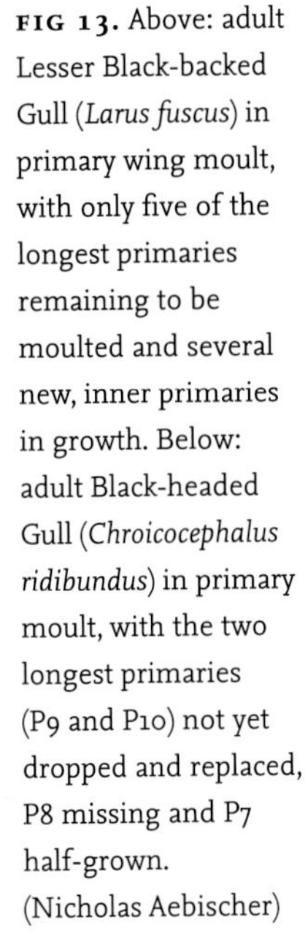

FIG 13. Above: adult Lesser Black-backed Gull (*Larus fuscus*) in primary wing moult, with only five of the longest primaries remaining to be moulted and several new, inner primaries in growth. Below: adult Black-headed Gull (*Chroicocephalus ridibundus*) in primary moult, with the two longest primaries (P9 and P10) not yet dropped and replaced, P8 missing and P7 half-grown. (Nicholas Aebischer)

primaries somewhere in their wing for about two-thirds of the year. This is a surprising pattern, because the efficiency of flight is affected by the loss of some of the primary feathers, particularly the longer outer ones. Most adults will have just completed the post-breeding moult when they start the next primary moult in December, which continues for several months. The information on moult in Franklin's Gull is based on a small number of specimens and requires further investigation, but the available evidence suggests that the post-breeding primary moult is spread over four months and the pre-breeding moult over a slightly shorter period. Carrying out two primary moults in a year is energetically and nutritionally costly, and no convincing explanation has been offered for it. It could be that exposure to more intense sunlight throughout the year (Franklin's Gulls winter south of the equator) causes the structure of the primary feathers to deteriorate quicker. However, this is not a convincing explanation, because other gull (and tern) species living in similar areas and making migrations across the equator appear to manage with a single annual primary moult (see also p. 337).

FOOD AND FEEDING

Gulls are usually classified as omnivores and scavengers, and animal material greatly dominates the diets of all species. The spectrum of food sources taken by gulls is much wider than in terns. Fish and invertebrates are the main food of many species, and small mammals are also taken by the large species. In addition, the larger gulls are often predators on eggs, chicks and even adults of other birds, and several consume eggs and chicks of their own species, although they usually avoid killing their own broods.

Typically, gulls obtain their food from the ground or near the surface of water. Most species can plunge from flight into water in an attempt to capture live food organisms within the top metre or so, but they lack the penetrating diving thrust of Northern Gannets (*Morus bassanus*) and some terns. Small gulls often pick food items from the water's surface, either while in flight or phalarope-like when floating on the surface. Gulls breeding inland feed on insects emerging from freshwater bodies and from flat terrestrial areas, and some even catch flying insects such as swarming ants and mayflies.

No gull species regularly and predominantly feeds on vegetable material, although some take berries and grain – for example, Herring and Lesser Black-backed gulls regularly search for animal feed on farmland and also consume grain in the spring and autumn. In some arctic areas, berries have been reported as a component of the diet of gulls, but this source of food is uncommon in more southern areas. That said, there are reports of berries being plucked off Hawthorn (*Crataegus monogyna*) bushes and even acorns still on oak trees being consumed by Black-headed Gulls. Many stomach analyses of gulls report small quantities of plant material, but these could have been accidentally ingested when picking up animal prey from the ground or even originate from the guts of the animals the gulls consumed. Many gulls will eat bread and other vegetable matter when offered it or when it is discarded by humans, and they often compete for food offered to waterfowl. The late Max Nicholson, who wrote *Birds and Men* (*New*

Naturalist 17, 1951), was one of many who frequently and regularly fed gulls in winter at lakes, in parks and on the riverside embankment in London in the first half of the twentieth century.

In the last century or so, several species of gull have started to utilise waste materials dumped by humans and they also follow behind tractors during the ploughing of agricultural land. Availability of foods associated with human habitation has attracted gulls to places such as landfill sites and fishing boats, particularly when nets are being hauled in or fish are being gutted at sea. In many areas across Britain, such sites have become a major food source for several species and have played a role in the growth of gull populations. As a consequence, some species are now considered pests and their numbers are managed. As refuse is increasingly being managed in ways that gulls can no longer exploit, and landfill sites previously used by gulls are being closed, this is likely to have an effect on the size of gull populations. Evidence of the magnitude of such changes is yet to be clearly identified, however, and the extent to which gulls can change their feeding habits and find new food sources remains unknown.

Scavenging on carrion or stealing food from other birds (kleptoparasitism) are other commonly employed methods of obtaining food. Kleptoparasitism in gulls is sometimes directed towards their own species and seems more frequent among male gulls. Smaller species of gulls join flock of Lapwings (*Vanellus vanellus*) and Golden Plover (*Pluvialis apricaria*), and steal prey items captured by the plovers, particularly when these are too large for them to swallow immediately. Sea ducks, particularly Eiders (*Somateria mollissima*), are attacked by large gulls when they surface with prey items in their bill, such as crabs, that need to be handled and processed before they can be swallowed. Indigestible items that are eaten by gulls as part of their diet are regurgitated as pellets, and offer a method of identifying some of the food consumed.

DRINKING

All animals need to replace fluid in their bodies; while some obtain sufficient water in their food, others will travel some distance to obtain it. Those species that spend long periods of time at sea or nest on small islands without available fresh water have no option but to drink seawater, although this causes physiological problems owing to its appreciable salt content.

A sailor crossing a large expanse of ocean in a small boat can use a solar still to produce drinking water, and gulls (and many other birds) have the

FIG 14. Urban Herring Gull (*Larus argentatus*) attracted to drink at a small source of fresh water. (Mike Osborne)

equivalent of this in the form of nasal glands, which are situated between the eyes. When they drink seawater, gulls use these glands to excrete a saline solution that has twice the concentration of the seawater, and by doing so reduce the concentration of salt in their bodies. Not infrequently, gulls are seen with water dripping from the end of their bill. These drops do not necessarily indicate that the bird has recently dipped its beak into water, but are the hypertonic saline solution excreted from the nasal glands. Using this method to remove excess salt from the body involves an energy cost for the birds, so it is not surprising that gulls show a strong preference for accessing and drinking fresh water wherever possible and are frequently seen congregating at freshwater pools and lakes. Herring Gulls at some urban sites even use small pools of fresh water when larger freshwater bodies are not available (Fig. 14). While Kittiwakes depend entirely on seawater and the healthy functioning of their nasal glands in their pelagic wintering areas, in the breeding season some will fly up to 2 km inland from their colonies to drink and bathe in fresh water. Those gulls that breed in colonies where no fresh water is locally available must depend on their nasal glands throughout the year.

VOICE

Most gulls are vocal, particularly in the breeding season, and this plays a part in their breeding behaviour. Calls are usually harsh and loud, and are used in territorial defence, as an alarm when birds are disturbed by predators or in greeting a mate. Quieter calls are used during courtship, and specific calls are also uttered, such as the wack-wack call of the Kittiwake during breeding, which tells its mate that it is not moving locally but is leaving and will be away some time on a feeding trip. Other specific calls are used during copulation, to maintain contact in flocks in flight, and to indicate that a food source has been found and that the calling individual is about to feed.

Studies using sonographs to analyse calls for several gull species have shown that there is a wide range of variation in the same call made by different individuals. Playbacks of the calls indicate that birds can recognise the call of their own mate but fail to respond to the same call when it is made by other individuals, suggesting that voice plays a part in individual recognition in addition to visual cues. Chicks of many gull species are able to recognise their parents at a relatively young age, and likewise the parents can identify their own chicks, apparently by their calls and actions – they respond differently to strange individuals. This is particularly important in species where the young wander or group together in a crèche. The Kittiwake is the exception in that adults appear to be unable to recognise their own chicks, although the usual isolated position of the nest on a vertical cliff face forces most chicks to remain in the same nest until they fledge. Occasionally, when two nests are built side by side, a Kittiwake chick will move across to the neighbouring nest and join chicks already there. If the wandering chick is placed back in its own nest, it soon returns to the neighbouring nest again, but such moves seem to be accepted by both pairs of adults.

ROOSTING

Gulls are inactive in the darkness of the night, spending these hours passively, although some species feed locally where there is strong artificial illumination, such as around working trawlers and in town centres. Most gull species spend the night in large flocks, with many roosting on water in sheltered bays at the coast, on reservoirs and on lakes, or on land in places that are free of predators, such as small islands.

Gull roosts are composed of single or mixed species, and may number thousands of individuals. Roost sites are often used year after year if their position excludes mammalian predators. Most species avoid roosting on water exposed to severe wave action, presumably because they do not possess the behavioural adaptations needed to cope with resting and sleeping while sitting on rough or breaking water. Sabine's Gull, Ross's Gull and the Kittiwake are totally oceanic outside the breeding season and are the only gulls that regularly spend the night far from shore. Kittiwakes will also roost offshore during the breeding season if they are not incubating or protecting small young in the nest. Before egg-laying, they leave the colony at, or long after, sunset and fly many kilometres out to sea in the dark.

CHAPTER 2

Black-headed Gull

APPEARANCE

The Black-headed Gull (*Chroicocephalus ridibundus*) is a small gull with a wingspan of about 85 cm. It is common both on coasts and inland, and while it is less abundant than the Kittiwake (*Rissa tridactyla*), it is much more familiar to the public thanks to its dark head in summer (Fig. 15, left), which gives rise to its common name. The hood is actually dark brown, not black, and while the eye is also dark, the white semicircle around the back of it draws attention to it when the bird is seen at close range. The species is frequently attracted to food in towns and even large cities, both inland and on the coast, but for more than half of the year adults lack the dark hood and are more difficult to identify. At that time, the white head has only small, dark grey patches behind the eyes, but the red bill and legs, and the light leading edge to the wing when in flight, aid recognition (Fig. 15, right). The sexes are similar, with the male being only slightly larger than the female.

FIG 15. Adult Black-headed Gull in summer plumage (left) and winter plumage (right). (Left, Alan Dean, right, Nicholas Aebischer)

Juveniles differ markedly from the adults. They have brown feathers on the wings, on the back of the neck and on top of the head, and these form part of a cryptic patterning that helps to conceal them in vegetation before they fledge. During the first year of life, they lack the bright red legs and bill of the adults, and, although present, the white outer edges to the wings are less obvious. First-years have a black band to the tail, but as this is also present in several other species of small gulls in their first year of life, it is a useful characteristic to age but not to species identification. In the first summer after hatching, the immature birds develop a dark hood later than the adults; this is usually incomplete and mixed with some white feathers. They also retain most of the dark band to the tail. Following the autumn moult, immature individuals entering their second year of life assume the adult plumage, including an all-white tail, but many still have less intensely red legs at this stage.

DISTRIBUTION AND HABITATS

The Black-headed Gull is widely distributed in Europe and breeds across Asia, and there is no evidence of subspecies within this large range. The species extended its western range in the twentieth century, reaching the Faeroes early on at an unrecorded date, Iceland in 1910 (50,000 adults now breed there) and Greenland in 1969. Breeding in North America was suspected as early as 1963 but not proven until 1977, when nesting by 14 pairs was recorded in Newfoundland. More recently, breeding has been suspected at other sites in North America, with recently fledged young being reported at several locations from Labrador to Massachusetts. Several young birds ringed in Iceland have been recovered during winter in Newfoundland and Labrador, but only one ringed individual of this species is known to have crossed from the mainland of Europe to the New World. It was ringed in the Netherlands and recovered 200 km off the coast of Labrador, so it seems likely that those seen in North America are mainly individuals reared in Greenland and Iceland.

Movement to North America has progressed in a series of stages rather by a sudden irruptive transatlantic movement from Europe, with birds breeding or reared in the Faeroes, Iceland and Greenland gradually moving west. This spread, together with a northern movement in Norway, Sweden and Finland, has been interpreted as a response to climate change. However, this theory ignores the spread of breeding to Italy (1960) and Spain (1960), and the fact that the increase within the species' core area in Europe during the twentieth century has probably been mainly responsible for the spread of the western limits of its breeding distribution.

Today, unknown numbers of Black-headed Gulls nest across Asia and some 4 million to 5 million adults breed in Europe, of which about 250,000 nest in Britain and Ireland. Those breeding here probably represent less than 5 per cent percent of the world population.

The Black-headed Gull is the commonest gull frequenting sports fields, landfill sites, harbours and areas around large shopping complexes, and the birds also seek food in urban streets, domestic gardens and even at bird tables during severe winter weather. In many areas, they are also the gull most likely to be seen following the plough in agricultural fields. The species achieved fame in literature as the character Kehaar in the well-known book *Watership Down*, written by Richard Adams (1972). Kehaar suffered a damaged wing and took refuge at Watership Down, eventually recovering and returning to his colony.

Numbers of Black-headed Gulls in and around towns and cities can be high, particularly in winter, but this has not always been the case. For example, until the end of the nineteenth century the species was considered scarce in central London, but the birds are now present here in their thousands in winter, where they are abundant in parks, sports fields, lakes and the river Thames, often seen competing vigorously for scraps of food.

The gathering of flocks of Black-headed Gulls around towns and cities varies through the year. Fig. 16 shows their density month by month within a radius of 5 km of Durham, where many were recorded on sports fields, pastures and the river Wear running through the city centre, or patrolling streets and shopping centres in search of discarded food. The gulls retreated to their colonies in

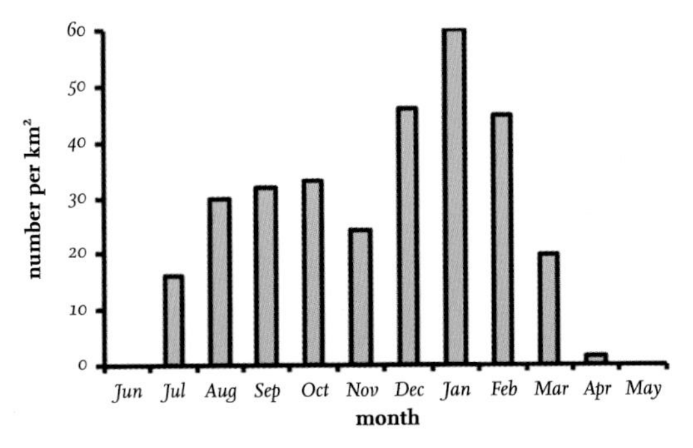

FIG 16. The seasonal variation in the numbers of Black-headed Gulls within a radius of 5 km of the city of Durham between 1982 and 1984, and in the absence of local breeding colonies. In each year, the gulls were absent in May and June, and the first individuals arrived in July, coinciding with the Durham Miners' Gala, when a considerable amount of food was dropped by the large numbers of people attending the event. Based on data collected by MacKinnon (1986).

March, and since none of these was within 40 km of the city, only an occasional immature bird was recorded in April and none of any age in May and June. There are many accounts of Black-headed Gulls frequenting other areas in winter and being absent there in spring and summer. It is obvious that the distribution of the birds in Britain becomes much more clumped during the breeding season and is limited by feeding range from colonies. At most colonies, first-year (non-breeding) Black-headed Gulls are also infrequent and those that are encountered are seen feeding with adults well away from colonies, commonly on intertidal estuaries.

HISTORICAL BREEDING STATUS IN BRITAIN AND IRELAND

The status of the Black-headed Gull in Britain and Ireland in or before the twentieth century is not clear owing to a lack of detailed information. On its website, the British Trust for Ornithology (www.bto.org) says that the species was a rare breeding bird in Britain in the nineteenth century, and the statement made by John Gurney Jr in 1919 that the species was in danger of extinction here cannot be true. Both opinions lack supporting evidence and are contradicted by the extensive egg collecting that occurred at that time (p. 52), and by the historical evidence accumulated by Michael Shrubb in his book *Feasting, Fowling and Feathers – a History of the Exploitation of Wild Birds* (2013). It is likely that the movement of colonies was misinterpreted as a decline in numbers, and that the 'disappearances' of the birds were probably movements of adults elsewhere and the establishment of new colonies in response to nest predation. Such movements of colonies are also frequent, and better known, in the Little Tern (*Sternula albifrons*), where they are closely related to poor breeding success induced by predators or by human disturbance.

There can be little doubt that the numbers of Black-headed Gulls breeding in Britain increased in the period between 1900 and 1973, as did those of many other gull species breeding here. However, Black-headed Gulls are not easy to count accurately. Colonies of very varying sizes occur both on coasts and inland, and the latter are often particularly difficult to count, with nesting taking place in marshes, on islands in lakes and tarns, and sometimes among tall or floating vegetation in boggy areas. In addition, colonies on moorland are often small and scattered, and new ones can remain overlooked for some time.

The first census of Black-headed Gulls in England and Wales was made in 1938, with further surveys in 1958 and 1973. The report of the census made in 1973 concluded that there were between 200,000 and 220,000 breeding adults, and

FIG 17. Map of the breeding distribution of the Black-headed Gull in Britain and Ireland in 2008–11. Reproduced from *Bird Atlas 2007–11* (Balmer *et al.*, 2013), which was a joint project between the British Trust for Ornithology, BirdWatch Ireland and the Scottish Ornithologists' Club. Channel Islands are displaced to the bottom left corner. Map reproduced with permission from the British Trust for Ornithology.

that since 1938 there had been an increase of 177 per cent, while the number of colonies containing more than 2,000 adults had doubled since 1958. However, part of the reported increase might be attributed to a better coverage of colonies in 1973. In Scotland, a census made in 1958 suggested that there had been little recent change in abundance, but that numbers had continued to be influenced by uncontrolled egg collecting. The national censuses of Britain and Ireland in 1970 and 1985 counted only coastal-nesting Black-headed Gulls and did not consider the appreciable numbers nesting inland. The census in 2000 included inland colonies for the first time and resulted in an estimate of about 336,000 breeding adults in Britain and Ireland, of which 44 per cent were nesting inland.

Between 1970 and 1985, a 71 per cent decline in coastal-breeding Black-headed Gulls in the Republic of Ireland was reported, but the numbers there had recovered by the time of the 2000 census. It is possible that an appreciable proportion of the missing birds in 1985 had not, in fact, died, but had moved to new breeding sites not visited at that time.

The general impression from these censuses is that the numbers of breeding Black-headed Gulls in Britain and Ireland increased during most of the twentieth century, and then numbers showed little overall change in Britain between 1985 and 2000. The meagre and incomplete data available since then hints that numbers have been maintained.

The BTO *Bird Atlas 2007–11* data, collected between 1968–72 and 2007–11, reported a reduction of 55 per cent in the number of 10 km squares in Britain and Ireland containing breeding Black-headed Gulls. Whether this reflects a decline in overall numbers is uncertain, as it could be the result of a trend towards fewer, larger colonies. Sample counts made at some coastal sites by the Joint Nature Conservation Committee (JNCC) show little evidence of a national change in numbers between 1986 and 2015, and this is supported by the national censuses in 1985 and 2000, which (although the earlier census was incomplete) did not suggest any major overall numerical changes within Britain and Ireland.

There has been a trend over time towards decreasing numbers of Black-headed Gulls breeding at inland sites, while numbers nesting at coastal sites have increased in England and Wales. However, this change does not seem to have occurred in Scotland. This movement probably reflects that birds breeding on coastal islands are often better protected from both humans and Foxes (*Vulpes vulpes*) than those breeding inland. The impact of the dramatic increase of Fox numbers in Britain in recent years is not frequently appreciated. Stephen Tapper (1992) found that numbers of this predator had increased fourfold between 1961 and 1990, and suggested that the effect of Fox predation on Black-headed Gull colonies may be greater and more widespread than that caused by American Mink (*Neovison vison*).

COLONIES

Colony sizes

The Black-headed Gull is a colonial breeder, nesting on a wide range of sites, including lakes, reservoirs, small moorland pools and tarns, marshes, sewage farms, clay pits, dunes, saltmarshes and industrial ponds. While numbers of nests in some colonies are easily counted, others are difficult to measure as the nests can be concealed by vegetation or lie on boggy ground, making access difficult or risking damage to the vegetation. Good vantage points do not exist near many colonies, and different methods are therefore needed here to estimate their size.

The basic unit used to measure the size of a colony is the number of nests, applying the realistic assumption that two adult birds are associated with each nest. In very large colonies, sampling of nest density has been used and then extrapolated across the total area of the colony as delineated from aerial photographs or detailed large-scale maps. At some sites, pair and hence nest numbers have been estimated through the less reliable method of counting numbers of birds in the air when disturbed and then using a conversion factor, such as dividing by 1.55. Most Black-headed Gull census counts have errors, which should be considered when making interpretations of the national status of the species and possible changes in abundance over time.

FIG 18. Part of the colony of Black-headed Gulls nesting with Sandwich Terns (*Thalasseus sandvicensis*) on Coquet Island, Northumberland. (John Coulson)

Data for years since 1980 on Black-headed Gull colony size recorded in the UK Seabird Colony Register allows for an in-depth assessment. While the species is essentially a colonial breeder, solitary breeding pairs on the day of counting comprised 10 per cent of the sites in Scotland and 8 per cent in England. Whether many of these pairs had been solitary for the entire breeding season is uncertain, because most of the nesting sites were visited by counters only once. It is possible that some single pairs were the sole remainders of a group present earlier in the breeding season that had suffered predation or disturbance, and had deserted the site prior to the observer's visit. Single pairs of Black-headed Gulls can probably breed without the need of stimulus of other pairs, although when they do so, they nest late and their breeding success is usually low.

In Britain and Ireland, the numbers of pairs of Black-headed Gulls estimated at breeding sites (subsequently referred to as colonies for convenience, and including sites with single pairs) between 1980 and 2015 ranged from only one pair to more than 10,000 pairs. Overall, a third of colonies in Britain contained between 10 and 100 nesting pairs (Fig. 19).

At the same time, the average size of colonies differed between Scotland and England, with 84 per cent in Scotland having fewer than 100 pairs (Table 8), while the comparable figure for England was a third lower, at 54 per cent. Colonies with 100 to 1,000 pairs were twice as frequent in England than in Scotland, while colonies of more than 1,000 pairs formed 13 per cent of all sites recorded in England, but only 2 per cent of those in Scotland.

These differences reflect the type of nesting sites used in the two regions, with the majority in Scotland being inland on upland sites and usually associated with relatively small waterbodies, while in England large coastal colonies were more frequent and were sometimes extremely large. The distributions of colony sizes in Wales and Ireland were intermediate between those in England and Scotland, with 72 per cent of colonies in Wales and 62 per cent in Ireland having fewer than 100 pairs (Table 8).

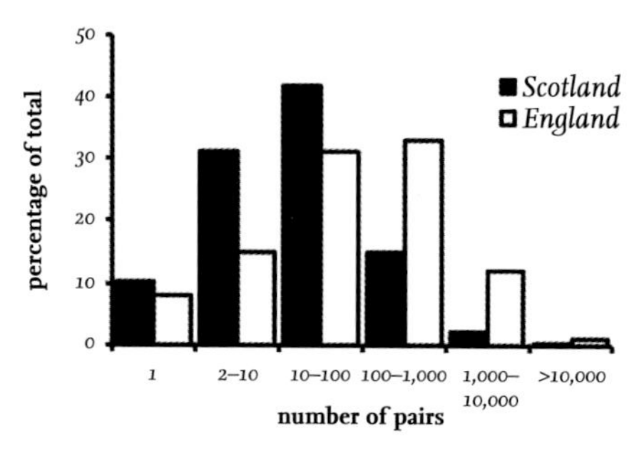

FIG 19. The size distribution of Black-headed Gull colonies (including sites with only one pair) in England and Scotland between 1980 and 2015, based on Seabird Colony Register data for 364 colonies surveyed in Scotland and 260 in England.

TABLE 8. The proportion of Black-headed Gull colonies in England, Scotland, Wales and Ireland with fewer than 100 breeding pairs, based on data from 1980–2015. Sites with a single pair present have been included.

	England	*Scotland*	*Wales*	*Ireland*
Number of colonies considered	260	364	50	82
Percentage with <100 pairs	54%	84%	72%	62%

TABLE 9. The percentage of Black-headed Gull colonies of different sizes in England and Wales in 1938, 1958 and 1980–2012. Sites with only one pair are included.

Period	*Total number of colonies recorded*	*Percentage of total by colony size (pairs)*				
		1–10	*10–100*	*100–1,000*	*1,000–10,000*	*10,000+*
1938	107	18%	47%	31%	4%	1%
1958	155	25%	41%	29%	5%	1%
1980–2012	260	24%	31%	33%	12%	1%

A comparison between the size of colonies reported in the 1938 and 1958 censuses made in England and Wales, and more recent data collated in the Seabird Colony Register, show progressive changes in colony sizes over time (Table 9). There has been an overall decrease in the proportion of colonies with fewer than 100 pairs since 1958, while the proportion and total numbers of large colonies with more than 1,000 pairs has more than doubled over the same period. Many colonies – particularly small ones in England and Wales – have disappeared for various reasons, including disturbance, predation and drainage, while others – particularly large ones – are now protected within nature reserves, although even a few of these have disappeared.

Persistence of colonies

While some Black-headed Gull colonies persist for many years, in general they are not as permanent as those of the Kittiwake or Herring Gull (*Larus argentatus*). Over the past hundred years, many Black-headed Gull colonies have disappeared while new ones have become established. Small colonies are particularly susceptible to decline, and this is especially true of those on upland moors, where the birds nest around small bog pools, reservoirs and lakes. A consequence of this is that some of these mobile groups can be missed during census work.

Various reasons have been suggested for the desertion of colonies. In some cases these are speculative, and include drainage, military activity, competition from increasing numbers of large gulls, aggression by Mute Swans (*Cygnus olor*) and Canada Geese (*Branta canadensis*), egg collecting, keepering, and predation by Foxes and American Mink. Rats and Foxes have been a persistent problem for Black-headed Gulls nesting at Blakeney Point in Norfolk, and no young were reared at the colony in 2000. In 2014, Fox predation at the same site resulted in very few chicks being reared by the 2,200 pairs present, while in 2016 an appreciable increase of rats caused the gulls to abandon the colony and, presumably, move elsewhere.

Ravenglass colony

The site of the very large Black-headed Gull colony on the coast of Cumbria at Ravenglass is owned by the Muncaster Estate. There has been continuous collection of eggs there since the seventeenth century, a practice that was recorded as being extensive in 1886. Del Hoyt has suggested the colony had an annual yield of 30,000 eggs over many years. Neither Michael Shrubb's searches (2013) nor my own have uncovered documentation to confirm or contradict this statement, but the colony was certainly used consistently as a source of freshly laid eggs for much of the twentieth century by the estate, and workers also took some young from time to time. The collection of eggs was well organised and was stopped each year on an agreed date, which left time for successful breeding by birds laying late or with repeated clutches. David Bannerman recorded that the estate collected 72,398 eggs at Ravenglass in 1941 and 24,568 in 1951, when a further 6,000 eggs were taken (presumably illegally) by others.

In 1954, the site was leased to Cumberland County Council as a local nature reserve and egg collecting ceased. The gulls were protected and studied in the ensuing decade by Niko Tinbergen's research group from Oxford University. During the period 1954–68, the colony size remained between 8,000 and 12,000 pairs, but during this time predation on eggs, chicks and adults occurred from time to time and breeding success was low. In 1968, a full-time warden was appointed, and he reported that 8,100 pairs nested. From then onwards, a marked decline continued each year; Neil Anderson's study in 1984 found only 1,300 pairs nesting, and the colony was totally deserted in 1985 (Anderson, 1990). Five pairs nested there in 1988 and three pairs in 1989, but none since (Fig. 20). Predation by mammals over several years, particularly Foxes, was thought to be the cause of decline and ultimate desertion. The abandonment of the Ravenglass colony probably involved a movement of Black-headed Gulls to several other colonies,

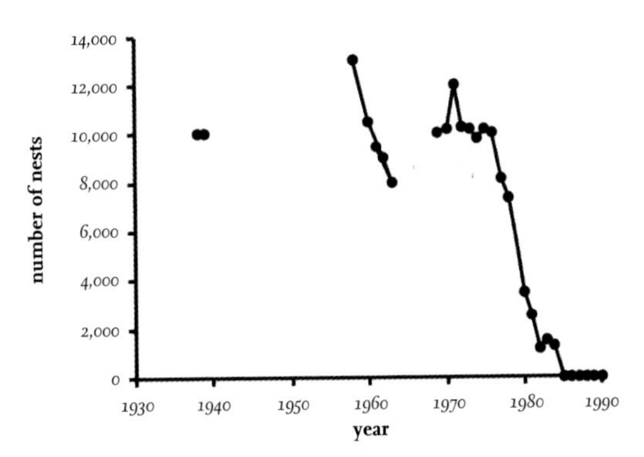

FIG 20. The numbers of nests (pairs) of Black-headed Gulls nesting at Ravenglass in 1939–90.

possibly those in the Ribble Estuary and Sunbiggin Tarn (see below), because no new or established colonies were nearer.

The history of the Ravenglass colony and other similar sites illustrates two important points. First, egg collecting over many years that was well organised and had an early cut-off had no adverse effects on the colony and did not lead to its desertion. Second, the creation of a nature reserve without effective predator management does not prevent colony desertion; indeed, in recent years there have been other instances of colonies within nature reserves disappearing.

Sunbiggin Tarn colony

The history of the large inland Black-headed Gull colony at Sunbiggin Tarn, near Orton in Cumbria, parallels that of Ravenglass. In 1891, a thousand eggs were said to have been collected from the edge of the colony. The site was estimated to contain 400 pairs in 1938, 1,200 pairs in 1958 and 2,500–3,000 pairs in 1973. The colony continued to grow and reached about 12,500 pairs in 1988 (see below), maintaining similar numbers in 1989, when Foxes were seen. Numbers of gulls then showed a noticeable decline in 1990, and only a few pairs bred in 1991. The colony persisted with a few pairs, and by 2000 had increased to 1,200 pairs, but by 2010 gulls had ceased nesting at the site altogether (Fig. 21).

Direct counting of all nests proved to be impossible at the large Sunbiggin Tarn colony because of the boggy areas within the tarn and the high vegetation, which would be damaged by trampling during a search for nests. A count of the colony was not obtained for the national census in 1985–87, and in 1988 a novel method was used to measure its size. This involved capturing a large sample of the breeding adults in late May and dyeing their tails yellow. A total of 1,010 adults were captured and marked during two weeks by attracting them with food placed about 1–2 km from the colony and then cannon netting them. When marking was completed,

FIG 21. Part of the large Black-headed Gull colony at Sunbiggin Tarn, Cumbria, during an exceptionally dry spring in 1984. (John Coulson)

the proportion of dyed birds in the colony was obtained by counts of 4,540 adults passing in and out of the site along flight lines. Of these, 186 – or one in 24.4 – had yellow tails, so multiplying the total number dyed by this proportion (24.4) gave a measure of the total number of breeding birds in the colony. This produced an estimate of 24,644 adults (rounded off to 12,500 pairs), identifying it as the largest inland colony in Britain at the time. Unfortunately, Clare Lloyd *et al.* (1991), in their book reporting the status of seabirds in Britain and Ireland in 1985–88, erroneously recorded this census as 25,000 *pairs* of Black-headed Gull, not 25,000 individuals.

Part of the colony at Sunbiggin Tarn could be viewed by people from the nearby road, and they often fed the gulls, which made attracting and capturing the birds in cannon nets much easier. The public's interest in the well-being and size of this colony was considerable, and not without humour. The investigators netting adult birds had to explain to onlookers what they were doing and that there would be an explosion when the cannons were discharged. On two occasions, the observers remarked that they were puzzled because they could not find a gull in their field guides with a yellow tail, and asked whether this was a rare species! Members of the public often also liked to guess how many gulls were nesting, generally coming up with figures that were widely lower than the actual numbers calculated – most likely because the extent of the colony could not be viewed from a single vantage point. Incidental to the main study, adults with yellow tails were seen feeding in the Lake District and several places on the river Eden up to 40 km from the colony a few days after marking.

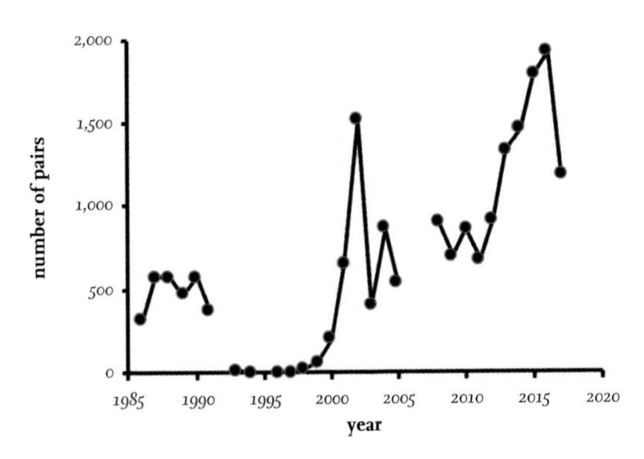

FIG 22. Numbers of Black-headed Gulls nesting on the Sands of Forvie, 1986–2017. Data mainly from the Seabird Colony Register.

Ythan Estuary (Sands of Forvie) colony

Black-headed Gulls have nested at the Sands of Forvie at the mouth of the river Ythan, near Aberdeen in north-east Scotland, for many years. In the late 1980s, about 500 pairs nested there annually, along with Sandwich Terns (*Thalasseus sandvicensis*), Common Terns (*Sterna hirundo*) and Eiders (*Somateria mollissima*). Fox predation became intense in about 1990, when breeding Sandwich Terns and Black-headed Gulls abandoned the area and the number of young Eiders declined. In 1994, steps were taken to reduce and then exclude Foxes from the breeding areas, and in 1998 the first Black-headed Gulls returned to breed. Numbers increased rapidly in the following years and reached more than 1,500 pairs in 2002. This level was not maintained, but numbers exceeded 500 pairs each year up to 2012, when the colony size increased annually until 1,921 pairs were recorded 2016 (Fig. 22).

Foulney and other colonies

The colony at Foulney in north-west England (Fig. 23) showed a pattern of decline spread over a few years in the late 1980s and early 1990s, before breeding there

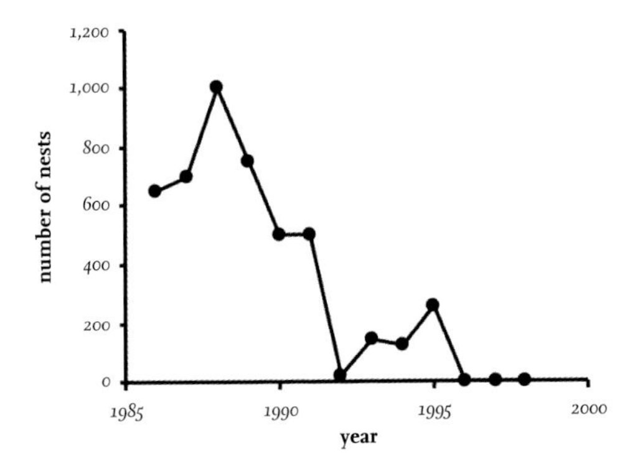

FIG 23. The decline and subsequent total desertion of the Black-headed Gull colony at Foulney, a coastal site in Cumbria. Data mainly from the Seabird Colony Register.

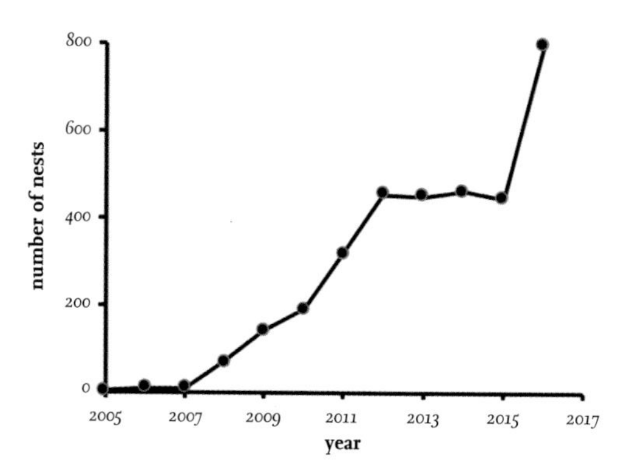

FIG 24. The rapid build-up of a Black-headed Gull colony at the RSPB reserve at Saltholme, near Stockton-on-Tees, between 2005 and 2017. Data mainly from the Seabird Colony Register.

ceased altogether sometime in or before 1996. As elsewhere, predation of eggs and nestlings was considered the main cause.

There are many other examples of Black-headed Gull colonies that showed appreciably smaller numbers each year before breeding stopped entirely. The national census in 2000 listed 12 large colonies that had disappeared since 1985, plus a further five that had lost more than 89 per cent of the maximum numbers of breeding pairs over recent years. In contrast, nine colonies showed major increases within a short time period, including the colony at Langstone Harbour in Hampshire, which in the 15 years between 1985 and 2000 increased more than 60-fold. Many new colonies had become established and subsequently grew rapidly, such as those at Hamford Water in Essex, Coquet Island in Northumberland, Larne Lough in Northern Ireland, Loch Leven in Scotland, the Ribble Estuary in Lancashire and Saltholme (Fig. 24), a nature reserve near Stockton-on-Tees, County Durham, that has been managed by the RSPB since 2007.

Reasons for colony declines

Egg collecting, particularly if stopped early in the season, has carried on year after year at several Black-headed Gull colonies without resulting in abandonment. And at Ravenglass, Neil Anderson's investigations (see above) ruled out a possible link between the decline there and contamination associated with the nearby Sellafield nuclear fuel reprocessing and decommissioning plant. In many areas – particularly the fenland of eastern England – colonies have been lost due to drainage and expanding agriculture, which have removed otherwise suitable nesting sites. However, it seems that the main cause of decline and disappearance of Black-headed Gull colonies is predation of eggs, young and sometimes adults. The presence and increase in numbers of mammalian predators, particularly Foxes and American Mink, have been associated with the decline of the colonies at Sunbiggin Tarn, Ravenglass and the Sands of Forvie. During his studies at Ravenglass, Hans Kruuk found that four Foxes killed 230 Black-headed Gulls in one night, and a total of 1,449 adult gulls were killed by Foxes in two consecutive breeding seasons. Studies made at the colony by Ian Patterson during the period of appreciable mammalian predation activity found that scattered and outlying pairs of gulls were entirely unsuccessful, while those in the central core areas, although having a very low success rate, did succeed in rearing a few young (Patterson, 1965).

Once colonies have started to decline, the impact and pressure from the same numbers of Foxes or American Mink produce a vicious spiral that has an increasing impact on the remaining birds. Another predator at Ravenglass was the Hedgehog (*Erinaceus europaeus*), which had developed the habit of consuming the flesh of live gull chicks that had not reached fledging age. Several of the colonies referred to in the past as sites where eggs were collected have now ceased to exist, and some authors have attributed these to uncontrolled egg collecting extending throughout the entire breeding season. It is ironic, therefore, that it was only after egg collecting had stopped at Ravenglass that the colony declined and then disappeared, presumably owing to the lack of efficient predator control.

When a colony disappears in a matter of a few years, it is unlikely that the adults have died, but rather that they have moved to other sites and colonies. The decline at Sunbiggin Tarn following predation resulted in some adults moving to Killington Reservoir alongside the M6 motorway in Lancashire, a move confirmed by a few ringing recoveries of adults previously marked at Sunbiggin Tarn. However, this colony contains nowhere near the numbers of birds recorded breeding at Sunbiggin Tarn, which implies that many adults from Sunbiggin probably moved considerable distances and to several different sites, as no other single large colony was established within 80 km of it at the time of its decline. This is also true of many small, ephemeral colonies on upland moorland in northern England and Scotland. A consequence of this is that mobile groups of Black-headed Gulls are often overlooked, and new colonies can grow rapidly through immigration in only a few years – as has happened at Saltholme (Fig. 24).

Exploitation of eggs and young

In the past, both the eggs and young of Black-headed Gulls were extensively exploited in Britain by humans. Michael Shrubb gives a detailed account of this in his 2013 book, *Feasting, Fowling and Feathers*, from which I have extracted some of the information given below.

Through the combination of Black-headed Gull numbers, their extensive inland breeding often close to human populations, and the relatively easy access to many nests compared to those of gull species that nest on coastal islands and sea cliffs, the eggs of this species have been exploited more extensively than those of other gulls in Britain and Ireland for the past 500 years and probably longer. In some places in the sixteenth century, nearly fledged young were collected by driving them into nets, and they were then housed and fed until required for human consumption. Some were purchased by the wealthy for 4–5d each, (equivalent to about £7 at today's prices) or given as gifts, just as we now give flowers. Oxford college records at the time also list Black-headed Gull eggs being bought for a fraction of a penny each.

In the seventeenth century, one landowner is said to have made a profit of £60 in one year (equivalent to about £7,000 today) by collecting and selling gulls' eggs. Exploitation of young birds probably decreased over time, but they continued to be sent to London markets nonetheless, and egg collecting became more frequent and extensive. Exploitation continued into the eighteenth century, and there are more records at this time of eggs being collected throughout England and Scotland. There are also records of gull colonies disappearing, the causes of which were attributed at the time to excessive egg collecting.

In the nineteenth century, egg collecting for human consumption throughout Britain and Ireland was extensive. In some areas it was organised by the landowners, but elsewhere it appears to have been uncontrolled. Well-established markets existed in London, and eggs collected in bulk in Norfolk and Hampshire were sent there for sale. That the Black-headed Gull was widely distributed as a breeding bird in southern England is evident from records of the large numbers of their eggs being collected and sold in the London markets. For example, in 1800 some 8,000–9,000 eggs were collected at Scoulton Mere in Norfolk, while in 1864 and later, 2,000 eggs a year were collected. From 1858, 700 eggs were taken each year at Hoveton in Norfolk, and in 1864 alone, 2,000 eggs were collected. According to *The Spectator* (17 April 1897), in 1834 at Stanford in Norfolk, a 'tumbrel-load' of eggs was collected in a day and sold for 3d a dozen.

It is probable that landowners enforced a degree of protection to the gull colonies in the nineteenth and early twentieth centuries owing to the value of the egg trade. In some cases, collecting was limited to taking eggs only from nests that held one or, occasionally, two eggs, to ensure that those collected had not been incubated, and at some but not all sites, collecting ceased on 15 June. As a result, clutches of three eggs were often left untouched, while cessation of collecting in June permitted some gulls to re-lay and breed successfully from late, repeat clutches.

An account of annual egg collecting in Ayrshire, Scotland, in the first 40 years of the twentieth century is given by Ruth Tittensor in a 2012 edition of *Ayrshire Notes*. The eggs were collected for local bakeries and domestic use, and even for egg fights between youths. Interestingly, the article includes a personal account of egg collecting by a person who later became a senior member of the Nature

Conservancy! Elsewhere, the sale of eggs also continued into the twentieth century. In the 1930s, for example, more than 200,000 gull eggs were sold annually at Leadenhall Market in London. Egg collecting increased extensively during the two world wars, with large numbers collected to supplement rationing. An archived Pathé News ciné film shows several Land Girls based at Muncaster Estate during the Second World War collecting baskets of eggs at the large Ravenglass colony and boxing them up in crates ready for dispatch to London. Hugh Cott describes eggs being on sale in Cambridge in 1951 at 9d each. A London market for Black-headed Gull eggs existed for the whole of the twentieth century, with some being sold as 'plovers' eggs'.

Today, despite current bird protection Acts, systematic collecting of Black-headed Gulls' eggs is still permitted in England by licence holders, although there is little general knowledge of the numbers of licences and eggs collected. The licences are issued annually by Natural England for collecting eggs for food consumption. As a result of a request for information, Natural England responded in 2016 by stating that 'Licences are only issued where evidence of hereditary/ancestral/traditional family rights to collect eggs can be given'. The public body also stated that it seeks 'to ensure that collection of eggs is sustainable in relation to numbers collected and colony size'. However, I still await a response as to how this is achieved and where the records are deposited, since they do not appear in the Seabird Colony Register or local county bird reports. An accurate census of a colony where many eggs are collected is often difficult to carry out because of the greater spread of laying and repeat clutches.

A freedom of information request granted by Natural England in 2016 revealed that it had issued licences recently for egg collection at six sites in England: the North Solent nature reserve in Hampshire; Lymington, Pylewell and Keyhaven marshes, also in Hampshire; Barden Moor in North Yorkshire; and the Upper Teesdale National Nature Reserve in County Durham. Currently, up to 40,000 eggs are collected annually and sent to London. These details are not publicised and I found no information in the published literature. A further freedom of information request to Natural England revealed that in 2015 it issued 24 licences to collect Black-headed Gull eggs in England (22 in Hampshire, one in Yorkshire and one in County Durham). Table 10 shows that the numbers of eggs collected in Hampshire remained at about the same level over the period 2009–15. Those taken in North Yorkshire increased 17-fold between 2011 and 2012, and since then the numbers collected have continued to increase, with more than 32,000 eggs taken in 2015.

No mention of these annual egg collections appear in recent county ornithological reports, while Natural England made the surprising statement

TABLE 10. The number of egg-collecting licences issued and Black-headed Gull eggs collected in Hampshire, North Yorkshire and County Durham in 2009–15, as reported by Natural England. The public body stated that data for earlier years have not been retained.

	2009	*2010*	*2011*	*2012*	*2013*	*2014*	*2015*
Hampshire							
Licences issued	24	22	22	22	22	22	22
No. of eggs collected	29,303	23,110	25,559	25,083	25,557	26,545	31,017
North Yorkshire							
Licences issued	1	1	1	1	1	1	1
No. of eggs collected	1,500	1,487	1,500	25,635	26,996	28,070	32,518
County Durham							
Licences issued	1	1	1	1	1	1	1
No. of eggs collected	45	55	25	25	25	25	25

that information prior to 2008 'has not been retained'. This public body should not destroy historical information on colony sizes and the numbers of eggs collected, and such data should be archived. Natural England also refused to identify the colonies from which eggs were collected, or to whom licences were issued, thereby preventing further research on these activities. How this quango carries out the annual monitoring of the colonies – as it claims to do – remains unknown, and why this information is not deposited in the national Seabird Colony Register, which it supports, requires investigation. It is likely that eggs laid by Mediterranean Gulls (*Larus melanocephalus*) are being inadvertently taken at some of the colonies for which licences are issued.

Natural England has indicated that the licences place restrictions on the dates of collecting, these normally covering the period 1 April to 15 May each year. They also restrict the length of time collectors can spend in the colonies on each visit. Natural England revealed that the County Durham licences authorise collection of eggs from three large colonies of gulls, which therefore must be in Upper Teesdale, but the quango is not willing to disclose which colonies are involved.

Black-headed Gull eggs collected in England are currently sold by up-market London stores and find their way onto the menus of several prestigious restaurants. The eggs are normally served hard boiled or lightly boiled and seasoned with celery salt, and a serving containing three eggs may cost anything from £7.50 to £22. One restaurant in 2015 was offering omelettes made from three gulls' eggs filled with lobster meat.

WINTER ABUNDANCE

Few British-reared Black-headed Gulls leave the region in winter, with most making only local movements. Of those that do migrate, a higher proportion move to Iberia from south-east England than those reared in north-east Britain (Fig. 25). In contrast, large numbers of Black-headed Gulls move from the Continent to winter in Britain and Ireland.

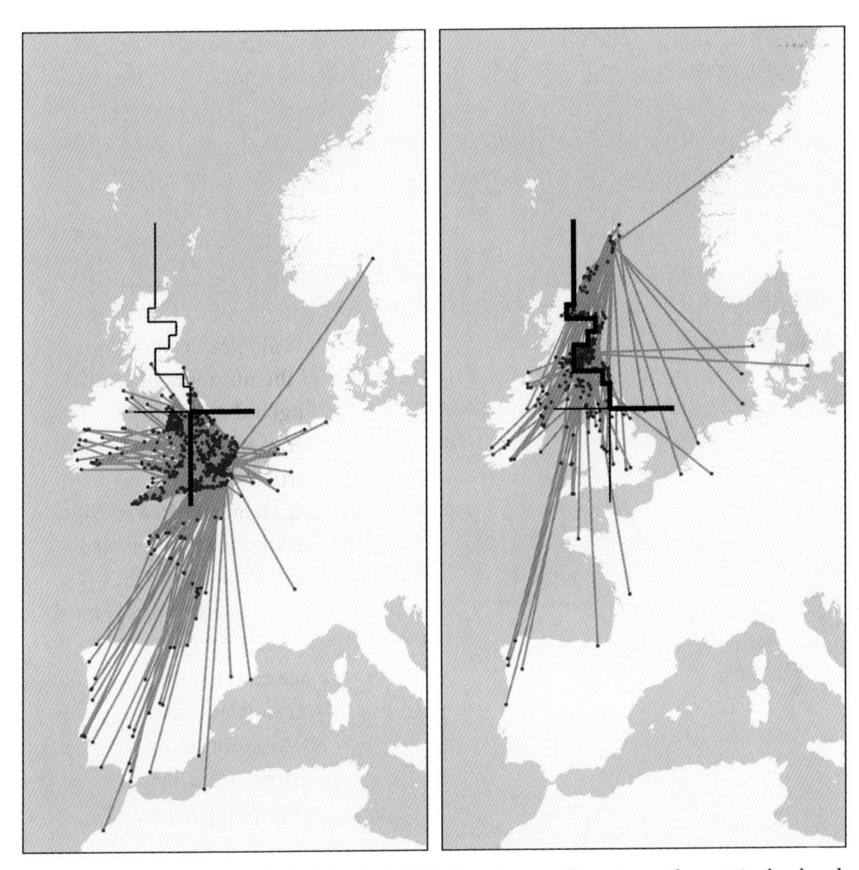

FIG 25. Left: Movements of Black-headed Gulls ringed as nestlings in south-east England and recovered outside the breeding season, based on 976 recoveries. Right: Similar mapping, but based on 447 recoveries of Black-headed Gulls ringed in north-east Britain. Note that fewer birds moved to Iberia from the north-east and that birds from the two regions have different winter distributions in Ireland. Reproduced from *The Migration Atlas* (Wernham *et al.* 2002), with permission from the BTO.

The BTO Winter Gull Roost Survey in 2004 estimated that more than 1.6 million Black-headed Gulls were using winter roosts in the United Kingdom and Northern Ireland., and since the total did not include all winter roosts, there are probably more than 2 million individuals wintering here. This number can be compared with 280,000 breeding birds in the United Kingdom and Northern Ireland, which with immatures added comes to about 360,000 individuals. As most of these remain here in winter, it suggests that some 77–82 per cent of wintering Black-headed Gulls recorded in 2004 had moved here from the Continent. Gabriella Mackinnon and I obtained a comparable figure of 71 per cent, derived from ringing recoveries up to 1980 (Fig. 26). Both percentages are rough estimates, but it is clear that a major proportion of the wintering Black-headed Gulls in Britain and Ireland originate from mainland Europe. Without new estimates, it is not known whether these numbers from the Continent have declined in more recent years as a result of the relatively mild winters.

Some readers may find it strange that, despite the satisfactory status and abundance of the Black-headed Gull as a breeding species in Britain, it has been Amber-listed here since 2015 as a species of conservation concern. This is based on the belief that numbers of wintering Black-headed Gulls in Britain have decreased by 33–50 per cent over the last 25 years, based on counts at winter night roosts. Since the British breeding numbers do not appear to have reduced (p. 46), a possible explanation is that a major decline has occurred in northern Europe – the area from which many of the British wintering gulls originate. However, there is little evidence of this apart from Denmark. An alternative explanation for the decline in winter numbers in Britain is that more individuals are remaining on the European mainland during the milder winters encountered in recent years. A preliminary analysis of recent winter ringing recoveries of Black-headed Gulls marked on mainland Europe suggests that the change of wintering areas is more

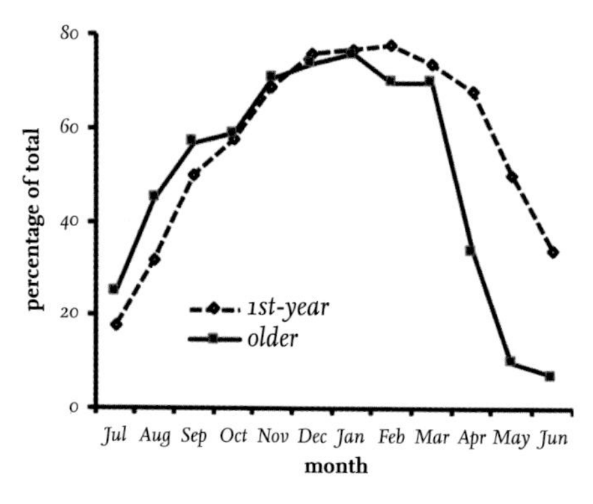

FIG 26. The percentage of Black-headed Gull recoveries reported in Britain and Ireland in each month that were originally ringed in mainland Europe. Note the trend for first-year birds to return to the Continent later. The recoveries of a few Continental birds in May and June may represent rings reported some time after finding.

likely to be the explanation. If this is correct and the species is not declining appreciably in Europe as a whole, but is modifying its wintering areas, then is it really a matter of conservation concern? In contrast to the assessment in Britain, the International Union for Conservation of Nature (IUCN) still regards the Black-headed Gull worldwide as a species of Least Concern.

Arrival of wintering birds

The numbers of Black-headed Gulls arriving in Britain and Ireland for the winter boosts the breeding season population by a factor of about four. These visitors come from a wide range of countries in northern Europe, with most starting their journey at locations around the Baltic Sea (Figs 27 and 28). Their arrival is spread over several months, from late July to early November (see Fig. 26).

The countries of origin of Continental Black-headed Gulls visiting Britain varies slightly with their final destination, with birds flying south-west or west to reach their wintering area (Figs 27 and 28). The exceptions are individuals reared in Iceland, some of which winter in Scotland and Ireland, but most of which avoid England and Wales. Other Black-headed Gulls reared in Iceland have been recovered from the east coast of North America, a route apparently taken by all of those reared in Greenland, although this wintering area is totally avoided by birds breeding in mainland Europe and in Britain.

Fig. 29 shows the number of recoveries of foreign-ringed Black-headed Gulls in Britain and Ireland for each month of the year up to 1985 in relation

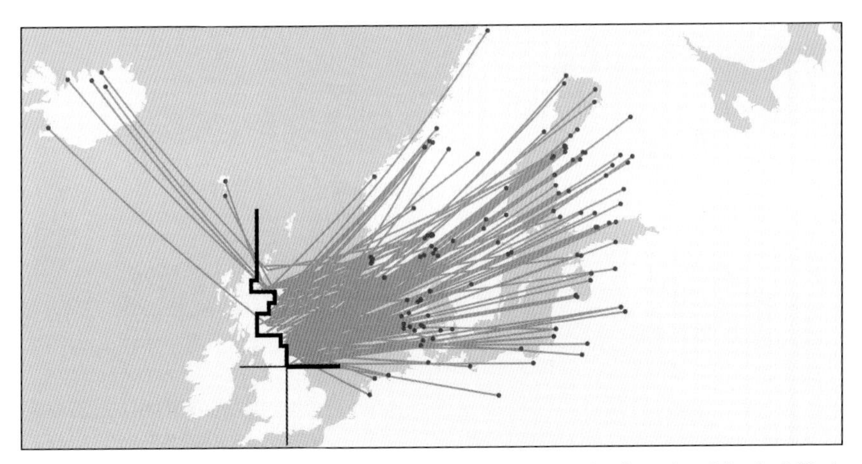

FIG 27. Black-headed Gulls ringed in winter in north-east Britain (to the right of the thick black line) and recovered abroad in a subsequent breeding season. Note the five Iceland recoveries. Reproduced from *The Migration Atlas* (Wernham *et al.* 2002), with permission from the BTO.

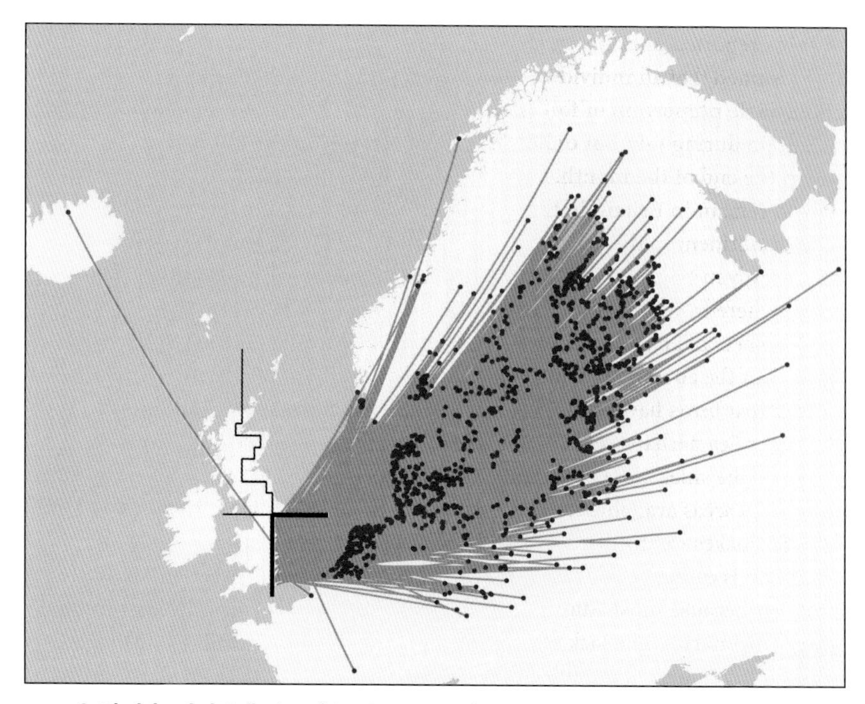

FIG 28. Black-headed Gulls ringed in winter in south-east Britain (to the right of the thick black line) and recovered abroad in a subsequent breeding season. Note the single Iceland recovery. Reproduced from *The Migration Atlas* (Wernham *et al.* 2002), with permission from the BTO.

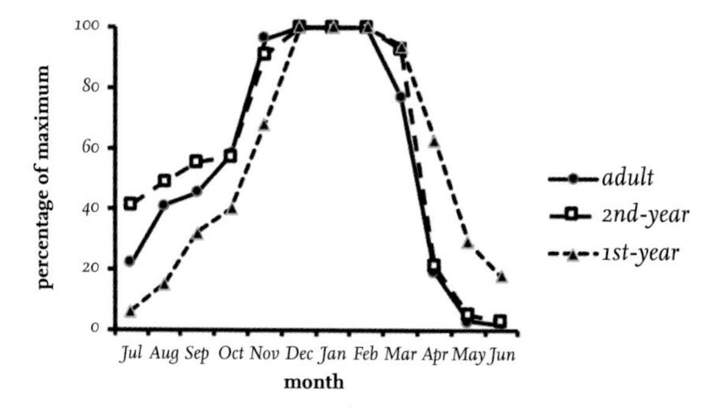

FIG 29. Recoveries in Britain and Ireland of Black-headed Gulls ringed in mainland Europe expressed as a percentage of the monthly recoveries between December and February. Note the high proportion of birds entering their second year of life (which moult early) reported in Britain in July, and the late departure of first-year birds in spring, with a few apparently remaining in Britain during the summer.

to those reported monthly from December to February, a period during which it is assumed that all individuals have arrived at their wintering areas. An appreciable proportion of foreign adults, possibly mainly failed breeders, arrive in Britain during July, but only a few young of the year have crossed the North Sea by the end of the month. A small proportion of first-year individuals reared abroad remain in Britain in the following summer, having failed to return to the Continent in the spring. These birds, entering their second year of life and already in Britain in July, are joined by more similar-aged immature birds moving here from the Continent.

Numbers of all ages cross to Britain in August. By the end of that month, about half the adults that winter here have already arrived, but only 15 per cent of first-year birds have arrived. The early movement from countries bordering the Baltic Sea perhaps reflects the less favourable feeding conditions there. For example, the small tidal changes in the Baltic markedly restrict the extent of inter-tidal areas available for foraging, while the warmer and drier continental weather makes earthworms less available there in late summer.

There is evidence of a lag in new adult arrivals in Britain in September, probably because most adults are still completing their moult and growing their longest primaries. The lack of, or reduction of, these long feathers must impair their ability to make long flights. A second peak of arrival occurs in November, and by the end of that month most of the adults that winter here have arrived, apart from a few birds that possibly turn up in December. The maximum numbers of Black-headed Gulls are present in Britain for only three months, from December to late February.

The timing of immigration of first-year and adult Black-headed Gulls in autumn to Britain depends on the country of origin. Dates after 1 July by which the first 25 per cent of recoveries of young birds and adults from abroad are recovered have been used as an index of the arrival times of first-year and adult Black-headed Gulls from different countries in Europe (Table 11). In general, adults and young birds arrive on different average dates, but this is not constant for each country of origin and varies by more than 60 days. Young birds arrive earlier than adults from countries closer to Britain, but the reverse applies for those from more distant countries. This suggests that first-year birds start their migration earlier than adults but move much slower, perhaps making a series of short journeys interspaced with stopovers at rest areas lasting a number of days. In contrast, adults appear to make much longer flights to their wintering area and arrive in Britain from European countries at less variable dates. The interpretation of the trend line in Fig. 30 suggests that first-year birds on migration take about 43 days longer than adults to travel 1,000 km.

TABLE 11. Date by which the first 25 per cent of adult and first-year Black-headed Gulls ringed abroad have been recovered in Britain in each 12-month period, starting 1 July. Based on MacKinnon and Coulson (1986).

Country of origin	*Distance to central England (km)*	*Number of adult recoveries considered*	*Date by which first 25% of adults recovered*	*Number of first-year recoveries considered*	*Date by which first 25% of first-years recovered*	*Difference in arrival dates between first-years and adults*
Belgium/ Netherlands	480	279	10 Sep	103	18 Aug	+23 days
Germany	720	150	27 Oct	57	7 Oct	+20 days
Denmark	770	187	14 Oct	80	12 Oct	+2 days
Norway	1,000	149	27 Sep	95	27 Sep	0
Sweden	1,400	167	30 Oct	73	30 Oct	0
Poland/ Lithuania	1,500	111	3 Sep	51	21 Sep	–18 days
Latvia/ Estonia	1,700	216	26 Oct	98	19 Nov	–24 days
Finland	1,900	231	12 Oct	141	1 Dec	–50 days

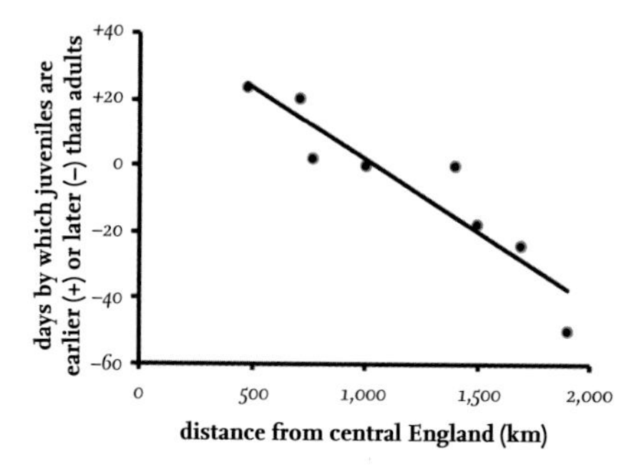

FIG 30. Difference in the dates of the first 25 per cent of recoveries in Britain of first-year and adult Black-head Gulls of Continental origin (data from Table 8) in each 12 months from 1 July, plotted against the distance from country of origin to central England. Note that the first-year birds from countries close to England arrived earlier than the adults, but that the converse was true for those travelling further.

Faithfulness to wintering areas

There is extensive and convincing evidence that some adult Black-headed Gulls from the Continent return to the same immediate wintering area year after year. The best example is a gull ringed in Poland, whose ring numbers were recorded in the same central London park over eight consecutive winters. However, such faithfulness to wintering areas is not absolute, as there are many instances of birds from the Continent wintering in Britain one year and then remaining in mainland Europe (usually in the Netherlands or Denmark) in the following winter (Fig. 31). It has been suggested, but not yet proven, that relatively mild conditions in western Europe in some recent winters have resulted in fewer Black-headed Gulls migrating here; this effect has also been reported for several species of waders and other waterbirds.

Evidence of the extent to which Black-headed Gulls use the same wintering area in successive years comes from individuals wing-tagged in north-east England, as

FIG 31. Black-headed Gulls ringed in winter in Britain and recovered more than 20 km away in a subsequent winter. Many individuals returned to Britain, but note the appreciable number that failed to do so in a subsequent winter and were recovered on the Continent. Reproduced from *The Migration Atlas* (Wernham *et al.* 2002), with permission from the BTO.

TABLE 12. The proportions of marked Black-headed Gulls that returned to the same wintering area the following year.

Age when marked	*Number marked*	*Number returning*	*Percentage returning*	*Percentage returning after adjusting for mortality*
Adult	593	285	48%	56%
Second-year	138	44	32%	38%
First-year	315	55	17%	24%

these had a high probability of being seen if they returned in the following winter. Most tagged birds were breeding on the Continent, and so the annual return would involve many individuals migrating considerable distances. In the winter after tagging, those that returned were seen on an average of five different days, so it is unlikely that many others were missed. The study detected that 48 per cent of adults and 32 per cent of second-year birds, but only 17 per cent of first-year birds, returned to the same area in successive winters (Table 12). Allowing for mortality in the intervening year (estimated at about 15 per cent for adults and 30 per cent for first-year birds), 56 per cent of the surviving adults returned to the same wintering locality, with 38 per cent of those in their second winter when marked returning for a second time, but only 24 per cent of those marked in their first year.

A total of 38 marked individuals that did not return to the study areas were seen by other observers elsewhere in the winter following marking. These were reported mainly on the east coast of Britain, south of the study areas. Two extreme cases were recorded; one bird 450 km away on the south coast of England, and the other 300 km north at Aberdeen in Scotland. In addition, one remained in Denmark. It is evident from this that only a proportion of individuals of all age classes return to the same wintering area, but that faithfulness is stronger (although not complete) in adults. The reason why some birds return while others fail to do so remains unknown, but is not primarily related to their sex.

Return of birds to the Continent

The return migration from Britain to the Continent is much more synchronised than the arrival and begins in late March, with most adults migrating in April (Fig. 32). First-year birds tend to depart later than adults and some do not leave until May, while others (perhaps 10 per cent of immature birds wintering here) remain throughout the summer. These birds join the feeding adults associated with colonies in Britain (since they leave their wintering feeding areas) and

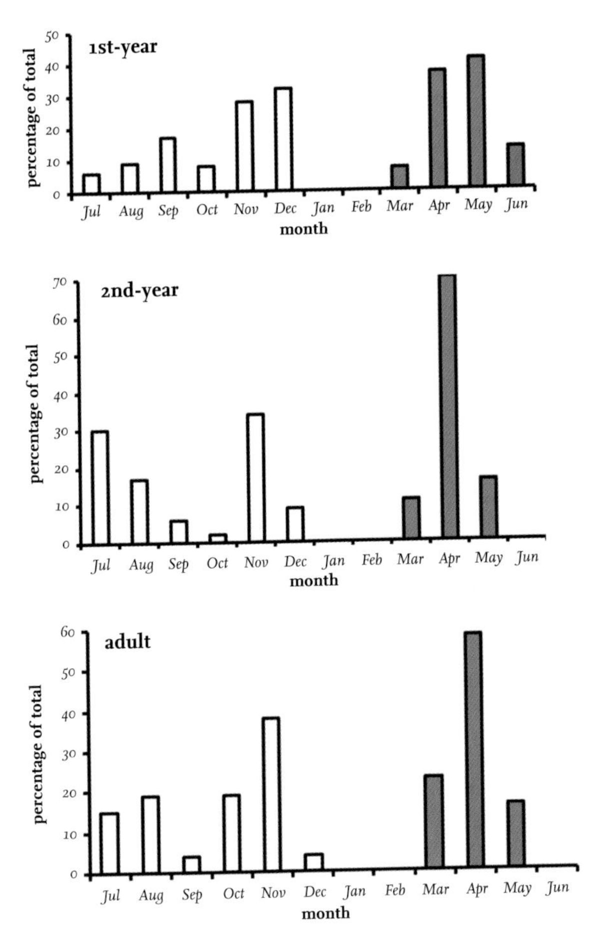

FIG 32. The month of arrival of Black-headed Gulls entering Britain from the Continent in autumn (white bars) and leaving in spring (blue bars), based on monthly change in numbers of ringing recoveries of foreign birds in Britain. Note that for all age classes, the spring departure is much more synchronised than the autumn arrival dates.

presumably do not return to the Continent until they have overwintered here for a second time.

By the end of July, numbers of second-years have arrived in Britain and joined the few that spent the summer here. These birds are not constrained by breeding and moult about four weeks earlier than adults, so this might explain their early crossing of the North Sea. More arrive in late October and early November, and many late arrivals originate from further away, east of the Baltic.

Distribution of Continental wintering birds

The Continental Black-headed Gulls that winter in Britain do not disperse randomly across the country, and are much more abundant in eastern and southern regions. The proportions of young gulls ringed in Britain or on the Continent and recovered from December to February in each of 10 regions of Britain and Ireland (4,267 recoveries in total) were used to produce an index of the proportions of geographical distribution in each area (Fig. 33). The percentages of Continental birds in these different regions ranged from 7 per cent to 80 per cent, based on at least 115 recoveries in each region and, in most cases, many more. The wide range of percentages obtained suggests that they are a close approximation to the actual proportions of birds of Continental origin. Ireland, Scotland and Cumbria had low proportions of Continental gulls, while the highest proportions occurred in south Wales, southern England, and eastern England from Yorkshire southwards. Overall, about 71 per cent of the wintering gulls in Britain and Ireland originated from the Continent – a high value, but consistent with the much higher overall numbers of Black-headed Gulls reported here compared with estimates in the breeding season.

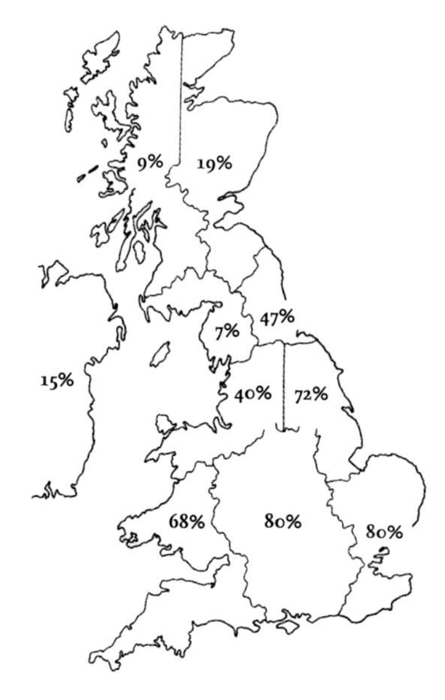

FIG 33. Estimates of the percentage of Continental Black-headed Gulls present in 10 regions of Britain and Ireland in winter based on numbers of recoveries of British and Continental ringed birds in each region between early December and the end of February.

Sex ratio of wintering birds

While the sex ratio of Herring Gulls and Great Black-backed Gulls (*Larus marinus*) captured in Britain during the winter approaches equality, that of wintering Black-headed Gulls shows a marked skew, with many more males being present. The data in Table 13 show that birds identified at breeding sites in northern England in May and June contained a minor excess of 108 males per 100 females. However, the sex ratio of wintering birds captured in England changed considerably, with a threefold excess of males in October and November, followed by the reversion of proportions to near equality between December to February. In the winter samples, most individuals were visitors

TABLE 13. The sex ratio of adult Black-headed Gulls cannon netted at landfill sites between October and February in north-east England, and near large colonies in northern England in May and June. Males were identified by a head and bill measurement of 82 mm or longer.

Month of capture	*Sample size*	*Males per 100 females*
Oct	138	345
Nov	168	282
Dec	172	102
Jan and Feb	206	102
May and Jun	1,112	108

from the Continent, and the skewed sex ratio in October and November suggests that either migrating males were arriving earlier than the females by an average of a month or more, or that males and females were feeding at different types of sites not sampled in those months, but were feeding at the same sites later in the winter. A total of 20 males and 11 females wing-tagged in the autumn were subsequently reported to have returned to the Continent, which gives some support for the suggestion that the bimodal pattern of the arrival time of Continental gulls (Fig. 32) represents males migrating earlier than females.

BREEDING BIOLOGY

Black-headed Gulls build their nest on the ground in areas with low-growing vegetation of varying density, ranging from sand dunes with much bare ground, to taller and floating vegetation around tarns and lakes. However, a few exceptions have been reported. At Seamew Crag, a small islet on Lake Windermere, Clive Hartley and Robin Sellers recorded several of the 50 pairs in the colony there nesting on top of low bushes over several years following 2009 (pers. comm.). In East Anglia in 1947, an entire colony numbering more than 300 pairs switched to nesting 2–3 m above ground level in young spruce trees, apparently in response to the flooding of their usual nest sites on the ground (Vine & Sergeant, 1948). Unlike some other gull species, there is only one old record of Black-headed Gulls nesting on a building, but in 2015 Robin Sellers found three small groups doing so, two near Perth and the other at Montrose, both in Scotland (pers. comm.).

Philopatry and colony faithfulness

Most Black-headed Gulls are two years old before they breed for the first time and a minority are a year older before they breed. Exceptionally, one-year-old individuals attempt to breed, although their success is very low. Many of those that

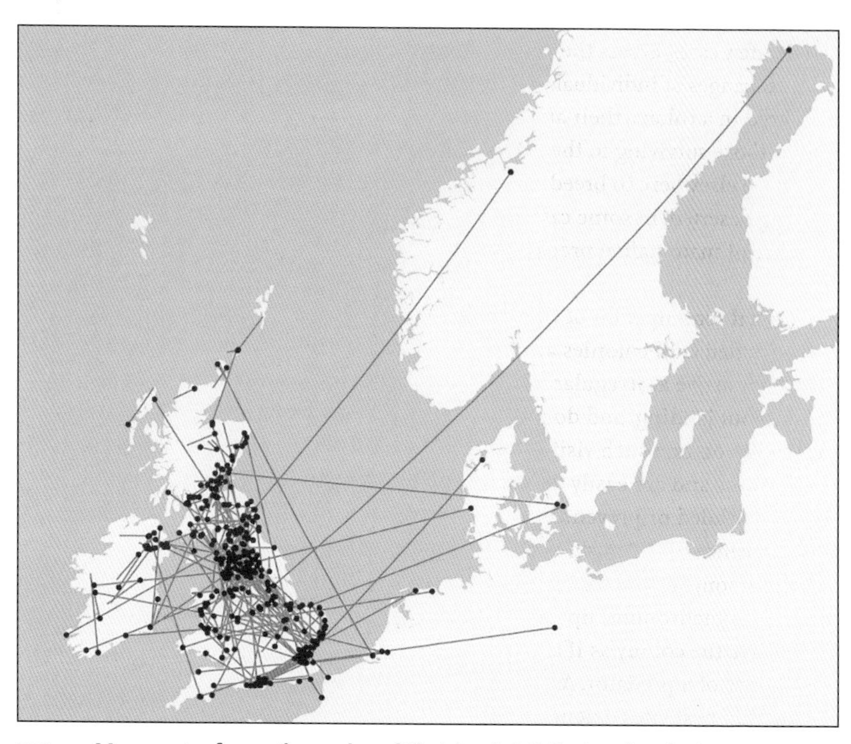

FIG 34. Movements of more than 20 km of Black-headed Gulls ringed as chicks in Britain and Ireland and recovered in the breeding season when of breeding age. Many returned to breed at or near where they were reared, but a similar proportion apparently moved to other colonies. A few moved to the Continent to breed, mainly between northern France and Denmark, but one moved to Germany, another to Norway and a third to northern Sweden. Reproduced from *The Migration Atlas* (Wernham *et al.* 2002), with permission from the BTO.

have survived to maturity return to breed in the colony in which they hatched as chicks (called philopatry), but others move to other colonies, often some distance away. The high proportion returning to the natal colony indicates that the young birds retain a good memory of where they were reared. However, it is easier to find marked individuals that have returned to their original colony than those that have moved elsewhere and consequently the extent of philopatry is often exaggerated. A realistic estimate of the extent of philopatry is obtained by measuring the proportion of those ringed as chicks and have reached adult age that were recovered during the breeding season less than 20 km from their natal site. In the case of the Black-headed Gull, about 60 per cent of the young that survive to breeding age are philopatric and the remaining 40 per cent move to other colonies and in a few cases, across the North Sea, to breed on the Continent. There have also been exchanges of individuals between Ireland and Britain (Fig. 34). Once adults have bred in a colony, their attachment to it becomes high, and at least 85 per cent of those surviving to the following year return to it to breed. The few adults that move elsewhere to breed are mainly from colonies that are in the process of being deserted, in some cases as a reaction to repeated nesting failure and the presence of mammalian predators.

Annual reoccupation of the colony

Black-headed Gull colonies are first visited by groups of adults in March each year. They arrive at irregular intervals during the morning, flying over the site without landing, and do not remain long at the colony or in the vicinity. Eventually, on one such visit some birds do land, but they exhibit a degree of nervousness and are easily disturbed. These visits become more frequent, but will be curtailed or prevented by cold and windy weather. As the days pass, the daily presence of the birds at the colony lasts much longer and spreads into the afternoon, but the colony is always vacated before sunset. The gulls usually leave by a synchronous 'up' or 'panic' flight, which sees them all suddenly rising high above the colony as if alarmed, yet without an obvious stimulus such as the appearance of a predator. As April progress, the amount of time the gulls spend at potential nest sites extends to the greater part of the day, but the colony is still deserted each night and reoccupied early in the morning, sometimes well before sunrise.

Pairs are formed early and courtship displays become common (Fig. 35). In the meantime, nesting material is collected locally and brought in to the colony. Birds nesting in some coastal colonies often collect substantial and untidy

FIG 35. Female Black-headed Gull (right) courtship-begging for food from a male. (Norman Deans van Swelm)

quantities of brown seaweed, while those at inland sites collect dry grass locally and carry it to the selected nest site to construct the nest. More material is usually added to the nest during incubation. Colonies and nesting sites are usually close to water, and nests are often substantial structures that raise the eggs above the local water levels.

The density of nests varies according to the size of the colony and the nature of the nesting site. Nests are 1–2 m or more apart where plant growth is sparse, but only 50 cm apart in dense vegetation, presumably because this tends to conceal the neighbouring pair, moderating the extent of aggression between close neighbours.

Eggs and incubation

Black-headed Gull eggs are brown with spots of a darker colour. They are laid in mid- to late April (the date of the first egg laid in different years at Ravenglass ranged between 12 April and 26 April), but even at this time the colony is still often deserted at night, with the adults moving to night roosts and leaving the early eggs unprotected. As clutches are completed and incubation begins, increasing numbers of birds remain in the colony throughout the night; it is at this point that predation by Foxes on incubating adults may occur.

In Britain, the peak of laying in Black-headed Gulls is reached at the end of April or in early May, which is earlier than in other gull species breeding here. Fig. 36 shows the date distribution for the first eggs in each clutch recorded by Ian Patterson at Ravenglass (1965). There is considerable laying synchrony by the majority of pairs, but there is often a distinct tail or a secondary late peak of birds that re-lay after losing their first clutch or delay laying owing to inexperience.

The typical clutch comprises three eggs, although two-egg clutches are common. Occasionally, four-egg clutches occur, but whether these are laid by

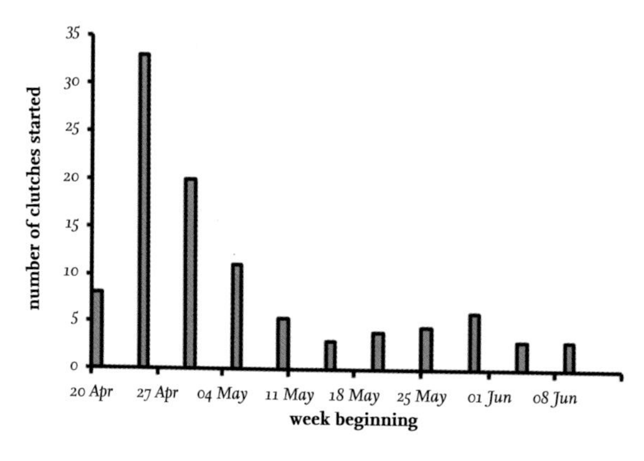

FIG 36. The spread of the laying season at Ravenglass, recorded by Ian Patterson in 1965. The secondary peak at the end of May/early June probably comprised repeat clutches produced by pairs that had lost their first clutch or by pairs of young, inexperienced birds.

only one female has not been investigated. Late-laying birds produce clutches of just one or two eggs. Average clutch size within a colony varies considerably, from 2.3 to 2.8 eggs. Nils Ytreberg (1956) recorded a high average of 2.9 eggs from 411 nests in Norway, but in a later year reported a smaller average of 2.62 eggs based on a sample of 100 nests (Ytreberg 1960). The average clutch size tends to be lower in years when laying starts late, such as in small colonies and those further north and in the uplands. Eggs are laid at variable times throughout the day and the interval between eggs is usually about two days.

Incubation of eggs and feeding of young is shared by both members of the pair. The eggs are covered by an adult irregularly while the clutch is still being laid, with the first egg probably protected rather than incubated. Obviously, incubation does not take place at night on occasions when the adults desert the colony in the evening. Intensive incubation starts at variable stages during laying of the clutch and often before the third egg is laid, but the brood patches of incubating birds are not fully vascularised for the first few days and initially the covered eggs may not reach a high enough temperature to facilitate development. This results in variable incubation periods, and the early start of incubation is sometimes enough to cause the third egg to hatch a day or so after the first two. Ivan Goodbody (1955) recorded incubation periods from the laying of the last egg of between 23 and 26 days at different nests.

Rearing young

A study by Roland Brandl and Ingrid Nelsen (1988) reported that Black-headed Gulls feed their chicks at intervals of about 45 minutes during daylight (but not at night), and that chicks receive at least 20 feeds each day. Each parent made five to six feeding trips per day, and obviously retained food from each trip for at least two feeding bouts. The authors found that this rate was mainly independent of the nestlings' age or brood size. Adults regurgitate food onto the ground for the chicks and often re-swallow any surplus.

The newly hatched young have cryptically marked down, and as they grow, the feathers are patterned in shades of brown. The chicks remain in the nest for about a week if undisturbed and are brooded by their parents for a decreasing length of time each day as they grow older. Brooding at night often continues for a day or two longer after daytime brooding has ceased. Chicks eventually move a short distance away from the nest and shelter in nearby vegetation, where they receive some protection from both adverse weather and predators. One parent remains at the nest site most of the time, feeding the chicks by regurgitation and defending them against neighbouring adults and attacking wandering young from other pairs.

Once chicks reach 20 days old, they develop an interest in searching the ground for items they can pick up and swallow. Although they probably find little that is edible this way, the behaviour seems to develop their ability to search for food. Young birds can fly when they are about 34 days old and leave the colony area soon after fledging, often accumulating in small groups in open areas nearby. There, they search for food, later joining feeding flocks of adults in fields or on the shore. There is no evidence that the young are accompanied or fed by their parents once they leave the colony, nor do they return to the nest site to be fed after their first flight.

Breeding success

Breeding success in Black-headed Gulls is highly variable and depends on whether the colony is protected from predators, including humans. American Mink, rats, Badgers (*Meles meles*), hedgehogs, Foxes and herons are major predators at some sites, and are suspected to have caused the desertion of several colonies. Intensive predation can result in very few young fledging from a colony. For example, in a sample area at Ravenglass studied in detail by Ian Patterson, 2,213 pairs each fledged an average of only 0.06 young in 1961, while in 1963, 2,290 pairs each fledged an average of 0.10 chicks (Patterson, 1965). This poor breeding success was mainly caused by Fox predation on eggs, young and adults late in the breeding season.

At colonies where predation has been typically much lower than at Ravenglass, Black-headed Gulls probably fledge about 0.6–0.7 young per pair each breeding season. On its website, JNCC reports an average production of about 0.6 young per pair annually based on a few sample colonies studied between 1987 and 2015, but in the best year 1.2 chicks per pair fledged. Studies of several colonies over many years in France reported an average of 1.4 young fledged per pair (Péron *et al.*, 2009). Once the last few young fledge, which is usually in late July or early August, colonies are rapidly deserted until the next breeding season.

FOOD AND FEEDING

Despite being small, the Black-headed Gull has been recorded feeding on a wide range of items, although in general these are smaller than those taken by the larger gulls. Animal material dominates, with invertebrates being by far the most frequently consumed food. Earthworms and the adult and larval stages of beetles and flies predominate in food obtained in fields. In coastal areas, marine crustaceans, molluscs and worms are frequently consumed when mudflats are exposed at low tide, or small food items are taken from the sea surface (Fig. 37). In addition to the consumption of animal matter, the gulls ingest a wide range of plant materials from time to time. Some of these are ingested incidentally with animal food, but seeds and grain are intentionally consumed – in autumn, for example, acorns are occasionally plucked from the tops of oak trees and Hawthorn (*Crataegus monogyna*) berries are picked off hedgerows by birds momentarily alighting on branches. A wide range of animal and vegetable materials, including bread, are taken around human habitation and at landfill sites. In some areas, bread and food scraps are regularly supplied by the public, particularly in parks and by lakes, and these form a major part of the winter diet in urban areas.

The gulls look for potential food either by flying 1–2 m over the ground or water and landing to pick up items, or (more frequently) by landing, usually as a dense flock in a field, where they spread out by walking over the ground. This latter habit is used extensively just after dawn, as soon as the gulls arrive from their night roosts and when earthworms are still active on or near the ground surface. Small fish and riparian insects are captured and consumed on or near freshwater sources. Insects such as swarming winged ants are caught in flight on calm days in late summer. At times, adults bring large quantities of hairy caterpillars collected from moorland to feed to their young, but the hairs of some of these species are irritants and many chicks regurgitate them in numbers, such that they sometimes litter the ground within the colony. Occasional small mammals and amphibians are consumed, but these form a minute part of the diet.

FIG 37. Black-headed Gulls feeding on small fish larvae. Note the unusual head position. (Norman Deans van Swelm)

Kleptoparasitism

Kleptoparasitism is common in Black-headed Gulls, whereby the birds steal from members of their own species and, particularly, flocks of Lapwings (*Vanellus vanellus*). Golden Plovers (*Pluvialis apricaria*) feeding on agricultural land in winter also regularly have their food stolen by the gulls. Lapwings and Golden Plovers feed mainly on earthworms and insect larvae obtained by actively searching the ground, and they are often joined by Black-headed Gulls (and sometimes Common Gulls, *Larus canus*). However, rather than searching for worms, some gulls just stand nearby or follow the plovers; when a plover finds a large food item that requires manipulation before it can be swallowed, it is rapidly and often successfully challenged by one of the gulls, and frequently the food item is stolen. In other cases, a plover may take off with the worm or insect in its bill, only to be closely pursued by a gull, which attempts to force it to drop the prey. When this happens, the gull takes the food as it falls or quickly turns to pick it up from the ground.

Such kleptoparasitism of Lapwings and Golden Plovers decreases the waders' own rate of food consumption, but this is partially mitigated by feeding at night under full moon conditions when the gulls are absent. The fact that Black-headed Gulls resort to kleptoparasitism suggests that they are not as efficient as plovers at searching and finding their own earthworms.

Kleptoparasitism by Black-headed Gulls appears to be a specialised feeding behaviour used by a minority of individuals, and experience is important in successfully obtaining food in this way. Even by their second year of life, young Black-headed Gulls have still not achieved the efficiency of adults when carrying out kleptoparasitic attacks (Hesp & Barnard, 1989).

In urban areas in winter, the tables are often turned on Black-headed Gulls. Those birds finding large food items that they fail to swallow immediately are frequently pursued by Common Gulls or even members of their own species, until they are forced to drop the food to the benefit of the pursuers. Such pursuits are particularly frequent in urban areas during severe winter weather, when large numbers of Common Gulls move from snow-covered high ground and congregate at much increased densities at urban feeding sites.

Feeding areas

In the breeding season, the feeding areas and ranges of Black-headed Gulls are limited by the location of the colony, and well-defined flight lines are often evident to and from large colonies. Many feeding adults forage within 10 km of the colony, although some will travel 40 km or more to favoured sites with a regular supply of food provided through the activities of humans. These may be fields where ploughing is in progress, landfill sites, large car parks (particularly near food stores), coastal picnic sites or riverbanks in towns where people feed bread, chips and other items to birds. Food supplements put out for farm animals are also exploited. Few Black-headed Gulls hunt far from the shore at coastal sites, and only rarely do they plunge-dive and submerge to capture fish.

In the 1960s, along the 18 km tidal reaches of the river Tyne (which was little more than an open sewer at the time), Black-headed and Common gulls were the commonest birds feeding in winter at sewage outfalls and on sewage items floating in the river. Once the sewage was piped separately to treatment plants to clean the river up, the numbers of Common Gulls reduced dramatically, but Black-headed Gulls remained abundant and were still able to find food in the river. This is because the outflow of filtered water passing from sewage plants into the river still contained small items, which were attractive to the Black-headed Gulls but not the Common Gulls.

The habit of large numbers of Black-headed Gulls following tractors ploughing fields is both widespread and spectacular, with many individuals competing for position just behind the plough blades, and diving to catch and consume worms or insects exposed when the soil is turned over. Similar feeding occurs when grass is being cut for hay. The introduction of grass cutting earlier in the year for silage attracts many Black-headed Gulls and has had a marked effect in some areas where the birds feed. Although silage production was introduced many years ago, it became far more extensive from about 1969, and by 1993 it had increased sevenfold as a means of producing winter cattle feed. Grass is now frequently cut twice a year for silage and much earlier in the season than for hay. This has allowed feeding by Black-headed Gulls on insects in these fields to be spread over a much longer period of the species' breeding season than was once the case, and presumably has been of benefit to the gulls.

Feeding at landfill sites in winter attracts flocks of hundreds and even thousands of Black-headed Gulls. It is a highly social activity and occurs at irregular intervals, often with a resting, inactive flock remaining nearby for long periods. From time to time, one or two individuals leaving the flock will then fly over the landfill working area and alight there or pick up a food item while in flight. This act of feeding by one or two individuals is a major stimulus to those birds in the resting flock, and the great majority will then leave the roost and stream over to the feeding site, land on the refuse and search for food – several hundred individuals may be highly active within a small area. Suddenly, feeding will stop, perhaps as the result of a loud noise, the approach of a person or the arrival of a lorry. On other occasions and for no apparent reason for alarm, the flock will rise as one and return to the roosting site, staying away from the landfill working area for many minutes or even hours.

Food-collecting techniques

Black-headed Gulls use several techniques when collecting food, summarised below:

1. *Aerial searching* Individuals patrol suitable areas, turning and swooping to pick up items discovered. Used on farmland and in urban areas.
2. *Walking over fields of short grass* The birds are usually spread out and search for food items on the ground.
3. *Feeding while floating on water* The birds swim buoyantly, turning rapidly to pick small items such as larval fish from the water's surface, phalarope-like (Fig. 37). Head-dipping into the water is less frequent and diving does not take place.
4. *Feeding frenzies* Many individuals converge on an appreciable food source, such as food discarded by the public, a freshly ploughed field or refuse at a landfill site. In all three cases, individuals are attracted from some distance by the erratic flight and calling of those birds already scrambling for food.
5. *Feeding on flying insects* This is most frequent in autumn, when winged ants take to the air on calm, warm days. The gulls 'hawk' after the insects, flying slowly and frequently changing direction.
6. *Consumption of items from trees, bushes and low plants* This is an infrequent method, with gulls approaching branches or low plants, landing on them momentarily, and plucking abundant insects from the leaves and thin branches. Similar methods are used to obtain acorns from oak trees and berries from Hawthorn bushes.
7. *Paddling in shallow pools on intertidal mud* This technique is used to disturb, detect and capture small invertebrates, including marine worms. While Herring Gulls also use this method on areas of short grass, this has not been reported in Black-headed Gulls.
8. *Kleptoparasitism* The birds steal food from Lapwings and Golden Plovers when the waders are feeding on soil invertebrates in pastures.

Use of winter feeding areas

The next sections are based on data gathered through extensive ringing and wing-tagging studies of Black-headed Gulls wintering in north-east England. Observing coloured leg rings on Black-headed Gulls was difficult in resting flocks and when they were feeding on grasslands, so to overcome this problem, plastic wing tags were used (p. 363). These proved more successful in identifying and following the behaviour of individuals throughout the day. The collaborative research involved identifying feeding areas and catching and marking birds prior to and during three years of intensive field studies on the species made, and was carried out by Gabriella MacKinnon at Durham University as part of her doctoral thesis (MacKinnon, 1986). It involved us in many cold early-morning starts in winter to reach the feeding areas and set cannon nets before the gulls arrived at about sunrise from their night roosts.

Most of the wing-tagged individuals studied in north-east England frequently returned to feed at the same sites during the whole winter (Table 14), but some moved away during this period. A marked difference in the behaviour of the gulls was detected between those using coastal sites and those feeding inland. At the coast, both adult males and females were particularly faithful to the same area throughout the winter, but at inland sites both sexes were less faithful, with adult females nearly twice as likely to move elsewhere than adult males. This difference between coastal and inland sites probably relates to the difficulty of accessing food in fields during periods of cold, frosty weather, which is less of a problem at coastal and intertidal areas as they are less susceptible to freezing. First-year birds were variable in the proportions that returned to feed at the same site, but overall, they were less faithful to sites than adults.

Sightings of wing-tagged birds that moved 10 km or more from the study areas in north-east England in the same winter and reported by other observers revealed that all moved south, remained on or near the east coast of England, and moved less than 100 km, apart from one bird, which moved more than 350 km. Of the 254 adults marked in winter in north-east England, 19 were seen in Britain during the breeding season. Nine of these birds were recorded within 50 km of the winter study areas, while 51 were recorded in colonies on the Continent, confirming that most of the marked birds studied bred outside Britain.

That some individuals return to the same feeding areas each day is remarkable – in some cases during the field studies, birds commuted up to 25 km

TABLE 14. The proportion of wing-tagged Black-headed Gulls in north-east England recorded in subsequent weeks during the same winter in a different feeding area. Data summarised from MacKinnon (1986).

Study area	*Adult males*		*Adult females*		*First-year*	
	Total sightings	**Percentage moving between areas**	**Total sightings**	**Percentage moving between areas**	**Total sightings**	**Percentage moving between areas**
Tyne (coastal)	865	1%	635	1%	—	—
Durham (inland)	648	15%	276	29%	198	32%

twice a day to and from a large overnight roost on land containing at least 10,000 birds. Observations of the Black-headed Gulls leaving their night roost in the early morning revealed that individuals from several parts of the roost left in successive waves spread over the 30 minutes before sunrise. Some of those leaving from different parts of the roost circled, formed flocks and departed inland. The birds arriving at sunrise at an extensive series of sport fields in Durham 25 km away gathered in one or two flocks, and invariably included several wing-tagged birds that were previously marked when feeding there. Yet some of these tagged birds (and probably the others) did not roost close to each other at night. How individuals dispersing from the night roost among thousands of others managed to gather into a flock returning to the same feeding grounds is a fascinating mystery to which I have no answer. How do birds regularly flying to the same feeding area manage to group together into a flock on leaving the large roost? This is an interesting aspect of behaviour that requires more research, which might be assisted through the use of small modern transmitters (p. 364).

Age structure in different feeding areas

First-year and adult Black-headed Gulls are not distributed in the same proportions in different feeding habitats within the same area. During the studies carried out in north-east England, first-year birds formed only 1–5 per cent of the Black-headed Gulls present near river mouths and on the coast

TABLE 15. The proportion of first-year Black-headed Gulls in winter counts near the mouths of four rivers in north-east and eastern England, and on the north-east coast. Proportions of first-year birds showed little variation between days or between years.

Site	*Total counted*	*Percentage of first-year birds observed*	*Total in cannon-net catches*	*Percentage of first-year birds in catches*
Mouth of river Tyne	13,466	5.0%	—	—
Mouth of river Wear	1,139	4.4%	—	—
Mouth of river Tees	1,877	4.1%	—	—
Humber Estuary	907	1.0%	—	—
Northumberland coast	1,040	3.2%	—	—
Durham county coast	4,552	5.0%	522	9.0%
Pastures and sports fields	15,196	22.1%	613	56.0%
Landfill (Jul–Nov)	6,780	11.2%	584	25.2%
Landfill (Dec–Mar)	5,428	20.2%	97	27.7%

(Table 15), and were more than four times as frequent feeding inland on pastures and grassland, averaging 22 per cent of individuals present. At landfills, first-year birds increased from 11 per cent in autumn to 20 per cent by midwinter. The reason for these differences is not known, but since birds feeding on the coast and those feeding inland used the same night roost, it would appear to be the result of variation in the feeding ability of birds of different ages (Table 16). This suggests that coastal feeding may require greater experience in order to become proficient, as has been shown in other studies of Black-headed Gulls relying on kleptoparasitism as a source of food (p. 78).

Interestingly, the cannon-net captures of Black-headed Gulls at landfill sites were composed of up to twice the proportion of first-year birds than those counted at nearby loafing sites before making a capture (Table 15). This probably indicates that the young birds take greater advantage of feeding frenzies and possibly have a greater risk of capture because of their less efficient feeding ability, which causes them to stay and feed for longer at landfill sites.

Feeding rates of Black-headed Gulls differ considerably between different sites (Table 16). Items are taken much more rapidly at sewage outfalls, as these sites offer an abundance of readily available food. Following the plough is less successful as it involves a degree of searching or gaining good positions, while walking over short grass on playing fields or pasture involves considerable searching before each food item is discovered. The figures in Table 16 do not take into account the size of each item consumed, although they were likely to be similar on fields and at coastal sites, and smaller at sewage outfalls.

TABLE 16. Feed rates measured as numbers of swallows per 12 seconds by Black-headed Gulls at different feeding sites. First-year birds were slower feeders than adults at all three sites. Data mainly collected by Gabriella MacKinnon and personal unpublished records.

	Adults		*First years*	
	Sample size	*Swallows per 12 seconds*	*Sample size*	*Swallows per 12 seconds*
Playing fields and pastures	47	0.15	26	0.04
Following the plough	21	1.10	21	0.43
Sewage outfalls	27	6.48	10	3.10
Landfill sites	59	2.20	67	1.41

Weight changes

The studies of Black-headed Gulls in north-east England confirmed that adult males are about 17 per cent heavier on average than females throughout the year, and that both sexes are 15–17 per cent heavier in winter than in summer (Fig. 38), reaching a peak weight in December when daylight is shortest. From February onwards, both sexes gradually lose weight, reaching their lowest weight in summer. Recently fledged first-year birds in July are much lighter than adults, but soon catch up, so that by September they have almost reached the weight of adults of the same sex. They then remain marginally lighter throughout the rest of the autumn and winter. The heavier weights in winter are, presumably, caused by an increased layer of subcutaneous fat beneath the skin, which reduces heat loss and acts as a reserve to tide individuals over difficulties encountered in feeding during severe weather.

The low weights of adults in summer does not seem to be a direct response to the stress and effort of breeding, because first-year birds, which do not breed, follow the same trend from November to March. Lower summer weights are achieved by adults by March, before they migrate and have started to visit their colonies. Between March and August, adults maintain the lower weight, and there is only slight evidence that breeding causes further weight loss. One effect of lower body weight in summer is that the energy cost of flying is reduced. This is at a time when food reserves protecting against cold weather and insulation to reduce heat loss are less necessary, and so there is no need for birds to carry the associated extra weight.

In the studies, only minimal differences were apparent in the weights of Black-headed Gulls caught in July and August just after dawn (before they have eaten their first feed of the day) compared to those caught in the afternoon

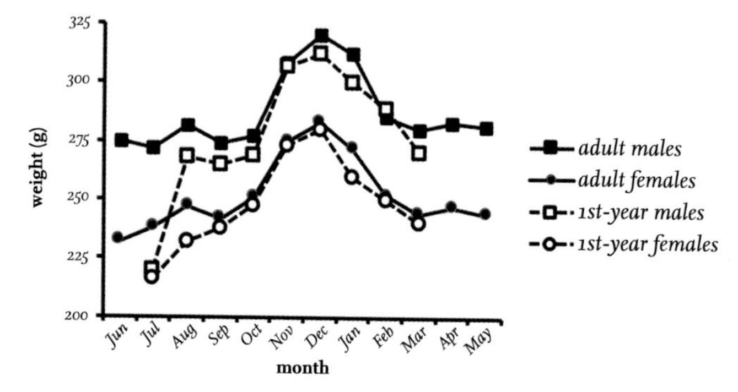

FIG 38. The monthly average weights of adult and first-year male and female Black-headed Gulls in north-east England. Based partly on data analysed by MacKinnon (1986) and on unpublished personal data, totalling 1,953 individuals. Note that few first-year birds were weighed in April, May and June.

(Table 17). However, as day length becomes shorter and the period when the gulls cannot feed increases, their overnight weight loss increases progressively. By December and January, the overnight weight loss recorded reached 18.3 g, which is an appreciable 6 per cent decrease of the afternoon body weight, at a time when body weight is at a seasonal maximum (Table 17 and Fig. 38).

There is a widely held view that winter is the period when birds face the highest risk of food shortages, leading to loss of weight and increased mortality. In the case of the Black-headed Gull, however, this is not so. Prolonged periods of hard frost do little to affect the average weight of the birds. Table 18 shows the deviation in weight of Black-headed Gulls from the monthly average in relation to the number of consecutive days with hard ground frosts before capture. Short periods of frosty weather had little impact on their weights, and even with more than four days of continuous hard frosts, their weights had declined by only 3.7 g, or about 1.2 per cent of the body mass expected at that time of year.

TABLE 17. Weight change between dawn and afternoon of adult Black-headed Gulls between July and January. Weights are the average for males and females recorded in north-east England. Significant overnight weight losses are indicated by an asterisk (*).

	Afternoon		*Dawn*		*Overnight loss (g)*
	No. of birds in sample	*Weight (g)*	*No. of birds in sample*	*Weight (g)*	
Jul and Aug	165	260.3	237	256.4	3.9
Sep, Oct and Nov	256	278.4	223	272.2	6.2*
Dec and Jan	405	298.3	135	280.0	18.3*

TABLE 18. The effect of consecutive days with hard ground frosts on the weight of Black-headed Gulls. Results are shown as the deviation from the average weight of adults of each sex in the month of capture in north-east England.

	Consecutive days of hard frosts			
	0	*1–2*	*2–4*	*5–7*
No. of samples	9	12	13	9
Change in weight (g)	+1.9	+1.6	−0.7	−3.7
Percentage change in weight	+0.6%	+0.5%	−0.2%	−1.2%

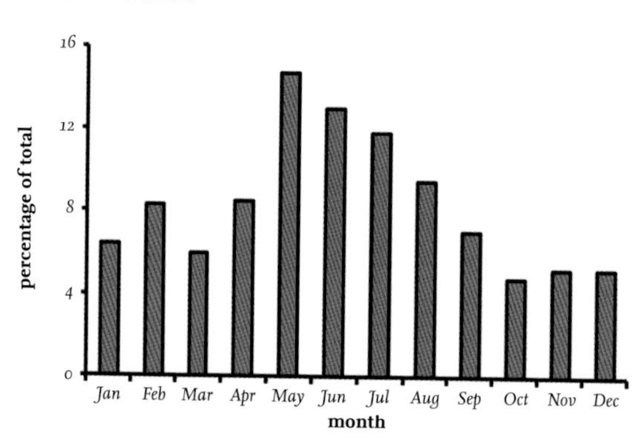

FIG 39. The percentage of the total number of ringed adult Black-headed Gulls reported dead in each month of the year. The distribution suggests that mortality is relatively low during the winter months and then peaks at about twice the winter rate during the breeding season (May to July). Figures are based on 2,575 recoveries of adults and include individuals recovered abroad.

With Black-headed Gulls achieving high body weights in winter and sustaining these during cold weather, this begs the question as to when the risk of mortality is highest in the species. This was first investigated by Jim Flegg and Christopher Cox in 1956–74 (Flegg & Cox, 1975), and the amount of related data have increased since then following the recovery of many more British ringed birds found dead from 1975 to 1984 (Fig. 39). In fact, the seasonal pattern of deaths has changed little since the original analysis, with the numbers reported dying each month during the breeding season (April to July) double those in each month between October and March. In the British climate, Black-headed Gulls are at a greater risk of being killed by predatory mammals while breeding (as found by Hans Kruuk at Ravenglass and Clive Craik in the Hebrides), than by food shortages in winter. While the birds are breeding and are attached to a particular colony site, their feeding areas are restricted. Agricultural fields planted with root or grain crops become unsuitable as feeding areas as the plants grow, while those where grass is allowed to grow tall for hay and silage production also become unsuitable until the first cuts are made. Once the grass is cut, the speed at which the fields are then visited by gulls strongly supports the belief that food may be in short supply during the breeding season.

ANNUAL SURVIVAL RATES

Considering that the Black-headed Gull is so abundant, there is relatively little information on the survival rates for members of the species. The first estimates were published in 1975 by Flegg and Cox, who suggested that the adult survival rate in the south of England was 76 per cent per year. This value was probably low because some recoveries of old birds were still to be reported, and in addition, doubts now exist about the durability of the rings then in use, which may have reduced the estimated survival rate (see below). More recently, Anne-Caroline Prévot-Julliard, Jean-Dominique Lebreton and Roger Pradel (1998) estimated a 90 per cent annual adult survival rate in France.

Gabriella MacKinnon and I used British BTO ringing recoveries to estimate the annual survival rates of Black-headed Gulls ringed as nestlings in the UK. The rate in the bird's first year of life proved to be low, followed by a constant rate of 84 per cent for the next 10 years of life (MacKinnon & Coulson, 1987). Thereafter, the annual survival rate declines progressively, with the oldest bird reported at 22 years of age; this is well below a maximum age of 33 years reported by Klaas van Dijk and Rob Voesten in 2014. There are two possible explanations for this pattern. One is that senility sets in after Black-headed Gulls reach about 10 years of age. The second is that the rings on the gulls start to be lost or become illegible after 10 years, resulting in fewer reports of old birds that are of this age in the population. We encountered a similar problem with aluminium and Monel rings used on Kittiwakes and Herring Gulls in the past, and examined both aluminium and Monel rings used by BTO ringers recovered during our studies on Black-headed Gulls in north-east England. We concluded that ring wear and loss was a problem and that a few individuals probably outlived the life of the rings used to mark them. As a result, survival rates were probably underestimated. In addition, based on information available for other seabirds, it is considered unlikely that senility occurs in 10-year-old individuals. We also found that Monel rings, which were introduced to overcome the wear found on aluminium rings, actually lost weight at a similar rate to the aluminium ones and were therefore little improvement.

Fortunately, methods have been developed that allow survival rates to be calculated based on age-truncated recoveries. MacKinnon and I used these in our studies (with help from Nicholas Aebischer and David Jackson) to determine annual survival rates for each year from 1950 to 1980 using recoveries of birds up to 10 years old (Fig. 40). This method excluded information on survival rates for the oldest 20 per cent of Black-headed Gulls living in the wild, but avoided bias produced by ring wear and whether senility was having any effect. Using this method on the accumulated BTO recoveries, annual survival of adults fluctuated around 83 per cent between 1950 and 1955, and then progressively increased to about 87 per cent from 1974 to 1980. The survival of first-year birds in the 1950s

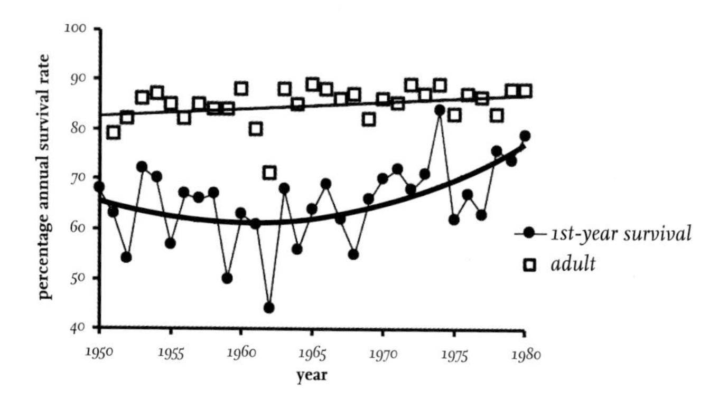

FIG 40. Estimates of the annual survival rates of adult and first-year Black-headed Gulls in Britain from 1950 to 1980. The adult survival estimates were based on birds recovered within 10 years of ringing to avoid bias. Each year runs from July to June, and the years shown relate to the start of each 12-month period. Both trend lines differ significantly from a constant survival rate over the period considered, and survival has gradually increased from 1950 to 1979. Note the low survival rate from July 1962 to June 1963, when adult survival declined to 71 per cent and that of the young fledged in 1962 decreased to 44 per cent. The winter during these 12 months was the most severe of the 30-year period.

was about 60 per cent and increased progressively to about 70 per cent during the 1970s.

The results show a number of interesting features. First, the annual survival rate has increased overall in both adults and first-year birds in more recent years. Second, the first-year survival rates show much more between-year variation. Third, in only one winter (1962/63, the severest during the range of years used in the study), do the birds show clear weather-related decreases in survival, with first-year birds managing only 44 per cent survival and adults 71 per cent survival in that year. And finally, the high (84 per cent) survival rate of first-years and above-average survival rate of adults in the 1974/75 winter coincided with the mildest weather at this time of year in England since 1869.

Regional survival rates

The survival rates of Black-headed Gulls ringed as pre-fledged chicks or as adults in different regions of Britain (Table 19) show little regional variation, and the small differences are all within the tolerances of the methods used. In general, the mortality rate in the first year of life is about double that of the birds once they become adults. However, the estimates for Ireland indicate that Black-headed Gulls here had both the lowest first-year and adult survival rates, and the

TABLE 19. Estimates of the annual survival rates of adult and first-year Black-headed Gulls, and life expectancy of adults, in Britain and Ireland according to the area in which they hatched. Data are restricted to those birds ringed between 1950 and 1980.

Region	*First-year*		*Adult*		*Adult life expectancy*
	Number	*Percentage survival rate*	*Number*	*Percentage survival rate*	
Scotland	303	60%	292	89%	8.6 years
North England	478	62%	623	83%	5.4 years
South England and south Wales	393	65%	593	84%	5.8 years
East England	336	67%	521	85%	6.2 years
West England and north Wales	535	61%	695	82%	5.1 years
Ireland	107	58%	130	78%	4.0 years
Total/average	**2,152**	**62%**	**2,854**	**84%**	**5.8 years**

lowest adult life expectancy. These figures are consistent with the appreciable current decline in adult numbers nesting there. The highest adult survival rate and adult life expectancies were in Scotland, which is consistent with the increase in numbers reported nesting there between 1985 and 2000.

MOULTING

There is surprisingly little information on the progress of the primary wing moult of Black-headed Gulls. The start of moult commences at the innermost primary, with the old feather usually dropped by adults in June (the earliest recorded date is 26 May), at a time many have already hatched chicks. Since the inner primaries are relatively short, the start of moult at this stage will have little effect on flight and the ability of parents to search for food for the young. After incubation is complete, the primary moult progressively spreads along the wing, until the last (outer) primaries are dropped in September and replaced by fully grown feathers by mid-October. In all, the primary moult takes three to three-and-a-half months from the first feather being shed to the longest primaries becoming fully grown.

The moult of the primary feathers by immature one-year-olds starts about a month earlier than in adults and is completed by September – a three-month period, which is generally shorter than that taken by the adults. The period when the outer and longer ninth and 10th primaries feathers are missing (August for first-year birds and September for adults) probably interferes most with efficient flight, and so this may inhibit long-distance migration until the new outer primaries have grown.

CHAPTER 3

Mediterranean Gull

APPEARANCE

Mediterranean Gulls (*Ichthyaetus melanocephalus*) closely resemble Black-headed Gulls (*Chroicocephalus ridibundus*) in size and shape, although during the breeding season the head is totally black (not dark brown, as in the incorrectly named Black-headed Gull, p. 43) and this extends further down the back of the head. Outside of the breeding season, the species is even more likely to be confused with the Black-headed Gull, because at that time both have white heads. However, the entirely white-tipped primaries of the adult Mediterranean Gull allow immediate separation. Mediterranean Gulls in their first winter resemble Common Gulls of a similar age and differ only in having a narrower black tail band, reddish legs and a paler mid-wing panel. Second-year Mediterranean Gulls are similar to the adults except for a small amount of black on the wing-tips.

FIG 41. Adult Mediterranean Gull in flight, with adult Black-headed Gull (*Chroicocephalus ridibundus*) further away. Note the colour and extent of the dark head. (Norman Deans van Swelm)

FIG 42. First-year Mediterranean Gull in flight. (Tony Davison)

FIG 43. Adult Mediterranean Gulls. (Norman Deans van Swelm)

DISTRIBUTION AND BREEDING STATUS

The Mediterranean Gull was rare in Britain throughout the first half of the twentieth century, with only 10 verified records of the species in England and none in Scotland, Wales or Ireland up to 1940. Records became more numerous in the late 1950s, and in 1968 the first pair bred in England, and numbers of breeding birds did not increase for at least 10 years. Now more than 1,000 pairs nest in Britain at 32 locations, with most in south-east England and the largest numbers in Hampshire. Breeding in Northern Ireland occurred for the first time in 1995 and in the Republic of Ireland (County Wexford) in 1996, but not until 2009 in Wales and even later in Scotland, although pairs were present there throughout the summer in some earlier years.

In the first 50 years of the twentieth century, the Mediterranean Gull was confined to the Black Sea and the eastern end of the Mediterranean, but its range gradually extended into central Europe and then western Europe. Breeding spread to Hungary in 1953, then to Germany, the Netherlands, Poland, Belgium and the south of France during the 1960s. The first pairs nested in Denmark in 1970, Italy in 1978, Spain in 1987 and Sweden in 2008. The exception to this comprehensive and recent pattern of expansion was a single bird that paired with a Black-headed Gull and bred unsuccessfully in the Netherlands in 1935. At the end of the twentieth century, some 2,500 pairs were breeding in western Europe, and by 2017 some 6,000 individuals were nesting in France, 4,000 in the Netherlands and 7,000 in Belgium. The main breeding areas are still around the Black Sea and eastern areas of the Mediterranean, but numbers breeding and wintering in western Europe are increasing rapidly.

When they first arrive in a new geographical area, Mediterranean Gulls frequently attach themselves to colonies of terns and other gulls, particularly Black-headed Gulls, although in one instance an individual was recorded in a Kittiwake (*Rissa tridactyla*) colony. In the early years of the colonisation of Britain and elsewhere, several mixed pairings with Black-headed Gulls, and in one case with a Common Gull, were recorded, but many of these breeding attempts failed, although a few hybrid young were reared. The fate of these hybrids is mainly unknown, but one that was ringed as a chick subsequently paired with a Mediterranean Gull and bred successfully.

The increase in records of Mediterranean Gulls in England is well illustrated by the analysis of records from the West Midlands for the period 1989–2010 (Fig. 44), which show a fourfold increase over the period. While this increase in records may have been influenced by the increase in numbers of observers in recent years, it mainly reflects the rapid rise in numbers of individuals visiting, which parallels the increase in numbers breeding in Belgium and the Netherlands. Fig. 45 shows the monthly distribution of records in the West Midlands between 1989 and 2005, where the months of arrival of individual Mediterranean Gulls have been tallied over the 17 years. Numbers were high in late winter, reaching a maximum in early spring. This was followed by a few records during the breeding season and then a rapid increase of records after June.

The breeding distribution of Mediterranean Gulls in Britain and Ireland in 2007–11 is shown in Fig. 46. This indicates that by that time the birds had already

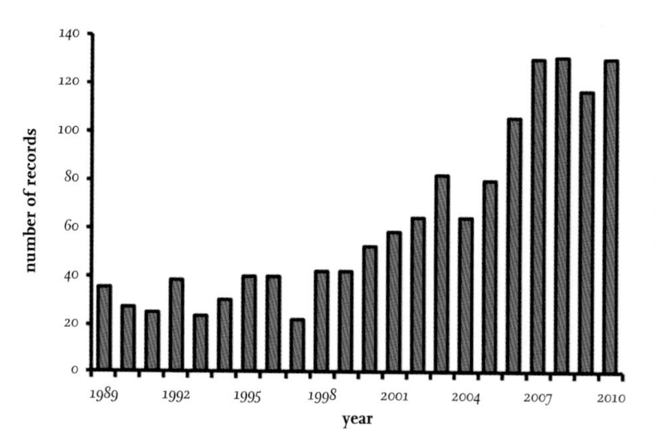

FIG 44. The annual number of records of Mediterranean Gulls recorded in the West Midlands in 1989–2010. Data based on Dean (n.d. [no date]).

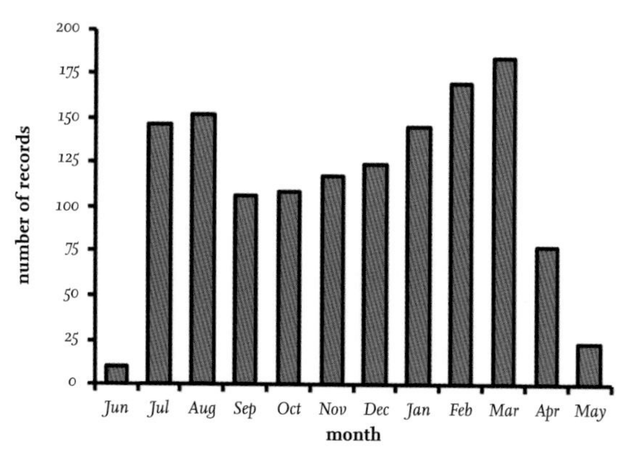

FIG 45. The monthly numbers of Mediterranean Gulls recorded in the West Midlands for the period 1989–2005. There are two peaks, one in July–August, which represents post-breeding dispersal and migration, and the other in January–March, as birds return to their breeding areas. In every month except for February and March, adults formed fewer than half of the records. Data based on Dean (n.d.).

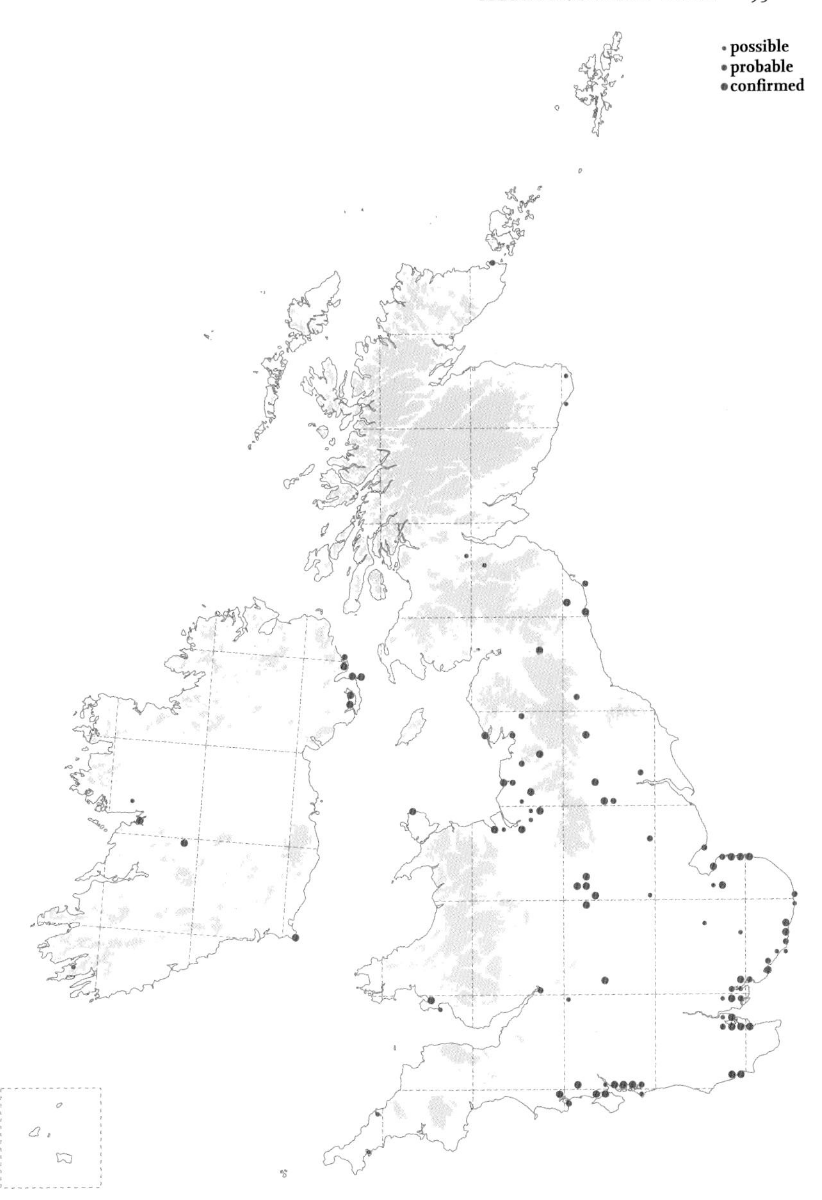

FIG 46. Mediterranean Gull breeding distribution in Britain and Ireland in 2008–11. Reproduced from *Bird Atlas 2007–11* (Balmer *et al.*, 2013), which was a joint project between the British Trust for Ornithology, BirdWatch Ireland and the Scottish Ornithologists' Club. Channel Islands are displaced to the bottom left corner. Reproduced with permission from the British Trust for Ornithology.

spread rapidly north and west from their restricted breeding area on the south coast of England in 2000.

The British Rare Breeding Birds Panel have been effective in collating records of breeding attempts, and their published data (including information to 2015) have been used to form Fig. 47, which shows the number of breeding attempts each year in Britain. Although numbers fluctuated in the early years, poor breeding success (caused in some cases by egg collecting) resulted in the panel at the time doubting that the species would become permanently established here. However, numbers of breeding pairs increased year after year up to 2012, forming a remarkable exponential curve, with the rate of increase remaining virtually constant and at the remarkably high level of 22 per cent per annum during the main period of growth. Records of non-breeding birds were widespread in each summer, presumably indicating individuals exploring new areas for potential colonisation. In the early years of the species' arrival in Britain and probably currently, this increase in numbers recorded must have been driven by immigration from Continental Europe (see below). A similar, very rapid increase and spread has occurred in Belgium and the Netherlands, preceding that in Britain by five to eight years. Wez Smith has reported that the Langstone and Chichester Harbour Reserves alone have had about 835 pairs breeding in 2017 and 1,736 pairs in 2018, and this only 20 years after the first pairs nested there in 1997.

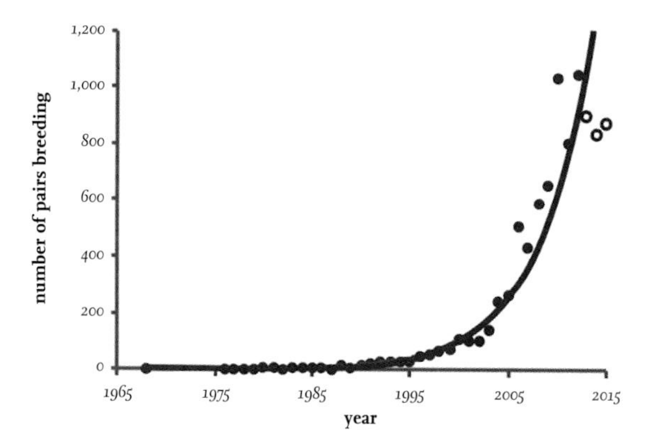

FIG 47. The number of pairs of Mediterranean Gulls breeding in Britain to 2015 as reported by the British Rare Breeding Birds Panel. The curve indicates a 22 per cent annual increase. The reported numbers deviated from the high rate of increase only in the period 2013–15 (white circles), when the increase appears to have ceased temporarily, because in 2016, there were at least 1,846 breeding pairs.

BREEDING

Breeding sites of Mediterranean Gulls are similar to those used by Black-headed Gulls (p. 49) and involve ground nesting at coastal and inland sites, which are usually associated with water to give some protection from mammalian predators. Pairs breed in dedicated colonies or join colonies of other seabirds, but they do not usually breed as isolated pairs. Sites are occupied quite late, sometimes not until late April. Egg-laying peaks in early May. In many cases, eggs that are lost are replaced during a further laying, but the extent of re-laying has not been documented in detail. The most frequent clutch size is three eggs, with replacement clutches tending to be smaller.

The arrival and spread of the Mediterranean Gull in Britain and on the near Continent will possibly produce competition with Black-headed Gulls in the next decade or so, since both have similar breeding and feeding habits during the breeding season. Such competition is likely to be to the detriment of the Black-headed Gull, since the Mediterranean Gull appears to have a broader diet, particularly in winter, and is more aggressive, as is evident by its ability to force its way into existing Black-headed Gull colonies.

Breeding success

Breeding success was low among the first Mediterranean Gulls breeding in England (Fig. 48). Many pairs failed to produce offspring and the frequent mixed pairings with Black-headed Gulls may have contributed to this low success rate. Rearing young was recorded in about half of these breeding attempts, but there was considerable year-to-year variation, particularly in the early years, when no

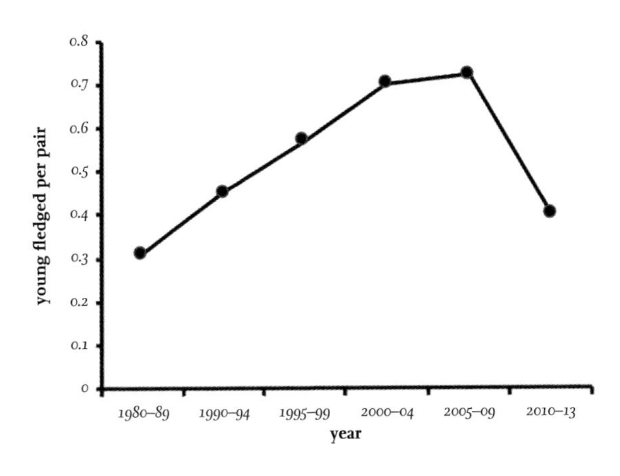

FIG 48. Estimates of the annual number of young fledged per pair by Mediterranean Gulls nesting in England between 1980 and 2013. The data do not take into account numbers of pairs present that failed to build nests or lay, and as a result these values should be regarded as maximum estimates.

young were known to have been reared. In contrast, 2005 was an exceptional year; for those pairs where breeding success was recorded, it reached the exceptionally high value of 1.4 young fledged per pair, which was 40 per cent higher than in any other year. To reduce the impact of year-to-year variation in breeding success, averages over a number of years have been used to create Fig. 48.

Breeding success was initially only 0.3 young per breeding pair and, while this increased in the 1990s, it was still insufficient to replace adult mortality and was much too small to support the 22 per cent annual increase in the numbers of birds reported breeding (see above). The higher level of 0.7 young per pair in the 2000s was mainly due to a higher breeding success in the larger colonies that had formed in the south of England by this time. For example, the colony of 57 pairs at Langstone Harbour, Hampshire, fledged 80 chicks in 2004, or 1.4 young per pair. At the same time, those breeding in small groups and mixed pairs elsewhere continued to show poor breeding success. Records of breeding success in Dutch and Belgian colonies, where Mediterranean Gulls nest alongside Black-headed Gulls, suggest that they breed appreciably more successfully than the Black-headed Gulls in the same colonies, in some cases rearing twice as many young per pair each year. The reason for this difference has not been identified, but the species response to predators is different and, for example, Mediterranean Gulls are more successful at defending their chicks against predation by Herring Gulls (*Larus argentatus*). In some colonies, Mediterranean Gulls predated eggs and small young of Black-headed Gulls and terns, a behaviour that is likely to increase in future years as Mediterranean Gull numbers increase.

It is possible to estimate the production of young required to achieve the 22 per cent annual increase without immigration recorded in Mediterranean Gulls in Britain (Fig. 47) by using the annual survival values reported by te Marvelde *et al.* (2009) of 84 per cent for adults and 55 per cent for immature birds in the 12 months following fledging. Considering 100 pairs of breeding birds, these adults suffer 16 per cent mortality in the following year, so 32 of the 200 birds will die and will need to be replaced by 32 new recruits to the breeding group. In addition, the annual increase (22 per cent) requires a further 44 new birds, so 76 young must survive to breed. The young breed for the first time when they are two years old, so after a mortality of 45 per cent in the first year and then a further 16 per cent mortality in the second year, there is a total mortality of 54 per cent before the birds reach breeding age. To produce 76 individuals that will survive to breeding age, 165 young must therefore be fledged by the 100 pairs, which is an average of 1.65 young fledged per pair. In reality, an even higher breeding success is needed, however, because the numerical recruitment is made by the surviving young produced two or three years earlier, when there were fewer adults.

In England, the annual production reached one fledged chick per pair only twice in the 33 years between 1980 and 2012, and the average was 0.55 young fledged per breeding pair. It is obvious from this, therefore, that the productivity of the species in England is nowhere near that necessary to achieve the 22 per cent annual increase, indicating that massive and sustained immigration is needed. Appreciable immigration has been recorded of young reared and ringed in colonies on the mainland of Europe.

IMMIGRATION AND MOVEMENTS

The 22 per cent annual increase in Britain requires more young recruits than those needed simply to replace the annual adult mortality, but not enough young are produced in Britain each year to achieve even this. The obvious conclusion is that the increase of numbers of Mediterranean Gull records in Britain and Ireland has been almost entirely driven by immigrants for at least the period 1968–2012. How long this high level of immigration continues and the numbers breeding in Britain increase at a remarkably high rate will remain unknown until more counts are made in future years. With a constant rate of increase, it follows that the actual numbers of immigrants arriving annually to breed in Britain in the early years of colonisation must initially have been small and then increased considerably and progressively from 1980 to the present time. The reason why immigration has progressively increased is not clear, but it is probably linked to particularly successful reproduction in mainland western Europe.

The origins of most immigrants are unknown, but ringing recoveries and, particularly, sightings of colour-ringed birds suggest that the majority were young birds reared in colonies in northern and western Europe, particularly France, the Netherlands and Belgium, where the species has also been increasing rapidly. However, some were reared further away, in Germany, Hungary and Poland, and single birds have immigrated from Italy, Serbia and Ukraine. Those birds breeding inland in central Europe must have originated from the long-established population centred on the Black Sea region, as this spread was recorded before the species began to breed in western Europe.

Large numbers of Mediterranean Gulls breeding in the Black Sea area winter in the west of the Mediterranean, along the eastern coastline of Spain and in Italy. In contrast, sightings in winter of colour-ringed individuals reared and marked in western Europe suggest that most of these birds winter along the Atlantic coast of Europe and northern Africa, and that they do not mix with the Black Sea birds.

Typically, new breeding areas are visited first by wintering individuals, and then a few remain in the summer and eventually some breed. Colour-ringed individuals readily and frequently move between the Netherlands, Belgium and Britain. One young bird ringed in Britain was found during winter on the coast of Morocco.

Mediterranean Gulls are highly mobile, and there are several records of individual adults changing their breeding areas. Some individuals move around extensively outside of the breeding season, while others often return again and again to the same wintering locality. Two examples of such behaviour by individuals with coloured and inscribed rings are shown in Tables 20 and 21, but there were long periods, including the breeding seasons, when neither bird was reported.

TABLE 20. The repeated sightings of Mediterranean Gull EP74563, which was first marked as an adult at Felixstowe in January 1996. It is not known where it bred each year, but it is likely that it nested in Belgium, Germany or the Netherlands and moved each winter to Suffolk.

Status	***Date***	***Location***
Adult ringed	27 Jan 1996	Felixstowe, Suffolk
Alive (ring read in field)	20 Dec 1998	Near Ipswich, Suffolk
Caught by ringer	25 May 2001	Kreekraksluizen, **Netherlands**
Alive (colour rings seen)	13 Feb 2002	Near Ipswich, Suffolk
Alive (colour rings seen)	16 Mar 2002	Zandvlietsluis, **Belgium**
Caught by ringer	31 Mar 2002	Berendrecht, **Belgium**
Alive (colour rings seen)	31 Dec 2002	Near Ipswich, Suffolk
Alive (colour rings seen)	21 Jul 2005	Folkestone, Kent
Alive (colour rings seen)	18 Sep 2005	Near Ipswich, Suffolk
Alive (colour rings seen)	14 Jan 2006	Near Ipswich, Suffolk
Alive (colour rings seen)	03 Feb 2007	Near Ipswich, Suffolk
Alive (colour rings seen)	10 Apr 2009	Molfsee, Schleswig-Holstein, **Germany**
Alive (colour rings seen)	08 Dec 2010	Near Ipswich, Suffolk

TABLE 21. The movements of an individual reared in Belgium and given an inscribed colour ring before fledging. Despite 13 sightings after ringing, the whereabouts of this bird between July and December each year and when it was in wing moult remains unknown.

Status	*Date*	*Location*
Ringed as a chick	Jun 2004	Belgium
Seen	Mar 2005	Lancashire
Seen	Jun 2005	Lancashire
Seen	Apr 2006	Calais, **France**
Seen	Apr 2008	Zeeland, **Netherlands**
Seen	Feb 2009	Lancashire
Seen	Apr 2009	Norfolk
Seen	Jan 2010	Lancashire
Seen	Mar 2010	Suffolk
Seen	Feb 2011	Lancashire
Seen	Mar 2011	Hampshire
Seen	Apr 2011	Hampshire
Seen	Jan 2012	North Wales
Seen	Mar 2012	Sussex

FOOD AND FEEDING

Much of the food consumed by Mediterranean Gulls is similar to the wide range of items taken by Black-headed Gulls both on land and from the surface of freshwater and marine areas. The species also uses similar techniques of surface dipping and ground searches to collect food. During the breeding season, most of the food consumed comprises invertebrates collected from terrestrial and intertidal sites, including many insects, while in some areas earthworms and marine organisms are consumed. Some individuals have been seen following behind farm ploughs, but most food is obtained by searching on grasslands and intertidal areas. Vertebrates other than fish are rarely taken, with the exception that some individuals take the eggs and young of other birds, including terns and Black-headed Gulls.

In winter, many Mediterranean Gulls in Italy, Spain, Portugal and north-west Africa follow fishing boats to feed on discarded fish and offal. Apparently, they frequently feed at night, a behaviour that differs from the feeding habits of the

Black-headed Gull. The species has also been recorded feeding at landfill sites, although not as frequently as Black-headed Gulls, and birds in England have been recorded feeding at pig farms. There are suggestions from the Black Sea area that breeding birds sometimes travel more than 70 km to feeding areas, but there is no evidence of such long-distance movements in western Europe.

CHAPTER 4

Common Gull (Mew Gull)

THE OLD BRITISH NAME for the Common Gull (*Larus canus*) was Mew Gull, which is descriptive of its call, and this name persists in North American literature, although the equivalents of 'storm gull' or 'fish gull' are used in many European countries. The new IOC World List of Bird Names has now selected the use of Mew Gull worldwide, but it will take time before this is accepted and becomes common usage.

APPEARANCE

In appearance, the adult Common Gull resembles a miniature version of the Herring Gull (*Larus argentatus*, p. 121), with a wingspan of about 1 m, similar grey wings and mantle, and a white head, underparts and tail, while the black wing-tips have obvious white mirrors (Fig. 49). However, it is much smaller, and can further be separated from the Herring Gull by its greenish legs and delicate yellow bill, which lacks the Herring Gull's red spot. In winter, the white head is streaked with grey (as is the head in many Herring Gulls) and the bill acquires a dark band near the tip (Fig. 50), which can cause confusion with the Ring-billed Gull (*L. delawarensis*; p. 339).

Immature individuals differ markedly in plumage from the adults, and young birds encountered in their first year of life might be mistaken for a different species. During its first year of life, the juvenile retains many of the brown feathers in the wing, while the distal half of the bill is dark, the tail has a broad dark band (absent in adults), the mantle has grey feathers and the black tips on the primaries lack white mirrors. Immature birds can be difficult to separate

FIG 49. Adult Common Gull. (Alan Dean)

FIG 50. Common Gull in winter plumage. Note the narrow black band on the bill, which is less distinct than in the Ring-billed Gull and the less bright legs. (Tony Davison)

FIG 51. Adult Common Gull in breeding season plumage. Note the bright yellow-coloured legs, which are much duller in the winter. (Thermos on fi.wikipedia)

from Ring-billed Gulls and Mediterranean Gulls (*Ichthyaetus melanocephalus*; p. 91) of a similar age without referring to finer differences in the plumage. Second-year, but still immature, birds show plumage that is very similar to adults, having lost the dark tips on their tail feathers, while the previously brown wing feathers are replaced with grey ones. In some second-year individuals, the black-tipped primaries have acquired white mirrors, but when present these are much smaller than those of the adults.

The oldest known Common Gull in the wild was 33 years and eight months old, and was still alive when the ring inscription was read. A second ringed bird was more than 31 years old when found dead.

WORLD DISTRIBUTION

The Common Gull breeds widely in the northern hemisphere, extending west from Iceland across the whole of northern Europe and Asia, to Alaska and western Canada. It is not quite circumpolar, however, because it is absent from most of Canada and the United States, where it is probably excluded by competition from the closely related Ring-billed Gull. In northern Europe, the Common Gull is an abundant breeding bird – about half the world population nests in Scandinavia. The southern limit of breeding in mainland Europe is reached in Switzerland, Belgium and France. Elsewhere in Europe, Common Gulls breed across northern Russia and are slowly spreading as a breeding species in most areas. They were first recorded nesting in the Faeroes in 1890, in Iceland in 1955 (where breeding had been suspected earlier), in Belgium in 1924 and in Switzerland in 1966. The species ceased breeding in the Netherlands in the nineteenth century, but returned in 1908 and has continued to breed there in increasing numbers.

The size of the world population of the Common Gull is uncertain, owing to the lack of reliable censuses throughout much of its range. Estimates vary from between 800,000 and 1.25 million adults, half of which nest in Scandinavia. Between 200,000 and 350,000 immature individuals are probably also present at any one time and should be added to the estimated numbers of adults.

DISTRIBUTION IN BRITAIN AND IRELAND

The Common Gull breeds abundantly only in Scotland, with the greatest numbers found in the central and northern regions (it is particularly common in Orkney and Shetland), but it is slowly expanding its range (Fig. 52). The first breeding record in Ireland was in 1934, and small numbers now nest in the west and north-east. Breeding in Wales did not occur until 1963 and currently there is only an occasional pair breeding in or close to Anglesey.

There were no records of Common Gulls nesting in England until the twentieth century (perhaps overlooked through confusion with Herring Gulls); the first records were between 1910 and 1913 on the Farne Islands in Northumberland, where a few pairs bred (apparently unsuccessfully). The species has bred on these islands on only one occasion since, despite intensive protection of seabirds breeding there. Breeding in England is currently restricted to a small number of birds regularly nesting on coastal shingle in Kent and Suffolk, and a few occasionally nesting on the coast in Cumbria.

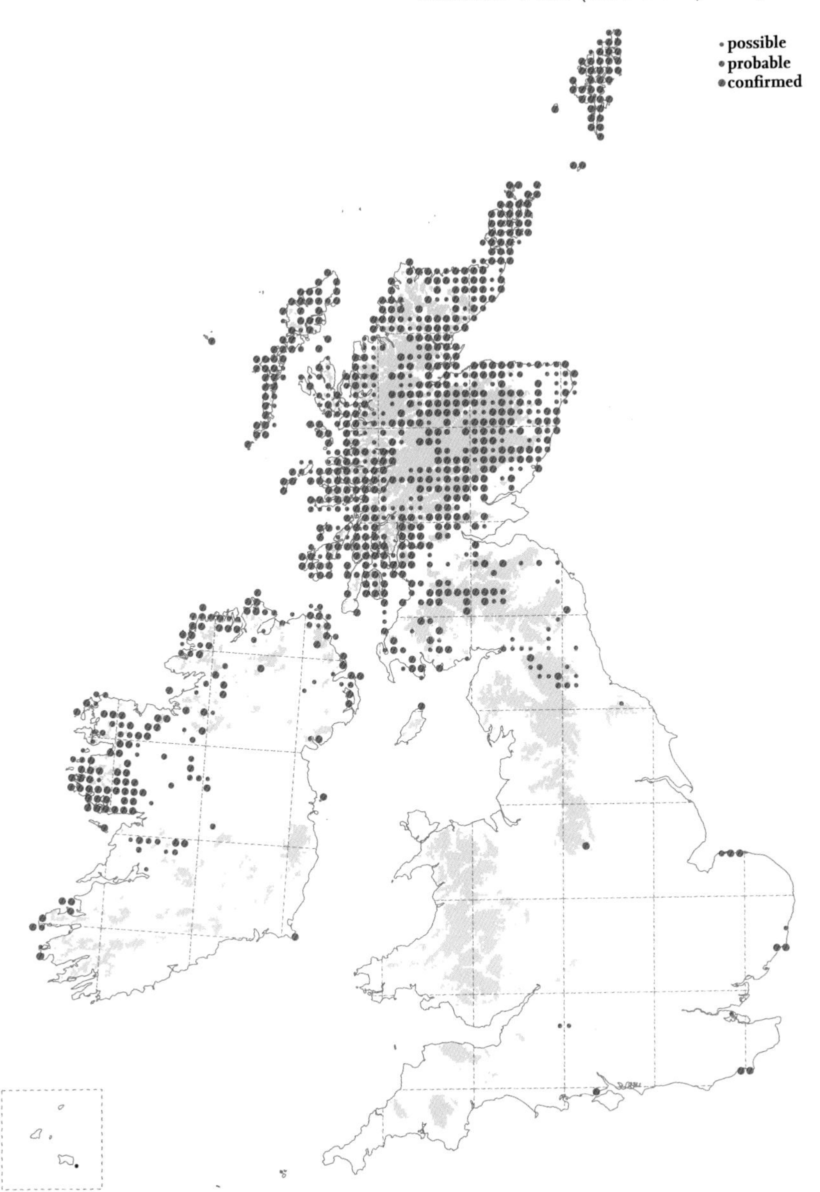

FIG 52. Map of the breeding distribution of the Common Gull in Britain and Ireland 2008–11. Reproduced from *Bird Atlas 2007–11* (Balmer *et al.*, 2013), which was a joint project between the British Trust for Ornithology, BirdWatch Ireland and the Scottish Ornithologists' Club. Channel Islands are displaced to the bottom left corner. Reproduced with permission from the British Trust for Ornithology.

During the last 50 years, sporadic nesting has occurred inland on moorlands in Yorkshire, County Durham and Northumberland, but these involved no more than a total of 10 pairs in any year and the sites have frequently changed. In the early 2000s, a few pairs of Common Gulls nested annually within a large lowland area in Northumberland where coal was being extracted from an opencast mine, but the site has now been closed and landscaped, and the gulls have left. Ringing has established that Common Gulls reared in Scotland seldom move into England, so it is possible that some of the sporadic breeding records in England could have been birds visiting from the Continent in winter, which then stayed on to breed.

The total number of breeding Common Gulls in Britain is uncertain, as the few very large colonies on moorland in Scotland are difficult to census and many solitary pairs nesting alongside rivers and lochs are easily overlooked. The latest estimate is that about 100,000 adults breed in Britain, which is a much lower figure than the number of Herring Gulls. Over the years, some sites in western areas of Scotland (and on the mainland of Europe) have been deserted as a result of predation by American Mink (*Neovison vison*). Currently, there is also a suggestion of minor declines in parts of Scotland, although some new sites have been colonised.

In winter, Common Gulls are abundant throughout Britain and Ireland, totalling 750,000 to 1 million birds, of which about 90 per cent are immigrants from countries bordering the Baltic Sea. The species has been put on the Amber List of species of conservation priority in the UK owing to concern that winter numbers here have declined. However, further evidence is needed to confirm a decline, particularly considering the difficulties of census counts during winter months.

SUBSPECIES

Four subspecies of the Common Gull have been described, although the status of these await more detailed investigation using modern techniques. The western European subspecies, which is the subspecies that breeds in Britain and western Europe, is *Larus canus canus*. Another subspecies, *L. c. heinei*, breeds in Russia; it is slightly larger than the nominate race and has a darker mantle and wings. No specimens of *L. c. heinei* have been collected in Britain, but the subspecies was placed on the British List by the BOU in 1994 based on three individuals among a sample of 250 Common Gulls trapped in south-east England (i.e. more than 1 per cent) that were considered to be *L. c. heinei* owing to their long wing length measurements. Since there is considerable overlap in the measurements of the *canus* and *heinei* subspecies, these three individuals would have to have been large males that were towards the extreme of the size range for the *heinei* subspecies. Shorter-winged males and females of this subspecies would not have been detected from their wing length, so it could be argued that the real proportion of *heinei* in this sample was possibly higher. The *heinei* subspecies has been reported as common in winter in southern Scandinavia and northern Germany, but the evidence in support of these claims has not been published.

Ringing recoveries also suggest that *Larus canus heinei* may occasionally occur in Britain, as a few birds ringed in winter in Britain have been recovered in the breeding season as far east as Vologda in Russia (longitude 46°E), from the area considered to be within the breeding range of the subspecies. A few birds ringed in winter in the Netherlands, Germany, Denmark and Sweden have also been recovered within what is regarded as the probable breeding range of *L. c. heinei.*

So far, data on the shade of grey on the wings and mantles of birds wintering in Britain have not been collected. The current evidence therefore suggests that a few Common Gulls from colonies in eastern Europe may regularly winter around the southern North Sea (including Britain). However, this conclusion would be more persuasive if the breeding distribution and variation within the *heinei* and *canus* subspecies were better known, along with how plumage shade and body size vary across Europe and western Asia. There is likely to be an area of overlap in Europe where both *L. c. canus* and *L. c. heinei* breed, but this has not yet been identified and needs to be defined.

Two further subspecies are *Larus canus kamtschatschensis* in eastern Siberia, and the shorter-billed *L. c. brachyrhynchus* in western North America (mainly Alaska). A case has been made, based on DNA evidence, that the latter should be regarded as a separate species, but so far this proposal has not been accepted. A field study has suggested that it is possible to identify the *kamtschatschensis* and *brachyrhynchus* subspecies in the field, but so far the suggested criteria have not been critically tested and they do not by themselves indicate whether these two subspecies are justified, or that only a cline is involved.

BREEDING

Common Gulls nest both inland and on the coast, and about half of those breeding in Scotland use inland sites. They nest as isolated pairs and in small colonies of fewer than 100 pairs, although a few colonies (or groups of colonies), such as those on the Mortlach Hills and Correen Hills of Scotland, have been estimated at several thousand or even tens of thousands of pairs. Numbers at these large colonies have, however, declined markedly in recent years and it is not known whether the missing birds have moved elsewhere or have died. Colonies are not usually visited by breeding birds until April, and even then, they usually leave each evening and roost elsewhere until around the time eggs are first laid.

Isolated pairs of Common Gulls nest along rivers and lochs, on shingle beaches and moorland, and, at a few places, on the gravel of railway tracks. Most nest on the ground, but there are records of nests in small trees, both here and on the Continent, and there have been several records of pairs breeding on buildings in Scotland and mainland Europe (Denmark and the Netherlands). The first record of breeding on a roof in Britain was reported near Inverness in 1969. By 1984, roof nesting had begun in Aberdeen, where several hundred pairs now breed, split into two groups: one in the city centre and the other on an outlying industrial estate. The gulls have also started to nest at oil installations surrounded by chain fencing, which fortuitously also provides protection from mammalian predators.

Breeding biology

There have been few published studies on the breeding biology of the Common Gull in Britain. The most detailed was made by Clive Craik (1999), who studied a colony of about 100 pairs on a small island off the west coast of Scotland for eight years between 1988 and 1995. He found that pairs nesting early were the most successful, with an average of 1.09 young fledged per pair in those that laid before 11 May, while those laying towards the end of the month failed to rear any young. This decline in success was caused by increased predation by Herring Gulls and Peregrines (*Falco peregrinus*) – in one year, 20 per cent of the large young were taken by the latter. American Mink reached the study island late in 1990 and no gull chicks fledged in the following two years, after which point the adults permanently deserted the site. Such desertions have also taken place elsewhere.

An intensive long-term study on breeding Common Gulls was carried out in Estonia for 38 years, initially by Sven Onno and continued by Kalev Rattiste (Rattiste, 2004, 2006). This involved colour-ringing many adults, allowing individuals to be identified and their survival and breeding success to be followed. Comparable studies have not been made in Britain and so it has been necessary to rely on the outstanding Estonian study for information on the breeding biology of this species.

The Estonian researchers found that incubation (as in most gulls) is shared by both members of the pair, with shifts lasting two to three hours. The start of incubation was variable, and as in many other gull species, it often started before the clutch was completed and produced asynchronous hatching. The time from the start of incubation to hatching was approximately 24 days, while fledging occurred after a further 35 days. The normal clutch size was three eggs, with a minority of females laying only two. Many, but not all, pairs laid a replacement clutch if the first was lost, but this was usually smaller, averaging 1.7 eggs and with most replacement clutches comprising only one or two eggs.

The Estonian study showed that breeding success increased with the age of the adults for at least eight years after they bred for the first time, and then reached a plateau before showing a slight tendency to decline in old birds that had bred in more than 12 years. Each first-time breeder fledged only a third of the number of young that was produced by those breeding for the eighth year. Marked birds that disappeared, and presumably had died, had bred less well in their last year of breeding, suggesting that their condition had been slowly declining. A similar effect has been reported for the Kittiwake (*Rissa tridactyla*; p. 231).

The researchers in Estonia suggested that the extent of the white mirrors on the wing-tips indicated the quality of individuals, and found that males with larger mirrors were less likely to change partners. In those males that did change partners, the female laid about four days later than in pairs that retained their mate from the previous year. Adult survival rates averaged 90 per cent, but dropped to 87 per cent in years with cold winters and rose to 93 per cent in years with mild winters.

The age at first breeding recorded in the study was variable, with 9 per cent of males breeding when two years old, 59 per cent breeding when three years old and 32 per cent delaying breeding until they were four years old. Females delayed the start of breeding longer than males, and 50 per cent did not breed until they were four years old. These differences suggest that adult males either suffered a higher mortality rate than females or that the sex ratio of chicks at fledging was unequal. Many of these effects are similar to those reported for the Kittiwake (p. 231).

MOVEMENTS

Common Gulls breeding in Scotland and Ireland disperse only short distances and most remain throughout life within the same country (Fig. 53), although a small number move from Scotland into Ireland. Only four birds have been recovered in mainland Europe: two in Iberia and two in France.

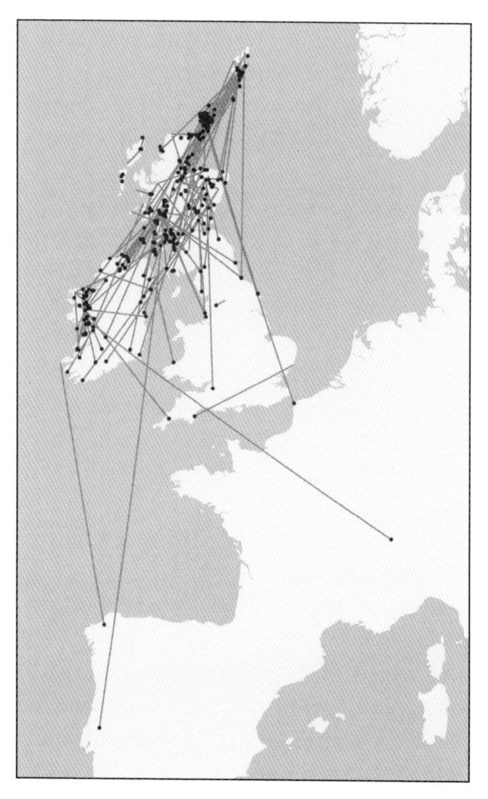

FIG 53. The movements of Common Gulls ringed as chicks in Britain and Ireland and recovered outside of the breeding season. The almost total absence of recoveries in Wales and England indicates that virtually all the numerous Common Gulls wintering in these countries are birds from the Continent. Reproduced from *The Migration Atlas* (Wernham *et al.* 2002), with permission from the BTO.

Birds breeding in Iceland, Denmark and the Netherlands are also mainly sedentary or move short distances into adjacent countries, and only a few make the sea crossing to Britain. In contrast, hundreds of thousands of Common Gulls migrate from Norway, Sweden, Finland and north-west Russia to winter in Britain and Ireland, and then return to the Continent to breed (Fig. 54). Many of those travelling from Sweden and Finland collect in Denmark for several days or weeks before crossing the North Sea to Britain in a single flight. Their arrival in Britain is rarely witnessed and it is possible this is because they arrive mainly at night.

There has been only one transatlantic ringing recovery. A nestling ringed at Devich'ya Luda, in the White Sea area of Russia, was recovered three years later in Newfoundland.

The departure of wintering Common Gulls leaving Britain and returning to the Continent begins with numbers aggregating in large flocks early in the day, often on inland lakes and reservoirs, and in sites where they have possibly roosted overnight. They drink, bathe, preen and are highly active and

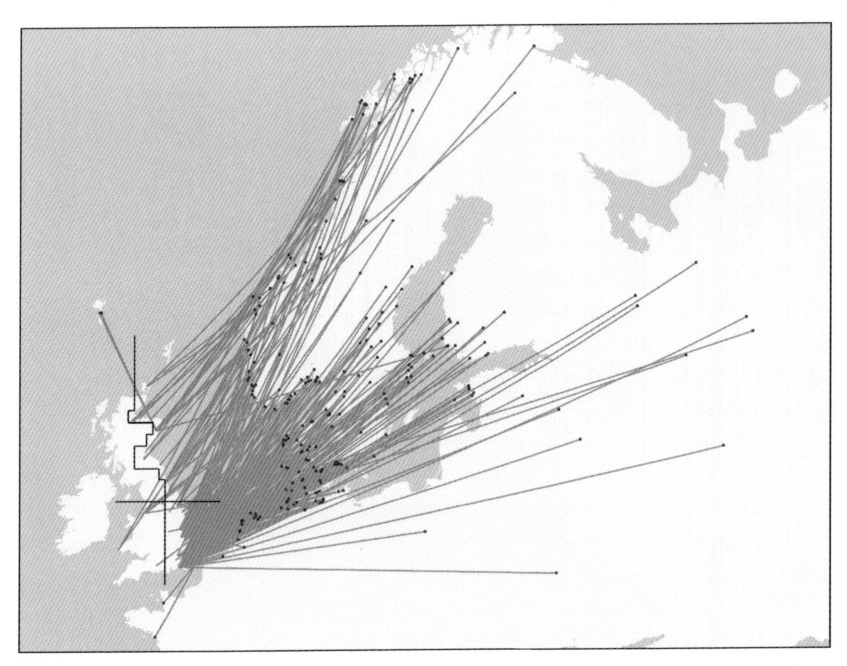

FIG 54. The recovery localities of Common Gulls ringed in winter in Britain and recovered abroad. Note that most were ringed on the eastern side of England and Scotland. Reproduced from *The Migration Atlas* (Wernham *et al.* 2002), with permission from the BTO.

vocal, but feed only infrequently. Groups leave the main flock throughout much of the day and move off in the same direction (usually north-east), and their flight is purposeful.

Common Gulls leaving southern Scotland in April to cross the North Sea were first observed by Ian Patterson, who saw flock after flock leaving the country's south-east coast at such a high altitude that many were difficult to see without an optical aid. Subsequent investigations by Bill Bourne using radar in north-east Scotland confirmed that these departing flocks continued over the North Sea in an easterly or north-easterly direction, but did so only in good weather and with prevailing westerly or south-westerly winds. Departures began at about 10.00 and were spread throughout much of the remaining daylight hours. Presumably, most birds make the North Sea crossing without stopping, which means that many reach the coast of Scandinavia in darkness (Bourne & Patterson, 1962).

High-altitude migration appears to be unusual in gulls and has been recorded only occasionally in Kittiwakes, but it is more frequent in several species of tern. Many Black-headed Gulls (*Chroicocephalus ridibundus*) also cross the North Sea, but some fly 1–2 m above the waves and whether others fly at higher altitudes has not been investigated.

In their first summer after hatching, young Common Gulls avoid visiting colonies, and ringing recoveries reveal that a small proportion of those reared on the Continent remain in Britain during the summer following fledging and probably remain there through the following winter, returning to the Continent for the first time in the spring when they are almost two years old.

One of the few investigations of Common Gulls in Britain was a study by Gabriella MacKinnon on an 8,000 ha study area around Durham. This revealed that a few birds arrived in late July and numbers increased progressively until November, and then arrivals declined slowly until March (Fig. 55), by which time most Black-headed Gulls had also left the area. Numbers then doubled in April, but these did not include marked individuals that had spent the winter in the area, and the increase was presumably caused by individuals moving from elsewhere in England prior to their return migration across the North Sea. Numbers declined rapidly, so that by the end of April only a few immature individuals remained into May, and none was present in June, although a few were present at that time in upland areas. In the study area, Common and Black-headed gulls often fed together, but arrived in the morning separately and from different directions from their night roosts.

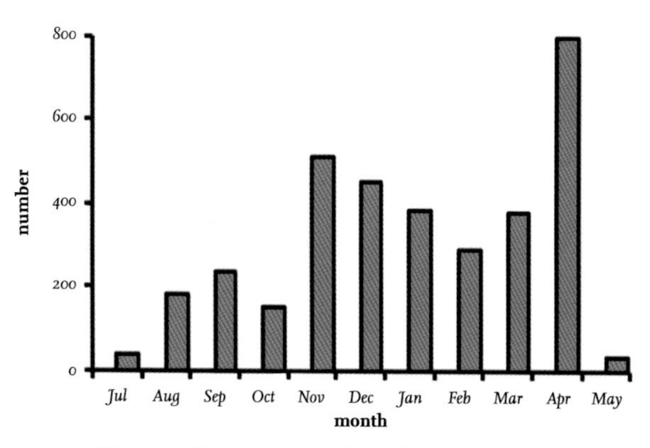

FIG 55. The monthly average number of Common Gulls in an 8,000 ha study area near Durham from July to May, based on counts made in consecutive winters and mainly on data collected by Gabriella MacKinnon. The increase between October and November reflects the arrival of winter visitors from the Continent, while that in April was presumably caused by individuals wintering elsewhere in England moving towards the east coast prior to departing for the Continent across the North Sea. No Common Gulls were present in the study area in June.

TABLE 22. The proportion of marked Common Gulls that returned to the Durham study area in the following winter. Data from MacKinnon and Coulson (1986).

Age at marking	*Number marked*	*Number returned*	*Percentage returned*
Adult	35	10	29%
Second-year	7	1	14%
First-year	22	2	9%

Common Gulls marked with wing tags (which made them easier to detect than marking with coloured leg rings) in north-east England in autumn and winter were mainly birds originating from mainland Europe. Some of these marked birds returned to the same area in England in the following winters (the first to return arrived in late July); this behaviour was age dependent, with 29 per cent of those marked when adult returning, but only 9 per cent of those marked in their first year of life doing so (Table 22). Faithfulness to the same wintering area was much lower than that found for Black-headed Gulls being studied at the same time and it was evident that many Common Gulls, and particularly the immature birds, wintered elsewhere in subsequent winters.

Sightings of eight wing-tagged Common Gulls that failed to return to the study area in the following year were reported by other observers up to 150 km further south along the east coast of England, and an exceptional individual spent its next winter further north, near Aberdeen in Scotland.

FOOD AND FEEDING

Common Gulls are omnivorous, with a diet composed mainly of insects, earthworms and other small food items such as grain obtained from agricultural land. Small crustaceans are foraged from intertidal zones, and small fish and aquatic invertebrates are caught from rivers and lakes. There are also occasional records of frogs, small mammals and young birds being swallowed, and some Common Gulls feed on dead animals killed on roads.

J. D. R. Vernon (1970a,b, 1972) made detailed studies of the many feeding methods of Common Gulls and drew attention to the importance of earthworms in their diet, while also reporting the frequent consumption of grain in autumn and spring. In addition to the items detailed above, the consumption of craneflies (daddy-long-legs) and their larvae (leatherjackets) on moorland in the breeding season is of considerable importance, because earthworms are few in number in these areas. In general, the feeding method used by Common Gulls is similar to that of Black-headed Gulls (p. 76). Flocks of gulls land together on pastures to search for earthworms and leatherjackets, but individuals soon spread out and walk over the ground as they forage for prey items. Most individuals walk in the same direction but they rarely interact. Common Gulls also join flocks of feeding Lapwings (*Vanellus vanellus*) and attempt to rob them when the waders capture earthworms, but kleptoparasitism is more frequently seen in Black-headed Gulls (p. 78) and Lapwings.

Until recently, many Common Gulls in winter used to feed at outfalls where sewage flowed into intertidal sections of rivers, but these sources of food have disappeared following improvements of sewage management (see below for details of the situation on the river Tyne at Newcastle).

In mild winter weather, Common Gull flocks feed on grasslands from sea-level to the highest cultivated areas in Britain, and even on moorland and upland pastures, providing the ground is clear of snow. As a result, the altitude range used by the species in winter and spring is much greater than that of Black-headed Gulls (Table 23), with the former more common above 305 m. During periods of extensive snow cover or hard frosts, Common Gulls feeding in upland areas move down to lowlands, river estuaries and grasslands near sea-level. Others join Black-headed Gulls in seeking food put out for domestic animals or patrol urban areas for food items left by people. This behaviour results in frequent kleptoparasitic and aggressive aerial pursuits of other gulls or corvids that have picked up food items too large to be swallowed quickly.

In autumn and spring, Common Gulls (and other gulls) are attracted to agricultural fields where active ploughing is in progress. They hover closely behind the plough and pick up items while on the wing, or land and run short distances over the ground to capture worms and insects exposed as the soil is turned over. They are also attracted to fields sprayed with slurry, presumably because the slurry stimulates earthworms and leatherjackets to rise to the surface.

TABLE 23. Comparison of altitude distribution of ground-feeding Common and Black-headed gulls in autumn and winter. Based in part on Vernon and personal data.

Altitude	*Percentage of Black-headed Gulls*	*Percentage of Common Gulls*
0–200 ft (0–61 m)	66%	25%
200–400 ft (61–122 m)	26%	27%
400–1,000 ft (122–305 m)	6%	25%
Above 1,000 ft (305 m)	2%	23%

Feeding on the tidal stretches of the river Tyne

Up until the late 1970s, when the river Tyne in north-east England was 'cleaned', huge quantities of particulate matter and untreated sewage from the million or so people living in the Newcastle area and waste from abattoirs were released at outfalls into its 24 km-long tidal reaches. The tidal stretches often became abiotic and anaerobic, caused by the large amounts of organic material flowing into them, and the water had a strong stench and was devoid of fish. In winter, however, more than a thousand Common Gulls regularly fed on this tidal stretch of the Tyne, mainly feeding near or at the outfalls.

In anticipation of a planned policy to clean the Tyne of this pollution, Gerry Fitzgerald and I (1973) carried out a series of gull surveys from 1969 to 1971 along 18 km of the tidal stretch of the river as passengers on the Port of Tyne Authority launch, which made daily inspections from the river mouth at South Shields to Newcastle. In 1993–95, several years after sewage and offal had been prevented from entering the river, Susan Raven and I (2001) repeated these surveys and found marked changes in the numbers of gulls feeding on the river.

By the winters of 1993–95, the numbers of Common Gulls present and feeding on the river had declined by 89 per cent from those present in 1969–70 (Fig. 56), but interestingly, the abundance of the Black-headed Gulls had not changed. Untreated sewage was now being piped directly to a preliminary treatment plant, where the particulate material was concentrated and transferred regularly to a ship (which, with typical Tyneside humour, was known locally as the 'honey boat'). This dumped the material in the North Sea off the river Tyne, just beyond the 3-mile (5 km) coastal limit.

The sites where the sewage pellets were dumped at sea attracted hundreds of Common Gulls; when this practice eventually ceased in 2000, Common Gulls in the area lost the major sources of their winter food. As a result, their numbers decreased from many hundreds to a maximum of 100 individuals at any one time over the whole 18 km surveyed length of the river. Such changes in feeding opportunities for Common Gulls must have occurred at the same time in many other large rivers in Britain as new legislation forced changes in sewage management throughout the country.

Unlike Black-headed and the larger gulls, Common Gulls rarely feed at landfill sites in Britain. In counts totalling more than 11,000 wintering gulls at landfill sites in north-east England, only five Common Gulls were recorded, and these were present only in severely cold weather. It is not clear why they avoided the landfill sites when these were so much favoured by other gull species.

Common Gulls feeding inland are represented by all age classes, with first-year birds averaging 13 per cent of the total recorded feeding on fields

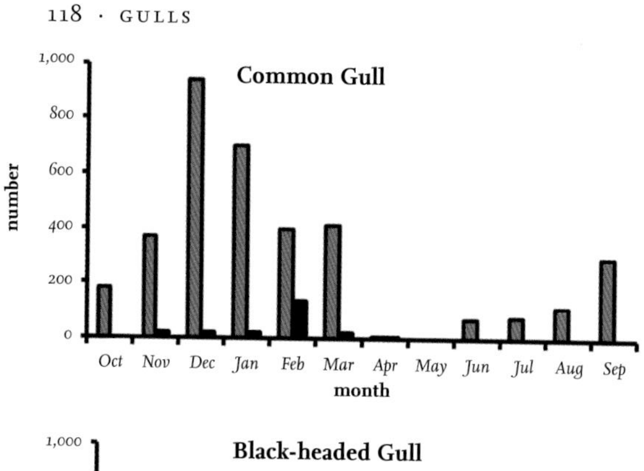

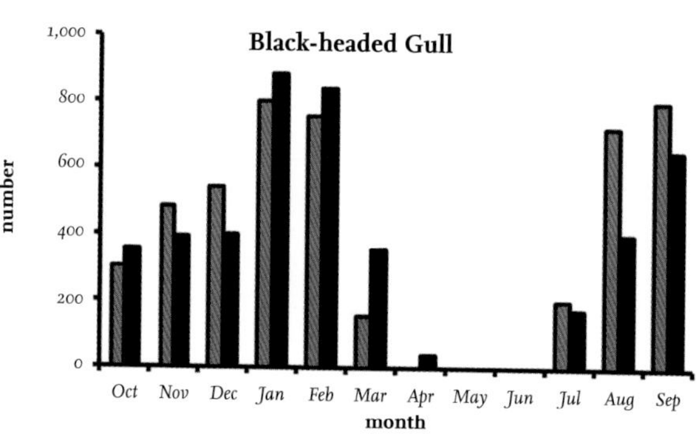

FIG 56. The numbers of Common Gulls and Black-headed Gulls present in each month along the tidal stretch of the river Tyne in 1969–71 (red bars) and 1993–95 (black bars). Note the huge reduction in Common Gulls but minimal change in numbers of Black-headed Gulls.

from August to January. In contrast, flocks feeding near the coast of north-east England included only 2 per cent of first-year birds (Table 24). Another study reported that Common Gulls present during the day on reservoirs and lakes included 22 per cent of first-year birds, but only 7 per cent of those feeding on

TABLE 24. The numbers and proportions of first-year Common Gulls at coastal sites and on inland fields in north-east England between August and March. From data mainly collected by Gabriella MacKinnon and also personal data.

Month	*Number observed*	*Coastal sites*		*Inland fields*		
		Number of first-years	*Percentage of first-years*	*Number observed*	*Number of first-years*	*Percentage of first-years*
Aug–Nov	1,107	41	3.7%	2,253	301	13.4%
Dec–Jan	2,493	42	1.7%	1,661	288	17.3%
Feb–Mar	216	12	5.6%	1,648	139	8.4%
Total	**3,816**	**95**	**2.5%**	**5,562**	**728**	**13.1%**

TABLE 25. The weight (g), standard errors and sample sizes (in brackets) of adult and first-year Common Gulls according to months of capture.

Month	*Adult males*	*Adult females*	*First-year males*	*First-year females*
Aug–Oct	392 ± 5 (30)	361 ± 6 (31)	394 ± 13 (4)	348 ± 12 (6)
Jan–Feb	436 ± 5 (26)	379 ± 7 (13)	446 ± 7 (13)	364 ± 12 (2)
Difference in winter	+44 ± 7*	+18 ± 9*	+52 ± 15*	+16 ± 17

* Indicate differences that are statistically significant.

nearby grassland were immature. These consistent differences suggest that many individuals in their first year of life feed in different ways and in different proportions to the adults.

Weight

Very little information exists about the weight changes of Common Gulls through the year. Birds that were captured in north-east England for ringing were also weighed and sexed, when females were found to be 13 per cent lighter than males on average (Table 25). First-year males did not differ in weight from adult males at any time, and adult and first-year birds of both sexes achieved their highest weights in January and February. There is a suggestion that weight gain in winter was higher in males than in females of both age classes. This pattern of weight gain suggests that the short, cold days of winter do not usually result in food shortages. The higher weights in winter are probably a response to lower environmental temperatures, with more subcutaneous fat reducing heat loss and acting as an energy reservoir should a period of extremely low temperatures occur.

NIGHT ROOSTS

Throughout winter, Common Gulls collect in large numbers to roost at night on lakes and reservoirs, and islands and other safe sites on the coast, including sheltered bays and harbours. However, they do not usually roost overnight on the open sea. Flight lines of Common Gulls indicate that some move up to 40 km from feeding areas to night roosts.

During the night, roosts on land are not full of immobile gulls with their heads tucked under a wing while peacefully sleeping, as many might expect.

While some Common Gulls do sleep for part of the time, there is continuous vocal chattering from the flocks all night, and the birds spend much time preening, walking about, having minor aggressive interactions with immediate neighbours and even walking to small pools of fresh water to drink. Just before dawn, the birds leave the roost in flocks – marked individuals have been recorded returning day after day to feed on the same grasslands. Interestingly, the flocks are composed of several dozen individuals, but how they assemble from among the thousands of individuals in the roost remains a mystery, particularly since marked individuals repeatedly recorded using a particular feeding area did not appear to group together at the night roost site.

CHAPTER 5

European Herring Gull

THE EUROPEAN HERRING GULL (*Larus argentatus*) ranks along with the Black-headed Gull (*Chroicocephalus ridibundus*; p. 43) and the Black-legged Kittiwake (*Rissa tridactyla*; p. 231) as one of the three most numerous gulls breeding in Britain and Ireland. In winter, virtually all Kittiwakes have departed into the Atlantic Ocean and numbers of Black-headed Gulls in Britain are swollen by a huge immigration from Continental Europe. In contrast, almost all Herring Gulls breeding in Britain and Ireland remain there throughout the year and only a relatively small number immigrate, from northern Scandinavia and north-west Russia.

APPEARANCE

The adult Herring Gull is readily identified by its large size (wingspan of 1.5 m and weight of 700–1,100 g) and its silvery-grey mantle and wings, the latter with black wing-tips that include obvious white spots, often referred to as mirrors. The rest of the body is white, although the head is often streaked grey in winter (Fig. 57). The legs are a flesh-grey colour and the bill is yellow with an obvious red spot on the lower mandible in both sexes. The eye has a golden-yellow ring, which gives the adult an impression of viciousness. Males are larger than females and have a more substantial bill. This gull is likely to be confused only with the smaller Common Gull (*Larus canus*; p. 103), which has greenish-yellow legs, and the Great Black-backed Gull (*L. marinus*; p. 209) and Lesser Black-backed Gull (*L. fuscus*; p. 179), where the wings of adults are black or very dark slate grey. Related species that breed in the eastern Baltic and in southern Europe closely

FIG 57. An adult Herring Gull, searching for worms and leatherjackets on an inland field. (Mike Osborne)

resemble the Herring Gull, but have yellow legs. These gulls are still uncommon in Britain, although their numbers are increasing rapidly.

There are four annual plumage stages in the immature Herring Gull, and at each moult individuals progressively acquire more of the characteristic plumage of the adult. Those in their first year look completely different from the adults, apart from their size (Fig. 58). They lack any white in the plumage and most of their feathers are brown with darker centres, the tail ends in an obvious dark band and the dark wing-tips lack mirrors. The bill is dark, as is the eye. First-year

FIG 58. Recently fledged young Herring Gull. (Mike Osborne)

FIG 59. First-year Herring Gull in flight. Note pale inner primaries, which separates it from a Lesser Black-backed Gull of the same age. (Mike Osborne)

Herring Gulls are very similar to first-year Lesser and Great black-backed gulls, and while identification can be made in the field, this requires experience and the detection of minor details. In particular, these include the markings and the colour shade of the inner primaries (seen in flight) (Fig. 59), the patterning on the tail, and the shade and degree of contrast of the colouring of body feathers.

Most second- and all third-year Herring Gulls have some light grey feathers on the mantle, and these make easier the separation from Lesser Black-backed Gulls of the same age, which have dark slate-grey or even black feathers. Traces of the dark tail bar remain until the end of the third year, by which time parts of the bill have become pale or even yellow. While there is variation in the rate at which individuals acquire full adult plumage, most Herring Gulls in their fourth year (when some individuals have started to breed) show all of the characteristics of the adults. Even then, however, some can still be recognised as sub-adults by the small size of the white mirrors and the more extensive black along the leading edges of the open wings, and some retain a dark area on the lower mandible next to, or replacing, the red spot.

The Herring Gull is highly vocal, which is well known to the public, particularly people living in or visiting coastal towns and cities, and also listeners of radio and television programmes, where their calls are frequently used to set the scene of a coastal location. The calls made by Herring Gulls are numerous and similar to those of Great and Lesser black-backed gulls, but the species can be separated with experience – Herring Gull calls are less harsh and slightly higher in pitch. The loudest call is probably the long-call (also known as the trumpet call; Fig. 60), which is frequently made when obtaining and defending a nesting territory during the breeding season. It is also made by sub-adults, but less frequently (Fig. 61).

FIG 60. Aggressive long call by adult Herring Gull acquiring winter plumage. (Mike Osborne)

FIG 61. Three-year-old Herring Gull, still lacking white mirrors and with some immature plumage feathers in the wing, giving a long call to threaten an intruding gull. (Nicholas Aebischer)

HERRING GULL RESEARCH

Scientific literature on the Herring Gull is vast and, more than any other gull, an in-depth monograph is well overdue on this species – the last one was written by Niko Tinbergen more than 60 years ago (Tinbergen, 1953). The species is the most thoroughly studied of all gulls in Britain and north-west Europe. Studies on birds called Herring Gulls in North America relate to what many now regard as a separate species, the American Herring Gull (*Larus smithsonianus*).

The first extensive research studies on Herring Gulls, starting in the 1930s, were carried out in Germany by F. Goethe and in Norway by Edvard Barth, followed by investigations by Rudi Drent, Arie Spaans and Kees Camphuysen in the Netherlands. In France, Jean-Marc Pons has added greatly to our knowledge of the biology of this species. In Britain, the study of Herring Gull behaviour was extensively studied by Niko Tinbergen and his co-workers Dick Brown, Richard McCleary and Nicolaas Verbeek in the 1960s and 1970s, mainly focusing on the large colony at Walney Island in Cumbria. Studies there have been continued by Sin-Yeon Kim, Maria Bogdanova, Pat Monaghan and Ruedi Nager, and currently by the BTO. Other studies have been made on islands in the Bristol Channel and on Skokholm by Mike Harris, Peter Ferns, S. J. Sutcliffe and J. W. F. Davis, while investigations on wintering Herring Gulls in the London area were made by Peter Stanley, Mark Fletcher and Trevor Brough, working for the Ministry of Agriculture, Fisheries and Food. Studies on urban nesting in Britain and Ireland were carried out by Pat Monaghan, Susan Raven, Peter Rock and me. Studies on Herring Gulls on the Isle of May were made by Jasper Parsons, Neil Duncan, George Chabrzyk, Margaret Emmerson and John Calladine. Others have also made contributions in studies at landfill sites and elsewhere.

The large gull colony on the Isle of May, in the mouth of the Firth of Forth in south-west Scotland, offered considerable opportunities for study of Herring Gulls and was a research site from 1966 to 1977 for a group working from Durham University. Here, a series of researchers ringed a large number of Herring Gulls of known age (16,500 chicks ringed). Studies were eventually interrupted by the massive cull of large gulls on the island made by the Scottish section of the Nature Conservancy Council (now the Scottish Natural Heritage), but not before the existence of so many marked and aged individuals allowed a much greater insight into the breeding biology of the species and then on the effects of the cull on the remaining adults, which by then were breeding at a much lower density. More recent information on this colony has been collected by the Centre for Hydrology and Ecology.

Following the Isle of May cull, the Durham group switched its research to gulls in north-east England. Urban nesting was investigated by Pat Monaghan, Susan Raven and me, and intensive studies on the origin of wintering Herring Gulls were also initiated. Large numbers of adults were marked with unique combinations of colour rings, initially in north-east England and subsequently also to south Scotland when Monaghan moved to Glasgow University. These marked birds led to studies on the use of landfill sites by gulls made by Neil Duncan, Callum Thomas, Jennifer Butterfield, Nicholas Aebischer, David Jackson and me, and they also facilitated an intensive study on the feeding behaviour of gulls at landfill sites by Susan Greig, supported by Pat Monaghan and me. All of these studies were ably supported by the technical help of Eric Henderson, Michael Bone, Kathy Flowers, John Richardson and many others, and continued for 20 years.

Owing to the vast literature arising from the numerous studies on the Herring Gull, I have had to be selective in this chapter in terms of coverage, and to retain a degree of balance in its extent in comparison with the other gull species. Consequently, I have concentrated particularly on the species' present and past abundance in Britain and Ireland, the long run of information gathered on the former vast colony on the Isle of May and its management, the effects of age on the species' biology and the role of landfill on its feeding behaviour.

DISTRIBUTION AND STATUS IN BRITAIN AND IRELAND

The distribution map for breeding Herring Gulls in Britain and Ireland, based on the 2007–10 national survey (Fig. 62), shows a more general distribution than that for the Lesser Black-backed Gull (Fig. 90). Herring Gulls are particularly abundant on the west coast of Scotland, where they are associated with the many small islands. In contrast to the species nesting on natural sites, urban nesting dominates along the south and north-east coasts of England and in much of Wales. Herring Gulls are much less abundant in Ireland and have declined there to a greater extent than in Britain; currently, there is only one large colony there and urban nesting is restricted to the Dublin area and a single locality on the south coast.

The numbers of Herring Gulls breeding in Britain and Ireland changed dramatically during the twentieth century. At some stage between 1890 and 1910, their numbers reached an all-time low, brought about by a century or more of intensive persecution and the collection of eggs for food. Improved methods of transport during the Industrial Revolution made many breeding areas more accessible, and this – together with improvements in firearms – drove the dramatic reduction in numbers of many seabirds, including Herring Gulls, Kittiwakes, Great and Lesser black-backed gulls and Great Skuas (*Stercorarius skua*), while Grey Seals (*Halichoerus grypus*) and Common Seals (*Phoca vitulina*) also declined markedly.

I have searched many published and unpublished records to try to obtain an estimate of the numbers of breeding Herring Gulls in England, Wales and southern Scotland in the early years of the twentieth century; my best estimate is that there were fewer than 200 pairs breeding in the whole region. It proved

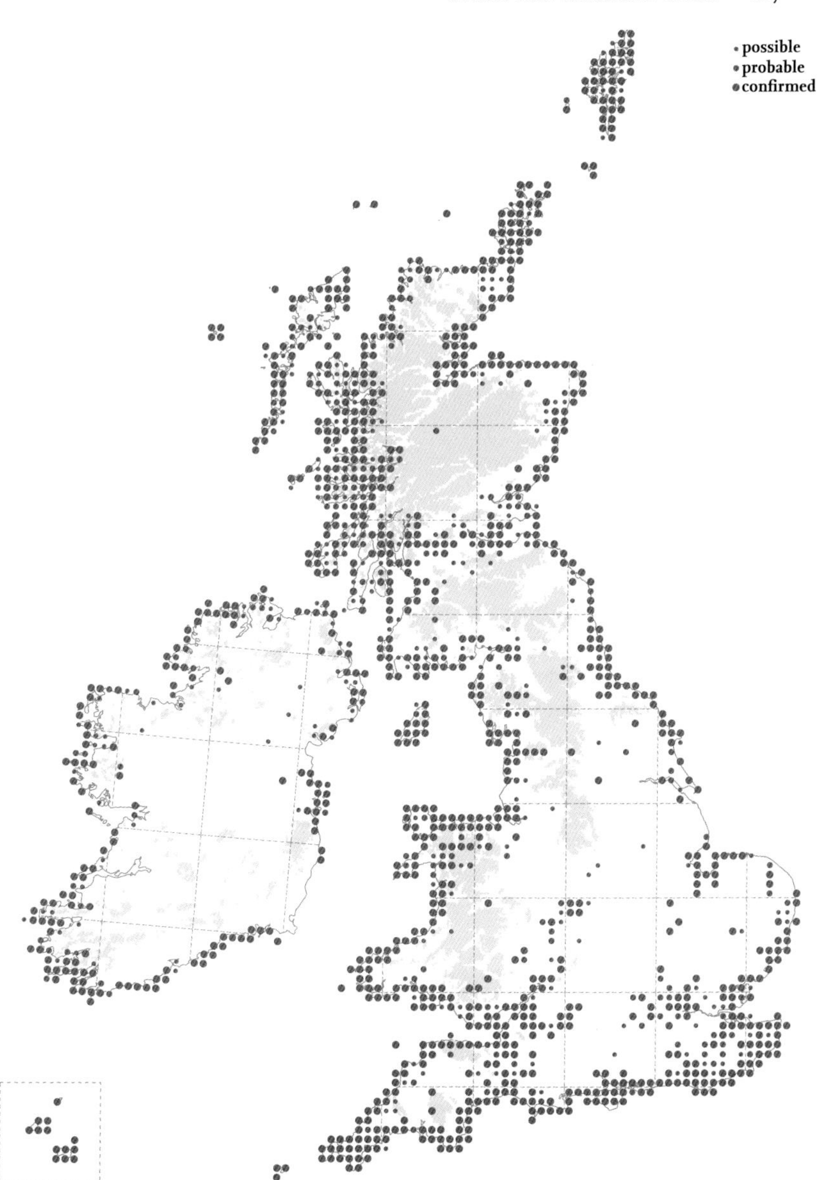

FIG 62. Map of the breeding distribution of the Herring Gull in Britain and Ireland in 2008–11. Reproduced from *Bird Atlas 2007–11* (Balmer *et al.*, 2013), which was a joint project between the British Trust for Ornithology , BirdWatch Ireland and the Scottish Ornithologists' Club. The Channel Islands are displaced to the bottom left corner. Reproduced with permission from the British Trust for Ornithology.

impossible to obtain information about numbers breeding on the northern Scottish mainland and islands, but there is no reason to believe that the situation there was much different from anywhere else in Britain, and it is realistic to assume that only small numbers of Herring Gulls bred throughout Britain and Ireland at this time.

Confusingly, several books written around the turn of the twentieth century described the Herring Gull in Britain as 'common', although it soon becomes evident that the word 'common' related to much smaller numbers at that time than it does now. For example, the discovery of six nests on the Farne Islands in 1912 resulted in the species being referred to as common in the accounts and records of the islands at that time.

The bird protection Acts introduced in the latter part of the nineteenth century were slow to come into effect, but eventually they resulted in a modest and then rapid recovery of many marine bird species during much of the twentieth century. In the case of the Herring Gull in Britain, the increase in numbers soon reached an impressive 13 per cent per year, a rate that resulted in their numbers doubling every six years. The growth was caused by a very high breeding success rate, which saw twice as many young surviving to breed than was necessary to replace those adults that had died.

The twentieth-century increase in Herring Gull numbers in Britain and Ireland

The introduction of legislation to protect birds in the late nineteenth century heralded the start of an extensive recovery of breeding seabirds in Britain and Ireland. The effects were remarkable, not only for Herring Gulls but also for other species that had reached low numbers. All started to increase at about the same time, although at lower rates. These recoveries had to be supported by abundant food, and presumably much of this had remained unexploited as their numbers declined during the nineteenth century. Additional sources of food also became available and were utilised at different times during the period of increase; these included more offal and rejected fish dumped overboard as bigger and more powerful fishing boats were introduced. Herring Gulls developed new feeding methods, including exploiting agricultural land and landfills, first near the coast and then eventually further inland. Several authors have attributed the entire population explosion of the Herring Gull to abundant food obtained at landfill sites (p. 167). However, this view is erroneous, as the gulls did not start to exploit this food source until the 1950s, when many small landfill sites, which burnt much of their refuse, were closed and large new sites were established.

When the first national seabird census was carried out in 1970, numbers of Herring Gulls breeding on the coasts of Britain and Ireland were estimated to have reached 690,000 adults. Increases in the numbers of any animal cannot carry on forever, and the Herring Gull population explosion ceased before the next census was carried out in 1985, when the estimate was 354,000 adults, or a 49 per cent decrease in just 15 years. Between 1985 and 2000, the numbers of Herring Gulls breeding on natural coastal sites declined further, but much less severely: by 17 per cent over the 15 years to 295,000 adults. There are no more national data available, apart from an index based on annual sample counts at Special Protection Areas organised by the JNCC, and these showed no meaningful change in the abundance of Herring Gulls at coastal sites between 2001 and 2015, although the average values hinted at a small decrease.

In contrast to the decline at natural sites on the coast, Herring Gulls have continued to increase rapidly as a breeding species at urban sites. In Britain, six species of gulls currently nest on buildings, with Herring Gulls being by far the most abundant and Lesser Black-backed Gulls coming in second. While Herring Gulls occasionally nested on buildings in England in the 1920s, the habit increased rapidly in the 1970s and numbers breeding in urban areas have increased dramatically since. More than 3,250 pairs of Herring Gulls were nesting in Aberdeen alone in 2000, when national numbers of urban nesters were estimated at 40,000 adults. However, a census of gulls nesting in towns is particularly difficult because many nest sites cannot be visited or even viewed from a distance. The use of a cherry picker (a hydraulic work platform) in Dumfries to improve and correct previous counts suggested that total urban nesting in 2000 may have involved close to 50,000 adults. Numbers nesting in urban areas have continued to increase, compensating in part for the decrease since 1985 in the numbers of gulls nesting at natural coastal sites.

The behaviour of Herring Gulls to nest on buildings in towns and cities is widespread on the Continent, with urban nesting common in the Netherlands, Belgium, France, Spain and Germany. While such breeding sites were originally coastal, there has also been a spread inland – for example, urban nesting is now taking place in Berlin. In the USA, the American Herring Gull breeds at several urban locations around the Great Lakes, while Glaucous-winged Gull (*Larus glaucescens*) has nested on roofs for more than 60 years in western USA and Canada.

Starting with the figure of 295,000 breeding adults reported in the national census in 2000, and then adding appropriate numbers of immature individuals in their first, second and third years, there must have been some 360,000 Herring Gulls in Britain in the late summer months of that year (assuming a breeding success rate of 0.4 young per adult, 20 per cent mortality in the first year, and 15 per cent mortality in the second and third years). Since ringing has shown that few British-reared Herring Gulls leave the country in winter and studies have estimated that numbers in winter increase by about 10 per cent through immigration from northern Scandinavia and Russia, there would have been a maximum of approximately 400,000 individuals wintering in Britain in 2000. This figure is appreciably less than the extrapolated estimate of 729,801 obtained from counts at a series of winter roost sites in England, Scotland and Wales in the 2003–06 period. The large discrepancy of more than 300,000 Herring Gulls is a concern and requires further investigation, but whichever numbers are used, it is difficult to understand why such an abundant species in Britain would be Red-listed, particularly when so many local councils consider it a pest in towns and cities. There is clearly a need for a new national census of the species, because the evidence currently available hints at the possibility that numbers of Herring Gulls in many parts of Britain may not have declined since the last national census in 2000.

The discrepancy above indicates that a word of caution is necessary regarding national censuses of gulls and other seabirds. Such work is extremely valuable and important, but it is also difficult to carry out and there are several sources of error. For example, Herring Gulls are surveyed by counting apparently occupied nests, but there is probably no date on which all breeding Herring Gulls have nests, and a proportion (perhaps as high as 10–20 per cent of adults) miss a breeding season and do not build nests, and so they are not included. In addition, the accuracy and reliability of totals reaching tens of thousands or more need to be determined, because currently they are used as if accurate to the last pair or nest. Sometimes, however, the total numbers have been rounded off to the nearest hundred, which attributes an accuracy far greater than is realistic, and yet the figures are still regularly used by both government and conservation organisations to determine the species' status and trends. Major changes in the abundance of species are undoubtedly correctly identified, but how large must an estimated change be in order to ensure it is real and not caused by limits in the census methodology?

The estimated decline of about 50 per cent in Herring Gull numbers in Britain between 1970 and 1985 was undoubtedly real and appreciable. A decrease in Ireland in the 15 years between 1985 and 2000 was also genuine, because it was confirmed by declines in each of the 16 Irish counties. However, there is no certainty that the decline really was 81 per cent and not, say, 65 per cent or 90 per cent. There are similar doubts over the accuracy of the 4–5 per cent decline in the totals for Herring Gulls in Britain reported over the same 15-year period (144,250 pairs in 1985, and 141,000 pairs plus at least an additional 10,000 in towns in 2000), particularly as counts at urban sites have been shown to be appreciably underestimated. As a point for discussion, has there been a change in the numbers of Herring Gulls in Britain in winter since 1985?

Dates of ringing recoveries in Britain strongly suggest that adult Herring Gulls are more likely to die in the summer, during and immediately after the breeding season, and not, as might have been suspected, in severe winter weather. This suggests that being tied to feeding grounds within range of a colony, plus the additional effort required to lay, incubate and rear young, may have become a particular stress in this species. However, this stress still has to be quantified, and it probably changes as new factors emerge to affect distribution and food supply. Around 1970, Herring Gull numbers were affected for the first time by three new factors: culling, food availability and botulism.

Culling

Extensive culling of adult Herring Gulls took place in England and Scotland from the early 1970s, and while Natural England have claimed this had little effect on numbers, it greatly underestimated the extent to which culling took place. No licence was needed to cull Herring Gulls at that time and no national records were kept of the extent of the culls carried out. Often, little or even no information of culls was made public and written records of some do not now exist. Indeed, I discovered by word of mouth long after the events that several more culls had taken place. Between 1970 and 1990, about 100,000 adult Herring Gulls were culled in Britain. This alone could have been sufficient to stop the population expanding, as the numbers were great enough to double the natural mortality rate of adults during that period. Some of these culls were large; that on the Isle of May in the early 1970s, for example (see below), carried out by the Scottish branch of the Nature Conservancy Council, reduced the numbers of breeding adult Herring Gulls on the island from 58,000 to about 7,000 birds over four years. The aimed reduction would have needed to remove some 51,000 individuals if carried out in a single year, but in the four years of the main culling, additional recruits arrived, and these were also culled to achieve and maintain the reduction, so that some 60,000 Herring Gulls were culled over the four-year period.

In the same decade, the RSPB killed large but unknown numbers of Herring Gulls nesting on several islands in England and Scotland for the benefit of terns, and the National Trust culled several thousand Herring Gulls on the Farne Islands in north-east England. In addition, local councils and industrial firms in several areas, the RAF at Lossiemouth and Carlisle, the Ministry of Agriculture,

Fisheries and Food, and dockyards at Rosyth (Scotland) and Barrow-in-Furness (Cumbria) all engaged in culling Herring Gulls. Landowners and United Utilities carried out a series of particularly large culls over many years at the Tarnbrook and Abbeystead colonies in Lancashire; some culling was said to be continuing during 2017. While thousands of Lesser Back-backed Gulls have been killed at these sites, an estimated 4,000 Herring Gulls and a few Great Black-backed Gulls were also killed.

In Wales, 100 Herring Gulls were accidentally culled on Skomer, an island off the south-west coast, between 1981 and 1985 during attempts to reduce the numbers of Lesser Black-backed Gulls. At the same time, many more adults died from botulism (see below), resulting in the considerable decrease of Herring Gulls breeding on the island (Fig. 63).

In 1991, I suggested that nearly 100,000 adult Herring and Lesser Black-backed gulls had been culled in Britain between 1972 and 1987. However, since then I have become aware of several more culls, and I now believe that more than 100,000 adult Herring Gulls alone have been culled in England and Scotland since 1970.

It is reasonable to believe that culling on this scale between 1970 and 1985 made an appreciable, but not the sole, contribution to the decline of coastal breeding Herring Gulls in England and Scotland. Is it a coincidence that, in the following 15-year period (1985–2000), when culling was taking place on a much smaller scale, Herring Gull numbers in coastal areas of Scotland were estimated to have declined by about a quarter while numbers in England increased?

A licence is now needed to cull Herring Gulls in England. The introduction of this restriction has had the effect of reducing annual numbers killed, but licences are still being issued and, in the last five years, more than 1,000 adult Herring Gulls were culled annually under licences issued by Natural England.

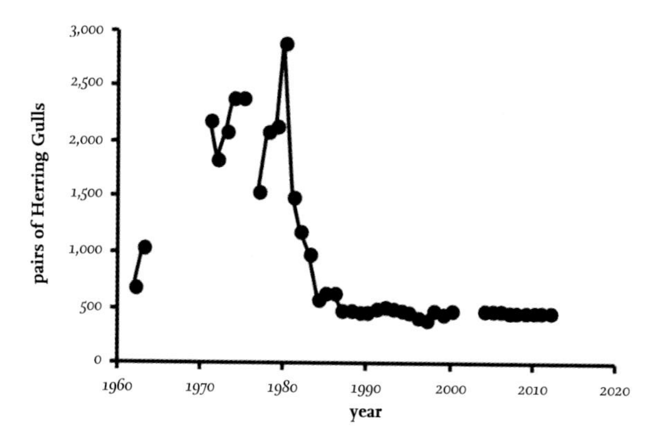

FIG 63. Numbers of Herring Gull pairs breeding on Skomer, an island off the south-west coast of Wales. From the early 1960s, numbers increased to a peak of 2,900 pairs in 1980, and then declined to about 500 pairs owing to high mortality of adults through culling and then from botulism.

Food availability and botulism

At the time of the decline of Herring Gulls in Britain, large landfill sites were numerous (methods of waste management to exclude gulls had not yet been extensively developed) and many gulls fed there. At the same time, botulism in gulls became more prevalent and widespread, killing many, although the actual numbers and proportions of coastal breeding Herring Gulls dying from this cause are uncertain because of the difficulty of identifying the presence of the toxin in dead birds. The time lag between ingesting *Clostridium botulinum* bacteria (which produce the botulism toxin) and death resulted in many gulls dying far from the source of the toxin, in night roosts or at sea. There has also been a suggestion that mortality from botulism poisoning was probably greater in Ireland than in England and Scotland, possibly because the higher rainfall in Ireland may encourage the survival of the bacteria.

Clostridium botulinum is associated with the presence of anaerobic water rich in organic detritus, and it has been identified as abundant in the majority of landfill sites that have been tested. Other sites not investigated may also have been possible sources. Landfill sites started to attract large numbers of gulls from about 1960, and they often had pools of stagnant, polluted water (Fig. 64)

FIG 64. Water collecting near the edge of a landfill site, often used by large gulls to drink, bathe and loaf, and is a site where botulism can be acquired. (John Coulson)

where many gulls drank, bathed and loafed after feeding on the dumped refuse. Consequently, this change in the feeding behaviour of large gulls probably brought them more frequently into contact with *C. botulinum*, increasing the frequency of individuals dying from botulism and overall mortality rates. Evidence that gulls acquired botulism mainly by feeding at landfill sites does not exist, but major sources of *C. botulinum* at other places where gulls regularly feed have been rarely identified in Britain and Ireland.

The paralysis produced by the botulinum toxin prevents gulls from flying, walking, drinking and feeding, and many probably died from dehydration rather than directly from the toxin. A series of paralysed gulls collected when suffering from botulism (confirmed from blood samples) recovered within a few days after water was frequently dripped into their mouths to reduce dehydration. After a day or so they would take food, and after a week they could stand and fly again, and were successfully released.

A feeding difference detected among Herring Gulls in north-east England in the 1980s and 1990s was that most urban nesting individuals rarely visited or fed at nearby landfill sites at any time of the year, and instead obtained their food at sea and in the intertidal zone. In contrast, many visiting gulls from Scotland and Scandinavia fed frequently at landfills. Whether this difference exists elsewhere is not known, but it is possible that it resulted in local urban breeding individuals being less at risk of developing botulism. If widespread, this behaviour also offers an explanation for the puzzling situation that numbers of urban breeding gulls have increased, while those breeding at natural sites have declined markedly.

The appreciable reduction in landfill sites that is currently taking place, and changes to the methods used to manage refuse, may already have reduced the risk of botulism occurring in gulls. Should the suggestion that botulism and culling played major roles in the decline in numbers of Herring Gulls in Britain and Ireland since 1970 be accurate, then the reduction of culling and the risk of contracting botulism may have already stopped this trend. Examining the available but incomplete data since 2002, there is no reliable evidence of a continued decline in breeding numbers of Herring Gulls in Britain. However, the decrease in the availability of offal and rejected fish discarded from fishing boats caused by fishing quotas and changing fishing methods has probably introduced a new factor adversely affecting Herring Gull numbers. This has already been suggested as the reason for the recent appreciable decline in numbers breeding on Canna, an island of the Inner Hebrides. There is a current need for a new national census, for without it, changes in the status of the Herring Gull since 2000 remain uncertain.

The Isle of May

The Isle of May in south-east Scotland became a national nature reserve in 1956, when the owners, the Northern Lighthouse Board, signed a nature reserve agreement with the Nature Conservancy Council. I supervised Durham University students carrying out research on Herring Gulls on the island for several years from 1966. This involved the ringing of approximately 16,500 chicks with metal and year-class colour rings, as well as the marking of adults with colour rings to identify individuals.

The first pair of Herring Gulls nested on the Isle of May in 1907 and the colony has been studied for many years. Fig. 65a illustrates the numbers breeding on the island since that time, with the plot showing an ever-steepening exponential curve. By using the logarithm of these numbers, the slope of the line is directly proportional to the rate of increase (Fig. 65b). It can also be seen that this rate of increase – 13.5 per cent per year – continued uninterrupted through

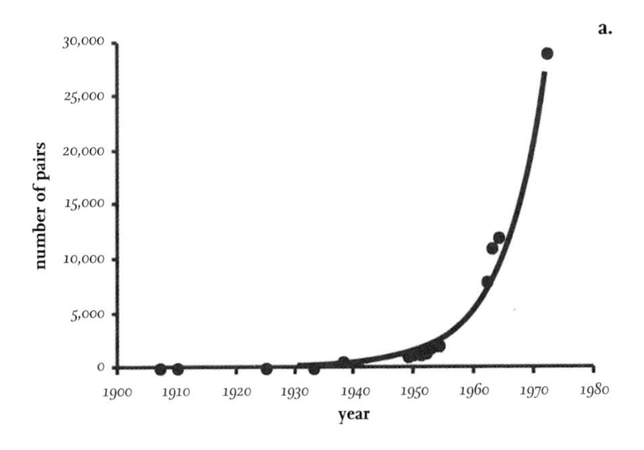

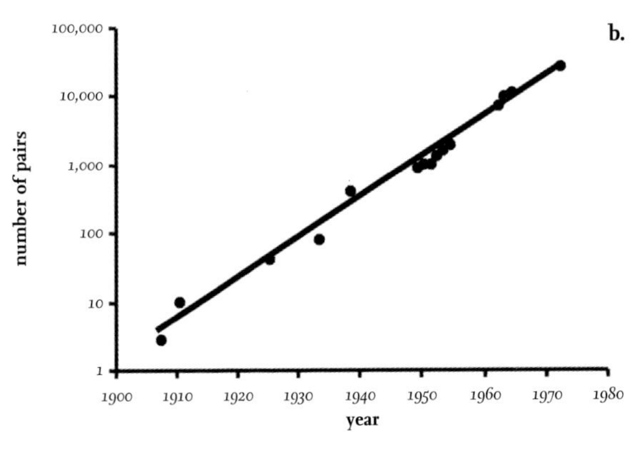

FIG 65. Graphs of the number of pairs of Herring Gulls breeding on the Isle of May, south-east Scotland, from 1907 to 1972, before culling took place. (a) the exponential increase; (b) the same data plotted on a logarithmic scale, with the best fit line indicating a 13.5 per cent annual increase.

the next 60 years. There is no evidence that the two world wars had an adverse effect on the population growth. Lesser Black-backed Gulls started to breed on the island in 1930 and their numbers increased at the even faster annual rate of about 16 per cent. However, the Herring Gull continued to be the most numerous gull there, and I estimated that by 1972 there were 29,000 pairs breeding at a high density on the island, along with an additional 2,300 pairs of Lesser Black-backed Gulls. High rates of increase of both species were also recorded elsewhere in Britain during the twentieth century.

Concerns were expressed that the high numbers and densities of large gulls nesting around the entire edge of the Isle of May were having adverse effects on other seabirds (particularly terns), soil and vegetation (particularly in areas with Puffin, *Fratercula arctica*, burrows) (Fig. 67). As a result, the Scottish section of the Nature Conservancy Council decided to carry out an experimental cull, to see if it was possible to reduce numbers of large gulls to 3,000 pairs. Both Herring and Lesser Black-backed gulls were culled, along with two pairs of Great Black-backed Gulls, by Nature Conservancy Council staff in the period 1972–76 using small narcotic-laced sandwiches placed in each nest. The total numbers were reduced by some 90 per cent.

FIG 66. Part of the large, high-density colony of Herring Gulls at Tarbet, part of the Isle of May, prior to the extensive cull. (John Coulson)

FIG 67. Damage to soil and vegetation caused by Puffins (*Fratercula arctica*) digging and Herring Gulls pulling up vegetation in an aggressive response to other gulls or for use in nest-building. (John Coulson)

During the main cull, the Nature Conservancy Council did not have enough personnel on the island to count the numbers of gulls killed, nor did they check those culled for rings. As a result, many gulls were incinerated without their rings being recorded (Fig. 68). Consequently, I offered to be present with a group of helpers on the island during most of the culls to count and examine the gulls

FIG 68. Bodies of culled Herring Gulls awaiting incineration without first being searched for leg rings. (John Coulson)

that were killed. We recorded the numbers culled prior to incineration for the Nature Conservancy Council and obtained information on approximately 1,500 ringed gulls. It was a most unpleasant task, but at least the information obtained from the ringing recoveries was of appreciable scientific value regarding the many birds that had been ringed as chicks and had subsequently returned to breed on the island, while gulls ringed elsewhere that had moved to the island were also recovered.

At least 45,500 adult gulls were culled on the Isle of May between 1972 and 1986, with the great majority killed in 1972, 1973 and 1974. This total is probably an underestimate, because a proportion of the gulls died at sea and most of these would have drifted away or sunk, with only a minority washed ashore during infrequent easterly winds. Twenty per cent fewer bodies were recovered on the island than the number of baits taken, but only 10 per cent was added to the total by the Nature Conservancy Council to allow for deaths at sea. This potential underestimation could raise the numbers culled to more than 50,000 Herring Gulls, about 700 Lesser Black-backed Gulls and four Great Black-backed Gulls.

By the end of the main cull in 1974, the density of nesting large gulls on the Isle of May had decreased fivefold, but the area of the island occupied by these birds had decreased only marginally, to about 90 per cent of that prior to culling. By 1977, the large gulls had been reduced to about 3,700 pairs. Further selective culling was necessary to clear selected areas completely, and by 1986 numbers were further reduced to 3,000 pairs. Thereafter, culling ceased, and the number of gulls increased only slowly, reaching about 3,000 pairs of Herring Gulls and 2,000 pairs of Lesser Black-backed Gulls by 2010. There was a marked change in the proportion of the two large gull species, with Lesser Black-backed Gulls increasing from 15 per cent pre-cull to 40 per cent by the time culling had ceased. Although no attempt was made to select Herring Gulls over Lesser Black-backed Gulls for culling, by 1979 the numbers of the latter nesting on the island had been reduced by 44 per cent to 935 pairs, far lower than the 80 per cent reduction achieved by that time for Herring Gulls. Presumably, Lesser Black-backed Gulls were less likely to take the lethal baits.

Great Black-backed Gulls bred on the Isle of May for the first time in 1962. They were represented by only three or four pairs in 1970, and although the species was not targeted, all individuals present were culled in 1972. The island was subsequently recolonised by Great Black-backed Gulls, and by 2014 there were 51 breeding pairs.

The Isle of May cull had some remarkable effects on the breeding biology of the Herring Gulls nesting there. Following the main culls between 1972 and

1974, the proportion of gulls breeding for the first time increased markedly and approached 50 per cent, which was more than three times that during the pre-cull period. However, their numbers were still far fewer than would have been expected from the large number of young that had fledged four and five years earlier, before culling started. It is therefore evident that the cull produced an additional effect, in that many prospectors about to breed for the first time (and which would have normally joined the undisturbed colony) moved elsewhere, perhaps influenced and dissuaded by the considerable disturbance produced by the culling. The high proportion of first-time breeders was maintained until at least 1978, and no colour-ringed adults marked before 1972 survived after 1974, suggesting that virtually all the older adults had been killed by then.

Following the initial intensive culls, the weight and wing length of new Herring Gull breeding recruits increased, and females of the same age laid larger eggs than previously, presumably because the much smaller numbers suffered less competition for food. For the first time on the island, three-year-old birds (mainly males) held territories (46 in 1976), and about half of these managed to pair and breed at this age. This contributed (although only in part) to the decline of the overall age at first breeding on the island by about a whole year. Interestingly, after the main culls, areas with densities of breeding adults lower than two pairs per 100 square metres failed to attract new recruits. Many of these low-density areas eventually became cleared of breeding gulls, sometimes without further culling simply because they were now unattractive to new recruits. Many of these changes to the breeding biology of the gulls can be regarded as density-dependent processes, which became obvious when the numbers and densities of birds on the island were dramatically reduced.

While the aim of the experimental cull was achieved, some aspects (aside from those directly associated with the task of removing large numbers of any dead animal) were disturbing. Large culls require the availability of adequate numbers of personnel, and in the case of the Isle of May cull, much valuable information would have been lost without the unrequested additional input from the Durham University research group. For example, the numbers of gulls killed on the island were initially estimated by the Nature Conservancy Council only from the number of baits that were apparently consumed, and body counts were not made until the support group specifically examined every culled bird before incineration. Not all gulls were dead when retrieved, and those that were not had to be dispatched in the field. In some cases, these were birds that had regurgitated the bait before it became fully effective. Only one other bird species was affected by the baits: Rock Pipits (*Anthus petrosus*) disappeared as a breeding species on the island at the same time, presumably because they fed on crumbs of the baits regurgitated by some of the gulls. A similar loss of Rock Pipits occurred on Coquet Island when culls of gulls were carried out there, but subsequently both islands were reoccupied by the species after a lapse of several years.

The reduction of gull numbers on the Isle of May benefited the vegetation, which soon recovered and spread into the many areas of bare soil, and Puffins and Eiders (*Somateria mollissima*) faced reduced predation by gulls. Terns later returned to breed on the island, but surprisingly not in areas previously occupied by gulls or terns. Since the large-scale reduction of large gulls numbers, culls of a few large gulls that regularly attempt to prey on tern eggs and chicks have been carried out, and areas of the island have been designated as gull-free zones, with any nests and eggs regularly removed.

While no attempt was made to select Herring Gulls over Lesser Black-backed Gulls in the 1970s Isle of May cull, Herring Gulls had been reduced greatly by the time it ended, declining from about 92 per cent to 80 per cent of the remaining breeding gulls. Since then, the numbers of Lesser Black-backs and Herring Gulls have slowly increased, but the latter species at slightly lower rate; currently, Herring Gulls form about 65 per cent of breeding large gulls on the island (Fig. 69).

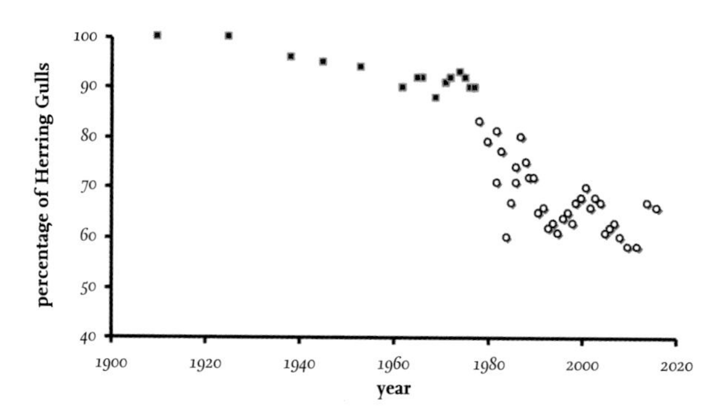

FIG 69. The percentage of Herring Gulls out of the total of Herring Gulls and Lesser Black-backed Gulls (*Larus fuscus*) breeding on the Isle of May from 1910 to 2016. The main culls on the island took place in 1972–74 and continued on a small scale in 1975 and 1976 and appeared to have a greater impact on Herring Gulls, which thereafter declined to about 65 per cent of the breeding gulls in 2016. Black squares indicate the percentage before and during the main culls, and open circles indicate the percentage once the main cull was completed. Data from 1982 kindly supplied by Francis Daunt (pers. comm.).

BREEDING

In southern areas of Europe, including England and Wales, many Herring Gulls remain near their breeding colony for much of the winter, visiting it from time to time and even roosting at night within the colony area. In late winter and early spring, the colony becomes occupied by many individuals on fine, calm days associated with high-pressure weather systems. In contrast, many breeding gulls in northern Norway do not return there until February and early March, but they start occupying territories soon after the snow cover has melted.

Niko Tinbergen (1953) found that most young Herring Gulls form their first pairs at 'clubs', places where gulls collect to roost during the day but that are not part of the breeding colony. Such pairing by young birds takes place long after the colony is reoccupied by older pairs. Only once they have formed a pair do they move into the colony and attempt to obtain and defend a territory there that will eventually contain their nest. Obtaining this territory is often a difficult task, as by this stage much of the colony is already occupied by older adults, and so many young birds nest within the lower-density areas at the periphery of the colony and lay later than established breeders. Density of a colony is important in determining the age at first breeding. On the Isle of May, no three-year-old males managed to breed until after the cull had appreciably reduced the density of breeding pairs.

Old adult males are the first to reoccupy the colony each year, often choosing the same territory or one close to that used in the previous year. Often, but not invariably, they pair with the same female as in the previous year, although the pairs do not regularly remain together during the winter and they reunite only by returning to the same territory in the colony as in the previous year. However, changes of mates commonly occur, and extremely few pairs remain with the same mate for life unless they survive for only a few years as adults.

Nest territory ownership is often challenged, and threats are very common. Many of these are non-combative gestures, involving pulling up plants or making threatening runs towards the intruding opponent. Actual physical contact does occur, when one bird may grasp the wing-tip of another by its bill and then drag it around. Disputes over ownership of territories in Herring Gulls rarely, if ever, result in major injuries. Once a secure pair bond has been formed, the female often joins in the defence of the territory.

In many colonies, Herring and Lesser Black-backed Gulls nest in mixed colonies. A study at Walney Island showed that nesting density of both species varied with the height of the vegetation (Hosey & Goodridge, 1980). Herring Gulls reached their maximum density on areas with short or medium-height

TABLE 26. The numbers of pairs of Herring Gulls and Lesser Black-backed Gulls (*Larus fuscus*) nesting per 10 m × 10 m plots in vegetation of different heights at Walney Island in Cumbria. There is considerable overlap between the two species, but Herring Gulls were most dense on medium-height and short vegetation, while Lesser Black-backed Gulls reached their highest density in tall vegetation. Based on data collected by Hosey & Goodridge (1980).

Vegetation	*Herring Gull (A)*	*Lesser Black-backed Gull (B)*	*Ratio B/A*
Tall	1.9	3.2	1.7
Medium	4.0	0.7	0.18
Short and scattered	3.4	1.9	0.56

vegetation, while Lesser Black-backed Gulls achieved their highest density when nesting in tall vegetation (Table 26). Surprisingly, most Lesser Black-backs avoided vegetation of medium height.

Occupying and retaining the nest territory, nest-building, incubation and protection of the young collectively restrict Herring Gulls to a much shorter period of daylight in which to feed, and this may contribute to their lower body weights of birds in the breeding season.

Nesting

Nest-building is often spread over 10 or more days before the first egg of a clutch is laid. The time involved in preparing and building the nest is much reduced in late breeding birds and those whose initial nest and eggs have been removed, and the nests of these pairs are usually less substantial. Nests are built on bare ground, areas of relatively short vegetation or cliff ledges. At a few colonies where trees were planted, Herring Gulls continued to nest between the trees, even when the trees attained a considerable height.

Nests on the ground (or roofs) are frequently sited alongside an upright object, such as a rock or, if on a building, a wall, parapet or ventilator. This provides some protection against adverse weather and restricts the direction of approach by intruding gulls. Nests are composed of plant material collected by both members of the pair. They are usually substantial structures with a central hollow to receive eggs, surrounded by a mound of vegetation that slopes away to ground level and can be up to 65 cm wide.

Nests at urban sites are frequently on flat or shallow, sloping roofs, while those on chimney stacks are almost always restricted to structures with two rows of chimney pots, which allows the nest to be placed in the space between four pots, or between the chimney and roof apex. Chimney stacks with a single row of pots (offering a nest site only between two adjacent pots) are rarely used. In many residential areas, the absence of chimney stacks or the types available play an important role in determining whether gulls nest there.

In Britain, laying starts in the last few days of April or the first few days of May, and the last pairs start their first clutches in late May or the first week of June, giving a spread of between five and six weeks within a colony. First-time breeders rarely replace lost clutches, but most other pairs do so, sometimes re-laying repeatedly, if necessary, until early July. Like other gulls, Herring Gulls rear only one brood in a breeding season.

Eggs and incubation

The most frequent clutch laid by Herring Gulls consists of three eggs; in a minority of cases, only two eggs are laid, and rarely a single egg is laid. The average clutch size in a colony is about 2.6–2.7 eggs and is highest among the early layers, progressively declining with date of laying, such that the majority of late-laying pairs produce only two eggs (Fig. 70). Early nesters whose eggs are removed tend to lay smaller replacement clutches that have similar numbers of eggs to those produced by young birds laying at the same time.

Clutch sizes laid by Herring Gulls progressively increase with their age. In a study of clutch sizes on the Isle of May (Table 27), the 12 pairs that included a three-year-old bird laid only one or two eggs, and the average number of eggs per clutch increased with age until birds reached seven or eight years old, when most pairs produced three-egg clutches. There was a small decline in clutch size in pairs that included a nine- to 11-year-old bird, which is the about the average age of most breeding Herring Gulls.

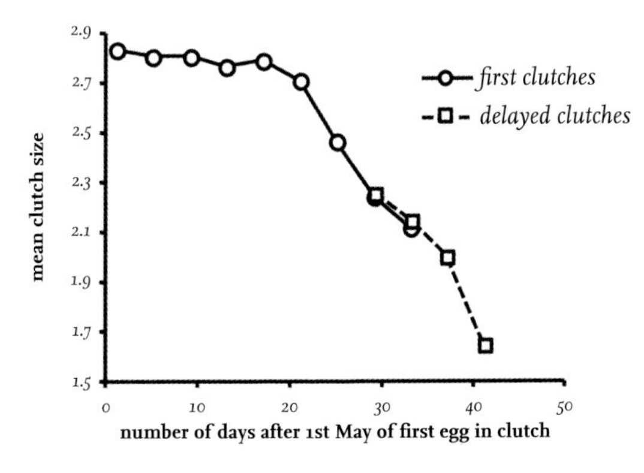

FIG 70. The seasonal variation in the mean size of first clutches of Herring Gulls (circles) and those laid later by breeders in replacement clutches (squares). The overlap in the trend lines suggests that clutch size is, at least in part, linked to the date of laying. Data collected by Jasper Parsons.

TABLE 27. The average clutch size laid by Herring Gulls of known age on the Isle of May, 1967–74.

	Age (years)						*Colony*
	3	**4**	**5**	**6**	**7–8**	**9–11**	
Number of clutches	12	23	64	141	163	51	632
Average clutch size	1.25	2.19	2.29	2.59	2.72	2.69	2.61

Eggs are laid mainly in the morning, and there are usually two days between successive eggs, but this can vary considerably – for example, the interval is greater in cold, stormy weather. The first two eggs in three-egg clutches are similar in size and volume, but the third egg is 10 per cent smaller in volume. This change in size of the last egg in a clutch is common in many other bird species (including Kittiwakes; p. 263), and is not the result of a shorter interval since the previous egg was laid (in fact, the interval between the second and third eggs is slightly longer). Instead, it appears to be associated with a gradual reduction in the material being transferred into eggs as the clutch approaches completion. This interpretation is supported by the second egg in a clutch being smaller than the first when only two eggs are laid, but is of a similar size to the first when a third egg is produced.

Four-egg clutches are extremely rare, and it has not been proven that a single Herring Gull has actually laid the four eggs; instead, it is likely that one or more eggs may have been contributed by a second female. The adults have only three brood patches and therefore the incubation of four eggs is unlikely to be successful.

The eggs of the Herring Gull are large (Table 28); in comparison, those laid by domestic Leghorn hens have a volume of about 55 ml. The shells have a mid-brown background and are heavily blotched with black and dark brown markings, a pattern that offers a degree of camouflage. The size of all eggs in the laying sequence tends to increase with the age of the bird, and the third egg is about 10 per cent smaller in eggs laid by adults over a range of ages (Table 29). As discussed above, a smaller last egg of the clutch in Herring Gulls has been recorded by many researchers, ranging from 8.7 per cent smaller by Barth to 11.0 per cent by Parsons. In addition, Parsons (1972) was able to show that, while the yolk is larger in bigger eggs, its size does not increase as rapidly as the total volume of the egg. As a result, while the third egg is about 10 per cent smaller, its yolk is only 6 per cent smaller.

TABLE 28. The size characteristics of Herring Gulls' eggs in relation to egg sequence in 455 three-egg clutches measured by Jasper Parsons (1972) on the Isle of May.

Egg number	*Length (mm) (L)*	*Breadth (mm) (B)*	*Volume (ml) ([0.47 × B² × L]/1,000)*	*Shape index (100 × B/L)*
First egg	69.4	48.5	77.9	69.9
Second egg	68.5	48.3	76.1	70.5
Third egg	66.2	46.9	69.3	70.8

TABLE 29. Size of first and third eggs in three-egg clutches laid by adult Herring Gulls of known age. Data from several unpublished sources (Neil Duncan, Jasper Parsons and John Coulson).

	Age (years)					
	4	*5*	*6*	*7*	*8*	*9–11*
Number of clutches	12	5	32	41	58	39
Volume of first egg (ml)	75.6	74.5	74.0	76.2	80.7	81. 2
Volume of third egg (ml)	67.7	68.2	69.1	68.4	70.9	73.1
Percentage decrease in volume of third egg	−10.4%	−8.5%	−6.6%	−10.3%	−12.1%	−10.0%

Incubation is shared by both members of the pair and is intensive, with eggs covered more than 95 per cent of the time. It is initiated by the female after she has laid the second egg and before the third egg is laid. Sometimes, low-intensity incubation begins at the time the second egg is laid and often results in synchronous hatching of the first two eggs in the clutch, but the third egg is usually laid two days later and hatches a day or so after the others. When incubation first starts, the capillaries supplying blood to the surface of the three brood patches are not fully exposed and the eggs do not reach as high a temperature as is achieved a few days later. Incubation lasts 27–30 days, the variation reflecting differences in the start of incubation and the frequency with which the eggs are left uncovered, which may cause their temperature to drop below the critical level required for development of the embryo. Incubation shifts last about two hours but are variable in length, ranging from 30 minutes up to five hours.

During the cull by the Nature Conservancy Council on the Isle of May in the early 1970s (see above), weights were recorded for 78 females that had died on the nest while incubating clutches of three eggs, all of which were measured. The females ranged in weight from 790 g to 930 g, but there was no correlation between egg size and female weight; large (heavy) females did not lay larger eggs.

Hatching success

In colonies not subjected to human or other forms of egg predation, or periods of severe weather, hatching success is 75–80 per cent. About 5–10 per cent of eggs are addled or the embryo dies, accounting for about a third of the hatching failures in colonies where egg predators are absent. However, in the presence of egg predators, much lower hatching success has been reported.

In a study on breeding Herring Gulls on the Isle of May made by Jasper Parsons (1975), hatching success peaked during the central part of the breeding season, and was 10–15 per cent higher when most other pairs were at the same stage of breeding and therefore laying synchronously than among birds that laid early or late (Fig. 71). Very little of this difference was caused by addled eggs and most was due to desertions and loss of eggs to predators, which were mainly other gulls and crows. This demonstrates an appreciable advantage to breeding when most of your neighbours are doing likewise.

Breeding success

There have been several studies of the average breeding success and productivity of Herring Gulls, both in North America (on the American Herring Gull, *Larus smithsonianus*) and in Europe; the results of some of these are listed in Table 30. Most values are between 0.6 and 1.0 young fledged per pair, which is appreciably fewer than the average clutch size of 2.7 eggs per pair (see above), indicating that about two-thirds of eggs laid fail to produce young that reach fledging age. The suggestion by Aonghais Cook and Robert Robinson (2010), based on modelling,

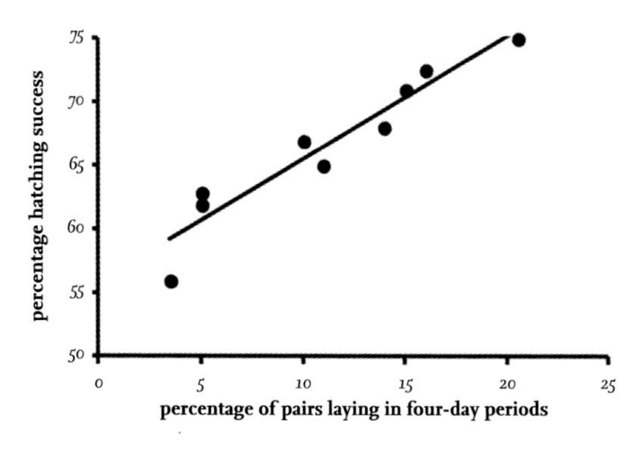

FIG 71. The relationship between the percentage of pairs of Herring Gulls laying on the Isle of May within four-day periods and their hatching success. Data from Parsons (1975).

TABLE 30. The average number of chicks fledged per pair of Herring Gulls at a series of sites in Europe and North America. Studies ranged from the late 1930s to the present time, and most figures are averages based on data collected over several years. Studies in the USA and Canada related to the American Herring Gull (*Larus smithsonianus*).

Location	*Number of chicks fledged per pair*	*Source*
Priest Island, north-west Scotland	0.92	Fraser Darling (1938)
Isle of May, south-east Scotland	0.85	Parsons (1971)
Skomer, south-west Wales	0.60	Harris (1964)
Walney Island, north-west England	0.91	Brown (1967)
Rooftops, north-east England	1.30	Monaghan (1977)
Average of several sites in UK	0.60	JNCC (2018)
Graesholm, Denmark	0.50	Paludan (1951)
Wilhelmshaven, north-west Germany	0.65	Drost *et al.* (1961)
Terschelling, Netherlands	1.20	Spaans & Spaans (1975)
Texel, Netherlands	0.88	Camphuysen (2013)
Kent Island, eastern Canada	0.91	Paynter (1949)
Mandarte, western Canada**	1.00	Vermeer (1963)
Eastern USA*	1.10	Kadlec & Drury (1968)

* American Herring Gull; ** Glaucous-winged Gull

that 1.3–1.5 chicks reared per pair are required to prevent a Herring Gull population from declining, would appear to be excessive. Many of the 13 field studies listed in Table 30 were made while the species was increasing and all but one reported fewer than 1.3 chicks fledged per pair.

Early-breeding Herring Gulls tend to lay more eggs per clutch, and while their percentage fledging success is not greater, they do rear chicks to fledging age (Fig. 72). Productivity continues to increase slowly for at least the first six years of breeding, but it is not known how the oldest birds (20–35 years old) perform because there is no way of ageing Herring Gulls (or other birds) apart from marking them when they are nestlings or in immature plumage. However, it is assumed that breeding success remains high for many, and few individuals live long enough to become senescent.

Seasonal variation in productivity was investigated on the Isle of May over four years and showed little change between years. Early-nesting pairs

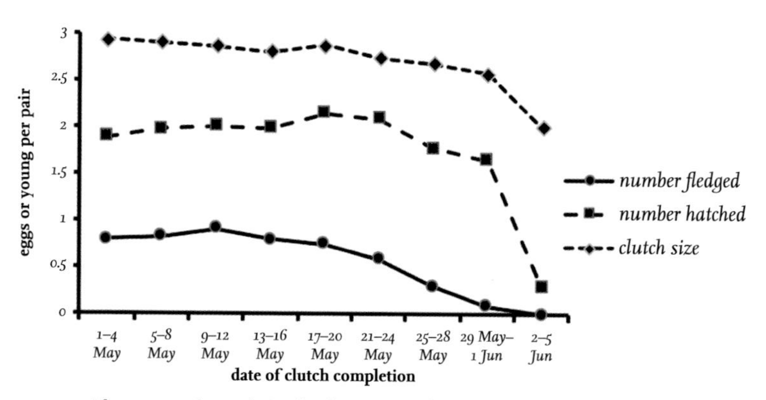

FIG 72. The seasonal trends in clutch size, number of young hatched per pair and number fledged per pair of Herring Gulls in relation to the date each clutch was completed on the Isle of May. Data pooled from 1968–70 and 1972, and based on a total of 1,086 clutches studied by Jasper Parsons, Margaret Emmerson and George Chabrzyk.

completing clutches up to 16 May laid large clutches and fledged an average of about 0.8 chicks per pair (Fig. 73), but in clutches completed after that date, production declined rapidly to the point that the 35 pairs laying for the first time that year and completing their clutch in June had almost total breeding failure. Poor breeding success in late-laying birds is reported in many bird species. The explanation normally given is that the birds are breeding when food has become less available (as in tits), or that they are young, first-time breeders.

Age of the adults is an important factor in determining breeding success within a colony. Young Herring Gulls breeding at three years of age were included in a study into breeding success on the Isle of May in the 1970s, and

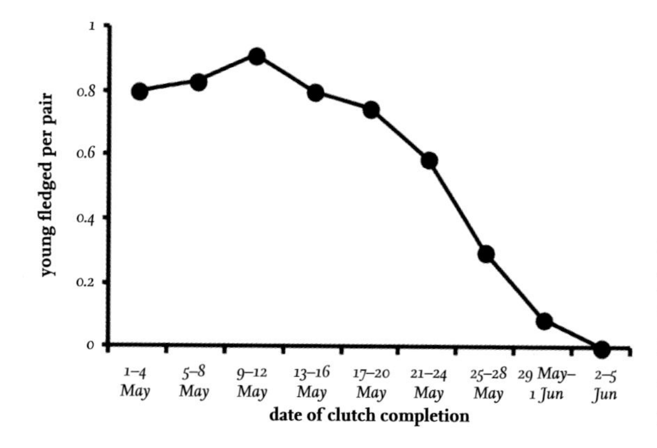

FIG 73. The variation in productivity of Herring Gulls nesting on the Isle of May in relation to date of clutch completion. Data as for Fig. 72.

TABLE 31. The breeding success of Herring Gulls of known age on the Isle of May, Scotland. Based on data collected in 1967–77 by the Durham University group.

	Age (years)							*Whole colony*
	3	**4**	**5**	**6**	**7**	**8**	**9–11**	
Number of pairs	12	31	76	145	75	76	75	755
Number of eggs	15	60	167	363	201	209	200	2,106
Number of chicks surviving 35 days	1	10	39	83	51	56	68	596
Mean clutch size	1.25	1.94	2.20	2.50	2.68	2.75	2.67	2.79
Chicks fledged per pair	0.08	0.32	0.51	0.57	0.68	0.74	0.91	0.79
Percentage of eggs producing fledged young	6%	16%	23%	23%	25%	27%	34%	28%

were found to have a very low success rate (Table 31). Many adults in the study were breeding for the first time at the age of four or five years (and a few did not breed until even older), but these had only 56 per cent of the hatching success of those aged nine to 11 years and 63 per cent of that of the colony as a whole.

Chick survival to fledging

A study on the Isle of May found that the mortality rates of chicks prior to fledging is highly age dependent, declining rapidly as the chicks become older and larger (Table 32). For the first five days after hatching, chicks are at risk from cannibalistic gulls. The risk of dying from starvation is small for the first two days because an appreciable amount of food in the form of yolk is retained from the egg, but after that chicks become entirely dependent on food supplied by

TABLE 32. The percentage of Herring Gull chicks on the Isle of May dying at different ages. Each percentage is based on the total that were alive at the start of each six-day period. Data collected by Jasper Parsons.

	Chick age (days)							
	0–5	**6–11**	**12–17**	**18–23**	**24–29**	**30–35**	**36–41**	**42–47**
Percentage dying	31%	22%	11%	8%	9%	6%	3.5%	3.3%

their parents. The amount of food required by young chicks is small, yet half of those that died in the first five days of life were underweight and their death appeared to arise from difficulties in persuading their parents to feed them, rather than the inability of the parents to collect enough food.

As discussed above, many studies have shown that the third egg of a clutch laid by Herring Gulls is appreciably smaller than the first two eggs in the clutch, and that this chick also suffers a high pre-fledging mortality; data from a typical study are shown in Table 33. Most of this differential mortality occurs in the seven days following hatching, after which it almost disappears.

These figures beg the question as to why the third chick to hatch suffers such a high mortality. Is it because it is smaller and has retained less yolk, or is it because the parents are less attentive to it when two (larger) chicks are also present? This question was answered by Jasper Parsons, who manipulated the hatching order by exchanging eggs between clutches. The result of this manipulation was that the third egg in the clutch hatched first; in a second experiment by Parsons, the first egg laid hatched last when exchanged, as if it were a normal third egg. Results indicated that two factors influenced high mortality in the third chick. When the last egg laid in the clutch hatched first, chick mortality in the first week of life halved, decreasing from 19 per cent in the control group of clutches to 9 per cent. When the first egg laid in the clutch was made to hatch last, chick mortality increased 2.5 times.

In another experiment, Parsons removed eggs laid early in the season, effectively delaying the laying date by 20 days. He then compared the breeding success of these delayed early layers with that of birds that naturally bred late in the season and suffered low breeding success (Fig. 74). The experimentally delayed birds, which eventually laid their clutches in the last days of May, had a very much higher breeding success compared with those that naturally laid their clutch at that time. The experimentally delayed early breeders retained the high breeding success of the normally early-laying birds used as controls (although their average clutch size was slightly smaller). Therefore, the poor breeding

TABLE 33. Percentage mortality of Herring Gull chicks on the Isle of May hatched from the first, second and third eggs laid in a clutch.

	Mortality day 1–7	*Mortality day 8–35*
First egg	22%	28%
Second egg	22%	35%
Third egg	58%	37%

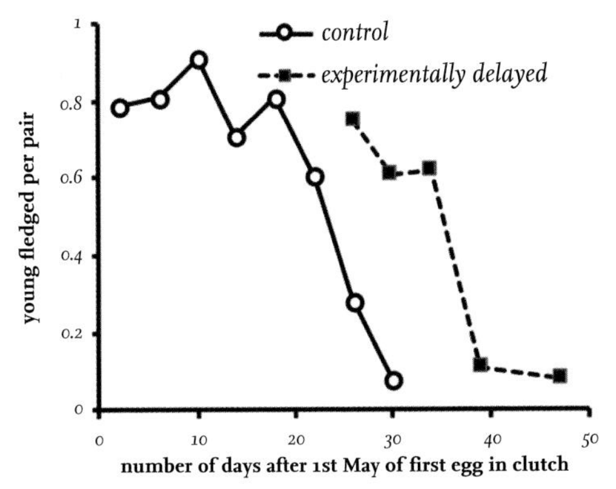

FIG 74. Number of young fledged per pair of Herring Gulls in relation to laying date. Circles show the normal situation, and squares indicate pairs that laid on 1–10 May but were delayed by about 20 days by repeated egg removal. Early-laying Herring Gulls that were delayed bred much more successfully than natural late breeders that laid and hatched chicks at the same time. The slightly lower productivity of the birds delayed until the end of May was due to their smaller average second clutch size, but this did not explain the poor success of those delayed until after 8 June (day 39). Based on Parsons (1973).

performance of naturally late-breeding Herring Gulls was not the result of adverse environmental conditions, such as food shortage or increased predation, but because they were poorer-quality individuals that were unable to breed as successfully. However, those individuals that were delayed longer, beyond the normal breeding period, did have a very low breeding success, probably because by then the colony had already started to disperse, potentially exposing these very late pairs to increased predation.

In some passerines, such as Great Tits (*Parus major*), the period of food abundance during the breeding season is very restricted, and if a pair does not nest at the optimal time, its breeding success is markedly reduced by food shortage. Results for Herring Gulls are different, however, and strongly suggest that a restricted period of abundant food available to feed to chicks is not the reason for the timing and synchrony of breeding in Herring Gull colonies.

Breeding success in relation to nesting density

The density of nesting gulls varies considerably within colonies. An analysis was made of the effect on the breeding biology of the density of neighbouring pairs nesting within a 4.5 m radius of the nest of the focal pair. Results showed that the average date of laying the first egg in each clutch was not earlier at higher densities. In contrast, clutch size increased progressively with nest density

(Fig. 75a), perhaps because more older gulls breed at high densities, but the percentage hatching success per pair peaked at an intermediate density (Fig. 75b), as did the production of fledged young (Fig. 75c). Those gulls nesting at a low density fledged only about half the number of young per pair compared with those nesting at the intermediate densities, while those at the highest density fledged 17 per cent fewer young per pair than those at the intermediate densities.

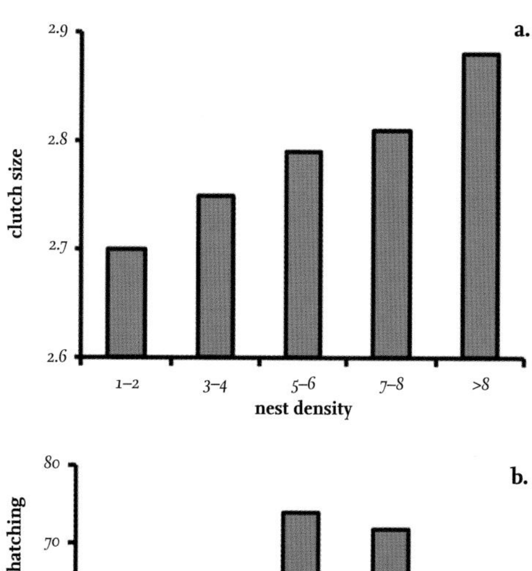

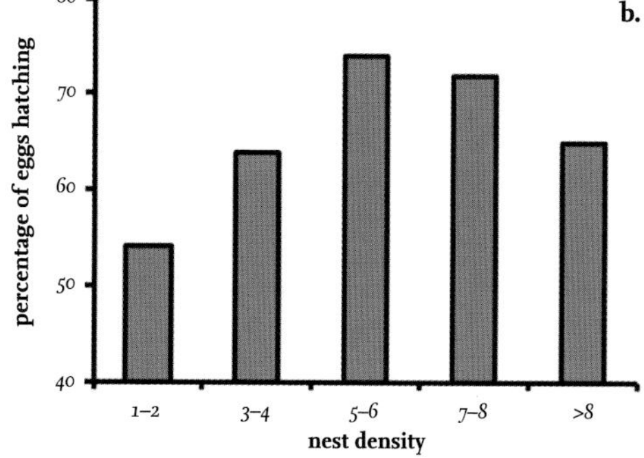

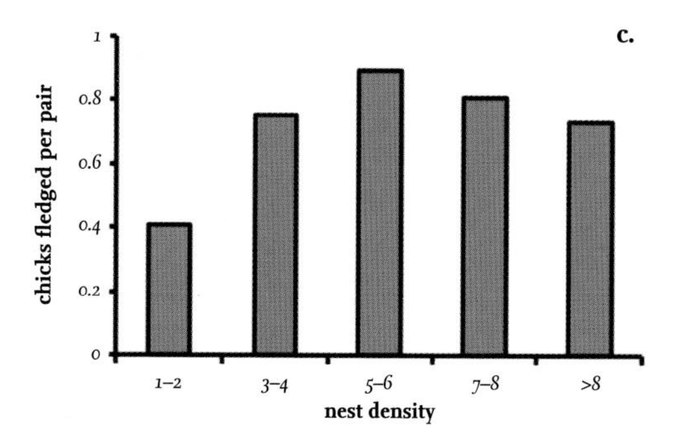

FIG 75. The relationships between density of Herring Gull nests and clutch size, hatching success and young fledged per pair. Density was measured as the number of nests within a 4.5 m radius of the nest of the focal pair. Clutch size (a) increased with density of nests, while hatching success (b) peaked at an intermediate density, as did the number of young fledging per pair (c). Pairs nesting at low density bred less well in all respects. These relationships could influence the density of breeding Herring Gull in colonies.

The birds nesting at low densities, including the periphery of the colony, probably included more young, inexperienced pairs, while at the highest density there was more intensive and frequent aggression between neighbouring pairs, and wandering chicks were more likely to be attacked and killed.

Breeding success in relation to brood size

A study carried out by Jasper Parsons over several years in the late 1960s on the Isle of May found little variation in fledging success of Herring Gulls that hatched different numbers of chicks (Table 34). Adults that hatched two or three chicks had almost identical percentage fledging success, but those with single chicks were slightly more successful, although not significantly so. These data suggest that rearing three chicks was not a major problem for the birds in the study and that enough food was available for them to do so. Broods of three young came only from clutches of three eggs, which was clearly the most productive clutch size per pair of adults.

Sex ratio of chicks

In a two-year study, Sin-Yeon Kim and Pat Monaghan (2006a) determined the sex ratio of Herring Gull chicks by DNA analysis and found that it did not differ from equality. In the first year, 54 per cent of the young from the first eggs in 49 clutches were male, and in the second year, 47 per cent of young in 118 clutches were male. There have been suggestions that the sex of eggs can be influenced in Yellow-legged Gulls (*Larus michahellis*) and Lesser Black-backed Gulls, possibly as a consequence of the body condition of the laying female, but confirmation of the extent to which this might occur is still wanting, and if it does occur, the deviation from equality would appear to be small.

Post-fledging

The behaviour of adult Herring Gulls when their chicks first fledge varies. In many cases, the bond between parents and chicks is soon broken, unless the

TABLE 34. The number and percentage of young fledged from broods initially of one, two or three chicks. Data recorded by Jasper Parsons on the Isle of May in 1967–68.

Number in brood at hatching	*Number of broods*	*Number of chicks hatched*	*Percentage fledged*	*Number of young fledged per pair*
1	124	124	41%	0.41
2	260	520	34%	0.68
3	360	1,080	35%	1.05

chicks return to their nest site territory, as some do. In other cases, the young and their parents fly together to feeding sites, with frequent contact calling between them. Their flight progress is slow and often involves gliding, presumably reflecting the less skilful flying abilities of the chicks. Most chicks are not fed by their parents once they leave the colony, but Tony Holley (pers. comm.) recorded some chicks aged up to 84 days returning in the evening to a colony in south-west England, food-begging and then being fed by adults; one exceptional juvenile was fed each evening for seven months. There is also a record by Bill Bourne of a fledged chick in Aberdeen begging and being fed by an adult in November.

The mortality rate of recently fledged Herring Gulls is high. Preliminary analyses of ringing recoveries suggest that about 40–50 per cent of fledged young Herring Gulls die during in their first year of life, but by the end of their first year their mortality rate decreases and approaches that of the adults, at between 7 per cent and 15 per cent a year. Recently fledged birds have not learnt the skills of feeding and swallow a wide range of food (and non-food) items. Few recently fledged chicks visit landfill sites and those that do tend to avoid the high-density feeding areas by keeping to the edges of flocks, or they remain as solitary individuals and patrol the surrounding areas in search of potential food items. Eventually, some of these young gulls develop successful attacks on older and smaller gulls. In winter, those present at landfill sites in the afternoon are inclined to remain there, searching out food for some time after most adults have left for the day.

Age at first breeding and adult survival rates

Herring Gulls only occasionally attempt to breed at three years of age and these are usually males. Most do so when four or five years old, and there are strong indications that a few individuals may not breed until they are six or seven years old, although it is difficult to be certain that these birds had not attempted to breed elsewhere when younger. A few Herring Gulls are known to live for more than 30 years in the wild (the oldest known was 34 years and nine months old when it died, tangled in nets), but a typical lifespan is between eight and 15 years, of which four to 11 years are spent as breeding adults.

There are several estimates of the survival rates of adult Herring Gulls, but earlier studies tended to underestimate the figure owing to rings being worn or lost before the longer-lived individuals died. More reliable estimates have been obtained by the annual identification of individuals with unique colour rings, assuming adults return to the same colony; these are included in Table 35. At first inspection, the survival rates appear to be similar, but this is not effectively so. When survival rates are converted to the average expectation of adult lifespan, and hence the number of times an average individual can breed, it is evident that this varies from 5.4 to 14.9 years for the figures listed. In long-lived species, a small change in the high annual survival rate greatly alters the number of potential breeding years and therefore is of major importance in determining whether the numbers of any species being considered are increasing or declining.

Table 35 also shows estimates of the number of chicks fledged in the lifetime of the adults. These figures are probably high, because no allowance has been made for adults that miss a breeding season. Presently, Herring Gulls need to fledge an average of about 0.7–0.8 chicks per pair each year to maintain their numbers. To keep the population at the same level, two chicks must reach maturity during the lifetime of each pair of adults to replace them when they die. This would require each adult breeding (living) for about 10 years, which is achieved only by adults with an annual survival rate of about 90 per cent. Unfortunately, the survival rate from fledging to maturity is poorly known, but it and/or the adult survival rates must have been even higher in the first 60 years of the twentieth century in Britain to permit the species to increase at 13 per cent per annum (p. 135).

Most studies have found a marked annual variation in survival rates, and a series of values over several years are required to obtain meaningful estimates. Two of the studies showed a significantly higher survival rate for adult females,

TABLE 35. Realistic estimates of the average annual survival rates of adult Herring Gulls and the corresponding expectation of adult lifespan and lifetime productivity.

Location	*Period*	*Average annual survival rate*	*Expectation of adult lifespan (years)*	*Young fledged per lifetime (assuming 0.8 chicks fledged per year)*
Isle of May, Scotland	1966–72	93.5%	14.9	11.9
Isle of May, Scotland	1989–94	88%	7.8	6.2
North-east England	1979–85	91%	10.6	8.4
South-west England	1991–2006	88%	7.8	6.2
Skokholm, south-west Wales	1965–70	93%	13.7	11.0
Skomer, south-west Wales	1986–2014	83%	5.4	4.2
Brittany, France	1983–90	88–92%	7.8–12.0	6.2–9.6
Netherlands	2006–11	83%	5.4	4.2
Arctic Canada	1998–2004	87%	7.2	5.8
Ontario, Canada	1981–97	91%	10.6	8.5

while one showed the reverse, and the others failed to detect a difference or the survival rates of the sexes were not considered separately.

Data on populations of animals are often put into the form of a life table; this includes numbers of each age class, to which the production of young can be added. Such a life table for the Herring Gull is provided in Appendix 2 (p. 442). Unfortunately, rarely is there sufficient available information to make the life table precise – for example, considering males and females separately, or the effect of age on productivity – but some of the outcomes obtained are of considerable interest nonetheless. For example, the life table shown in Appendix 2 highlights the appreciable numbers of immature Herring Gulls, which are rarely shown in data of species numbers because these are based on the counts of nests of breeding birds. The life table suggests that, at the end of the breeding season, about half of all Herring Gulls are immature; by the start of the next breeding season, they represent 37 per cent of all individuals. As a result, numbers from census work giving breeding numbers probably need to increase by 50 per cent to obtain the actual numbers of immature and mature individuals.

MOVEMENTS

Fig. 76 shows the extent of movements of young Herring Gulls from where they were reared to where they were, in most cases, breeding. There is considerable movement within Britain and to the east of Ireland, but of greater interest are the movements across the North Sea to the Netherlands, Germany and Denmark, the single movements to France and Spain, and the exceptional movement of one individual to Iceland.

In the early days of bird ringing, there was great interest in records of recapture following considerable migrations at sites where the birds had been reared and ringed – the Swallow (*Hirundo rustica*) is a good example of this. This ability to return to the place of rearing is described as philopatry, indicating a return to the 'land of the father', and many examples have been recorded over the years. It soon became commonly accepted that most bird species, including gulls, were philopatric.

Niko Tinbergen, in his remarkable book *The Herring Gull's World* (1953), commented on the few cases of Herring Gulls then known to have returned to breed in the colonies in which they had been reared, and mentioned two that had moved to a nearby colony. On this evidence, he believed that the great majority of young Herring Gulls were philopatric. Much more data have been accumulated since! These show that many young Herring Gulls have been found four or more years later breeding in the colony in which they were reared, but few researchers have acknowledged how much more difficult it is to detect those birds that move elsewhere. To find birds that return involves searching one colony, but locating those that move requires similarly intensive searches in numerous other colonies.

Jasper Parsons gave the young reared in each of three years on the Isle of May distinct year-class colour rings, and many of these birds were subsequently seen breeding on the island. However, a few were reported or recovered elsewhere, either breeding, or of breeding age and presumed to be breeding. In a separate study, Pat Monaghan discovered that many of these Isle of May chicks, when adult, were breeding in South Shields and Sunderland, but the magnitude of the movement away from the Isle of May was revealed when 34 marked chicks from the island were culled as breeding adults on the Farne Islands in 1976, and a further nine on Coquet Island in the same year.

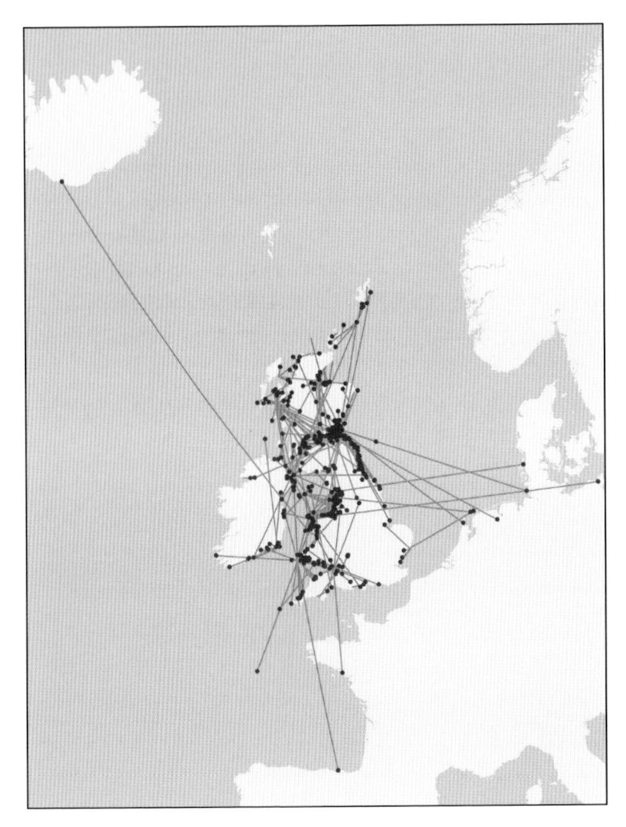

FIG 76. Records of Herring Gulls ringed as chicks in Britain and Ireland and recovered in the breeding season when of breeding age and more than 20 km from where they were reared. Reproduced from *The Migration Atlas* (Wernham *et al.* 2002), with permission from the BTO.

A measure of the degree to which young birds return to their immediate natal area can be gleaned from the many ringed birds culled on the Isle of May in the early 1970s (p. 135). The island was divided into eight zones, and numbers ringed and later culled in these areas were collected. This allowed the proportion that had returned to the same sub-area to be determined and compared to a value of about 12.5 per cent if they recruited into any part of the Isle of May at random. Table 36 shows the results based on three year classes of Herring Gulls ringed on the island. Birds fledged in three different years showed almost identical proportions (around 65 per cent) returning to the area in which they were reared, or five times the expected value had they returned to the area by chance. Clearly, many young Herring Gulls recognise their natal area. Unfortunately, the intensity of culling was so great on the island that there was no time to sex the culled birds before they were incinerated, and it is not known whether this intense philopatry varied between the sexes.

Most of these gulls had been reared in 1967 and so were of breeding age before the Isle of May cull began in 1972, and many that moved to breed elsewhere chose to do so before the cull and its associated disturbance commenced.

A result of the Herring Gull cull on the Isle of May was that large numbers of birds that had been ringed by year class as chicks were killed and the few that remained could be readily counted. In Table 37 the numbers of colour-marked chicks estimated to be alive in 1973, after the cull, are compared with the combined total of colour-marked birds culled and numbers culled and known to be alive still on the island. In all three years of ringing, there was a discrepancy between those accounted for and those estimated to be still alive, suggesting that the differences were birds that had moved elsewhere to breed. Pooling the data for the three years suggests that 45 per cent of the surviving birds were elsewhere. This can be only an approximate value, but it gives an indication of the proportion of individuals that are not philopatric.

TABLE 36. The number of Herring Gulls ringed in eight areas of the Isle of May and recovered in the same area as breeding adults during the early 1970s cull.

Year ringed	*Number culled in eight areas*	*Number culled in natal sub-colony*	*Percentage in natal sub-colony*
1966	414	278	67%
1967	508	332	65%
1968	342	215	63%
Total	1,264	825	65%

TABLE 37. An estimate of the percentage of Herring Gulls ringed on the Isle of May that were believed to be still alive and present on the island immediately before culling took place.

Year ringed	*Number fledged*	*Estimated number alive in 1973**	*Total of cohort culled or seen breeding on the island after culls ended*	*Percentage survivors returning to the island*
1966	3,943	1,040	735	71%
1967	4,811	1,494	870	58%
1968	3,896	1,423	590	41%

* Based on life-table estimates of annual survival rates.

National ringing recoveries in the breeding season of Herring Gulls ringed as chicks in Britain and Ireland and recovered when old enough to breed give a measure of the extent of philopatry. In a total of 4,051 reported recoveries, 766 (19 per cent) moved more than 20 km from the place of ringing and therefore were unlikely to have been philopatric. It is probable that the degree of philopatry varies with conditions in colonies, and that the high density on the Isle of May resulted in a higher rate of emigration.

It should not be assumed that Herring Gulls nesting elsewhere than in their natal colony have become lost. An interesting study in Belgium by Harry Vercruijsse on Herring Gull chicks that were individually colour-ringed found that the immature birds visited several colonies and all of those surviving to breeding age then visited their natal colony. Of these individuals, some stayed to breed, while others moved away and bred elsewhere. This suggests that potential breeding recruits made a judgement about the suitability of the natal colony, and perhaps chose another colony because they assessed that it had better prospects. Rather than having an inbuilt mechanism requiring them to return to the natal site to breed, the birds were making a choice. How this choice is made and how the assessment is made is not known for certain, but it is likely that competition for sites plays a part. There is also some evidence that young, prospecting gulls first visit colonies late in the breeding season and are attracted by those where there are young fledging, which is also an indication that the colony is likely to be a safe nesting place in a subsequent year.

There have been only a few ringing recoveries of Herring Gulls reared in Britain and Ireland that in winter crossed the North Sea or reached south-west France, Spain and Portugal (Figs 77 and 78). These represent a very small proportion of the large number of ringing recoveries within Britain; indeed more than 99 per cent of recoveries of British-reared Herring Gulls are within

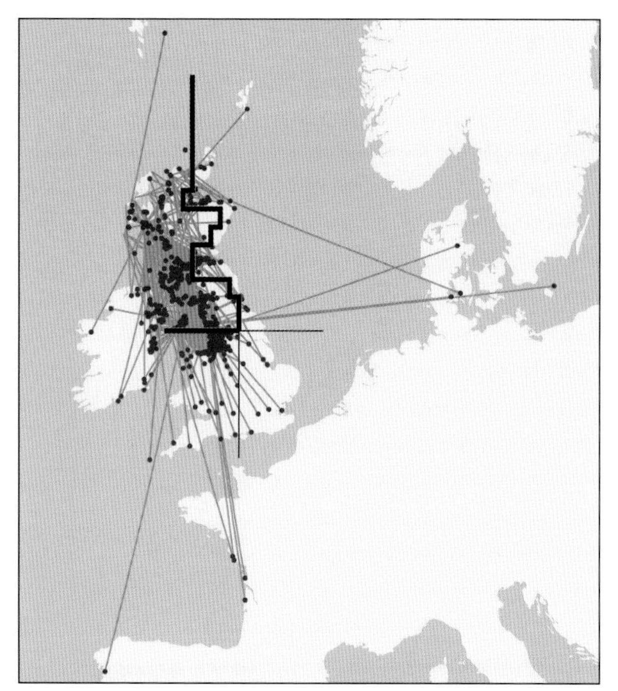

FIG 77. Movements of Herring Gulls ringed as chicks in north-west Britain and recovered outside the breeding season. The black line marks the area within which the chicks were ringed. Reproduced from *The Migration Atlas* (Wernham *et al.* 2002), with permission from the BTO.

Britain. Some of these birds remained close to the colony in which they were reared, while others made modest movements, typically heading south from their natal colony. Recently fledged Herring Gulls in Scotland show a greater tendency to move into England by following the coastline, with a minority of individuals making a cross-country movement. Adult Herring Gulls breeding in England, Wales and Scotland make only short dispersal movements, and these are often of short duration; many move only 100 km or less from the colony in which they breed. For example, a colour-ringed female breeding at South Shields in Tyne and Wear visited Scarborough in Yorkshire (about 100 km away) year after year, but she remained there for no more than one to two months each year before returning to her breeding area.

One- and two-year-old Herring Gulls do not visit breeding colonies in the summer and instead many stay in or near the areas where they spend the winter. Three-year-old birds start to visit colonies, mainly arriving late in the breeding season and some weeks after the established breeders have arrived.

Herring Gulls reared in southern Norway, Denmark, the Netherlands and Belgium show similar local movements towards the south and tend to follow the coastline; only a few cross the North Sea to winter in Britain. In contrast,

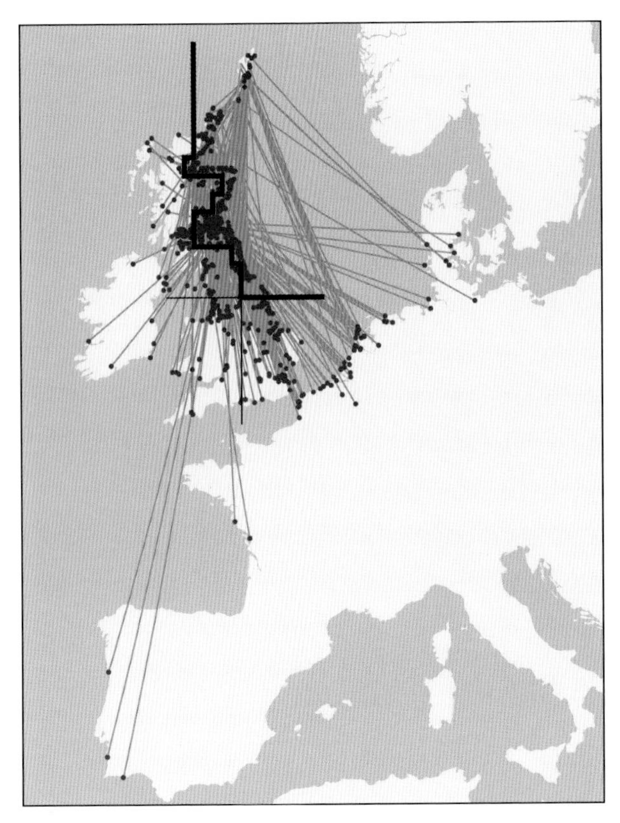

FIG 78. Movements of Herring Gulls ringed as chicks in north-east Britain and recovered outside the breeding season. The black line marks the area to the north-east within which the chicks were ringed. Reproduced from *The Migration Atlas* (Wernham *et al.* 2002), with permission from the BTO.

Herring Gulls breeding in northern Norway and north-west Russia migrate considerable distances and many winter in countries bordering the North Sea, including east and south England and Scotland (Fig. 79). These large, dark birds increased the numbers wintering in Britain by 10–20 per cent, but most did not leave Scandinavia until they had almost completed their primary moult (Coulson *et al.*, 1984a). They arrive in late September and October in eastern Scotland and England, where some remain during the winter, while others move inland through north and central England to winter in the London area, where many feed at landfill sites. In contrast, few of these birds winter in Wales, Ireland or on the western side of England and Scotland.

An estimate of the numbers of Herring Gulls present in the West Midlands during two-weekly periods was made by Alan Dean (n.d; Fig. 80). Appreciable numbers were present from only the beginning of November to the end of February, with a very marked peak during January. These numbers do not

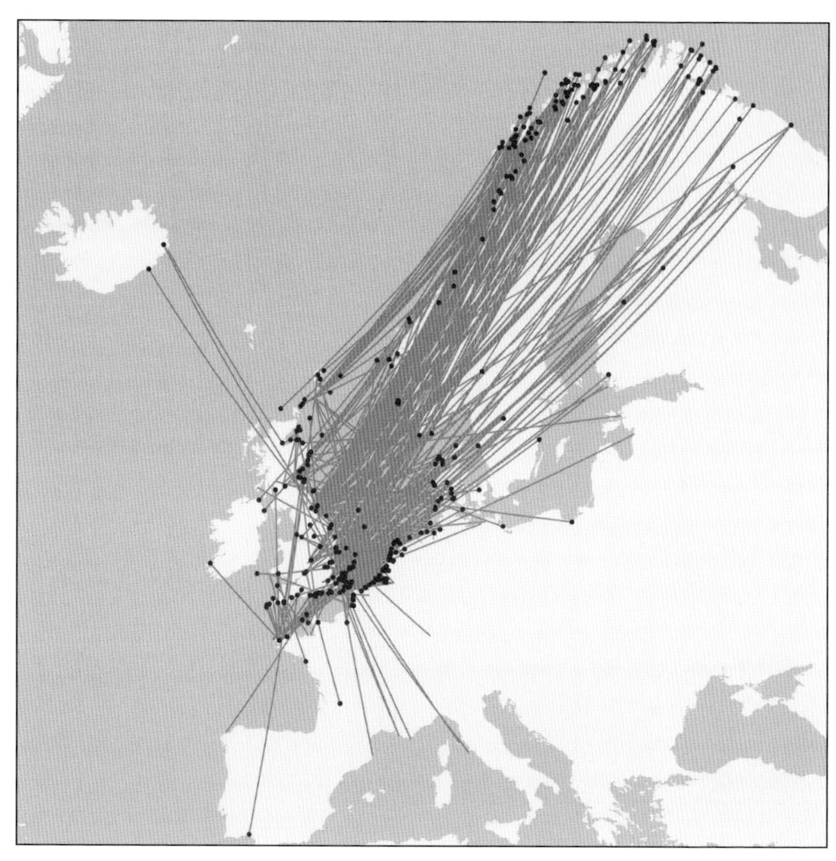

FIG 79. The exchanges of Herring Gulls between Britain and Scandinavia that were not ringed as chicks in Britain and Ireland. Most of those from Scandinavia were winter visitors to eastern Britain. Note that few reached western Britain and only two have been reported from Ireland. Reproduced from *The Migration Atlas* (Wernham *et al.* 2002), with permission from the BTO.

correspond to the overall numbers present in Britain from October to the beginning of February, which do not change markedly, and so the January peak must be caused by the redistribution of Herring Gulls within Britain, possibly with many (including birds of Scandinavian origin) moving inland to feed at landfill sites.

The numbers of Herring Gulls present in County Durham and the south side of Tyne and Wear in 1988–90 were estimated from ratios of colour-ringed birds seen at various sites, including landfill (Fig. 81). The figures are approximate but illustrate the gradual increase in numbers in the autumn and the build-up of

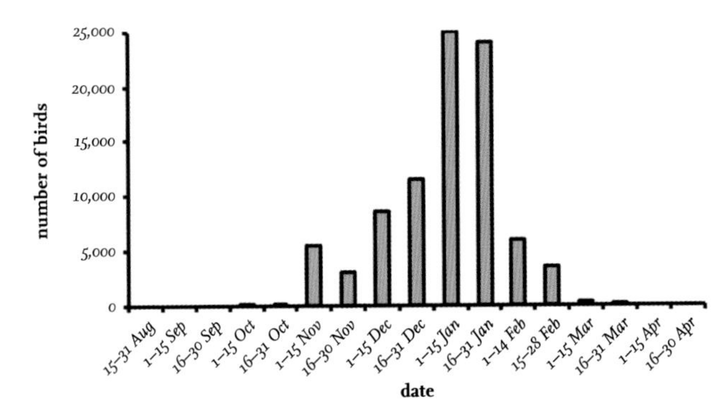

FIG 80. The estimated numbers of Herring Gulls wintering in West Midlands based on data collected by Alan Dean (n.d.).

large numbers from late October to early February, similar to the situation in the West Midlands. Most arrivals were individuals that had moved south from eastern Scotland, while additional individuals from northern Scandinavia arrived in October. Particularly noticeable was the sudden and synchronous departure of individuals from north-east England in the first two weeks of February, which involved both Scottish and Scandinavian breeders. A similar sudden departure was seen in the West Midlands, suggesting that a return towards breeding areas at this time may be widespread, involve huge numbers of Herring Gulls.

Following the large-scale colour-ringing programme of adult Herring Gulls visiting north-east England (p. 171), many of these birds were subsequently found breeding in colonies during the summer, and they formed two distinct groups. The majority were breeding at sites along the entire east coast of Scotland and

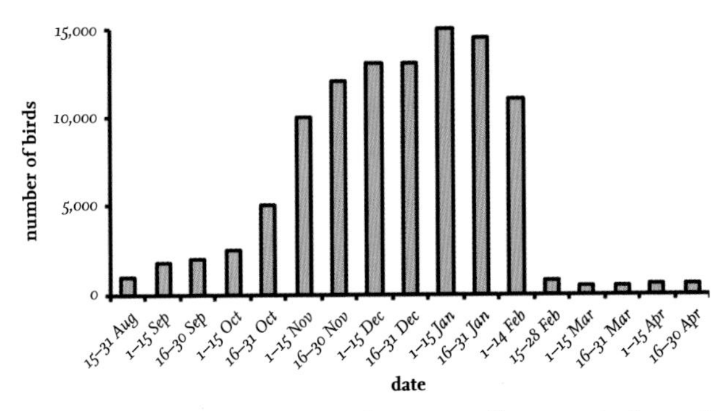

FIG 81. The estimated numbers of Herring Gulls present in County Durham and south Tyne and Wear from mid-August to the end of April, based on field counts in 1988–90 of the numbers and proportions of marked birds. From late February to mid-August, there were about 200 local breeding birds and a small number of immature individuals. On any working day, less than 25 per cent of the gulls were at landfill sites at any one time.

as far north as Shetland, while the minority were large, dark-winged individuals discovered breeding in northern Norway and north-west Russia. Many individuals of both groups returned to north-east England in subsequent winters.

The Scottish breeding birds visiting north-east England in winter did not remain in a specific wintering area but were much more mobile, making what is best described as a grand tour as they moved around a large wintering area in east and south-east England. In this manner, there was a regular and continuous passage of Scottish birds into and through north-east England. The first adults from Scotland (possibly failed breeders) started to arrive in north-east England as early as late July, and others continued to arrive until the end of December. As many adults were colour-ringed, it became obvious that most of the Scottish birds stayed only a few weeks in one area. For example, most adults caught and marked in October had left the area by November and did not reappear until the following October, when many returned for just a few weeks, only to move on again out of the large study area. They were replaced by other adults, which then also made relatively brief stays, and these exchanges continued in November and December. The extent of this mobility was remarkable. In an attempt to mark most of the 300–600 Herring Gulls making daily visits to Coxhoe landfill site in Durham, 2,000 adults were captured and marked there between September and December in one year. Of the several hundred gulls sighted at the landfill in early January, only five were marked individuals and the remaining were unmarked. Searches at sites within a 60 km radius found only two more marked birds.

Adult males from Scotland tended to arrive in north-east England earlier in the autumn than females, and they also returned earlier to Scotland – some having done so during November, at a time when some adult females were still moving south into England. Consequently, the sex ratio of Herring Gulls in north-east England changed appreciably and progressively from September to January (Table 38). Sightings of many of these birds in subsequent years confirmed that this change in sex ratio was not caused by females avoiding landfill sites in the autumn, and nor did adult males avoid landfills in winter.

The Scottish and northern Scandinavian Herring Gulls that were still present in England in January all left and returned to their breeding areas over a period of less than two weeks in early February. This mass departure was preceded for several days by more social and vocal activity, and with a reduced inclination to feed. Surprisingly, such synchronised mass departures and movements of thousands of Herring Gulls have not been observed, and departures did not start when the birds were at feeding sites during the day. It is possible that they set off directly from their overnight roosts at daybreak. Several individuals leaving for Norway were seen at their colonies just a few days after being seen in England

TABLE 38. The sex ratio of colour-ringed adult British Herring Gulls seen in autumn and winter in north-east England. Most originated from colonies in eastern Scotland.

Month	*Sample size*	*Males per 100 females*
September	427	120
October	291	114
November	546	68
December	389	44
January	730	43
February	210	51

in February, and may have made non-stop flights across the North Sea. Such a journey would take about 17 hours, in which case the gulls would have reached the Norwegian coastline in darkness.

FOOD AND FEEDING

There have been extensive studies during the breeding season on food and feeding in Herring Gulls in Britain and elsewhere. These confirm that the species is omnivorous and consumes a wide range of terrestrial and marine animal food, although it is not as extensive as the Great Black-backed Gull diet, nor does it include such large items. Animal food – both vertebrate and invertebrate – dominates. Fish, either caught or obtained from human fishing activities, makes up most of the diet, while plant material is occasionally consumed. In contrast, less is known of the food eaten during winter. Grain obtained from farmland is consumed in autumn and spring, while in some places in the Arctic and the Netherlands (but not in Britain), berries are eaten in the late summer. In many areas, much of the food is obtained on intertidal areas, where the gulls feed mainly on marine invertebrates, including sea urchins, crabs, starfish, cockles and mussels. In a few places, Herring Gulls have learnt the technique of collecting a shelled bivalve mollusc, then flying above a road or rocky area, half-hovering and repeatedly dropping the item onto the hard surface until the shell is broken, allowing the soft inner parts to be consumed. This skill is not widespread and takes a long time to acquire, as indicated by the much lower success of immature gulls, which often fail to drop the shells onto a hard surface.

In many areas, Herring Gulls feed at sea. The species is the most numerous gull feeding on offal and rejected fish thrown overboard from fishing vessels off the coast of north-east England, although this feeding habit is assumed by Lesser Black-backed Gulls off the Netherlands and south-east and north-west England, by Great Black-backed Gulls in western Britain, and by Gannets (*Morus bassanus*) and Great Skuas in parts of Scotland (Table 49, p. 222). All these species outcompete the Herring Gull when they attempt to feed together. While more often seen scavenging behind fishing boats, Herring Gulls are capable of plunge-diving to shallow depths to capture small fish, often sandeels and young Herring (*Clupea harengus*).

A hundred years ago, Herring Gulls were rarely encountered inland, but over the century they have been increasingly reported feeding on pastures and in fields that are being ploughed (Fig. 82). Feeding on pastures occurs more often in the early morning, when the soil and grass are still wet from overnight rain or dew. The Herring Gulls search for, and catch, earthworms and insects (particularly leatherjackets and beetles) at or near the ground surface. They also paddle on the spot in pastures, as if marking time, which presumably stimulates the earthworms to move, so that the gulls can detect them. Similar paddling is used in shallow pools of water on intertidal mudflats, and is a technique employed by several other gull species to disturb potential prey.

Herring, Lesser Black-backed, Black-headed and Common gulls are all attracted to fields that are being ploughed. They fly close behind the plough until a food item is exposed, and then scramble to capture the prey. Numbers of gulls

FIG 82. Herring Gulls following a plough to obtain invertebrate food. (John Coulson)

of several species, including Herring Gulls, frequently search for food in late spring and summer in fields while grass is being cut for silage or hay, presumably because insects and small mammals are disturbed or exposed and then readily captured. Kleptoparasitism is frequently used by individual Herring Gulls in these situations, pursuing members of their own species or smaller gulls if these catch a food item that cannot be swallowed immediately. Kleptoparasitism is also used in the marine environment, when one or more Herring Gulls join a flock of feeding Eiders on the water, waiting until a duck surfaces with a large food item in its beak and forcing it to give up the prey.

The marked increase of Herring Gulls nesting in towns in recent years has led to some feeding in the streets, but the importance and value of this is frequently exaggerated. It has, incorrectly, been assumed that Herring Gulls were first attracted to nest in towns because of the food they discovered there. However, I observed Herring Gulls as they first started to nest in towns in north-east England in the 1950s and 1960s, and they did not feed in the streets then but regularly moved to the shoreline or out to sea to feed. The habit of obtaining food in towns developed secondarily and years after they bred there for the first time. Discarded food left in streets eventually became a source of food for a few individual gulls, and was then encouraged by people deliberately offering them items such as bread, potato chips and the like. Eventually, this led to a few gulls becoming specialists, developing the technique of snatching food from the hands of unsuspecting people (see box on p. 168). Later, urban nesting gulls eventually discovered – and exploited – food in refuse bins and in plastic sacks left out overnight for removal, resulting in rubbish becoming strewn across streets. Even now, most gulls nesting in urban areas obtain only a very small proportion of their food within towns, and typically move tens of kilometres to the coast or to farmland to feed.

Many wildlife protection organisations and local councils still believe that urban nesting gulls are present in towns because of the food they find there, and therefore advocate that preventing gulls obtaining food in streets and elsewhere in towns will decrease their numbers. They encourage the use of gull proof containers and request that people do not feed the gulls. While these actions tend to keep towns tidier, the often-repeated idea that they will reduce or stop gulls nesting there is flawed. Gulls nest in towns primarily because eggs and chicks reared on roofs are usually safe, and hence breeding success is usually higher than when nesting in natural sites.

Feeding at landfill sites

Despite appearances to the contrary, Herring and Black-headed gulls do not normally feed continuously at landfill sites. At many sites, feeding by Herring

Urban issues

The invasion by, and problems associated with, Herring Gulls nesting in urban areas have resulted in the species being regarded as a pest in some places. This status developed because, as numbers increased, reports increased of gulls swooping at humans in defence of an unfledged chick on nearby buildings or one that had fallen from its nest. The more numerous gulls were also very noisy in the early hours of the morning. Their attacks, which usually involve swooping close over a person's head from above or behind at more than 50 kph, are intimidating. From time to time, large gulls do strike people, but this is rare. Very occasionally they have drawn blood by pecking at a person as they pass overhead, but people are more often struck by the trailing feet, which usually do less damage. Some of these fast fly-by diving attacks are accompanied by defecation, and the targeting seems to be remarkably accurate – in some areas, residents have had to protect themselves by carrying open umbrellas even on fine days!

As a result, complaints about gulls nesting in towns are numerous, ranging from their droppings soiling pavements, windows and the sides of buildings and even pedestrians, to the noise they make during courtship and fighting to gain or retain territories at early hours of the morning. Parked cars left for a day or so can be covered with gull droppings, and on some cars the faeces cause the body paintwork to blister. In Whitby, such irritations led one publican to post a notice reading 'Keep Whitby tidy; eat a gull a day'. Of course, such action would contravene the current bird protect legislation, but it is but one of many expressions of irritation caused by Herring Gulls nesting in urban areas.

In some coastal resorts and inland centres, a few individual Herring Gulls specialise in snatching food from people. This habit is not new – people were being robbed by Herring Gulls in Llandudno in the early 1950s – but it is increasing. In the city of Aberdeen, Herring Gulls stand on top of street lights along the main road leading from the fish docks, and when a lorry passes by loaded with uncovered boxes of fish, an individual often swoops down to pick up a meal. In the mainly enclosed Inverness railway station between 2012 and 2014, one individual Herring Gull frequently flew in through the train entrance and landed on top of a stationary train. It waited, often many minutes, until a catering trolley with food for passengers was wheeled to the train, and then it swooped, picked up a packet of sandwiches and left the same way it came in. In the south of England, one Herring Gull was reported regularly entering a shop, where it would pick up a packet of food displayed near floor level and then walk out with it into the street. Fortunately, these patterns of specialised, novel behaviour repeatedly used by individuals are not copied by the great majority of Herring Gulls nesting in towns.

Claims that Herring Gulls have attacked and killed pets, including a tortoise and small dogs, are infrequent, but are quickly blown up by the media. In July and early August every year, the press and television stations frequently report complaints about Herring Gull attacks. In one town, postmen refused to deliver mail to houses with gull chicks on their roofs, and some people have said they are scared to go outside. In effect, the media do much to increase public fear of gulls breeding alongside the human population.

Gulls takes place for short periods and only when no refuse lorries are unloading, and adds up to just 30–60 minutes during the day. At some, there is a lack of food for gulls when the site opens in the morning because refuse deposited the previous day has been covered, and there is a lag of an hour or two before the first loads of refuse are collected and taken to the landfill site. The high level of human activity at the site then disturbs feeding except during the half-hour after midday, when incoming vehicles are prevented from tipping during the lunch break and the bulldozers and compactors leave the site.

This midday feeding by gulls is so regular at many landfill sites that the birds accumulate in large numbers on a nearby field or a flat, undisturbed area less than an hour beforehand and wait for activity at the work face to cease. Once work stops, the gulls usually pour onto the area in a feeding frenzy. On other days, the gulls make no attempt to feed immediately and instead remain on the loafing area for up to 20 minutes, until finally a feeding frenzy suddenly develops and persists until vehicles return to the site at about 12.30.

When more landfill regulations were progressively introduced during the second half of the twentieth century, they required refuse to be covered by soil each day. Few gulls could feed at the sites after they closed for the day (usually at 16.00), after noon on Saturdays, or on Sundays and public holidays, when they were closed all day. Herring Gulls arrived at the landfills on all these days but left early on Saturdays, as soon as the freshly deposited refuse was totally covered. On Sundays, they arrived, circled the landfill site and passed on, usually without landing, presumably responding to the lack of activity and exposed refuse.

A landfill site at Aberdeen proved to be exceptional. Here, Herring Gulls fed for much of the day and did not fly off to avoid incoming vehicles or their drivers when they left their cab to check that their load had been emptied – the gulls often fed almost at their feet. This lack of consistency in the responses by the gulls at different landfill sites is marked, but at each it is usually consistent between days and in general depends on the level of vehicle traffic visiting during working hours.

Almost all the gulls present at landfills feed in a dense frenzy, often involving hundreds of individuals. Only rarely does a gull feed alone at the work face, although a few adult females and immature birds might search filled areas that have been covered days and weeks previously. Immature Herring Gulls are under-represented at feeding frenzies, usually forming less than 10 per cent of the gulls waiting to feed, but many choose to avoid the highly competitive situation and instead feed elsewhere.

The intensive social feeding that takes place at landfill sites is unusual, and it is difficult to understand why active feeding is so often restricted in time and is an all-or-nothing behaviour, with solitary individuals rarely feeding there.

The gulls give an impression of acute wariness and display a considerable initial reluctance to begin feeding at sites. The activity is eventually triggered when one or two individuals leave the loafing area nearby, fly over the refuse and land. There is then a mass movement to feed by most of the roosting flock within a minute. The feeding frenzy is often initiated by Black-headed Gulls, but they are soon displaced by the larger gull species and Herring Gulls in particular. But why are gulls so reluctant to start feeding at landfill sites? Because it is a relatively recent habit, it may be that they have not yet adapted a more efficient response. They respond as if the landfill site is dangerous and are suspicious of approaching until one or a few individuals have demonstrated that it is safe. The only time similar mass behaviour is observed in Herring Gulls is when they exploit a shoal of fish near the surface or when nets are hauled in by fishing boats, but in these situations they do not show such hesitation before starting to feed.

The intensive study of marked Herring Gulls feeding at landfill sites in north-east England made by the research group I led revealed several interesting fact about the behaviour of the gulls. While numbers of Herring Gulls tend to be similar on successive working days at landfills, only 30 per cent were present on consecutive days and many were replaced daily by different individuals. Searches within the area revealed that only 5 per cent of the missing marked gulls had moved to another landfill site. They also showed that, on average, each bird visited a landfill site for only one-and-a-half days per week, spending the rest of the time feeding elsewhere, presumably out at sea. Of course, there were exceptions. Two individuals (out of more than 200 recorded) fed at the landfill site on all five working days in a given week, and some large, dark-winged colour-ringed individuals from northern Scandinavia frequently attended the same landfill site on three consecutive days of the working week. At the other extreme, some individuals seen in the area were not recorded again at landfill sites, although they were seen elsewhere in the area.

Landfill sites are generally thought to be attractive to gulls because they are sources of plentiful food, and occasionally this can be so, such as where waste from processing chickens was dumped at one particular site. However, it is usually difficult for individuals to find food, and in Herring Gulls it requires years of experience before they become efficient in exploiting hidden or partially concealed food items among the mass of paper, plastic, wood, building waste and other deposited materials. This difficulty is also suggested by the small proportion of immature and, particularly, first-year gulls feeding at landfills, their low efficiency in obtaining food there and their need to revert to kleptoparasitism. In contrast, mainly immature individuals gather when food is dumped on the beach at Arbroath and at other fishing ports (Fig. 83).

FIG 83. Large group of almost entirely immature Herring Gulls feeding on shellfish dumped on the tideline at Arbroath. (John Coulson)

Effect of age on feeding behaviour at landfill sites

Studying the feeding behaviour of individual Herring Gulls at landfill sites is difficult owing to the short periods the birds spend feeding during the day and the large numbers involved at such times. We solved this problem by videoing gulls feeding at the work face of a landfill site and replaying the film frame by frame, following different individuals on each occasion. This also allowed colour-ringed birds of known age to be identified and their feeding behaviour followed. Susan Greig used this method to make a detailed study of the behaviour of 119 different colour-ringed adult Herring Gulls of known sex, a smaller number of colour-ringed immature birds, and a large sample of unmarked immature birds that could be aged by their plumage. Five age categories were used. Birds in adult plumage could not be aged but included some that had been marked as adults several years previously and were at least eight to 12 years old.

Because the active movement of individuals feeding at landfill often caused them to move out of view, 15-second observations were used as the unit of feeding behaviour, resulting in several hundred values being obtained for each age and sex category. Feeding behaviour was recorded in a series of categories, particularly the frequency of pecking at material on the landfill, the frequency with which items were swallowed, and when kleptoparasitic attacks were made or received.

Fig. 84a shows the rate at which individuals pecked at, and swallowed, food items at landfill sites. From this, it is evident that, irrespective of their age, the birds

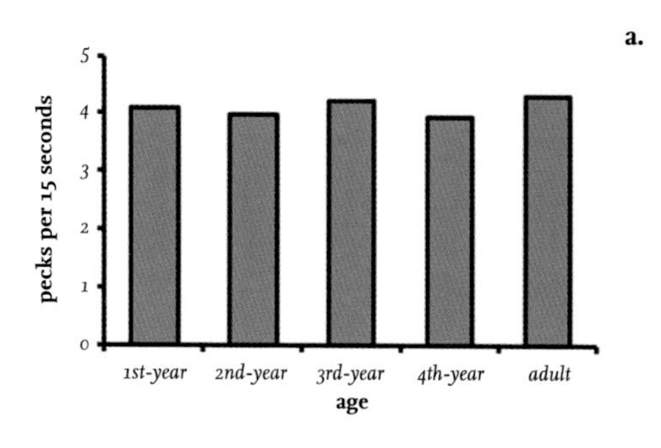

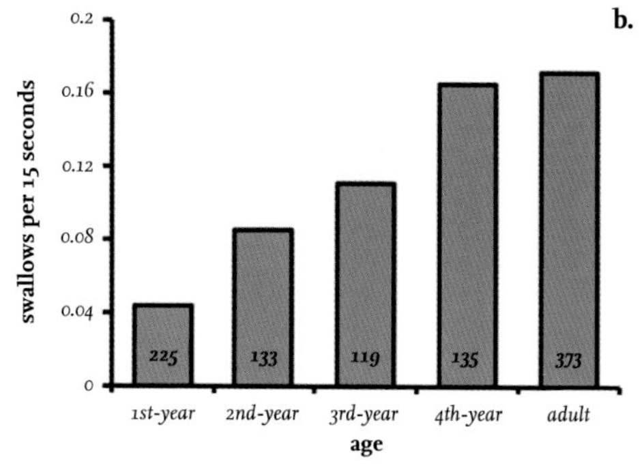

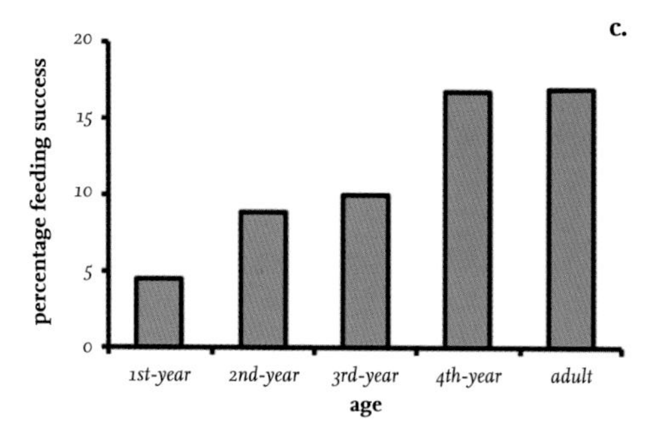

FIG 84. Feeding behaviour of Herring Gulls of different ages at landfill sites: (a) the rates of pecking; (b) the rates at which food items were swallowed (the sample sizes are given in each column); (c) the success of finding food. Based on Greig *et al.* (1983).

tended to peck at potential items at a similar rate. Other aspects of feeding varied markedly with age, however. For example, adults swallowed food at about four times the rate of first-year birds, while birds in their second and third years were intermediate, but fourth-year individuals were able to swallow food at the same rate as adults (Fig. 84b). It is noteworthy that immature individuals had a low rate of food swallowing despite identical rates of pecking (searching) for food items by all age classes and were therefore appreciably less efficient at locating food at landfill sites (Fig. 84c). About a third of the food obtained by first-year birds was achieved by displacing or robbing other gulls that had already discovered a food source, but even dependence upon kleptoparasitism did not prevent them from feeding appreciably less successfully than older birds. This low success rate of young gulls explains why far fewer first-years are usually observed at landfill sites, and that it takes four years for birds to achieve the success rate of full adults at such sites.

Effect of sex on feeding at landfill sites

Differences in the feeding behaviour of adult male and female Herring Gulls recorded at landfill sites were considerable (Table 39). Males moved less rapidly, but despite this, they found and swallowed food items at a faster rate than females. This greater success of males is partially achieved by being more aggressive, gaining access more readily to areas with more food, and attacking and stealing food appreciably more frequently from other individuals. Adult males made aggressive attacks on other gulls three times more frequently than

TABLE 39. Differences in the feeding behaviour of adult male and adult female Herring Gulls recorded at landfill sites. The rates were recorded per 15 seconds, but the averages are expressed here per minute. Data based on Greig *et al.* (1985).

	Adult males (n = 414)	*Adult females (n = 389)*
Pecks per minute	15.0	19.8
Walking steps per minute	24	32
Swallows per minute	0.72	0.64
Attacks on other adults per minute	0.18	0.07
Attacks on immature birds per minute	0.76	0.01
Percentage of food obtained by attacking	16%	5%

adult females. Despite immature Herring Gulls being less abundant than adults, males attacked them four times more frequently than they attacked adults, while females did not discriminate between adults and immature individuals.

Cannibalism

A few Herring Gulls, like other large gulls, are cannibalistic during the breeding season, feeding on the young of their own (and other) species. This apparently occurs only in large colonies, probably because in a small colony there are few unprotected chicks at any one time, and cannibals require large numbers of nesting gulls for it to be an efficient method of feeding. A detailed study on cannibalism was carried out by Jasper Parsons on the large Herring Gull colony on the Isle of May before it was reduced in size (Parsons, 1971a). He found that few adults were regularly cannibalistic, estimating that only one in every 500 birds fed in this way. The cannibals were usually males that were well-established breeders in the colony, and they continued the habit of taking young in successive breeding seasons. The chicks taken by the cannibals were small, usually under a week old, and were snatched from or near a nest and quickly swallowed in flight. However, one cannibalistic gull that was studied in detail consistently took chicks that were too large to swallow in flight, and instead brought them back alive to its nest site before killing and consuming them.

In his study area, Parsons marked hundreds of chicks with soft plastic rings as soon as they hatched, each coded with an individual reference number for the nest and the order of hatching in the brood, and he accurately mapped the position of all nests. Subsequently, tags from the chicks consumed by cannibals were recovered from regurgitated pellets at the cannibal's nest site (Fig. 85).

The effect of cannibalism considerably reduced the production of chicks. Of 1,415 young that hatched in the study area, 43 per cent fledged. Among those that did not survive, 23 per cent were killed by cannibals and a further 3 per cent died from head injuries that probably resulted from attempts by the cannibalistic adults to capture them. The remaining 31 per cent that died did so from other causes, including attacks by neighbouring adults when chicks trespassed into their territory. Chicks taken by cannibals were selected from nests at the lower average density of 2.11 within a 3 m radius of a focal nest, compared to an overall density of 2.74 in the colony. Rarely, a parent ate one of its chicks (a behaviour called kronism), although it was not known if these had already been killed by other causes; only 15 cases were recorded, compared with 329 young consumed by cannibals.

FIG 85. Map of Parson's (1971a) study area on the Isle of May, showing where two cannibal Herring Gulls obtained chicks from within the colony, and returned to their nests.

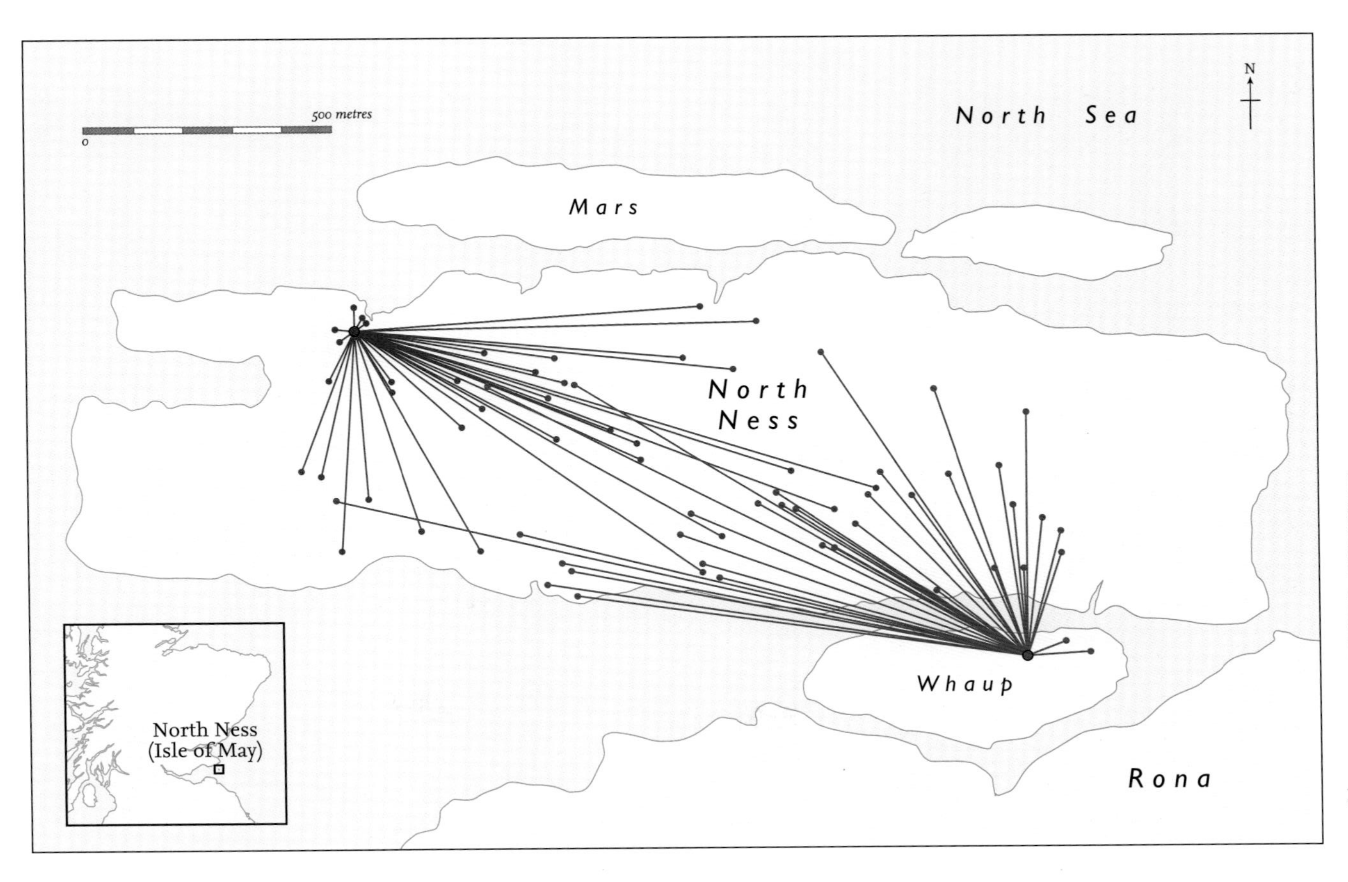
500 metres
0
North Sea
N
Mars
North Ness
Whaup
Rona
North Ness
(Isle of May)

TABLE 40. The percentage of young Herring Gulls taken by cannibals at a study area on the Isle of May in relation to the order in which they hatched within a brood. Data from Parsons (1971a).

Hatch sequence of chicks	*Number marked and at risk*	*Number taken by cannibals*	*Percentage taken by cannibals*
First	611	30	4.9%
Second	509	27	5.3%
Third	295	26	8.8%

The cannibals took almost twice the proportion of chicks hatched third in the brood than those hatched first (Table 40), suggesting that parental guarding became less effective as successive chicks hatched and started to move from the nest into nearby plant cover. However, because fewer third-hatched chicks were available (some pairs laid only two eggs), similar numbers of first-, second- and third-hatched chicks were consumed. Chicks were taken throughout the breeding season, further reducing fledging success, which declined with date; the cannibals exaggerated this trend but were not the main cause (Fig. 85).

A comparison of areas with and without cannibalism on the Isle of May showed that, in both, the highest fledging success was achieved by early breeders but that the cannibals reduced the potential fledging success from an average of 0.80 to 0.67 young per pair, and the greatest effect was on the early and late breeders (Fig. 86). Whether the cannibals had any effect on the overall size of the Isle of May colony prior to the large-scale cull is doubtful, and competition for nesting space, changes in the high adult survival rate, immigration and the appreciable emigration of young gulls may have been more important factors.

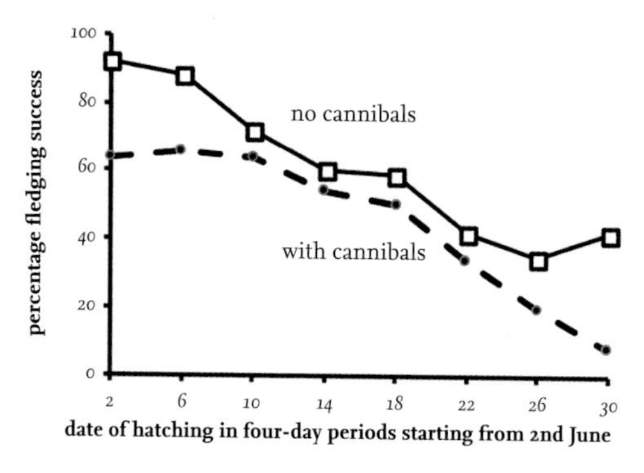

FIG 86. The impact of cannibals on the fledging success of Herring Gulls at a study area on the Isle of May. Fledging success was highest in early-laying pairs and progressively declined in the later layers. Note the greater effect of cannibals on the early- and late-laying pairs. Data from Parsons (1971a).

TABLE 41. The proportion of young taken by a single cannibalistic Herring Gull at a study area on the Isle of May in relation to date and numbers of young chicks available. Data from Parsons (1971a).

	9–15 June	*16–22 June*	*23–29 June*
Number of chicks at risk	144	375	113
Number of chicks taken by cannibal	9	9	8
Percentage	6.3%	2.4%	7.1%

TABLE 42. The level of cannibal predation by adult Herring Gulls through the breeding season on the Isle of May. Most predation occurred on chicks less than a week old. Data from Parsons (1971a).

	Before 5 June	*5–9 June*	*10–14 June*	*15–19 June*	*20–24 June*	*25–29 June*	*After 22 June*	*Total*
Number of chicks at risk	99	201	289	443	260	84	22	1,398
Number of chicks predated	22	22	21	34	29	23	7	158
Percentage of chicks predated	22%	11%	7%	8%	11%	27%	32%	11.3%

A detailed study of one cannibal showed that it took similar numbers of chicks per week throughout the breeding season (Table 41), despite there being far more chicks available in the peak week of hatching (commencing 16 June) than earlier or later. Chicks hatched at the peak of the breeding season were less at risk of predation by cannibals, whose needs were more than met at this time by availability. This pattern of predation existed throughout the colony as a whole, with those chicks hatching at the peak between 10 June and 19 June suffering the lowest predation rate (7–8 per cent), compared to much higher rates (22 per cent and even higher) for both the earliest and latest hatching chicks (Table 42).

In colonies where cannibals are present, there is therefore a clear advantage for pairs to lay and hatch chicks when most of the other pairs are at a similar stage of breeding (Fig. 86). This is an excellent example in support of the idea that the advantage of synchronised breeding is that predators are swamped with available food at peak times, thereby reducing the risk to individual chicks. Frank Fraser Darling originally proposed the theory in his book *Bird Flocks and the*

Breeding Cycle (1938), but at that time it was not fully supported by the data he was able to collect in the field.

In his Isle of May study, Parsons also found that cannibals rarely took chicks within a radius of 15 m of their own nest, but instead concentrated on nests further away (Fig. 86). This behaviour presumably had a survival value for the cannibals' own chicks, having the effect of preventing them from killing and consuming their offspring. An interesting situation arose with the cannibal that selected chicks too large to swallow in flight carried them back to its own nest site, where they were killed and consumed (see above). When this bird's own chicks hatched, it appeared reluctant to kill any of the chicks it caught at its nest site, yet continued to capture them and bring them back. The consequence of this change in behaviour was that, in addition to its own brood of two chicks, seven other live chicks were present at the nest site and at least some of these additional individuals were fed. A few days later, during a severe thunderstorm, eight of the chicks died, but one – an adopted chick – survived and went on to fledge!

CHAPTER 6

Lesser Black-backed Gull

APPEARANCE

The Lesser Black-backed Gull (*Larus fuscus*) is a large gull that commonly breeds in Britain and Ireland. It is only slightly smaller than the Herring Gull (*L. argentatus*; p. 121), but has proportionately longer wings, which makes it look more streamlined. Adults have a very dark grey (almost black in one subspecies) mantle and wings, and yellow legs.

FIG 87. Adult male Lesser Black-backed Gull. (Tony Hisgett, Wikimedia Commons)

FIG 88. First-year Lesser Black-backed Gull. Note the dark inner primaries, which distinguish it from a Herring Gull (*Larus argentatus*). (Nicholas Aebischer)

In juvenile and first-year birds, the inner primaries are the same shade of brown as the outer ones. Herring Gulls of similar age have paler inner primaries and other minor differences, but these are variable and do not allow certain differentiation in the field. The legs of Lesser Black-backed Gulls in their first year are greyish brown, similar to those of young Herring Gulls, and the colour changes gradually to yellow over several years. The bills of first-year birds are dark and then gradually lighten, but do not become yellow with a red spot until the bird is mature at four or five years of age. Older but still immature individuals show progressively more extensive dark grey feathers on the mantle and wings, but lack white mirrors on the wing-tips, and they retain part of the black tail bar for the first three years of life. By the time they are four years old, they have acquired adult plumage and yellow legs. Some breed for the first time at this age, but a few may be up to seven years old before they do so.

DISTRIBUTION IN WESTERN EUROPE AND THE NORTH ATLANTIC

The Lesser Black-backed Gull has changed both its distribution and abundance during the last century. In France, the species was originally restricted to

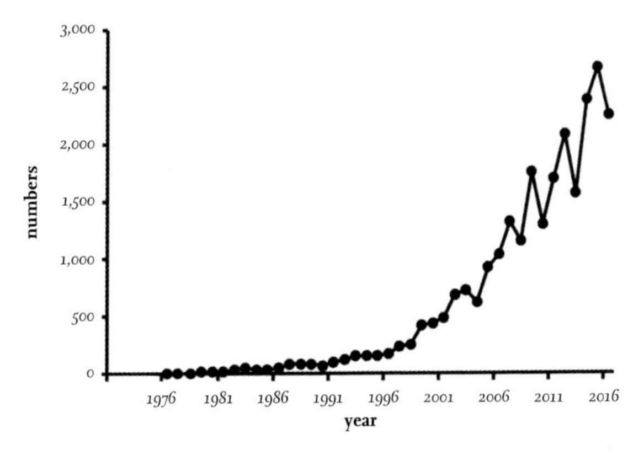

FIG 89. The approximate numbers of Lesser Back-backed Gulls recorded annually in eastern USA and Canada from 1976 to 2016 based on the Audubon Christmas Bird Count. While the recent numbers are probably exaggerated owing to increased observer participation, the rise in the numbers of gulls since 2000 has been dramatic nonetheless.

Normandy and then spread to Brittany and elsewhere early in the twentieth century. It colonised Spain in 1971, and it started to breed in Portugal in 1978 and the Canary Islands in about 2000. It first bred in the Netherlands in 1926 and Germany in 1927. It spread to the Faeroes and had started to breed in Iceland by 1920. It bred for the first time in Greenland in 1986 (confirmed in 1990) and has since increased there to about 1,000 pairs. In the last few years, hundreds of adults have started to winter along the eastern coasts of Canada and the USA (Fig. 89) and one bird, paired with an American Herring Gull (*Larus smithsonianus*), reared young in New Hampshire, USA, in 2007 and continued to do so for several years. If the Lesser Black-backed Gull is not already breeding in eastern Canada (young individuals believed to be hybrids have already been reported in Labrador), it seems likely that this will soon occur.

DISTRIBUTION IN BRITAIN AND IRELAND

Historic distribution

Little is known about the distribution and numbers of Lesser Black-backed Gulls in Britain and Ireland during the nineteenth century. In part, this is due to confusion with Herring Gulls and Great Black-backed Gulls (*Larus marinus*). Numbers were certainly low at the end of the nineteenth century and the start of their recovery is not well documented, although it is evident that the population increased and new colonies were formed in many areas during the twentieth century. For example,

FIG 90. Map of the breeding distribution of the Lesser Black-backed Gull in Britain and Ireland in 2008–11. Note the many inland sites. Reproduced from *Bird Atlas 2007–11* (Balmer *et al.*, 2013), which was a joint project between the British Trust for Ornithology, BirdWatch Ireland and the Scottish Ornithologists' Club. Channel Islands are displaced to the bottom left corner. Map reproduced with permission from the British Trust for Ornithology.

the Lesser Black-backed Gull did not start breeding on the Isle of May until 1930, 23 years after Herring Gulls first nested there, but since then their numbers have increased annually at a rate of about 14 per cent per annum (similar to that of the local Herring Gulls), until both species were culled in the early 1970s when they became too numerous on the island. In Britain as a whole, the species increased in many areas, particularly at coastal sites, and this continued until at least 2000 (more than 10 years after Herring Gulls had ceased to increase), when more than 223,000 adults were recorded breeding in Britain and about 10,000 in Ireland.

Current distribution

The distribution of breeding Lesser Black-backed Gulls reported in the 2000 national census (Fig. 90) reveals a strong western bias in the species' distribution in Wales, England and Scotland. Large colonies on the eastern side of Britain are restricted to the Firth of Forth area of Scotland and south-east England. Urban nesting is even more restricted, with numbers in south-west England and the Clyde region of Scotland, but few on the east coast. The distribution in Ireland is also more numerous on the west and north coasts, where most colonies number fewer than 500 pairs and urban colonies in 2000 were present in only two locations. Since this national survey, numbers at large colonies in England have declined due to predator activity at Orford Ness, Rockcliffe Marsh and Walney Island, and through extensive culling at Tarnbrook Fell (Abbeystead, Lancashire), Skomer and the Farne Islands; an exception is the large increase in numbers breeding on the Ribble Estuary in Lancashire.

Numbers nesting at many urban sites have increased and more towns have been colonised, but the extent of these changes has not been quantified since 2000. The current trend in the population of the Lesser Black-backed Gull in Britain is unknown and a new census is now long overdue, as evidence that it should continue to be classified as a species of conservation concern is lacking. The fact that only small numbers of Lesser Black-backed Gulls nest along much of eastern Scotland and England is not explained by the lack of suitable feeding areas and instead coincides with the absence of islands along much of that coastline. Those breeding on the east coast use a series of islands in the Firth of Forth (particularly before they were extensively culled) and the Farne Islands; numbers used to nest on Coquet Island, Northumberland, but were reduced on two occasions by culling and are kept low through management. Then there is a long gap to the south, with no suitable sites for Lesser Black-backed Gulls until the Outer Trial Bank in the Wash (a little-known island built in the 1970s for experimental water storage and later abandoned) and Orford Ness in Suffolk, both of which have been large colonies. Lesser Black-backed Gulls now also nest on buildings in several places along the east coast of Scotland and England, but they are invariably few and occur in much smaller numbers than Herring Gulls.

Lesser Black-backed Gull numbers increased rapidly at Orford Ness following their colonisation of the spit, reaching 5,500 pairs by 2000 (others report even larger numbers). Where these birds came from is unknown, as there was no large colony in south-east England from which they could have originated, and the possibility that many moved from Belgium and the Netherlands cannot be excluded. Since 2000, numbers have declined to only a few hundred pairs, apparently due to the arrival of, and predation by, Foxes (*Vulpes vulpes*). As a consequence of the predation, many of the breeding adults moved elsewhere. Some, including a few marked adults, moved locally to Havergate Island and to Lowestoft, but these account for only a small proportion of the total that left Orford Ness. It is suspected that the remainder moved to join more distant colonies, including those in Belgium, France and the Netherlands, as well as the Outer Trial Bank in the Wash. The Lesser Black-backed Gull is infrequent along much of the south coast of England between Kent and Cornwall, although Herring Gulls are common here. A few breed on the mainland coast of northern Scotland, and many in Orkney and Shetland.

Lesser Black-backed Gulls are often more numerous than Herring Gulls in colonies on the western side of Britain, and this pattern tends to be repeated at urban sites in both the Midlands and western areas. There are historic records of Lesser Black-backed Gulls nesting inland on moorland and bogs in greater numbers than Herring Gulls, but there are currently few such colonies in Britain. The largest was at Abbeystead on Tarnbrook Fell in Lancashire, with an estimated 16,000 pairs in the 1970s, but culling has now markedly reduced numbers there to less than a tenth of this. In the recent past, large inland colonies existed on Flanders Moss in central Scotland and on an island on Loch Maree in north-west Scotland. Unfortunately, gulls no longer nest at these sites, probably due to human disturbance at Loch Maree. Extensive drainage and tree planting-programmes have made much of the former breeding area at Flanders Moss unsuitable and Foxes have reached the area. Relatively small numbers of Lesser Black-backed Gulls nest in Ireland, with about a third of the 5,000 pairs breeding at inland sites.

COMPARISON WITH HERRING GULL NUMBERS AND DISTRIBUTION

Despite much overlap, there are obvious differences in the geographical distribution of colonies of Herring and Lesser Black-backed gulls in Britain and

Ireland. On natural nesting sites, Lesser Black-backed Gulls mainly use relatively flat areas with moderate vegetation (Fig. 91), often located on islands or in isolated areas not normally visited by Foxes or Badgers (*Meles meles*). Herring Gulls nest on similar sites, but they use a wider range of habitats, which include more areas with sparse and very short vegetation, broad cliff ledges, the tops of isolated rocky stacks and steep slopes on sea cliffs. The distribution of Herring and Lesser Backed-backed gull breeding areas is remarkably different, considering that they often nest side by side. The data in the 2000 national census in Britain and Ireland has been used to make a comparison of their distribution, as a more recent new national census has not yet been carried out.

Lesser Black-backed Gulls started to nest on urban sites in Britain more recently than Herring Gulls and currently show a different urban geographical distribution. While both species nest together in many towns, the proportions of the two species can differ markedly. Herring Gulls nesting on buildings are usually much more numerous in eastern England and Scotland, where few or even no Lesser Back-backed Gulls are involved. On the western half of Britain, and particularly in south-west England and south-west Scotland, the reverse situation occurs, with Lesser Black-backed Gulls dominating and fewer Herring Gulls nesting in urban areas.

FIG 91. Lesser Black-backed Gulls (and one Herring Gull, *Larus argentatus*) nesting in tall vegetation. (John Coulson)

Overall, the proportion of Lesser Black-backed Gulls among the large gulls nesting on roofs in Britain appears to have increased. In 1970, they formed about 5 per cent of large gulls nesting on roofs; this figure had risen by 1976 to 11 per cent, by 1994 to 19 per cent, and by 2000 to 35 per cent – seven times the proportion in 1970. No recent data on the proportions are available.

The proportions of Lesser Black-backed Gulls and Herring Gulls nesting on buildings and roofs also varies markedly throughout Britain. In 2000, in the extreme north of Scotland, Inverness, Nairn, Banff, Angus, Dundee and Berwickshire Lesser Black-backed Gulls accounted for less than 2 per cent of urban nesting gulls, while in Aberdeen the figure reached 4 per cent. In contrast, in most areas in south-west Scotland Lesser Black-backed Gulls formed the majority of nesting gulls, and in Glasgow the species comprised 98 per cent of nesting gulls and were the only species recorded nesting in Cumbernauld, to the north-east.

Similar variation occurs in England, with Herring Gulls dominating urban areas along the east and south coasts. In 2000, they accounted for more than 90 per cent of nesting gulls in most coastal towns and cities with the exception of Suffolk, where 72 per cent of the urban gulls were Lesser Black-backs. The situation changes dramatically in Avon and Gloucestershire, where 87 per cent of the urban gulls were Lesser Black-backed Gulls. In Wales, only 18 per cent of urban nesting gulls were Lesser Black-backed Gulls, while in Lancashire and Cumbria, about 35 per cent of the nesting gulls were this species. In Ireland, 28 per cent of the small total of urban gulls were Lesser Black-backed Gulls. The reason for these variations in the proportion of large gulls is not immediately evident. Both Herring and Lesser Black-backed gulls nest in the same towns and often on the same roofs or neighbouring chimney stacks. In some areas, Lesser Black-backed Gulls appear to show a marginal preference for extensive flat roofs common on industrial estates, but in other towns these sites are occupied by Herring Gulls or by both species nesting together. Comparisons of types of nest sites used in towns and cities in northern England and southern Scotland (where Lesser Black-backed Gulls were more numerous) with those dominated by Herring Gulls have failed to reveal differences.

A further insight has been obtained by comparing the proportions of the Lesser Black-backed and Herring Gulls nesting on natural rural sites around the coastline (Fig. 92). This confirms major differences in many areas, and reveals that the proportions of Lesser Black-backs are often much higher in western areas.

The proportions of Lesser Black-backed Gulls nesting at natural sites in the administrative counties of Britain and Ireland vary widely, from 0 per cent to

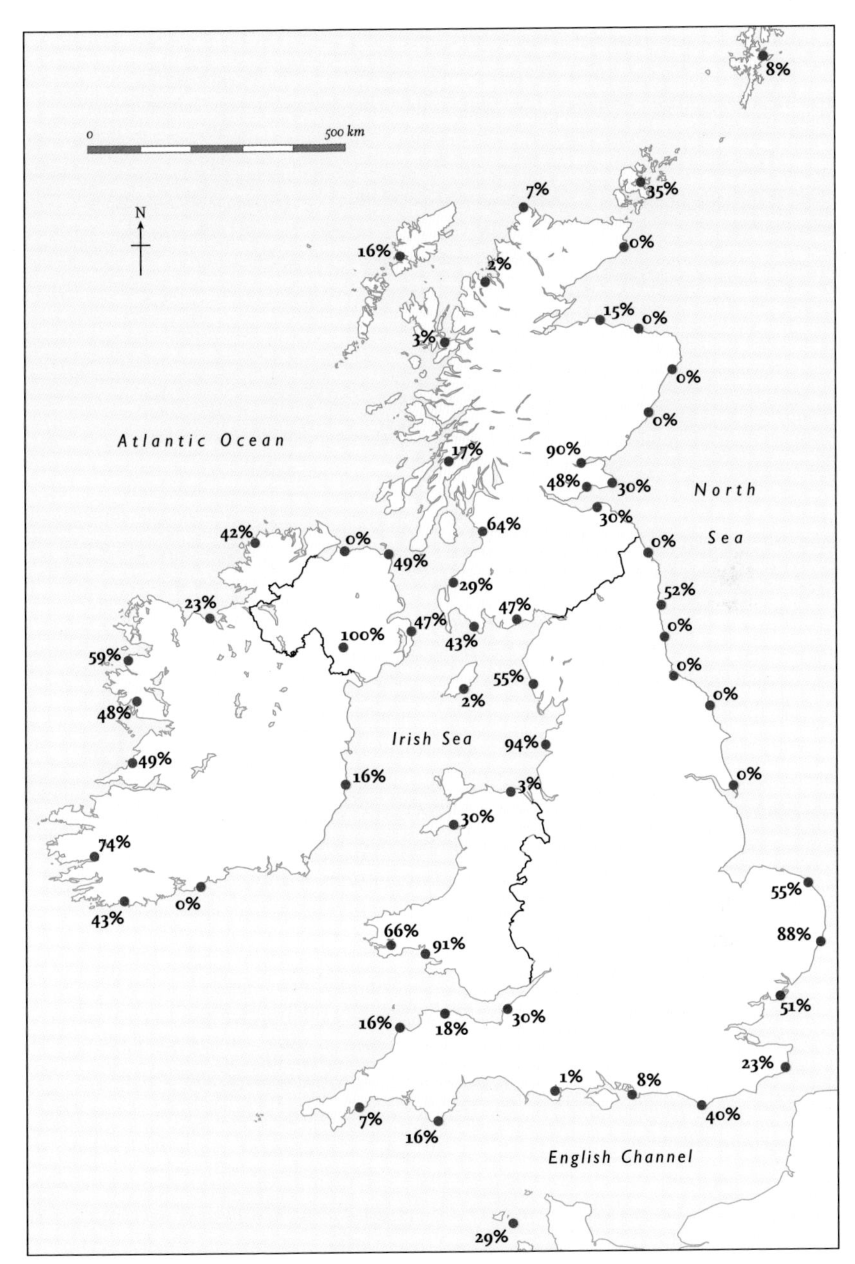

FIG 92. The percentage of Lesser Black-backed Gulls among the total numbers of Herring and Lesser Black-backed gulls in coastal districts of Britain and Ireland in 2000.

TABLE 43. The percentage of Lesser Black-backed Gulls among the total number of Herring and Lesser Black-backed gulls nesting at natural rural sites in administrative counties in Britain and Ireland, compared with those breeding at urban sites within the same administrative areas. Mainly based on data collected in 2000 and records in the Seabird Colony Register.

Number of administrative areas studied	*Percentage of Lesser Black-backed Gulls nesting at natural sites (range and average)*	*Percentage of Lesser Black-backed Gulls in nearby urban areas*
7	0% (average 0%)	1%
10	1–9% (average 4%)	5%
7	10–29% (average 21%)	17%
7	30–59% (average 48%)	50%
5	60–99% (average 75%)	61%

95 per cent, and these figures have been compared to the proportions at urban sites within the same areas. As is obvious from Table 43, there is a positive correlation between the two values obtained for the same administrative areas. Where there is a low proportion of Lesser Black-backed Gulls breeding at natural nesting sites, there is a similarly low proportion breeding in nearby towns. The proportions of Lesser Black-backed Gulls in natural areas have probably persisted for many years, while proportions among gulls nesting in nearby towns have been established more recently.

There are two obvious possible explanations for this relationship. First, the individuals that have moved into towns may have been determined by the proportion currently nesting or historically breeding at local natural sites (and where the numbers were often much larger). This infers that the proportions in towns have been an overflow of birds that were reared or had been breeding at natural sites in the recent past. A second explanation is that the species composition within an administrative district has been determined by the availability of feeding areas preferred by Herring or Lesser Black-backed gulls. A third possibility, that an area with no suitable natural sites for many nesting Lesser Black-backed Gulls also had few suitable nesting sites in nearby towns and cities, can be rejected since no major differences in the sites used by the two species in urban areas have been identified. An examination of the proportions of Lesser Black-backed Gulls and Herring Gulls nesting in natural and urban sites in larger regions of Britain and Ireland does not indicate this relationship between natural and urban sites (Table 44), suggesting that the correlation is a local effect. The table also reveals that in Ireland and Wales,

TABLE 44. A comparison of the proportion of Lesser Black-backed Gulls among the total numbers of Herring and Lesser Black-backed gulls nesting in natural sites in major regions of Britain and Ireland, compared with the proportion of these species found in urban sites in the same regions. Based on data collected in 2000 for the national census.

	Percentage of Lesser Black-backed Gulls among those nesting at natural sites	*Percentage of Lesser Black-backed Gulls among those nesting in urban sites*
Scotland	23%	40%
Ireland	34%	28%
England	45%	55%
Wales	60%	18%

Lesser Black-backed Gulls are under-represented in urban areas compared to other parts of the region.

That young Herring Gulls reared on natural coastal sites will frequently move to breed in urban areas was shown in their movement between the Isle of May and urban colonies in Tyneside and Wearside (p. 157). However, similar movements by young Lesser Black-backed Gulls have not been recorded, perhaps because far fewer have been colour-ringed at natural nesting sites and not because it does not occur.

SUBSPECIES – A TAXONOMIC NIGHTMARE

Lesser Black-backed Gulls breeding in Europe are often regarded as comprising three subspecies. Those breeding in the eastern Baltic belong to the subspecies *Larus fuscus fuscus* and have wings as dark as those of Great Black-backed Gulls. The subspecies breeding in Britain is *L. f. graellsii* and was named by Alfred Brehm in 1857 from specimens collected there. It was separated from *fuscus* on the grounds that the adults have slate-grey wings with black only on the wing-tips. In 1922, birds breeding in the western Baltic were described and named as *L. f. intermedius* by Eiler L. Schiøler. As the name implies, the average shade of grey/black on the wings and mantle of this subspecies lies between those of the other two.

A major taxonomic problem relates to the subspecies *intermedius* and *graellsii*, both of which breed in western Europe and are very similar, differing only in the average darkness of the slate-grey coloration on the mantle and wings of individuals. Since it was named, the subspecies *intermedius* has been ignored by many authors, including Harry Witherby *et al.* in their *The Handbook of British Birds* (1940), David Bannerman (1962) and the British Ornithological Union in 1971, all of whom combined its alleged geographical distribution within that for the subspecies *graellsii*. More recently, *The Birds of the Western Palearctic* (volume 3, 1983) accepted the subspecies, but without question or comment, while the *Handbook of the Birds of the World* (volume 3, 1996) recorded *intermedius*, but with a surprising and discontinuous distribution, stating that it breeds in Denmark, the Netherlands, southern Norway and in north-east Spain, but apparently not in France, Belgium or Germany. In 2000, the European Bird Census Council, again without new research, listed *intermedius* as breeding in Belgium, Denmark, Germany, the Netherlands, and parts of both Norway and Sweden, while the subspecies *graellsii* was restricted to breeding in Britain, Ireland, France, Portugal, the Faeroes, Iceland and Greenland. One author stated that the Lesser Black-backed Gulls breeding at the southern edge of the species' range in the Ebro Delta of north-east Spain belong to *graellsii*, another stated that they were all *intermedius*, and yet another source confidently stated that *graellsii* was restricted to Britain and Ireland. Such confusion is indicative of the dubious nature of the two subspecies.

Those who have examined the three subspecies in detail using mitochondrial DNA analysis have cast doubt on the validity of *Larus fuscus intermedius*. George Sangster and his co-workers (1999) found that *intermedius* was not diagnostically distinct, could not be identified in the field and did not justify subspecies status. In 2002, a study made by Dorit Liebers and Andreas Helbig, which included an examination of the genetic make-up of *graellsii* and *intermedius*, reported that they could find little difference between these two subspecies, and they lumped them together in their further investigations.

To an observer in the field, there is appreciable variation between individuals in the shade of grey on the wings of Lesser Black-backed Gulls, and even the shade on the wings of the same individual can vary according to the direction and intensity of light and the position of the bird in relation to the sun. Colour shade also differs in an individual that has newly grown primary feathers when compared with faded, older feathers that are about to be moulted and are almost a year old. Shades in museum specimens may have altered slightly, depending on how they have been stored over time, while photographs are unreliable unless a standard shade chart is included in the same exposure, because the darkness of the grey on the wings can be adjusted.

Even within a single colony, there is appreciable variation between individuals in terms of the darkness of the grey on the wings and mantle. Many experienced ornithologists in Denmark, the Netherlands and Belgium are uncertain as to which subspecies are breeding in their countries. In 2013, one author summarised the confused state of opinion in the Netherlands by recording the differing opinions expressed by 15 Dutch ornithologists, pointing out that some believed the Dutch birds were *intermedius*, others believed they were *graellsii* and two suggested (but this was not proven) that both occur together in an extensive area of overlap. It is worth noting that there are several records of Lesser Black-backed Gulls ringed as chicks in Britain (the type locality of *graellsii*) that have crossed the North Sea to breed in colonies in the Netherlands and Belgium (where *intermedius* is alleged by some to breed). Cases of exchanges in the opposite direction are also known, indicating gene flow across the North Sea.

Lesser Black-backed Gulls breeding in the western region of Europe would be better represented by a single subspecies, *graellsii*. This includes birds breeding in the countries around the North Sea coastline and at the western end of the Baltic Sea, and extending further west to the Faeroes, Iceland and Greenland, and perhaps involving a gradual change in the average darkness of the wings running south-west to north-east across northern Europe. The concept of the cline, introduced by Julian Huxley in 1942, is used to describe the situation where the characteristics of a species gradually and progressively change as a result of varying selection pressures over its geographical distribution. Such gradients are often indicative of appreciable gene flow and are known in several seabird species, such as Puffins (*Fratercula arctica*), Common Guillemots (*Uria aalge*) and Black-legged Kittiwakes (*Rissa tridactyla*). The concept of clines would seem a better fit than subspecies for the situation of Lesser Black-blacked Gulls in western and northern Europe, with the shade of grey on the wings and mantle progressively darkening from England and France towards southern Scandinavia. If individuals cannot be identified to subspecies in the hand or in the field then, as Sangster pointed out, naming it is of no value and hinders scientific progress.

Reappraisal of the status of the two subspecies is not just a taxonomic exercise. It is important because *graellsii* has been listed as a subspecies of conservation concern. If the subspecies *intermedius* was rejected, *graellsii* would then be much more abundant and widespread in Europe and hence its conservation listing would need to be revised.

As mentioned above, the nominate subspecies *Larus fuscus fuscus* breeds in eastern Scandinavia and north-west Russia, and is characterised by having wings and a mantle that are as dark as those of the Great Black-backed Gull. It is quite distinct when compared with *graellsii*, and some believe it should be given species status because it has different feeding habits and migrates inland to south-east

Europe and then to central Africa, whereas *graellsii* migrates along the western coast of Europe and to the west coast of Africa.

The subspecies of Lesser Black-backed Gulls in western Europe show considerable variation in the timing of the moult of their primaries. Many of the birds breeding in Britain start losing and replacing the inner primary feathers while they still have unfledged chicks, and most have completed replacing the primaries by the end of October, although a few do not do so until January. As with other gulls, immature individuals go through their primary moult about a month earlier than the adults (see also p. 36).

Further taxonomic problems exist with yellow-legged, dark-winged gulls breeding east of the White Sea in Siberia, which are very similar to *graellsii* but are now called *Larus fuscus heuglini* or *L. f. taimyrensis* by some, and are even considered by others as a separate species, *L. heuglini*, with *L. f. taimyrensis* possibly regarded as a hybrid. An additional subspecies, *L. f. barabensis*, occurs in Asia. In the past, these gulls were all described as subspecies of the Herring Gull rather than the Lesser Black-backed Gull, and specimens in the zoological museum in Moscow are still labelled as such, illustrating the extent of confusion that exists with these forms. The migration route taken by *heuglini* appears to be in a south-easterly direction, allowing the individuals to winter in the Indian and Pacific oceans, so few could be expected to visit western Europe. Neither has been recorded in Britain, but that is not surprising since *heuglini* looks like a slightly large *graellsii* and I doubt whether it could be reliably identified in the field away from northern Russia. Confirmation of a record here would depend on a ringing recovery or on a DNA sample obtained from an individual. To have forms of Lesser Black-backed Gulls with similar slate-grey wings (*graellsii* and *heuglini*) separated geographically in Europe by the black-winged nominate race creates an interesting problem, and the taxonomic relatedness of the group is in need of detailed investigation.

BREEDING

The Lesser Black-backed Gull is an intensely colonial species. Colonies can contain several thousand pairs and often include other species, usually Herring Gulls. Single pairs of Lesser Black-backs rarely breed in isolation, although they are encountered breeding in colonies of Herring Gulls.

Although both Herring Gulls and Lesser Black-backed Gulls have very similar breeding biology, adult Lesser Back-backed Gulls occupy nesting territories later in the year. This difference follows through to egg-laying, with Lesser Black-backed Gulls both starting and reaching the peak of laying four to seven days later on average than Herring Gulls at both natural sites and in towns. The spread of laying for both species extends through May and into the first half of June, with re-laying of lost clutches by both species continuing into early or even mid-July.

The eggs of Lesser Black-backed Gulls are well camouflaged, with heavy and variable blotching of dark colours. The most frequent clutch size is three eggs, with an appreciable minority of two-egg clutches laid by established breeders, while an occasional single-egg clutch is produced by late-laying birds. The average clutch size is usually between 2.70 and 2.87 eggs, and varies only slightly between years. As in other gull species, the final egg of the clutch is often long and narrow, and has a smaller volume than the earlier eggs. The eggs are usually laid at two-day intervals, but sometimes intervals of three days occur. Incubation takes about 27 days, with the third egg often hatching a day or so after the others, presumably because effective incubation starts shortly after the second egg is laid. The survival of chicks hatching from the third egg is often low, presumably because they are smaller at hatching, have smaller yolk reserves and suffer competition for food from their larger siblings.

Breeding success

Hatching success is highly variable. In some colonies, such as those in the Netherlands studied by Kees Camphuysen, 70–85 per cent of eggs hatched but predation by other gulls on chicks (cannibalism) was frequent, resulting in only 0.26–0.46 young fledged per pair in a series of years. Low breeding success is also common in other large colonies, including Skomer in south-west Wales, where an average of just 0.35 young fledged per pair between 1990 and 2015, and in only three of those years did the rate exceed 0.6 chicks per pair. In contrast, colonies where there is no cannibalism, predation by mammals or human disturbance have a productivity up to the time of fledging that frequently exceeds one chick per pair, such as that found at Tarnbrook in three years of study. A statement made in 2005 that Lesser Black-backed Gulls that laid three-egg clutches invariably managed to rear three chicks to fledging in Bristol is puzzling, as it makes no allowance for any losses during incubation or addled eggs (which elsewhere is in the range of 5–10 per cent of the eggs laid), nor any mortality at all during the pre-fledging period.

With the high adult survival rate among Lesser Black-backed Gulls, fledging 0.7 young per breeding pair each year is probably sufficient to replace the annual loss of adults. Equally, persistent low productivity below this level on Skomer and in neighbouring colonies offers a ready explanation for the progressive 50 per cent decline in numbers of Lesser Black-backs there between 1990 and 2015.

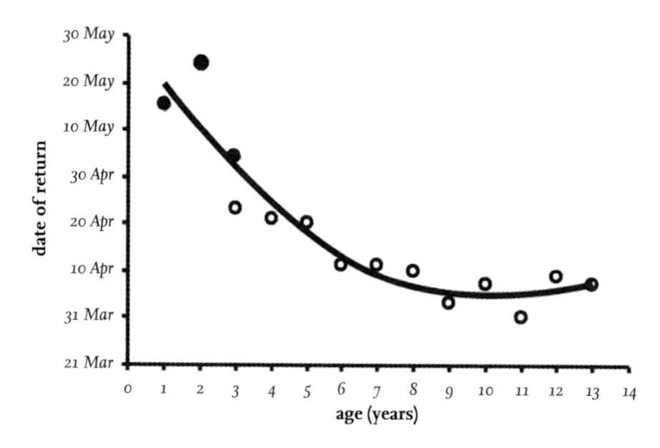

FIG 93. The influence of age on the date of return of Lesser Black-backed Gulls to the Zeebrugge colony in Belgium. The filled circles are birds that did not breed, while the open circles are birds breeding at that age. Birds aged six years and older returned at very similar dates. Based on data by Bosman *et al.* (2013).

Annual reoccupation of breeding colonies

Age affects both the timing of the annual return of adult Lesser Black-backed Gulls to their colony and the timing of egg-laying. A detailed study in Belgium in 2013 (Fig. 93) found that older individuals returned to the colony site first in early April, while three-, four- and five-year-old individuals arrived three weeks later. The few one- and two-year-old immature birds that did visit the colony arrived during the second half of May, some seven weeks after the arrival of the older breeding birds. Results suggested that a quarter of the three-year-olds that visited the colony remained and bred there in the same year. This is a remarkably high proportion to breed so young and is not supported by studies elsewhere.

Philopatry

A proportion of Lesser Black-backed Gulls return to breed close to the place where they hatched, a behaviour called philopatry (Fig. 94). I investigated birds ringed as chicks on the Farne Islands in Northumberland and subsequently culled there as breeding adults. Despite being ringed on seven different islands, 56 per cent (241 of the 428 individuals recovered) had returned to breed on the same island on which they had been reared. A further 26 per cent had moved a short distance to the next nearest of the seven islands. These results and similar studies on other gulls elsewhere, in which the birds return to their natal island many years after being reared there, are impressive, and have led to generalisations that most seabirds are highly philopatric.

However, many other ringed Lesser Black-backed Gulls in the study did not return to the Farne Islands, but had moved elsewhere to breed; while these birds are much more difficult to locate, many examples now exist. In the case of the breeding adults culled on the Farne Islands and Isle of May, and at Tarnbrook, there were also many that had been marked as chicks elsewhere in colonies some

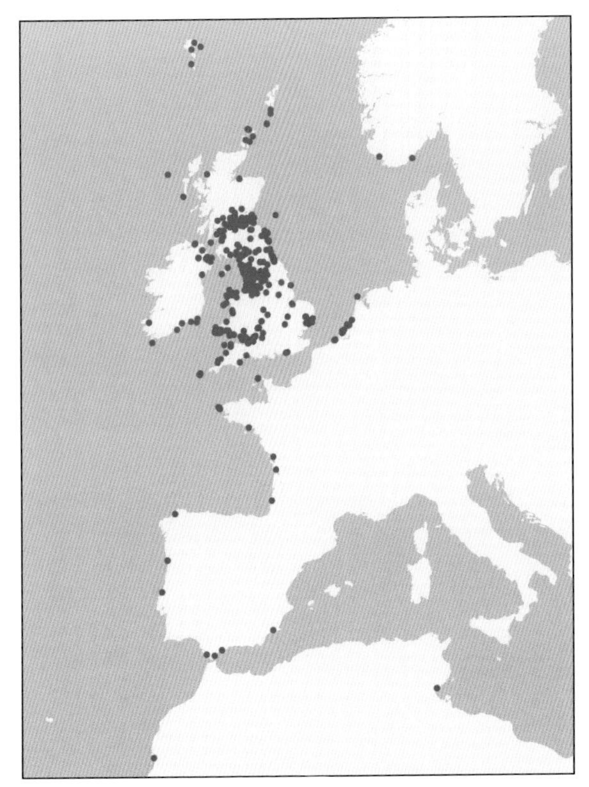

FIG 94. The recovery locations of Lesser Black-backed Gulls ringed as nestlings in Britain and Ireland and recovered as adults in the breeding season. Many of these birds returned to breed within Britain, but only a minority returned to where they had hatched, while some moved to Norway, France, Belgium, the Netherlands and even the Faeroes. It is uncertain whether those reported from Iberia and North Africa were breeding, but these areas have been recently colonised. Reproduced from *The Migration Atlas* (Wernham *et al.* 2002), with permission from the BTO.

distance away, and in one case, a bird ringed as a chick in Belgium moved to north-west England to breed. The proportion that return to their natal area is variable and its intensity has probably been exaggerated in the past. A balanced estimate based on ringing recoveries reported by the public is that only about a quarter of those reaching adulthood are highly philopatric, while the majority move elsewhere to breed, perhaps influenced by their sex (with males often more highly philopatric than females), feeding conditions and the extent of competition within the natal colony. The appreciable proportion of birds that are not philopatric suggests that preventing Lesser Black-backed Gulls from breeding successfully by removing eggs and chicks in a colony will not have much effect on numbers, because many of the adults that join the colony as breeding birds have probably been reared elsewhere.

The modest proportion of individuals returning to breed where they were reared raises an interesting question. If many individuals are not philopatric, then how did those that move choose a colony in which to breed? One theory is that young gulls visit several colonies late in the breeding season of the year before they breed for the first time, and that they are particularly attracted to areas where there are many well-grown chicks present. This would be a sound strategy, because a colony that is successful in one year is likely also to be successful in the following year and would therefore be a good place to breed. Further exploration and testing of this idea will require information about the young birds' movements, perhaps using miniature recorders attached to two- and three-year-old individuals to record in detail where and when they explore colonies; so far, this difficult investigation has not been made.

MIGRATION

Less than 100 years ago, Lesser Black-backed Gulls were summer visitors to Britain and all departed between August and October to winter in Spain, Portugal and north-west Africa, reaching as far south as Senegal (Fig. 95, left). The return migration to the UK took place in February and March. This migratory behaviour changed in the 1940s, when a few began to winter in Britain, particularly in south-west England, where they were seen feeding at landfill sites. In January 1953, 165 individuals were recorded in England and Wales during a census of wintering gulls in Britain. Ten years later, in January 1963, more than 6,900 were recorded, and in January 1973, 25,000 were reported. In 1983, 44,000 Lesser Black-backed Gulls were counted in Britain during the winter, and it was estimated that as many as 70,000 were probably wintering in Britain and Ireland at that time. The great majority wintered in south-west England, while only 10 were reported in north-east England, along with a few in Wales, the Clyde area of Scotland, and the east and south of Ireland. Those birds wintering in Britain were mainly adults, but some immature birds might have been missed through confusion with young Herring Gulls in the poor light of winter afternoons, when counts were made as the gulls arrived at their roosts.

This increase in numbers of Lesser Black-backed Gulls wintering in Britain coincided with the development of the Herring Gull habit of feeding at landfill sites (p. 167). The availability of this new winter food source was also exploited by Lesser Black-backs, which presumably must have been an important factor facilitating the change in their migratory behaviour. In 1980, Robin Baker analysed ringing recoveries of individuals ringed at Walney Island in north-west England, and was the first to try to quantify this change in behaviour. He suggested that 80 per cent of adults were wintering in Britain at the time, although he probably exaggerated the extent because he did not

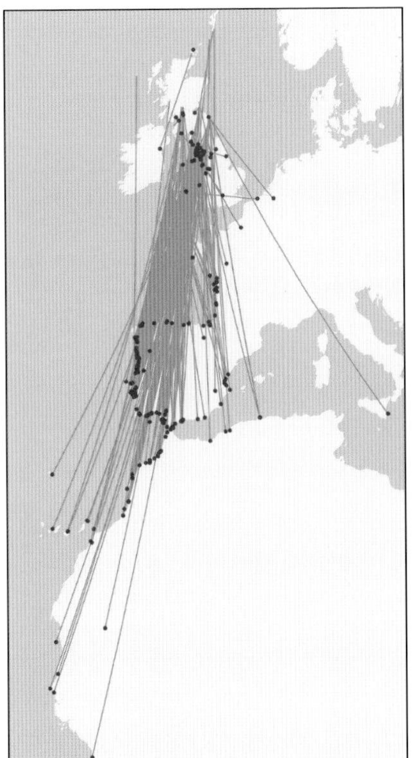

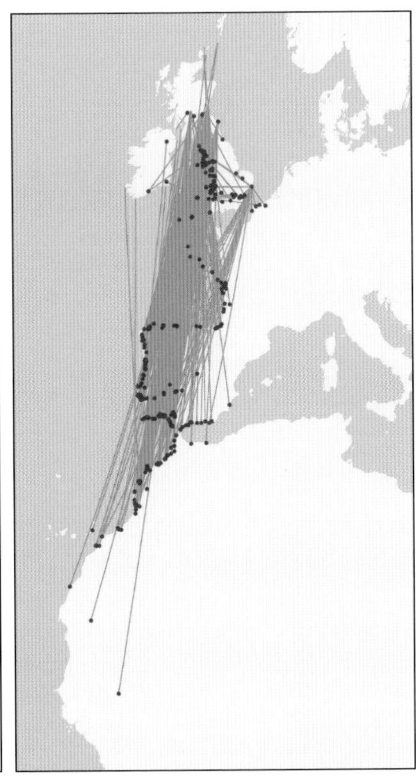

FIG 95. Left: winter recoveries of Lesser Black-backed Gulls ringed in Britain and Ireland up to 1969. Right: winter recoveries of Lesser Black-backed Gulls ringed in Britain and Ireland between 1970 and 1997. Despite increases in birds wintering in Britain in recent years and in the overall population, the numbers moving to Iberia and north-west Africa may not have changed. Reproduced from *The Migration Atlas* (Wernham *et al.* 2002), with permission from the BTO.

allow for the lower reporting rate of ringed birds from Iberia and North Africa. It is probable that the numbers moving from Britain to winter in those regions did not, and still have not, decreased at all, and that the increasing numbers wintering in England approximate the numerical increase of this species breeding in Britain.

Studies on marked adult Lesser Black-backed Gulls breeding in Belgium and the Netherlands in recent years indicate a different situation there, with only a few birds remaining in the Netherlands in December and January (less than 16 per cent) and even fewer in Belgium. The adults at that time of year were more than 1,000 km south of their breeding colonies on average, wintering mainly in south-west France and Iberia. This resembles the pattern of migratory behaviour in the Lesser Black-backed Gull in Britain in the 1950s, at a time when there were few, if any, breeding in Belgium. There is a need for a new and comparative study of the extent of past and present migratory movements of Lesser Black-backed Gulls from colonies around the North Sea and in the western Baltic, particularly as more landfill sites close and this major source of winter food disappears.

Many of the ringing recoveries of Lesser Black-backed Gulls reared in Britain and reported in winter in Africa are birds in their first year of life, suggesting that immature individuals migrate further than adults. However, this impression is contradicted by several reports that the Lesser Black-backs seen in Morocco are mainly adults. Is this difference due to the higher mortality rate of the young birds, producing more recoveries in Africa, while the adults tend to survive the winter and die in the summer months at or near their breeding colonies, or do the first-year and adult birds visit different areas in Africa? To answer this, more information is clearly needed about numbers and age distribution of Lesser Black-backed Gulls wintering in North Africa.

The annual migration of Lesser Black-backed Gulls leaving Britain and other countries around the North Sea mainly follows the Atlantic coastlines (Fig. 96), although inland records in Iberia are becoming more frequent – possibly because there are both more inland refuse sites and more observers. Only a few British ringed Lesser Black-backed Gulls have been reported from the western Mediterranean, and it is uncertain whether these made inland crossings of France or Iberia, or moved through the Straits of Gibraltar.

The autumn migration of Lesser Black-backed Gulls breeding in countries around the North Sea is usually slow, with many stops on the way (in contrast with

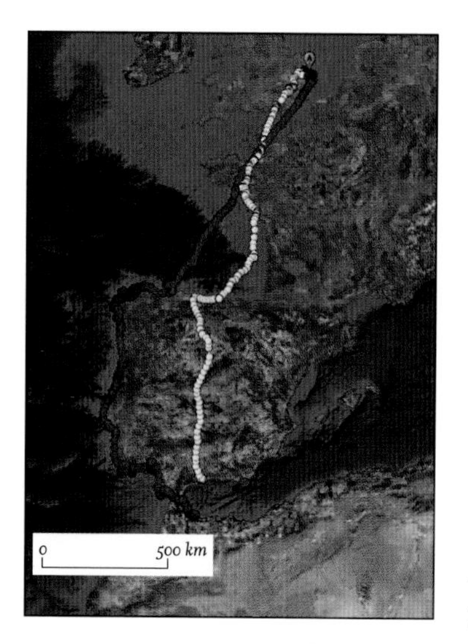

FIG 96. The outward (red) and return (yellow) flights to and from the wintering areas of an adult Lesser Black-backed Gull tagged by Viola Ross-Smith at Orford Ness. Image from BTO research funded by the UK Department for Business, Energy and Industrial Strategy as part of their Offshore Energy Strategic Environmental Assessment programme. Reproduced with permission of Hartley Anderson.

the rapid migration across Europe of birds breeding in the east of the Baltic; see below). Migration from the North Sea area often starts in August, and a few, possibly failed breeders, move as early as July, although only low numbers appear to reach southern Iberia and Morocco before December. The impression gained from sightings of marked individuals is that many take a 'grand tour' of feeding localities as they move south along the French, Spanish and Portuguese Atlantic coastlines, rather than making a single long flight to reach a specific wintering locality and then staying there for several months. These slow and progressive movements are similar to the smaller-scale autumn and winter movements of Herring Gulls from Scotland into England (p. 164). In both species, the individual birds obtain knowledge of many feeding sites *en route*, and in both cases, the return migration in the spring to the breeding areas is rapid and probably made as a single return flight in the second half of February or the first two weeks of March.

Recoveries of the young, immature individuals ringed in Britain indicate that some of these birds winter in Iberia and remain there throughout the following summer (and even longer in some cases), returning to Britain only in the late spring when nearly two, and in some cases three, years old.

Some Lesser Black-backed Gulls from more northern countries that pass through Britain during their migration originate from Iceland and the Faeroes, while those from south Norway pass through only the south of England. Recoveries of birds ringed in Iceland suggest that many of these migrate earlier, moving further and more rapidly than British birds (probably because they are compelled to make a long sea crossing). It is likely that many Lesser Black-backed Gulls breeding in Greenland do not migrate to Europe but instead move south-west and winter on the Atlantic coast of North America as far south as Florida. So far, there is only one record of a Lesser Black-backed Gull reared and ringed in Iceland recovered in North America, and none has crossed the Atlantic from the rest of Europe.

Most individuals of the subspecies *fuscus*, breeding in the eastern Baltic and northern Scandinavia, migrate inland in a south-easterly direction through central Europe to reach the eastern end of the Mediterranean. Some birds then move south to the large lakes of eastern Africa and to the Congo Basin. In contrast to the slow autumn migration of British Lesser Black-backed Gulls, these very dark-winged individuals move rapidly – often up to 400 km per day – and so probably do not stop to feed on more than one occasion before they reach the Mediterranean. On their return spring migration, the birds move slower and have more stop-overs, but still average more than 180 km a day. As a result of their migration route, black-winged adults believed to belong to the subspecies *fuscus* are infrequently reported in Britain – there have been only two recoveries here of individuals ringed in the subspecies' breeding area. There is considerable reluctance to accept field sightings of *fuscus* in Britain because it is believed that there is a zone of interbreeding between these birds and members of the lighter-winged subspecies, which can produce very dark-winged individuals (p. 190).

FOOD AND FEEDING

The Lesser Black-backed Gull is an omnivore and a scavenger, and like the Herring Gull, it consumes a very wide range of foods. Marine fish often predominate in the diet of birds breeding at coastal colonies and are a major component of the food fed to chicks at many of these sites (Table 45), but there is considerable variation and those breeding inland obtain their food elsewhere. The gulls catch some fish during shallow dives into shoals that approach the surface, but most are obtained by attending fishing boats, particularly when nets are being hauled in, when unwanted fish are discarded or when fish are being gutted. Landfill sites are an important source of food when and where

TABLE 45. Composition of food, estimated by weight, consumed by Lesser Black-backed Gulls in the breeding season.

	Fish	*Marine invertebrates*	*Insects*	*Earthworms*	*Birds*	*Mammals*	*Plants*	*Food waste*
White Sea, Russia	56%	15%	10%	?	12%	0%	6%	0%
Farne Islands	77%	3%	1%	?	0%	0%	0%	19%
Skokholm	80%	0%		12%	?	?	4%	4%
Bristol Channel	28%	6%	6%	8%	6%	1%	10%	34%
Walney Island	35%	18%	2%	++	0%	7%	2%	35%
Dumfries (15 km inland)	3%	2%	35%	c. 32%	0%	4%	2%	22%
Netherlands	86%	13%	1%	?	0%	0%	?	?
Netherlands – food fed to chicks	78%	17%	5%	?	0%	0%	?	?
Netherlands (30 km inland)	0%	0%	++	++	0%	0%	0%	++

++ Indicates earthworms were frequent but could not be quantified.
? Indicates not considered.

it is available, but their importance should not be overestimated. The closure of the local landfill site at Dumfries, for example, had no detectable effect on breeding success, and numbers of Lesser Black-backed Gulls nesting in the town continued to increase, with invertebrate food collected on agricultural land becoming more important.

Surveys in the North Sea show that some Lesser Black-backed Gulls travel far from land to follow fishing boats, although their density declines progressively as distance from the shore increases (Table 46). In the Netherlands, the detailed comparative study of Herring Gulls and Lesser Black-backed Gulls made by Kees Camphuysen in his thesis revealed a very marked difference in the food and feeding areas used by these species, which has developed since the latter became numerous in that country. These differences are illustrated in Table 47, which shows that the Lesser Black-backed Gull feeds predominantly at sea and behind

TABLE 46. The change in the numbers and proportions of Herring and Lesser Black-backed gulls seen during marine transects at different distances from the Dutch coastline and made during the breeding season. Based on data in Camphuysen (2013).

Distance from shore	*Herring Gull numbers per 100 km*	*Lesser Black-backed Gull numbers per 100 km*	*Lesser Black-backed Gulls per Herring Gull*
0–5 km	229	230	1.0
5–10 km	107	221	2.1
10–25 km	37	96	2.6
25–50 km	10	85	8.5
50–100 km	2	35	17.5

TABLE 47. The percentage of samples containing food from different sources and obtained from Herring and Lesser Black-backed gulls breeding at the same colonies in the Netherlands. Based on data in Camphuysen (2013) and the examination of more than 5,000 food samples for each species.

Source of food	*Herring Gull*	*Lesser Black-backed Gull*
Marine	23%	87%
Intertidal	74%	3%
Terrestrial	23%	32%
Human waste, landfill, etc.	13%	6%

fishing boats, while the Herring Gulls, despite their name, do not regularly feed on Herrings (*Clupea harengus*) but mainly utilise intertidal sites, such as shallow and tidal areas of the Wadden Sea, where they take appreciable amounts of invertebrates as food. This marked distinction in the feeding areas of the two species is also evident in south-east England, but not in most other areas and it does not apply to those using inland breeding sites. The British name Herring Gull suggests that, in the past, this was the gull attracted to fishing boats around most of Britain, and this is still true for most of the country today.

In some areas, young birds – particularly ducklings, seabirds, waders and gamebirds – are frequently consumed by Lesser Black-backed Gulls. Such predation appears to be carried out by individuals specialising in feeding on a particular species or by using specific methods, similar to those individuals that have developed techniques to 'prey' on humans holding sandwiches or ice creams at coastal resorts and in towns (see box, p. 168). Such specialists develop a consistent behaviour pattern of obtaining food and obviously benefit from past experience, returning day after day to the same feeding locality. Lesser Black-backs also prey upon small mammals and they sometimes scavenge road kills of both mammals and birds.

Breeding on moors and preying on young Red Grouse (*Lagopus lagopus*) and wading birds has made the species unwelcome in some upland areas, particularly where large colonies have developed, such as at Abbeystead and Tarnbrook in Lancashire. On these moors, large and increasing numbers of nesting gulls excluded Red Grouse and wading birds from breeding on many hectares of heather moor extending far beyond the limits of the colony, which was the initial reason for culling gulls nesting there.

The Lesser Black-backed Gull is one of several species that frequently feed at landfill sites in many areas, but in other regions (usually on the eastern side of Britain) it is only an infrequent visitor. For example, in north-east England, of more than 5,000 large gulls captured at landfill sites by cannon netting, not a single Lesser Black-back was caught and only two were seen out of many thousands of gulls counted while feeding at these sites over several years.

Studies of the food consumed by Lesser Black-backed Gulls while nesting some 15 km inland from the coast in Dumfries, south-west Scotland, found that they rarely fed in the marine environment during the breeding season and that neither marine invertebrates nor marine fish were consumed. Most of their food was collected from agricultural areas, where invertebrates – particularly earthworms, beetles and leatherjackets (cranefly larvae) – were frequently consumed, or at landfill, with individual gulls moving up to 40 km from the colony to feed. This inland feeding on agricultural land during the breeding season has often been overlooked because much of it occurs in the early morning, between dawn and 07.00, when dew is often still on the ground and earthworms are active at or near the surface. In addition, the gulls spread out in different directions and far from their colony to feed, so they are usually at low density on agricultural land. For this reason, they may go relatively unnoticed except when they are attracted in numbers to fields where ploughing or grass cutting for silage is in progress.

In most past studies, the high and frequent consumption of earthworms taken by Lesser Black-backs from agricultural fields has been overlooked or underestimated because the invertebrates are rapidly digested and leave only minute chaetae in regurgitated pellets, which are visible only under high magnification. Since each earthworm has large numbers of chaetae (four bundles on most body segments), those recovered in pellets or regurgitated samples obtained from gulls cannot realistically be converted into numbers of individual worms consumed, particularly since many chaetae are passed through the gut and are not deposited in the pellets.

The commonly held view of the public and local authorities is that urban nesting gulls obtain much of their food within town boundaries, but this was not supported in my studies at Dumfries and towns in north-east England. These found that the birds sourced only 5–10 per cent of their food from items left lying in streets or from sacks of rubbish left out overnight for collection in the morning. Many local councils in Britain have assumed incorrectly that, by reducing food waste in towns, the gulls can be starved and their numbers reduced. The error in this is obvious when one considers that several urban colonies each have well over 2,000 or more breeding large gulls, all requiring food each day – obviously, many of these birds need to feed far from their nesting sites in towns and cities.

Little plant material is consumed by large gulls, apart from grain found in fields and cooked materials such as bread and potato chips inadvertently made available. Other plant material is probably accidentally ingested or originates from the intestines of consumed animal prey. However, in the Kandalaksha region of north-west Russia, berries were found in 83 per cent of the stomachs of individuals of the dark-winged subspecies *Larus fuscus fuscus* that were examined in late summer. There are also records of Lesser Black-backed Gulls eating crowberries (*Empetrum nigrum*) on the island of Texel in the Netherlands, but in most breeding areas in Britain berries are not readily available.

The development of offshore wind farms has raised concerns over whether the moving rotor blades are likely to kill seabirds. The British Trust for Ornithology has made an extensive study of the feeding movements at sea of

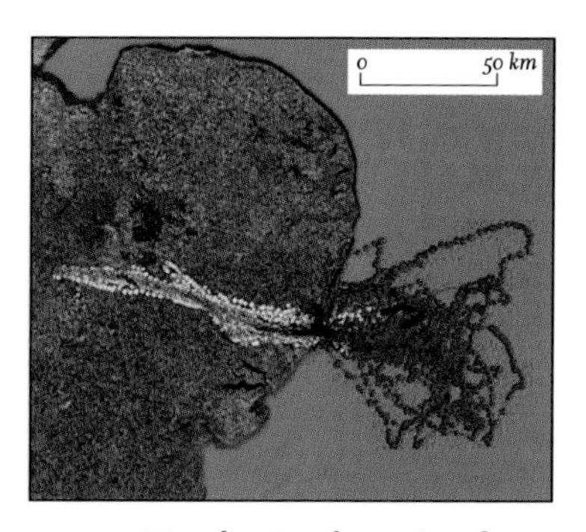

FIG 97. Map showing the routes of a series of feeding trips of a male (red) and a female (yellow) Lesser Black-backed Gull marked with GPS tags and tracked from Orford Ness during the breeding season. The black area at sea indicates the proposed position of wind farms when the trips were recorded. Data from Ross-Smith *et al.* (2014a). Image from BTO research funded by the UK Department for Business, Energy and Industrial Strategy as part of their Offshore Energy Strategic Environmental Assessment programme. Reproduced with permission of Hartley Anderson.

Lesser Black-backed Gulls nesting at Orford Ness in Suffolk by using a logger attached to the leg, which records at frequent intervals the bird's position and the height at which it is flying (Ross-Smith *et al.*, 2014a). These studies covered three successive breeding seasons and produced much new information about the feeding habits of the species (Fig. 97). As an example of the information that can be obtained by using loggers, the main findings are summarised below:

1. For 89 per cent of the time gulls were at sea, they were flying less than 20 m above it – i.e. within the range of the moving rotor arms of turbines.
2. The gulls tended to fly lower over the sea than over land, and the height was not influenced by wind speed or direction.
3. Feeding areas at sea varied from year to year, and the gulls spent more time in the areas of wind farms when they had young chicks than at other times of the breeding season.
4. Males spent more time near the wind farms late in the breeding season, but this difference was not seen in females.
5. Feeding trips during the breeding season usually lasted between three and eight hours, and individuals ranged up to 85 km.
6. Most individual Lesser Black-backed Gulls fed both at sea and on land, but a few fed only on land and were never recorded going to sea.

MOULTING

The timing of the primary moult in the subspecies *graellsii* (and including the alleged subspecies *intermedius*; p. 189) has been studied in detail by Peter Stewart in adult gulls captured by the Severn Estuary Gull Group in south-west England. He found that it commences in many individuals in early June, when most are hatching chicks. However, there is considerable variation, with some individuals not starting their moult until early July or even later. Once started, the moult continues, with the successive loss and replacement of primaries extending into late autumn; the earliest recorded date of completion of moult was October. The causes of variation in the start of the primary moult in individual Lesser Black-backed Gulls in Britain is unknown. In most cases, the longest primaries are shed in the autumn, which must reduce flying efficiency and may explain the slow migration of those birds moving to winter in Iberia. In contrast, individuals belonging to the subspecies *fuscus* do not usually begin the primary moult until they have reached their wintering area in October, and the replacement of their primary feathers can continue until April. Unfortunately, however, there are exceptions to the patterns described above.

POPULATION DYNAMICS

Having spent much of their lives since fledging well away from breeding areas, marked Lesser Black-backed Gulls are not usually seen in colonies for the first time until they are three or four years old. Even then, they are usually present only as prospecting individuals, arriving late in the season when breeding pairs are already incubating eggs or have chicks, and usually remaining on the periphery of the colony. In Britain, few of these young birds breed before they are four years old (although several individuals breeding when they are three years old have been reported in Belgium) and some are five years old before they attempt to breed for the first time, while a few may even be six years old before they breed.

Lesser Black-backed Gulls are long-lived, with the oldest recorded individual surviving in the wild for a few days short of 35 years. However, this bird was exceptional and would represent the longevity of about one in 1,000 individuals. Surviving for a further 10 or 11 years beyond adulthood would be a typical average life expectancy.

The average annual survival rates of adults have been determined in several studies, with the results revealing that the values are spread over a considerable range. No doubt they vary and fluctuate from area to area and also over time. In the Netherlands, Kees Camphuysen calculated that adult Lesser Black-backed Gulls had an annual survival rate of 87 per cent a year (compared with 83 per cent for Herring Gulls in the same colony). Data from the Isle of May (Centre for Ecology and Hydrology) and from a similar study colony in Belgium both indicated a 91 per cent annual adult survival rate, and these last two figures correspond to an average expectation of life of 11 years once the birds reached adulthood.

The numbers of Lesser Black-backed Gulls breeding on Skomer, south-west Wales, increased between about 1960 and 1993, and then declined progressively to the present time (Fig. 98). Survival data collected annually from 1978 to 2015 on the island gives an average annual adult survival rate of 88 per cent, which translates into individuals surviving for an average of eight breeding years. However, the annual survival rate varied over this considerable period, with a 92 per cent rate recorded from 1978 to 1994, then an appreciable decrease to 82 per cent per year in the following 10 years, followed by an increase to 87% per cent per year in the period 2004–15 (Fig. 99). While the change in the annual survival rate from 92 per cent to 82 per cent may seem relatively small, in effect it reduced the average lifespan of adults – and therefore their number of breeding years – by more than half, from 12 years to about five years. The numbers of Lesser Black-backed Gulls breeding on Skomer have clearly been affected by

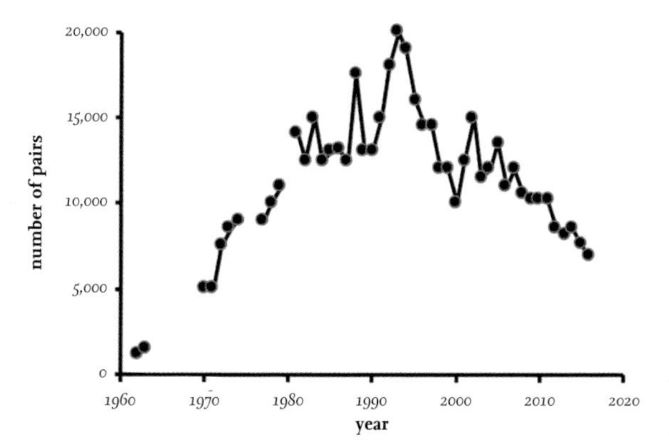

FIG 98. The estimated numbers of pairs of Lesser Black-backed Gulls breeding on Skomer from 1962 to 2016.

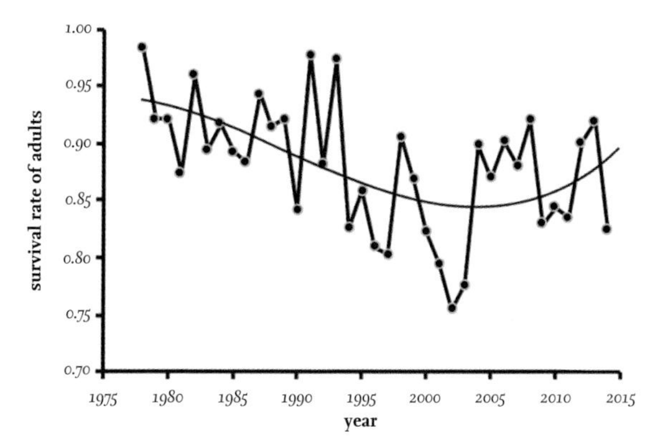

FIG 99. The annual survival rates of adult Lesser Black-backed Gulls breeding on Skomer, as determined and reported by the Wildlife Trust of South and West Wales in a study supported by the Joint Nature Conservancy Committee. The thin line is the smoothed trend line over the period 1978–2015.

changes in adult survival rates, but the causes of these changes are unknown, except for the increased mortality in the 1978–94 period being attributed to botulism.

Studies on marked birds in several colonies have shown that an appreciable proportion of adult Lesser Black-backed Gulls that had bred in a previous year missed one or more breeding seasons, although this figure appears to vary considerably from year to year. John Calladine and Mike Harris recorded that the proportion was at least 33 per cent in 1993 and 1994 on the Isle of May, and that an appreciable number of these birds missed two consecutive breeding seasons. In another study, 25 per cent of adult males and 17 per cent of females with previous breeding experience missed a breeding year.

Some pairs of Lesser Black-backed Gulls in three different British colonies even built nests but did not lay eggs. For example, this behaviour was recorded in 49 per cent of the pairs that built nests on Skomer in 1996, and was never less than 11 per cent in any year between 1991 and 2015. There have also been cases reported where pairs built several nests but laid in only one. These occurrences of pairs building more than one nest were too infrequent to provide a complete explanation for the high frequency of empty nests.

What causes these missed years is uncertain. It could be influenced by a shortage of one sex, or it could be that the individuals are in poor condition and so opt out of breeding to avoid the additional effort required to incubate and rear young, thereby increasing their chance of being able to breed successfully in future years. Lesser Black-backed Gulls need to rear about 0.5–0.6 chicks per year to fledging to replace adult losses, but allowing for some adults missing a breeding season, they may actually need to fledge about 0.7–0.8 young per breeding attempt. Since 1987, Lesser Black-backed Gulls breeding on Skomer have fledged an average of only 0.3 young per pair, which has probably contributed to the decline in numbers breeding, particularly if other colonies in the area had similarly low breeding success and decreasing adult survival rates.

Based on the dates of discovery of dead Lesser Black-backed Gulls carrying a ring, many more adults die in the summer months than in the winter. Unbiased evidence is difficult to obtain, but counts of dead adults found along beaches in the Netherlands revealed that numbers peaked in June and July (similar to Black-headed Gulls; p. 86), while the adults were breeding. This is also the time that weights of Lesser Black-backed Gulls are low, which could be interpreted as an adaptation to the need for great flight efficiency because more flying is needed to obtain the extra food demanded by the chicks, or that breeding is a stress caused by the need to defend a territory, eggs and young. An interesting study by Ruedi

Nager and others found that female Lesser Black-backed Gulls that were made to lay more eggs had poorer survival rates, and that those that survived bred less well in the following year (Nager *et al.*, 2001).

The main information missing from the population dynamics of the Lesser Black-backed Gull is the survival rate from fledging until breeding for the first time. As in most gulls, survival in the first year is high, but the precise survival level after the first year until breeding age is reached is still uncertain, although is often assumed to be similar to that of adults.

CHAPTER 7

Great Black-backed Gull

APPEARANCE

The Great Black-backed Gull (*Larus marinus*) is the largest gull in the world, with a wingspan of more than 1.5 m. In older literature and occasionally in current accounts, it is called the Greater Black-backed Gull, presumably to be consistent in separating it from the Lesser Black-backed Gull (*L. fuscus*; p. 179), with which it was confused in the nineteenth century.

Adults are readily identifiable by their large size and their black wings and mantle, whose colour is as intense as on the wing-tips. In addition, they have a substantial, powerful bill (particularly so in the males), and the flesh-coloured legs prevent confusion with the Lesser Black-back, which has yellow legs (Fig. 100). The adult's bill is yellow with a red spot on the lower mandible, as in

FIG 100. Fourth-year Great Black-backed Gull (left) and similar-aged Lesser Black-backed Gull (*Larus fuscus*). (Nicholas Aebischer)

the Herring Gull (*Larus argentatus*) and Lesser Black-backed Gulls. The voice is gruff and deeper than that of other gulls, and once it becomes familiar, the bird's presence can readily be identified when flying and calling overhead among flocks of mixed gull species.

Juveniles and first-year birds (Fig. 101) have the typical flecked, camouflage plumage of immature large gulls, but the background colour is grey (rather than brown as in Herring Gulls) and it has a more distinct chequerboard patterning. Caution is necessary, however, because some first-year male Herring Gulls are similar in size to female Great Black-backed Gulls of the same age.

The bill is totally dark in recently fledged birds, and a small, light area at its base becomes more extensive during the second, third and fourth years. The second- and third-year birds progressively acquire more adult plumage; they already have white heads and breasts at this age, but still retain some immature feathers on the wings and mantle, despite having acquired some totally black feathers. Fourth-year birds have still not assumed full adult plumage (Fig. 100), and while the bill is now mainly yellow, there is often a dark spot on the lower mandible that in full adults turns red.

FIG 101. First winter Great Black-backed Gull. (Nicholas Aebischer)

DISTRIBUTION

The Great Black-backed Gull is restricted to the North Atlantic and adjacent seas. It breeds in northern Europe and along part of the east coast of North America, and its distribution includes Greenland, Iceland, the Faeroes and as far north as Svalbard and Bear Island. In Europe, it is mainly a coastal breeder, nesting along the northern seaboard from north-west Russia to northern France, which it colonised in 1920, at the same time as it spread to Svalbard and Denmark. In North America, the species has spread north to Nunavut in Canada and south on the east coast of the USA as far as North Carolina. Numbers worldwide have been estimated at about 340,000 adults.

The species is sedentary over much of its range, including North America, Greenland, Britain and Ireland. However, birds breeding in north-west Russia and many from the Norwegian coast migrate in winter into the North Sea and the adjacent countries. In general, the species is increasing and slowly expanding its range on both sides of the Atlantic, helped by the fact that it is now not hunted except in Denmark.

The BTO census in Britain and Ireland in 2007 showed that breeding Great Black-backed Gulls have a marked western distribution, with concentrations on islands of various sizes in the Hebrides, Orkney and Shetland, and on the mainland of northern Scotland (Fig. 102). There is also a strong western distribution in Ireland and in Wales, from where the range extends into south-west England. At that time, some 34,000 adults were reported breeding in Britain and Ireland. In the past, Great Black-backed Gulls used to breed inland, but these colonies were reduced or exterminated by upland sheep farmers and gamekeepers, who considered the gulls a threat to livestock, and only 40 were nesting inland in 2000.

In the last 20 years, the species has increased its range and spread along the east and west coasts of Scotland, into north-west, north-east and south-east England. This expansion was delayed by culling of large gulls in the Firth of Forth, particularly on the Isle of May in Scotland, the Farne Islands in Northumberland and at Abbeystead in Lancashire, where small numbers of breeding Great Black-backs were killed during culls of other gulls. In addition, numbers were killed on Skomer and Skokholm, off the south-west coast of Wales, to reduce predation on Manx Shearwaters (*Puffinus puffinus*). The species has also spread within Ireland, particularly along the south and north-east coasts, and it now breeds (albeit in small numbers) around most of the coastline.

New Great Black-backed Gull breeding areas are usually established by a single pair joining existing colonies of Herring Gulls or Lesser Black-backed Gulls, followed by an increase of numbers in later years. This occurred at the large

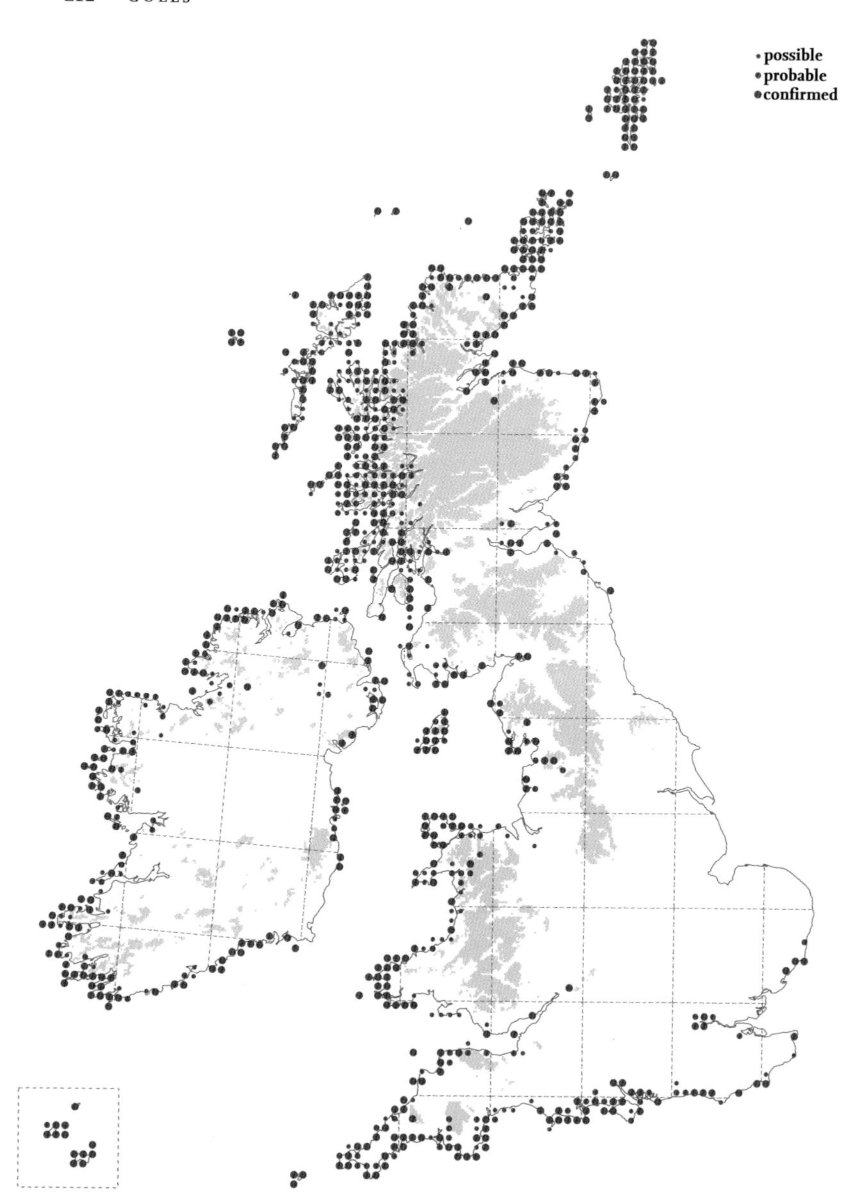

FIG 102. Map of the breeding distribution of the Great Black-backed Gull in Britain and Ireland 2008–11. Reproduced from *Bird Atlas 2007–11* (Balmer *et al.*, 2013), which was a joint project between the British Trust for Ornithology, BirdWatch Ireland and the Scottish Ornithologists' Club. Channel Islands are displaced to the bottom left corner. Map reproduced with permission from the British Trust for Ornithology.

inland gull colony at Abbeystead and has been recorded on the coast at many sites, including the Isle of May, the Farne Islands, Orford Ness and Havergate Island in Suffolk, Rockcliffe Marsh in Cumbria and on the Ribble Estuary in Lancashire.

Biometrics

Biometrics of Great Black-backed Gulls recorded in four studies are shown in Table 48. From this, it can be seen that the average wing lengths of Great Black-backed Gulls of both sexes breeding in Wales, Norway, north-west Russia and North America are nearly identical. This is in contrast to Herring Gulls – and Kittiwakes (*Rissa tridactyla*) and Puffins (*Fratercula arctica*) – which conform to Bergmann's rule and are appreciably larger in arctic Europe than in Britain. One effect of this is that males of the large Herring Gulls breeding in northern Scandinavia are of similar size to female Great Black-backed Gulls breeding in

TABLE 48. The biometrics of adult Great Black-backed Gulls breeding in Norway, north-west Russia, Wales and on the east coast of North America, compared with those breeding in Scandinavia and captured in winter in north-east England.

	Males				*Females*			
	Number of birds measured	*Weight (g)*	*Wing length (mm)*	*Bill depth (mm)*	*Number of birds measured*	*Weight (g)*	*Wing length (mm)*	*Bill depth (mm)*
North-east England (winter)	201	1,845	501	26.2	223	1,472	477	23.4
Wales (breeding)	95	1,713	499	—	108	1,486	478	—
North-west Russia	116	1,829	—	—	93	1,488	—	—
Norway	36	1,806	499	—	42	1,407	474	22.8
New Brunswick, Canada	94	—	498	26.5	120	—	470	24.1
Newfoundland and Sable Island, Canada	8	—	501	26.2	7	—	470	23.6
Ontario, Canada	15	—	502	27.4	9	—	482	24.3
Maryland, USA	?	—	505	—	?	—	478	—

the same area, which could potentially result in greater competition between them than in Britain. The similar size of Great Black-backed Gulls over much of their range meant that the birds caught in winter from unknown breeding areas could be sexed with about 98 per cent accuracy based on their head and bill measurements and, if necessary, confirmed by wing length. Head and bill length alone correctly sexed 95 per cent of individuals.

Historical distribution in Britain and Ireland

During the eighteenth and nineteenth centuries in Europe and North America, the Great Black-backed Gull was persecuted and its numbers were much reduced. Feathers were used in the hat trade and for bedding, and in upland Britain the fully grown birds were also considered a threat to lambs and gamebirds and were consequently killed, while some birds were shot for no reason other than that they were suitable targets. As a result, the species was extirpated from appreciable parts of Britain.

At the beginning of the twentieth century, numbers of Great Black-backs in Britain and Ireland probably reached their lowest level. A recovery then began, although it came about much later and was much slower than for other gull species because persecution continued in many areas. Distribution had been reduced to three isolated groups: one in south-west Wales and south-west England, including the Scilly Isles; one in the west of Ireland; and a third (and by far the largest) group on the islands and mainland of western and northern Scotland. As numbers increased slowly through most of the twentieth century, the species gradually returned as a breeding bird to a few areas from which it had been extirpated.

By 2000, more than three-quarters of breeding Great Black-backed Gulls in Britain were in Scotland. There were still extensive areas – including much of the east coast of Scotland and England, north-west England and south-west Scotland – where they were few or absent, but several of these have since been colonised. Despite this recovery, the national census in 2000 hinted that a small decline in the totals breeding in Britain and Ireland might have occurred since 1985, although this perceived decrease may simply reflect an incomplete coverage of the census and the considerable difficulty in making accurate counts of the species.

Great Black-backed Gulls have begun to nest on buildings in towns, but more recently and in much smaller numbers than Herring and Lesser Black-backed gulls. In 1976, only seven pairs were recorded nesting on buildings at three localities in Britain. By 1994, the total had increased to 11 pairs at 10 sites, but numbers in Cornwall were not recorded at that time. By 2000, numbers nesting at urban sites had increased to 83 pairs at 26 sites, with most in Devon and Cornwall, but seven urban nesting pairs were reported in Scotland and a single pair nested for the first time on a building in Wales. Although no national census has been carried out since, other towns and cities in Scotland and England now have nesting Great Black-backed Gulls, including Dumfries in south-west Scotland, Edinburgh in south-east Scotland and several towns in Cumbria. Pairs often select a higher point for nesting than other large gulls, such as flat roofs, on top of lift shafts or on other similar structures that are raised above the main roofs of the town, similar to their tendency to use raised areas at natural sites.

BREEDING

Great Black-backed Gulls tend to lay one or two weeks earlier than American Herring Gulls nesting in the same area (Fig. 103), with many of their clutches started during the second half of April and early May. The usual clutch size is three eggs, and as two-egg clutches are less frequent than in Herring Gulls, this results in a slightly larger average clutch size compared to that species, with 2.9 eggs laid in Wales on average and 2.8 eggs in North America. As is the case with many other gull species, lost clutches are usually replaced. Incubation from

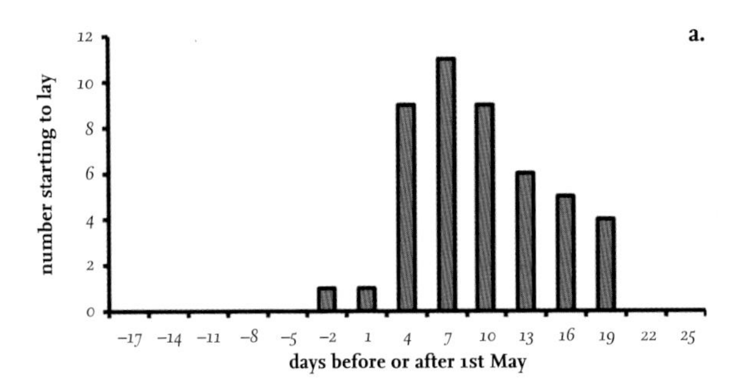

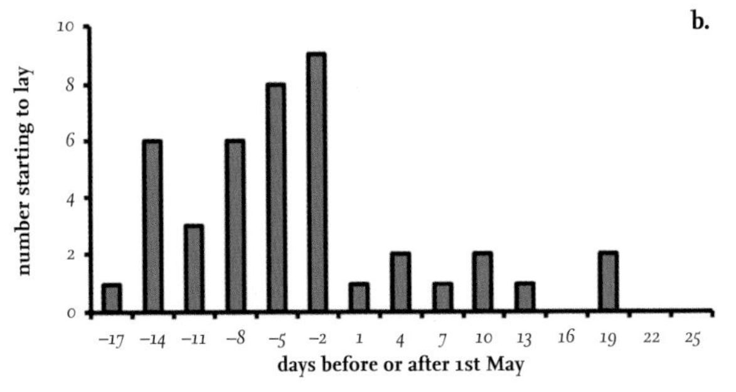

FIG 103. The date on which the first egg of each clutch is laid by (a) American Herring Gulls (*Larus smithsonianus*) and (b) Great Black-backed Gulls breeding in the same colony. Data from Erwin (1971).

completion of the clutch takes about 27 days. Incubation and feeding of the young is shared by both parents, but neither the frequency of feeding nor the pattern of duty sharing by the sexes have been investigated.

The growth of the chicks follows a typical pattern for gulls. After a slow start (with the chick relying on, and benefiting from, some yolk still present after hatching), there is an appreciable period of constant increase in weight each day of about 42 g, the largest daily weight gain recorded for any gull species. The third stage sees progressively lower daily growth rates, until a maximum weight is reached just before fledging (Fig. 104), which is slightly below the adult weight and persists for several weeks or even months. It is not recorded how long the young are fed by their parents after fledging, but in some cases they return from time to time to the nest site and are fed there by the parents. Owing to the large size of the adult Great Black-backed Gull, growth and development during pre-fledging are spread over a long period of about 50 days before sustained flight is achieved, giving a longer fledging period than in any other gull species. If undisturbed, breeding success is often high, with an average of more than one chick fledged per pair; in some areas, averages of up to 1.4 fledged per pair have been reported.

There is little information on the age of first breeding of this species. Some accounts state that breeding occurs at four years of age, but no details are given. However, if the species is like other large gulls, there is probably considerable variation, with some individuals being five or even six years old before they nest for the first time.

Apart from detailed research of their feeding habits, Great Black-backed Gulls remain the least-studied breeding gull species in Britain. In part, this is because they breed in relatively inaccessible sites and at low densities. The species is seen as a top predator that has adverse effects on other birds, and rather than encouraging study, this has often resulted in the removal of eggs and culling

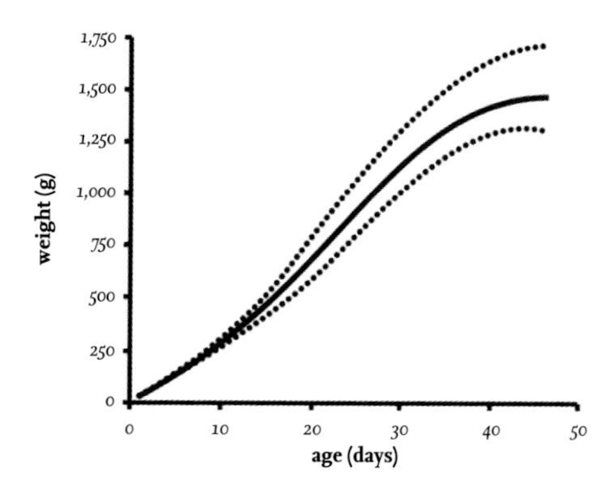

FIG 104. The typical growth rate of a Great Black-backed Gull chick (indicated by the continuous line) and the limits of normal variation, some of which is probably caused by the sex of the chick (indicated by the dotted lines). Some chicks return to the nest site and are fed after fledging, but it is not known how regularly or frequently this occurs. Adult males weigh 1,820 g on average, and females 1,500 g.

of adults in an attempt to benefit other species. It is also the least colonial of British gulls, often breeding as isolated pairs on small islands that are difficult to access, or within colonies of other large gulls. Studies to assist management have begun on the breeding biology and survival rates of this species on Skomer and Skokholm, leading to a change in conservation policy on these islands following many years of culling and egg removal.

In some places, the Great Black-backed Gull does nest colonially, but at lower densities than other gulls. In Britain, colonies seldom number more than 100 pairs; the largest colony was reported by P. G. H. Evans in the early 1970s on North Rona, a small island 80 km off the north-west coast of mainland Scotland, with about 2,000 pairs. However, numbers there had declined to 1,000 pairs by 2000 and then to fewer than 200 pairs within the past 10 years. Copinsay, in Orkney, was reported to have just over 1,000 breeding pairs in 2000, but numbers have since also declined there.

On the east coast of North America, colonies of Great Black-backed Gulls are often larger than in Britain and a few of these have been subjected to detailed studies. One investigation showed that breeding success was appreciably higher in pairs nesting on the edge of colonies compared to those in the centre, which is the opposite to findings in several other species of gulls. It was also reported that their breeding success was twice as high when their nearest neighbours were American Herring Gulls (*Larus smithsonianus*) rather than other Great Black-backs. In both cases, the differences were attributed to the aggressive and cannibalistic nature of the Great Black-backed Gulls.

MOVEMENTS

Recoveries of nestlings ringed in Britain show that most make only modest dispersals of a few hundred kilometres within the country, although a few marked in north-east Britain did leave the country, crossing the North Sea. Four of these birds moved to the Faeroes, one reached Iceland and two moved to Ireland (Fig. 105). Of those marked in south-west Britain or in the south of Ireland, a few moved to the Bay of Biscay and only two travelled further and were recovered in Spain (Fig. 106). Despite the fact that appreciable numbers have been ringed on both sides of the Atlantic, there are only two recorded transatlantic crossings. These were a young bird reared and marked on the east coast of Canada, which was recovered in Portugal, and an adult ringed in Newfoundland as 'fully grown' in July 2006, which was found dead near Plymouth, England, eight months later.

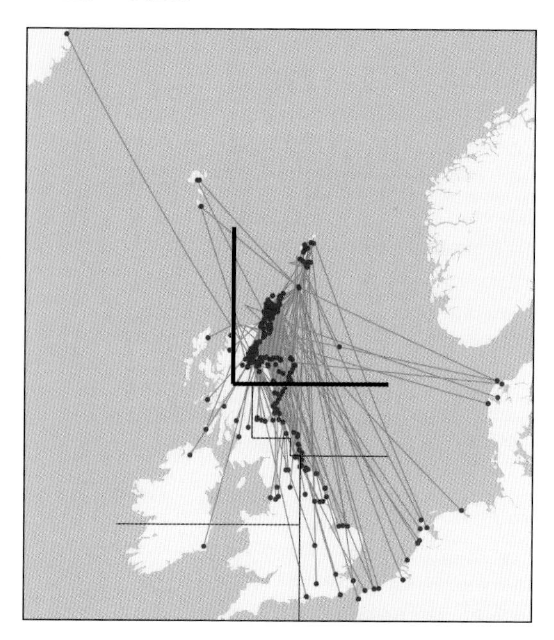

FIG 105. The movements of Great Black-backed Gulls ringed in north-east Britain during the breeding season and recovered in the non-breeding season. The bold black line shows the area within which they were ringed. Reproduced from *The Migration Atlas* (Wernham *et al.* 2002), with permission from the BTO.

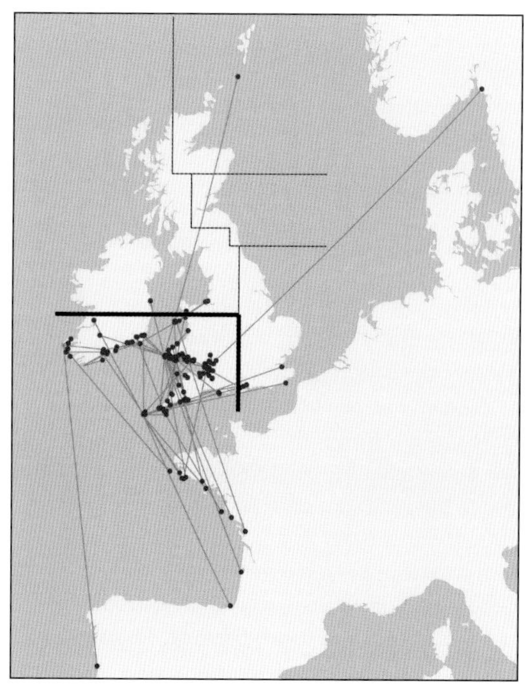

FIG 106. The movements between the breeding season and non-breeding season of Great Black-backed Gulls ringed or recovered in SW Britain and southern Ireland. The bold black lines show the areas of ringing. Reproduced from *The Migration Atlas* (Wernham *et al.* 2002), with permission from the BTO.

FOOD AND FEEDING

Apart from feeding occasionally on berries in summer and early autumn within the Arctic Circle, the Great Black-backed Gull's diet comprises entirely animal material composed of the greatest range of foods recorded for any species of gull. They are opportunists, and exploit the whole range of food consumed by Herring and Lesser Black-backed gulls; in addition, as a top predator, they consume large items that are not normally taken by the other species, including a wide range of fish, birds and small mammals.

Most studies have found that fish is the most important component of the diet, and judging from the species and sizes, most are obtained as discards from fishing vessels. That said, in some areas of the North Atlantic the gulls follow shoals of large predatory fish and marine mammals to capture fish driven to the surface by these marine hunters. Presumably, Great Black-backed Gulls must have caught most of their food in this way in the distant past. Where the gulls breed in large colonies of other seabirds, fish are less frequently consumed and are replaced by young and adult seabirds and Rabbits (*Oryctolagus cuniculus*), which can form a major part of their diet at some sites during the breeding season. The Great Black-backed Gull is capable of killing and consuming an animal half its own size. Such large food items are dismembered, whereas smaller items – such as fish and young seabirds – are usually swallowed whole and indigestible parts are later regurgitated in pellets.

Historical and recent accounts have claimed that adult Great Black-backed Gulls kill fully grown sheep. These reports are almost certainly a misinterpretation of the situation where the gull is seen standing near or on a sheep that has died from other causes, and confirmed cases of gulls actually killing adult sheep are lacking. Great Black-backed Gulls will feed on the carcasses of large animals such as a sheep or seals, and are known to have attacked animals that are already seriously ill, such as lambs and gulls of their own species. In 2015, the media reported that a 'giant gull' was attacking adult sheep in Ireland, with the story occupying many columns of newsprint for several days. Observations of what actually occurred were not reported. The suspicion that large gulls, and Great Black-backed Gulls in particular, are a threat to sheep and gamebirds in the uplands, whether real or otherwise, accounts for their disappearance from many former inland breeding areas where they once bred in the past.

On islands such as Skomer in Wales, where Great Black-backed Gulls breed alongside other species, adults prey during daylight on a broad spectrum of seabirds. In addition, Manx Shearwaters and Storm Petrels (*Hydrobates pelagicus*) are killed at night and in the early morning when they visit burrows and nest sites near the nesting gulls. Many of the shearwaters killed are prospecting young birds that are preparing to breed for the first time, and the extent of this mortality has been a cause of concern. A survey funded by the Joint Nature Conservation Committee found that 92 per cent of Great Black-backed Gull territories on Skomer contained remains of shearwaters at the end of the breeding season (Fig. 107), and that each pair of Great Black-backed Gulls was killing an average of eight Manx Shearwaters and three Rabbits each year. Adult Puffins were also captured and consumed, both on land and at sea, but numbers could not be determined. The diet of the Great Black-backed Gulls nesting on Skomer in 2013 was grouped into 11 types of food (Fig. 108). From this, it can be seen that fish, birds and Rabbits formed substantial parts of the diet. However, food regurgitated by Great Black-backed Gull chicks before they fledged showed that much more fish was present than in the adult diet. Presumably, the young were being fed with food obtained on feeding trips, while predation on seabirds represented food captured and consumed by adults in the colony mainly for their own maintenance.

In an attempt to protect Manx Shearwaters, numbers of Great Black-backed Gulls were culled on Skokholm between 1949 and 1985, and on Skomer during the 1960s and 1970s. As a result of these management policies, numbers of Great Black-backed Gulls nesting on Skomer were reduced from about 280 pairs in 1960 to 25 pairs in 1984. An outbreak of botulism in the early 1980s also contributed to their decline, but since then numbers on Skomer have gradually recovered to about 100 pairs. Numbers breeding on Skokholm were reduced to about five pairs in 1968, but since culling stopped in 1985 they have increased to 92 pairs. The large numbers of shearwaters nesting on these islands appear able to tolerate the current level of predation by Great Black-backed Gulls without declining.

Over much of the year, Great Black-backed Gulls are offshore feeders and frequently in attendance near fishing boats, particularly while nets are hauled in

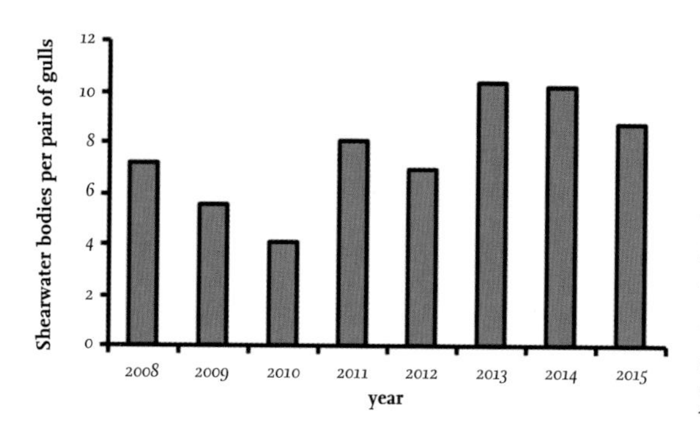

FIG 107. The number of Manx Shearwater (*Puffinus puffinus*) found per pair of Great Black-backed Gulls on Skomer, south-west Wales, in 2008–15. Data from the JNCC website.

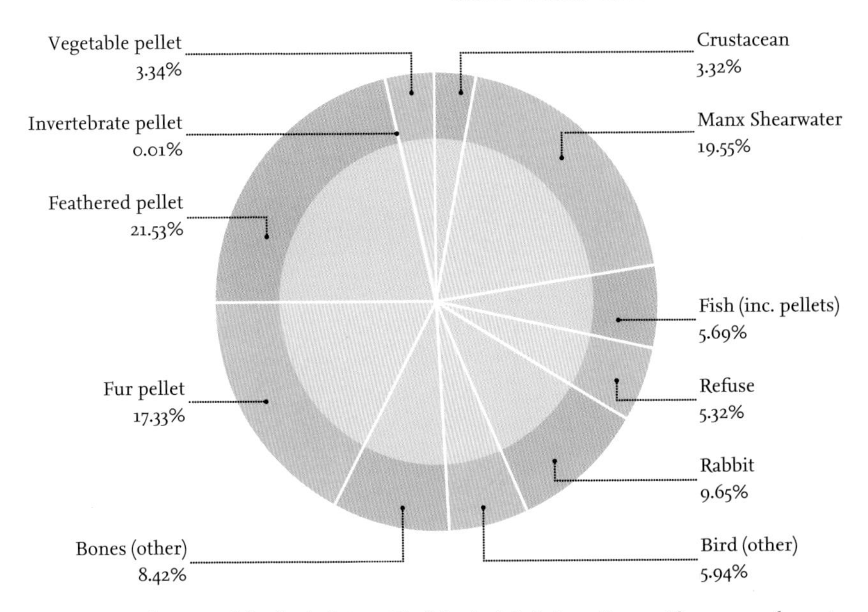

FIG 108. Evaluation of the food of Great Black-backed Gulls breeding on Skomer, south-west Wales, in 2012. The numbers are estimates of the percentages of the total food obtained from these organisms. Data from the JNCC website.

and unwanted fish are discarded. While Herring Gulls and Lesser Black-backed Gulls are often abundant around inshore fishing vessels off the British coast, on some occasions they are outnumbered by Great Black-backed Gulls. Up to 1,000 Great Black-backs have been recorded around single fishing boats off Shetland and in the Irish Sea (Table 49), but in contrast they are infrequent around fishing boats off the east coast of England in the summer.

During the breeding season, the Great Black-backed Gull occurs further offshore than the other large gulls, although in much reduced densities beyond 25 km from the nearest European coastline. Birds breeding in Newfoundland have been recorded travelling 100 km and more from land to feed near fishing boats working in areas of the Grand Banks. In winter, individuals have occasionally been reported up to 200 km from land, but these records are restricted to the continental shelf, which is still essentially associated with the fishing industry. The limited distance Great Black-backs move offshore (which prevents the species from being considered pelagic) is probably determined by its strong preference to roost at night on land or in sheltered waters along the coastline, although there are a few records of birds remaining offshore and feeding at night within the North Sea.

TABLE 49. The average numbers of seabirds attending each trawler fishing for white fish off Scotland and in the Irish Sea in summer and winter. Data mainly based on Hudson & Furness (1989) and my unpublished data for north-east England.

	Shetland, summer	*Irish Sea, summer*	*Irish Sea, winter*	*Clyde, Scotland, winter*	*North Rona, summer*	*East Scotland, winter*	*North-east England, summer*
Great Black-backed Gull	234	213	251	6	4	15	1
Lesser Black-backed Gull	6	2	<1	3	1	2	0
Herring Gull	30	5	16	241	90	87	45
Black-legged Kittiwake	3	75	78	24	35	0	56
Northern Fulmar	485	23	3	<1	155	1	1
Northern Gannet	9	11		6	5	15	0
Great Skua	12	<1	<1	<1	0	0	0

Commercial fishermen discard fish from their vessel for a variety of reasons: because the fish are below the minimum legal or marketing size; because fish escape from nets as these are hauled in; because there are restrictions on the percentages of catch that can be landed; and because of fish quotas. Some fishermen attempt to maximise their financial returns by keeping only fish that will sell for the highest price and discard the rest. As part of the reformed Common Fisheries Policy, catches of undersized quota fish may no longer be discarded in European waters. Instead, all of the catch must be landed and counted against the quota. This new rule was implemented in 2015 and will become more widespread and more wide-ranging up to 2019, including part of the Atlantic Ocean, and the Baltic, North and Mediterranean seas. These changes are likely to have a major impact on several seabird species, particularly on numbers of large gulls and Northern Fulmars (*Fulmarus glacialis*). The current dependence of Great Black-backed Gulls on commercial fisheries is likely to affect them now and into the future, as a result of major changes in the intensity and methods of fishing, and particularly the new regulations. Already, declines in numbers of Great Black-backed Gulls have been reported from parts of the east coast of Canada, associated with decreased commercial fishing activity. On Canna in the Hebrides, numbers of breeding Great Black-backed Gulls declined by 80 per cent between 2000 and 2009, a reduction attributed to much lower fishing activity and smaller landings of fish at nearby ports. A national census is needed to explore whether such declines have occurred elsewhere and that the reported changes in numbers are not the result of some other cause, such as disease.

There are few records of Great Black-backed Gulls catching fish at sea apart from at fishing vessels, although this may be because there are few observers offshore. Some individuals are attracted to, and feed on, surface shoals of sandeels when these are also attended by other gull species, and they will plunge from the surface or make shallow dives from flight after other fish, although the depth of the dives appears to be less than 1 m.

As mentioned above, small food items are also consumed by Great Black-backed Gulls and the variety of items they take is almost endless. Refuse and carrion are also consumed, and kleptoparasitism (parasitism by theft) on smaller gulls and even their own species is common, particularly by males. Many prey are captured alive, but the birds are also major carrion feeders, exemplified by the frequent occurrence of this species feeding on fish offal at sea, at fish docks and at landfill sites (see below).

The diet of Great Black-backed Gulls breeding on Cape Clear Island, Ireland, was investigated by Neil Buckley (1990). He reported that it comprised mainly fish and birds, as on Skomer (see above), but in the absence of Rabbits and shearwaters, 90 per cent by weight of the food was fish and auks, the former being mainly obtained from fishing boats. Other foods represented only 10 per cent of the diet and included small mammals, crabs and goose barnacles.

Breeding areas of Grey Seals (*Halichoerus grypus*) where pups are born in late autumn and early winter are frequented by Great Black-backed Gulls (often together with Grey Herons, *Ardea cinerea*), which feed on the afterbirths and, less frequently, on dead young seals. The species is undoubtedly opportunistic, exploiting whatever is available.

Feeding at landfill sites

Although the Great Black-backed Gull is rarely the most numerous gull attending landfill sites (Table 50), and numbers vary markedly from day to day, it is the dominant species when present because of its sheer size and aggression. Some male Great Black-backs feeding at landfills frequently use kleptoparastism as the main method of obtaining food. Instead of rummaging through the refuse, they frequently stand on a raised site within the feeding area watching out for other gulls taking large food items they cannot swallow quickly. The forager is then attacked and pursued by the Great Black-backed Gull, which is usually successful in stealing and swallowing the item. Susan Greig studied this kleptoparasitic method of feeding by using video recordings (p. 171) and found that the average rate of swallowing food was higher in Great Black-backed Gulls than in Herring Gulls of all ages. This was mainly achieved through kleptoparasitism, which was successful in 99 per cent of 333 attacks on individuals of other gull species and much more

TABLE 50. The average numbers of gull species present at one landfill site during a working day in north-east England (Northumberland and Durham) in winter (October to the end of January) and summer (April to July). Note that the three year classes of immature Great Black-backed Gulls and Herring Gulls (*Larus argentatus*) formed only a small proportion of the total number of individuals of those species in winter.

	Average, winter	*Maximum, winter*	*Summer average*
Great Black-backed Gull adults	32	301	0
(first-year to third-year, immature)	(3)	(24)	
Herring Gull adults	278	950	5
(first-year to third-year, immature)	(72)	(170)	(32)
Lesser Black-backed Gull	0	0	0
Black-headed Gull	162	525	0
Common Gull	<1	5	0

so than the fewer attempts made by Herring Gulls. The latter were much more abundant on all occasions and received many attacks from Great Black-backed Gulls, but attacks between two Great Black-backed Gulls (often a male attacking a female) also occurred and, surprisingly, were more than twice as frequent than would have been expected. Herring Gulls usually attacked their own species to steal food and rarely attacked Great Black-backs, but when they did so, the attempt almost always failed. Presumably, Herring Gulls soon learn to recognise Great Black-backed Gulls and regard them as unprofitable targets for kleptoparasitic attacks.

Frenzy feeding by gulls at landfill sites rarely extended uninterrupted for more than 20 minutes. Studies by Greig found that the feeding rate of Great Black-backed Gulls was high during the first 10 minutes but then declined by 30 per cent (based on 820 records), whereas it decreased by only 10 per cent in Herring Gulls when feeding continued beyond 10 minutes. Presumably, the exposure (and hence availability) of large food items declined rapidly during the feeding frenzy, which involved many gulls, and this affected the feeding methods of Great Black-backed Gulls more than Herring Gulls. Only when new material was dumped, or the existing material was disturbed by the compactor, did the feeding rates increase again. This rapid removal of suitable food items and the subsequent decline in the feeding rate may explain why many feeding frenzies ended spontaneously within 20 minutes, even when there was no human activity or other disturbance at the working face.

There is no doubt that where large numbers of Great-backed Backed Gulls nest within seabird colonies (and these are few), their impact on the breeding success of their own and other species is considerable. Prey species involved include Puffins, Manx Shearwaters, Kittiwakes, Northern Fulmars and even other larger gulls. Dead chicks of their own species are frequently consumed by large Great Black-backed chicks in the last two weeks before they fledge.

In areas of Scotland, such at the mouth of the river Ythan near Aberdeen, and at several localities on the east coast of Canada and the USA, studies have revealed that predation by Great Black-backed Gulls (and other large gulls) is responsible for the high mortality of Eider (*Somateria mollissima*) ducklings and is believed to have caused declines in local numbers of Eiders. Kleptoparasitism by Great Black-backed Gulls on flocks of Eiders diving for food is also common. The gulls sit on the water near a flock of feeding ducks and wait for one to surface with a large prey item that cannot be swallowed underwater (such as crabs, whose legs must be removed by the Eider before it can swallow it), before attacking the duck and usually forcing it to relinquish its prey.

Feeding behaviour in winter

Regular winter counts of Great Black-backed Gulls visiting inland landfill sites in north-east England have revealed that the numbers present varied dramatically from day to day, and much more so than the numbers of Herring Gulls and Black-headed Gulls (*Chroicocephalus ridibundus*). For example, a single bird present on one day was replaced by more than a hundred on the next day, and then numbers decreased to three on the following day. These large fluctuations were first noted by Pat Monaghan during her study of gulls feeding at the inland Whitton landfill site in County Durham. Later counts at a series of landfills in north-east England confirmed the frequency of these fluctuations, with, for example, between one and 298 individuals present at a site on five consecutive days in the same week. Two examples from Coxhoe landfill in County Durham of this extreme variation occurring within a week are shown in Fig. 109. Counts at other landfill sites in the area and day roosts on the coast showed a close positive correlation with the numbers of Great Black-backed Gulls at Coxhoe: when numbers were low there, the missing birds were not visiting other landfill sites, and on some occasions were reported moving offshore to feed behind fishing boats.

Great Black-backed Gulls in north-east England roosted at the coast each night and chose at dawn whether to go to sea or move inland in search of food at landfill sites. Susan Greig used a large database of daily counts collected by others at inland landfill sites in north-east England to identify the environmental factors that correlated with variations in numbers. She found that the most important relationship was wind strength during the night, which was positively correlated with large numbers of Great Black-backed Gulls visiting landfill sites on the following day.

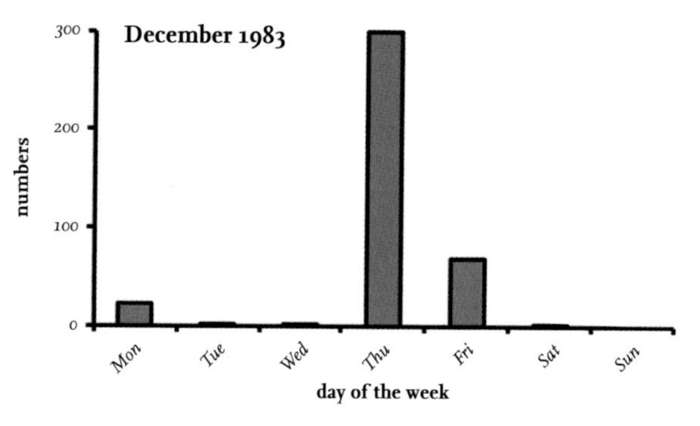

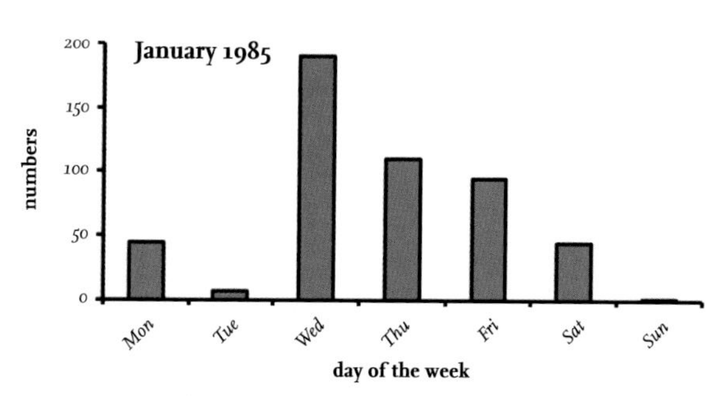

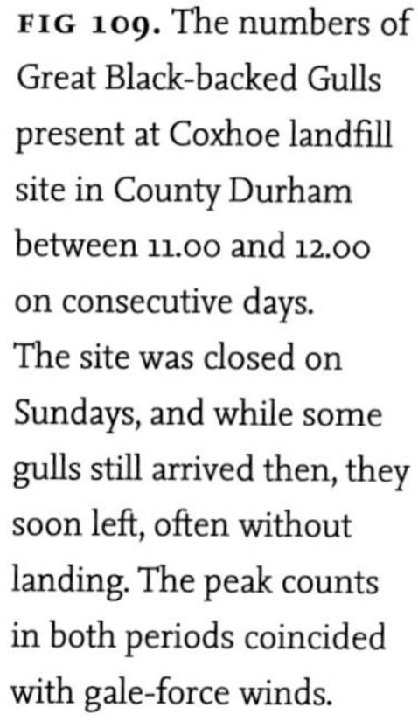
FIG 109. The numbers of Great Black-backed Gulls present at Coxhoe landfill site in County Durham between 11.00 and 12.00 on consecutive days. The site was closed on Sundays, and while some gulls still arrived then, they soon left, often without landing. The peak counts in both periods coincided with gale-force winds.

Gales kept many fishing boats in harbour, which significantly reduced scavenging opportunities for gulls. Most of the Great Black-backed Gulls arrived at inland landfill sites within two hours of sunrise, and presumably they used the wind speed at night or at dawn as an indicator of the amount of possible offshore fishing activity, responding accordingly. This relationship suggested that feeding on discarded fish and offal behind boats at sea was the birds' preferred method, and that they chose to visit landfill sites only when fishing activity at sea was likely to be unproductive. In contrast, Herring Gull numbers at landfill sites were only marginally affected by strong winds and the birds tended to occur in similar numbers each day.

STUDIES ON MARKED INDIVIDUALS

In a study in north-east England, numbers of wintering Great Black-backed Gulls were captured at six landfill sites by cannon netting and each adult was given a unique combination of coloured rings that allowed them to be individually identified in the field (Coulson *et al.*, 1984a). From the recoveries, various aspects of the birds' behaviour and population dynamics were revealed.

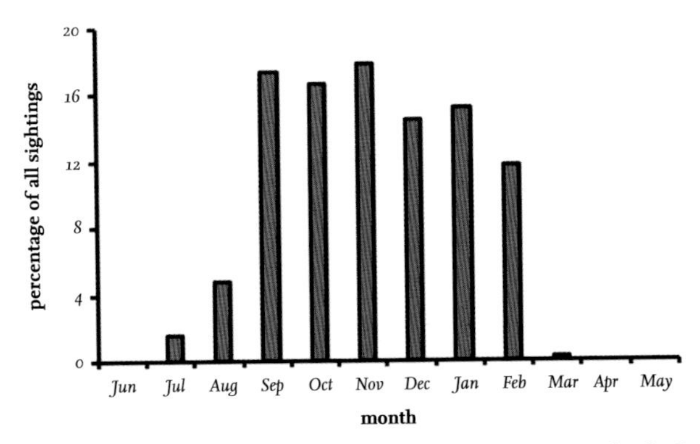

FIG 110. The percentage of all sightings in north-east England of colour-ringed adult Great Black-backed Gulls in each month (based on 461 field records), starting from June and following the winter during which they were marked. The majority of birds departed synchronously from the wintering area during the second half of February and the movement was virtually completed before March.

Breeding areas

Sightings and ringing recoveries of Great Black-backed Gulls in the breeding season identified their breeding areas. A total of 48 were reported from sites along the entire coastline of Norway and two were seen in north-west Russia, but only one was found breeding in Britain, in northern Scotland. From this, it appears that the great majority of wintering Great Black-backed Gulls in north-east England were migrants from Scandinavia and north-west Russia. All the adults visiting north-east England arrived in early September and left abruptly in the middle or end of February (Fig. 110), although a few immature individuals remained in some summers.

Visits to landfill sites

Great Black-backed Gulls visiting landfill sites near the coast were captured in almost equal proportions, but those caught at inland sites were dominated by females by more than two to one (Table 51). This difference in behaviour was confirmed by counts of the sex of unmarked birds, with females consistently more numerous than males at inland sites; subsequent sightings of the marked birds confirmed this finding. The reason for this difference in the behaviour of the sexes is not immediately evident, but it may indicate a greater preference of males to feed at sea.

In all, 424 Great Black-backed Gulls were captured at landfill sites in north-east England, and it was surprising that 81 per cent were full adults and only

TABLE 51. The numbers and percentages of adult male and female Great Black-backed Gulls colour-ringed and then subsequently seen at coastal or inland landfill sites in north-east England.

	Number of males	*Number of females*	*Total*	*Percentage males*	*Percentage females*
Captures at coastal sites	124	149	273	45%	55%
Captures at inland sites	46	105	151	30%	70%
Sightings at coastal sites	189	179	368	51%	49%
Sightings at inland sites	50	144	194	26%	74%

8 per cent were in their first year of life (Table 52). Anne Hudson and Bob Furness found consistently high proportions of adults, averaging about 90 per cent, feeding behind fishing boats off northern Scotland. There are several possible explanations for these proportions. First-year (and older immature birds) may have been less inclined to move to north-east England from Norway, but there is no evidence of this. In any case, even smaller proportions of birds in their second, third and fourth winters were recorded, and so the habit of using different wintering areas would have had to persist for the first four years of life. Another possibility is that adults were more likely to be captured, but this explanation was rejected because field counts of unmarked birds at both landfills and coastal roosts invariably found that first-year birds formed less than 10 per cent of the Great Black-backed Gulls present. A third possibility is that the captures were representative of the population. If this was indeed the case, then the breeding success was low, with 80 adults (40 pairs) producing only eight young that survived to their first winter, equivalent to just 0.2–0.3 young fledged per pair each year. Great Black-backed Gulls in Britain often fledge more than one chick per pair and usually have a higher productivity than Herring Gulls breeding at the same sites. For example, between 1996 and 2015 on Skomer, Great Black-

TABLE 52. The age distribution of 424 Great Black-backed Gulls captured at landfill sites in north-east England, 1978–81.

	First-winter	*Second-winter*	*Third-winter*	*Fourth-winter*	*Adult*	*Total*
Number	34	15	17	15	343	424
Percentage	8.0%	3.5%	4.0%	3.5%	80.9%	

backed Gulls averaged 1.2 chicks fledged per pair and the figure fell below 1.0 in only three of these years. However, low productivity has been recorded by Clive Craik in western Scotland due to predation by American Mink (*Neovison vison*), while on Canna in the Hebrides, only 0.1–0.3 young were fledged per pair each year between 2001 and 2006, but then productivity improved to an average of more than one per pair between 2007 and 2015.

Adult survival rates

A return rate of 82–84 per cent a year was obtained for adults using the same wintering area in north-east England (see below), suggesting that the survival rate would have been even higher had any marked adults been missed or changed their wintering area. The value above suggested that, unless numbers of Great Black-backed Gulls were declining, the adult survival rate must have been high. Urs Glutz von Blotzheim and Kurt Bauer suggested a 93 per cent annual adult survival rate for this species, and rates of between 78 per cent and 92 per cent were estimated recently at a colony in Newfoundland. An adult survival rate of 90 per cent was obtained for one year on Skokholm. There is a need for further estimates on the survival rates of this neglected species, but it is obviously long-lived. While the oldest ringed individual lived for just over 29 years in the wild, which is lower than the longevity records for most other species of large gulls, this could simply be because fewer Great Black-backed Gulls have been ringed, or that there is a lower viability of rings placed on this large bird and that some rings are lost before long-lived individuals die.

Winter site fidelity

Most of the colour-ringed adults marked in winter in north-east England returned to the same wintering site year after year and remained there throughout the whole season. Of the adults originally marked between September and November, 83 per cent were seen again in north-east England a year later, although surprisingly the return rate was lower for those initially captured and marked in December and January (Table 53). This lower rate suggests that some of these individuals may already have been moving north from more southern wintering areas when they were caught, although the main departure did not occur until mid-February (Fig. 110).

Of the adults that were seen in the following winter, 80 per cent were found at the sites where they had been initially ringed, suggesting that they benefited from experience and knowledge of food sources in their chosen wintering areas. Of the 20 per cent of individuals that were seen, but at a site other than that at which they had been ringed, none had moved further than 30 km.

TABLE 53. The proportion of Great Black-backed Gulls colour-ringed in north-east England and seen there in a subsequent winter.

Month of ringing	*Total marked*	*Seen in a subsequent winter*	*Not seen again*	*Percentage seen in a later winter*
Adults				
Sep & Oct	92	75	17	82%
Nov	112	94	18	84%
Dec & Jan	139	86	53	62%
First-years				
Sep–Jan	34	0	34	0%
Second- and third-years				
Sep–Jan	32	10	22	31%

This behaviour contrasted with that of the 34 birds that were in their first year when marked. None was seen again in northeast England in a subsequent winter, and presumably the majority had moved to winter in other areas, although some would have died. Some of those marked in their second or third years of life returned, but they did so at a much lower rate than full adults. The return of some adults to the same wintering area has been recorded in several gull species, but most of these records were of a specific individual. Where overall rates of return do exist for other species, they show much lower rates than reported here for the Great Black-backed Gull.

CHAPTER 8

Black-legged Kittiwake

THE BLACK-LEGGED KITTIWAKE (*Rissa tridactyla*; Fig. 111), here simply referred to as Kittiwake for convenience, is the most numerous gull in the world and has a circumpolar distribution in the northern hemisphere. It is closely related to the much less numerous Red-legged Kittiwake (*R. brevirostris*), which breeds on only a few islands in the Aleutian

FIG 111. Kittiwake and half-grown chick on nest. (Mike Osborne)

chain in the Pacific Ocean. Both species nest on cliffs and show several adaptations for this (see box on pp. 254–6), which justifies the separate generic name, *Rissa*. The hind toe on each foot is reduced to a small lobe, presumably because the birds have had little need to walk at any time of the year. This has given rise to the scientific name *tridactyla* for the Black-legged Kittiwake, and to a common name meaning 'three-toed gull' in several European countries. In Britain, the common name describes the *kittiwaak* calls of the birds when they reunite at the nest and both display.

Both kittiwake species have shorter legs than most other gulls, which lowers their centre of gravity. This presumably assists the birds in withstanding the buffeting of strong updraughts around sea cliffs where they nest. Both species are totally oceanic outside of the breeding season.

APPEARANCE

The Kittiwake breeds around the mainland coasts and islands of the North Atlantic (subspecies *Rissa tridactyla tridactyla*) and North Pacific (subspecies *R. t. pollicaris*). The subspecies *pollicaris* has slightly more extensive black on the wing-tips and is marginally darker grey on the wings and mantle. The length of the hind toe and the absence of a claw in *pollicaris* are often mentioned as two differences between the subspecies, but this distinction was based on a very limited number of specimens and it is neither a consistent nor reliable characteristic. In the North Atlantic, the size of Kittiwakes is progressively larger towards the north of the species' breeding range; such a gradual change is called a cline. A similar situation occurs in the Puffin (*Fratercula arctica*), and is presumably an adaptation to environmental temperature and the greater need to reduce heat loss from the body in colder areas. It is not clearly established which of the subspecies breeds in the few colonies along the north coast of Siberia, nor the route (east or west) these birds take to wintering areas in the Atlantic or Pacific oceans.

The Kittiwake is a small gull with a wingspan of about 1 m; adults superficially resemble several other gulls that have grey wings and white bodies, but differ in having black legs, a waxy-yellow bill and triangular black wing-tips that lack white 'mirrors' (Fig. 112). Its dark brown eye gives the impression of a docile animal. As in other gulls, the sexes are similar, with the male being only slightly larger than the female. The Kittiwake's habit of breeding on narrow ledges on coastal cliffs also aids identification, as does its highly characteristic greeting call (Fig. 113).

FIG 112. Adult Kittiwake in flight. (Mike Osborne)

FIG 113. Pair of Kittiwakes engaging in a greeting ceremony. The *kittiwaak* call gives the species its English common name. (Rob Barrett)

FIG 114. Full-grown Kittiwake chick with adult. Note the white streaks of droppings on the side of the nest. (Mike Osborne)

FIG 115. Adult Kittiwake (left) and juvenile Kittiwake (right) in flight. Note the rounder wing shape in the juvenile, caused by shorter ninth and 10th primaries. (Mike Osborne)

In their first year of life, young Kittiwakes differ markedly from adults, to such an extent that for some years they were thought by some museum workers to be a different species (Fig. 114). Unlike the adult, the neck of the juvenile has a black collar, the tail is slightly forked tail and has a black terminal band, and there is a broad black 'W' marking spread across the wings (which is also present in the immature stage of several other small gulls). The wings are shorter and seem broader in juveniles than in adults (Fig. 115). The bill is black and the legs are also black at fledging, but the leg colour soon changes to leaden grey during the first winter, only to revert to black in older individuals.

DISTRIBUTION

The Kittiwake breeds extensively in arctic areas in the Atlantic and Pacific oceans. In the Atlantic, it breeds as far south as northern France. There is one small colony in north-west Spain and until recently a few birds nested on an island off the coast of Portugal. One factor affecting their southern breeding distribution is the risk of mortality of young in nests exposed to strong sunlight and high temperatures. Both the nesting sites in Iberia were in gullies and on ledges on north-facing cliffs that did not receive direct sunlight.

Distribution in Britain and Ireland

The Kittiwake is widely distributed as a breeding bird around most of the coastline of Britain and Ireland, with the largest (yet decreasing) numbers in Orkney and Shetland, on the east coast of Scotland and on the north-east coast of England (Fig. 116). This is a very different distribution from that at the end of the nineteenth century, when there were only two colonies on the east coast of England (the Farne Islands and Bempton). As late as 1954 in their book *Seabirds* (*New Naturalist* 28), James Fisher and Ronald Lockley could list only 16 colonies in England, but by the end of the century, there were at least 46, and this increase has been accompanied by a considerable increase in the size of the colonies present 60 years earlier. Similar, but much less well documented, increases have occurred in Scotland, Wales and Ireland during the twentieth century.

Kittiwakes are an intensively colonial species, with nests often built close together on narrow cliff ledges, even those that are too narrow for Common Guillemots (*Uria aalge*). Not only is breeding colonial, but the collection of material to build the nest is also often a highly social activity, with a hundred or more Kittiwakes flocking to a damp or wet place on neighbouring cliffs to collect mud or vegetation (*see* Fig. 134), and then forming a stream of birds flying to and from the nest sites.

In 1900, Kittiwake colonies in Britain were located on a small number of high cliffs with the sea immediately below. Most nests were on the lower half of the cliff, in places where they were protected from human disturbance. Since

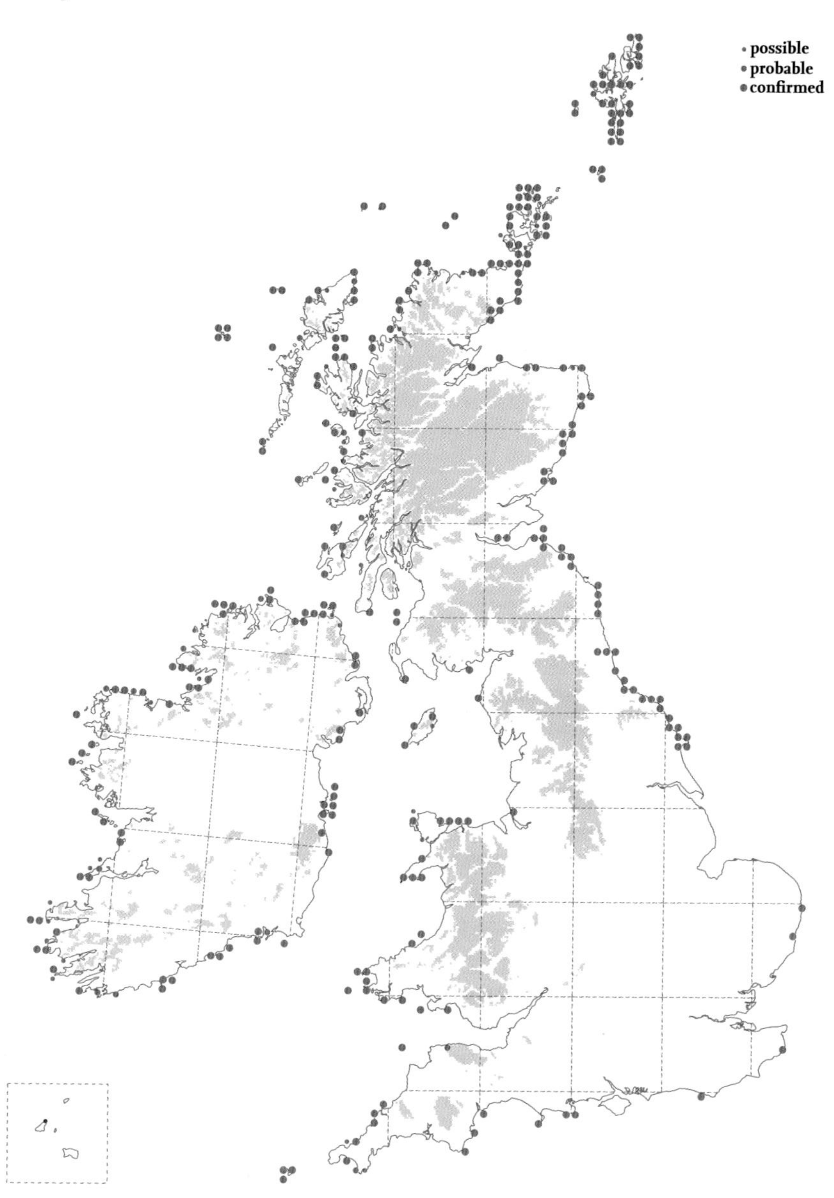

FIG 116. Black-legged Kittiwake in Britain and Ireland 2008–11. Reproduced from *Bird Atlas 2007–11* (Balmer *et al.*, 2013), which was a joint project between the British Trust for Ornithology, BirdWatch Ireland and the Scottish Ornithologists' Club. Channel Islands are displaced to the bottom left corner. Map reproduced with permission from the British Trust for Ornithology.

1930, newly formed colonies in Britain have been on lower cliffs, as illustrated in Fig. 117, which shows the heights of cliffs used by colonies of Kittiwakes before 1930, in 1960 and in 2000. The situation pre-1930 relates to the few long-established colonies. As Kittiwake numbers increased during the twentieth century, more new colonies were formed, many of which were on lower cliffs; 20 per cent of colonies are now on cliffs less than 20 m high. This trend of using lower cliffs continues today, and the birds are often protected from human interference.

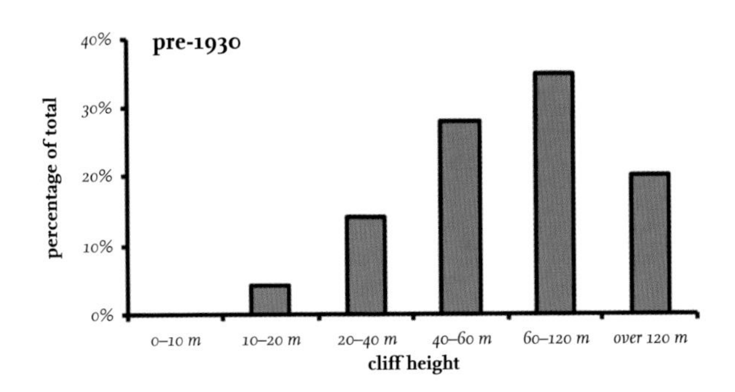

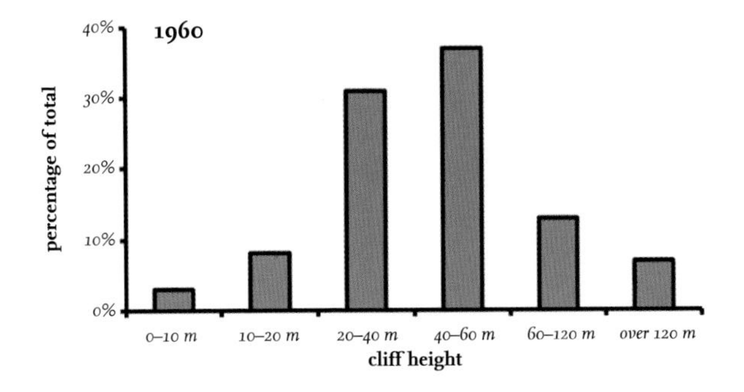

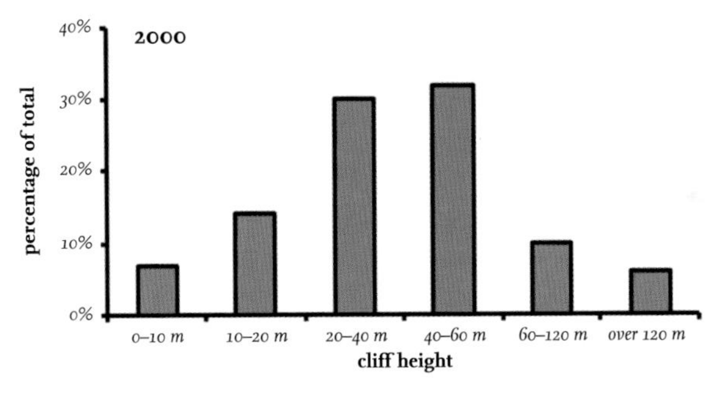

FIG 117. The height of cliffs used by breeding Kittiwakes in Britain before 1930, in 1960 and in 2000. Based on those colonies for which it was possible to obtain reliable measurements.

The changing habit of nesting on lower cliffs led to Kittiwakes accepting the sides of buildings with window ledges as acceptable sites. The first such colony on a building was at Dunbar in east Scotland (Fig. 118), followed by another in North Shields (Figs 119 and 120). Kittiwakes now nest on a low cliff and the remains of the wall of an old castle at Dunbar, and do not differentiate between the two sites (Fig. 121).

Colonies on buildings and other man-made structures occur elsewhere in Britain (Fig. 122), and in Norway and Alaska, while gas and oil platforms in the North and Irish seas, and even a navigation post in Sweden (Fig. 123) are used. Other structures used for nesting by Kittiwakes include the ferry terminal on the river Tyne near North Shields (Fig. 124).

Kittiwakes have even nested on the ground in Denmark, building nests first on a boulder beach and later spreading onto adjacent sand dunes. Unfortunately, this colony no long exists because it was invaded by rats. There are also two records of Kittiwakes nesting on sand dunes in Norfolk, among breeding Sandwich Terns (*Thalasseus sandvicensis*), but both of these attempts failed as the nests and chicks were buried by drifting sand (Fig. 125).

FIG 118. The first Kittiwake colony in Britain on a building: the grain-drying factory in the harbour at Dunbar. (John Coulson)

FIG 119. The west side of the brewery warehouse Kittiwake colony at North Shields, Tyne and Wear. (John Coulson)

FIG 120. Close-up of nesting Kittiwakes on one of the windows seen in Fig. 119. (John Coulson)

FIG 121. Dunbar harbour cliff and remains of Dunbar Castle, south-east Scotland. The inset shows nests on part of the castle. (Becky Coulson)

FIG 122. Wall specifically built for nesting Kittiwakes after demolition of the yacht club building on which they formerly nested at Lowestoft, Suffolk. (John Coulson)

ABOVE: **FIG 124.** Kittiwakes nesting on the roll-on roll-off gantry at North Shields, Tyne and Wear. (John Coulson)

LEFT: **FIG 123.** Navigation tower with nesting Kittiwakes, Sweden. (John Coulson)

FIG 125. Kittiwake nesting on a sand dune, Norfolk. (John Coulson)

POPULATION ESTIMATES

Despite worldwide information on the Black-legged Kittiwake being limited in recent years, the International Union for the Conservation of Nature (IUCN) moved it to its Red List in January 2018 and have expressed the view that the species is faced with the risk of global extinction. It is also considered a species of conservation concern in the UK and Ireland. The data leading to these decisions have not been published, although the IUCN suggests that the Kittiwake has declined globally by 40 per cent since 1970. This figure must be an informed estimate rather than based on fact, however, since many colonies in the species' worldwide distribution have not been counted twice since 1970. In any case, the Kittiwake must still be the most numerous gull in the world and currently is nowhere near extinction.

Numbers of Kittiwakes in Britain and Ireland reached a low at the end of the nineteenth century as a result of human exploitation for food, plumage and sport, but from about 1900 numbers began to increase in the surviving colonies. However, it was not until after 1930 that any new colonies formed. Numbers increased continuously for most of the rest of the century, with a fourfold increase between 1930 and 1985. Between 1959 and 1970, almost all Kittiwake colonies increased in size and more new ones were established. Total numbers reported in 1985 in the national census of Britain and Ireland showed a modest increase since 1970, recording about half a million breeding pairs.

Increases in the population of any animal species cannot go on forever, and this is true for the Kittiwake. By 1985, some colonies in Britain and Ireland were showing a decline for the first time that century. Between 1970 and 1985, 31 administrative areas (counties and regions) in Britain and Ireland still showed an overall increase in Kittiwake numbers, but 18 others recorded a decrease. This was the first indication that the expansion was beginning to slow down. The main areas of decline were in northern Scotland and in some colonies adjacent to the Irish Sea. By 2000, only 18 administrative areas continued to show an increase in breeding Kittiwakes, while 30 reported declines, so the influencing factors had become more widespread. Between 1985 and 2000, total numbers were estimated to have declined by more than 16 per cent.

There has been no national census of Kittiwakes breeding in Britain since 2000, but recent counts at many colonies show an overall decline in several regions. This has been particularly marked in the Northern Isles, where several colonies have declined by more than 70 per cent since the 2000 census, giving rise to much publicity and concern. Declines have also been reported in parts of the Scottish mainland, in Ireland and on the south coast of England. However,

TABLE 54. The estimated numbers of breeding pairs of Kittiwakes in Britain and Ireland since 1920.

Period	*England**	*Wales*	*Scotland*	*All Ireland*	*Total*
1920	Est. 7,000	—	—	—	Est. 50,000–60,000
1959	30,400	—	—	—	Est. 170,00–180,000
1969–70	50,596	6,891	346,097	44,383	448,000
1985–86 (revised)	84,700**	8,771	359,425	44,220	497,000
2000–04	77,329	7,293	282,213	49,160 (Eire 36,100)	416,000
2015	Est. 65,000	Est. 6,000	Est. 85,000	Est. 32,000 (Eire 22,000)	Est. 188,000

* Includes the Isle of Man and Channel Islands.
** The numbers reported for Bempton and Flamborough Head were 83,700 by Lloyd *et al.* (1991) and 85,095 by Mitchell *et al.* (2004). There is reason to suspect these figures represented the numbers of birds and not pairs, and they have been revised accordingly. The original data and the figures were taken from numbers reported by a third party, who could have misinterpreted the counters' figures. A count for this colony in 1970 was 30,800 pairs, and it would be unique for such a large colony to increase so rapidly to the claimed 85,095 pairs by 1985–86 and then decline markedly by 2000.

there are some areas on the east coast of Scotland and England where the species has shown only small declines since 2000, and some east coast colonies have remained unchanged or increased in size. Approximate estimates of the numbers of breeding pairs in Britain and Ireland are shown in Table 54. While the figures for 2015 include considerable potential error and a new national census is required for more accurate counts, it would appear that Kittiwake numbers in Britain and Ireland as a whole have reduced to levels similar to those in 1959.

The Joint Nature Conservation Committee website lists the changes since 1999 or 2000 at 25 Special Protection Areas (SPAs) where Kittiwakes breed, the majority of them in Scotland (Appendix 4; p. 446). Changes are reported as the percentage decrease annually to allow for the varied interval of eight to 15 years between counts. All but one of the 25 listed SPAs in Scotland showed a decrease, ranging from 15 per cent to 1 per cent per annum (Fig. 126a).

Counts since 2000 obtained from the Seabird Colony Register for 15 colonies on the east coast of Scotland showed that 12 (80 per cent) had declined. In

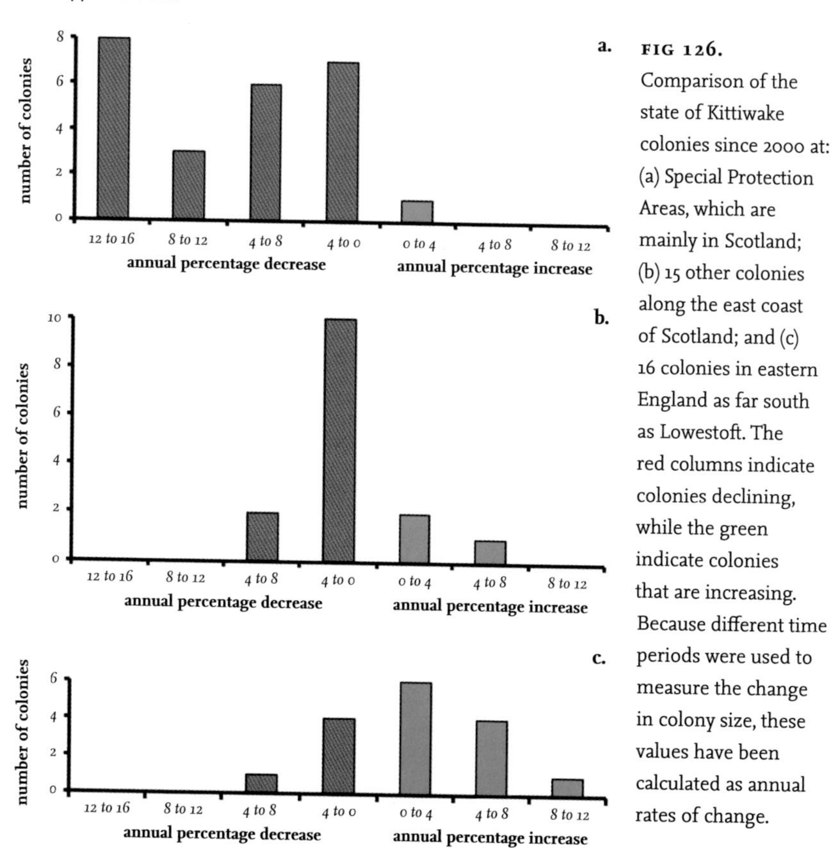

FIG 126. Comparison of the state of Kittiwake colonies since 2000 at: (a) Special Protection Areas, which are mainly in Scotland; (b) 15 other colonies along the east coast of Scotland; and (c) 16 colonies in eastern England as far south as Lowestoft. The red columns indicate colonies declining, while the green indicate colonies that are increasing. Because different time periods were used to measure the change in colony size, these values have been calculated as annual rates of change.

contrast, counts at 16 colonies along the east coast of England revealed that only 31 per cent were in decline and 69 per cent were growing (Fig. 126c).

There are considerable differences in the rates of change between some colonies that are situated close to each other. This is puzzling, since birds from nearby colonies often overlap in feeding areas, which may extend 50 km or more from land (p. 290). A hint of the underlying cause became apparent when the annual rate of change in size for each colony on the east coast of England between 2000 and 2015 was correlated with the colony sizes in 2000 (Fig. 127). This showed that there is a highly significant correlation between colony size and rate of change, with small colonies still increasing in size while larger ones tended to decline. This relationship is interesting, because it suggests that small colonies may be more attractive to recruits than large ones, perhaps because there is less competition for sites. This is also supported by earlier evidence that, in the

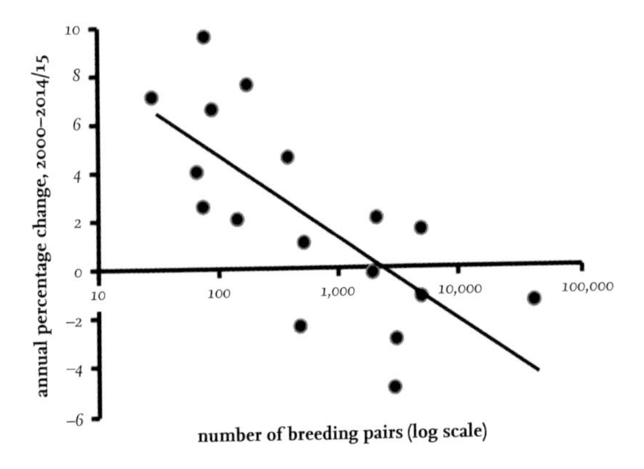

FIG 127. The change in size of Kittiwake colonies in east and south-east England between 2000 and 2014–15, plotted against the size of the colony in 2000 on a logarithmic scale. Large colonies have tended to decline, while smaller ones increased in size over the same period. This relationship with colony size explains much (85 per cent) of the variation in the rates of change in the size of colonies.

1960s, small Kittiwake colonies grew at a faster rate than large ones (Fig. 128), as evidenced by the growth over time of individual colonies.

In the Irish Republic, an almost complete census was carried out in 2015. This showed a 32 per cent decline in Kittiwakes over the 15-year period since the 2000 census and affected colonies around most of the country except for the south-west sector, where numbers have remained much the same. The largest declines occurred in the south and west of Ireland, but despite this, several new colonies have formed there. In Northern Ireland, both increases and decreases have been reported for individual colonies, but the total numbers do not appear to have declined below the 1985 level.

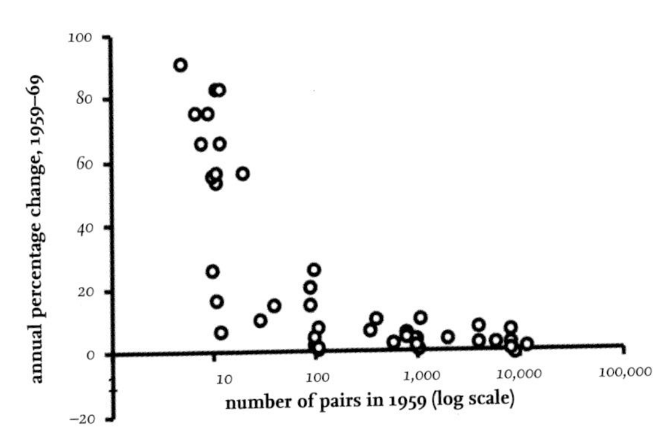

FIG 128. The annual rate of change in Kittiwake colonies in relation to their size over the period 1959–69. Almost all colonies with fewer than 100 nests grew at an appreciably higher rate than those with more than 500 nests. Based on the results from the 1959 and 1969 national censuses of British Kittiwake colonies.

The variation in extent to which numbers of occupied Kittiwake nests have changed between different colonies suggests that the decline is driven by food shortage during the breeding season (p. 446). This is confirmed by very poor breeding success in many colonies, with fledging appreciably lower than the critical level of about 0.8 chicks per pair. The situation is complicated by new evidence that some experienced adults are missing one or more breeding seasons and so are not included in the nest census, but this would appear to be of relatively minor influence compared with other evidence of an appreciable decline in numbers.

Numbers in winter

In winter, and in December in particular, there are relatively few records of Kittiwakes around the British coastline. However, sightings have increased in recent years, with a few adults visiting fish docks and harbours. These are probably not local breeding birds, since those seen at the mouth of the river Tyne in north-east England over many years never included colour-ringed birds from nearby colonies, while several Norwegian and one French-ringed bird were identified there. The BTO Winter Atlas of Birds tends to exaggerate the abundance of Kittiwakes around Britain in winter, because local breeding birds often start returning to inshore waters and the vicinity of their breeding colonies in January and February, and so may be included in the winter records. In the North Sea in December, Kittiwakes are at a relatively low density of about three birds per square kilometre, and a figure of 255,261 adult Kittiwakes present in the entire North Sea in December 2009 can be little more than a good guess about birds from unknown breeding areas.

BREEDING

The breeding biology of the Kittiwake has been comprehensively studied, perhaps more so than any other gull, as a result of long-term investigations of colour-ringed adults at Middleton Island in the Pacific Ocean, in France and at several sites in Britain. Much of the following information has been derived from these studies.

Annual reoccupation of the colony

In the 1950s and 1960s, some Kittiwake colonies in Britain were reoccupied by adults in early January, and in one year the birds were already back on nest sites on Christmas Day. Since then, the return time to colonies each year has become progressively later, and it is now often early March before the adults arrive. The reason for this later return is not yet known, but it is possible that feeding conditions in the Atlantic have changed and that this has had the effect of delaying the return.

The numerous marine molluscs that live on the surface of oceanic waters are an important winter food source for Kittiwakes, and these have reduced in number in recent years, apparently caused by a slight increase in the acidity of the seawater, which in turn has increased the rate of erosion of the calcium carbonate in their shells. While this change has not increased the winter mortality rate of Kittiwakes, it may have retarded the build-up of their body condition prior to their return to the colonies. The timing of the return to colonies in the High Arctic is much later than in Britain, being delayed until the end of April or even early May. Presumably, these birds remain longer in the Atlantic before returning to their colonies.

The annual reoccupation of Kittiwake colonies is always a gradual process, spread over two months within an individual colony. At first, the colony is occupied for only one or two hours in the morning by older birds. When these leave the colony each day, they fly directly out to sea, continue beyond the horizon and spend the rest of the day out of sight of land, presumably feeding well offshore. As the days go by, the colonies are occupied for longer periods of time each day and the number of birds present slowly increases.

Young birds about to breed for the first time arrive at the colony several weeks after the older birds, and in some cases a few adults have already built nests and laid eggs by the time they turn up. This behaviour means that the older birds have first choice of nesting sites, and each bird changing its mate is likely to pair with an equally old individual. When arriving late, first-time breeders must find unoccupied sites within the colony. These are usually of poor quality, at the edge of the colony or where breeding adults in the previous year have since died, leaving a site unoccupied. Other young Kittiwakes, aged two or three years old, that have not yet bred also visit the colony. They do not occupy nesting sites until late in the season, when most of the breeding birds have large chicks. These birds may pair and even engage in low-intensity nest-building, but mating and egg-laying do not occur (Fig. 129).

In Britain, the last Kittiwake chicks fledge in the first week of August, which in many years coincides with adults deserting the colony. The adults start their departure with 'panic flights', which see all of them synchronously pouring off the cliff and flying out to sea. The birds often return to their nests soon after, only to repeat the panic flight a few minutes later. Eventually, the adults fail to return, and the colony remains empty until the following year.

FIG 129. A young adult Kittiwake landing with nesting material in July, when most nests contain half-grown chicks. Note the long black edge to the outermost primary, which indicates that the bird is probably two years old. (Mike Osborne)

These panic flights are similar to those that occur when a large predatory bird appears near the colony, but there is no sign of predators at the end of the breeding season. Panic flights also occur when the colony is first reoccupied in the spring, but these soon cease as more Kittiwakes arrive and individuals gain confidence in visiting the cliffs for the first time in several months.

From the time that Kittiwakes first visit the colony in the spring and until the first eggs are laid, the adults vacate the colony overnight. At first, they leave early in the day, usually by midday, but the departure becomes progressively later, until by the end of April adults are staying in the colony until several hours after sunset and then arrive back before first light in the morning. Presumably, departing from the colony in the dark is not a hazard because there are no major obstructions when flying over the open sea.

As numbers increase at the colony in spring, up to a quarter of the birds at any one time are in pairs. This is the time of maximum noise, with the returning member of the pair, and then both, engage in mutual 'kittiwaaking', which often spreads to birds on neighbouring nests (Fig. 130).

FIG 130. Kittiwake returning to its mate (top right), with all the neighbouring adults joining in the greeting ceremony and calling with open bills. (Mike Osborne)

Age at first breeding

Some Kittiwakes start to visit breeding colonies when two years old, most of them arriving late in the breeding season when many breeding birds already have eggs or young (Fig. 129). Visits are usually restricted to fine summer days, and the prospectors are absent on days with strong winds and rain. Third-year Kittiwakes begin visiting the colonies earlier in the year, but still later than established breeders. Some form pairs and breed, but others do not breed until they are four or five years old. Based on a sample of 124 males of known age, the average age at first breeding was 4.2 years, while 50 females breeding for the first time had an average age of 4.0 years. Thus, many bred for the first time a year later than might have been expected, based on when they acquire full adult plumage. Most young birds pair with a partner of similar age, but there have been cases of old females that had already bred for 16 years pairing with young males that had not bred previously. Unlike younger females, these older females retained their previous nest site. Since males of all ages with previous breeding experience regularly do this and are unwilling to move, the older females were able to attract only a young male breeding for the first time as a partner.

Adults skipping breeding years

Evidence of adult Kittiwakes with previous breeding experience missing a breeding season and not pairing or building a nest comes from three sources.

1. Study of marked birds

In my study of colour-ringed adults at the North Shields, Tyne and Wear colony from 1954 to 1990, some adult Kittiwakes that had bred in a previous year then missed a breeding season, although they survived and did breed in a later year. Some individuals that skipped breeding for a year visited the colony but were not seen regularly and many did not form pairs, while others were not seen in the colony at all that year. Kittiwakes that missed a year included both sexes, although the behaviour was twice as frequent among females than males, involving about 10 per cent of females and 5 per cent of males in any year. It occurred every year, but declined from about 8 per cent of adults each year from 1954 to 1960 to 4 per cent from 1981–90 and without individual years with a markedly higher proportion missing breeding. The cause of this behaviour remains unknown, although in a minority of cases it was associated with a failed breeding attempt in the previous year or the loss of the mate from the previous year during the winter. It was not caused by a shortage of potential mates.

2. An exceptional year with major shortage of females

An exceptional situation occurred at the Marsden colony in Tyne and Wear in 1997, when during the chick rearing season large numbers of the adult Kittiwakes died as a result of an algal neurotoxin and about 70 per cent of those killed were adult females. In the following year, 1998, there was an obvious shortage of adult females in the colony, with about 35 per cent of adult males occupying sites but failing to obtain partners or building nests. The sex ratio was restored in 1999 and later years, with most adults breeding, so this event of high non-breed was restricted to a single year.

3. Evidence of recent and high non-breeding at colonies in the Firth of Forth

At eight of the nine Kittiwake colonies in the Firth of Forth, south-east Scotland, the number of nests were recorded annually between 1986 and 2016 (the only colony excluded was at Dunbar, which was counted from 2000 to 2007). The data gathered from these counts permit a detailed analysis of numerical changes.

The total of pairs at the eight colonies (excluding Dunbar) shows that numbers increased from 1986 to 1989 (and for many years previously), but that a progressive decline began in 1989 and continued for the next 28 years, with a decrease of 5 per cent per year over the whole period (Fig. 131). The rate of decline of individual colonies from 2000 to 2016 shows considerable differences (Table 55) and totals for individual years showing an overall large decline were 1994 (–24 per cent), 1998 (–34 per cent) and 2013 (–27 per cent), but within the next two years there were increases of 27.5 per cent (in 1996), 13.0 per cent (in 1998) and 49.5 per cent

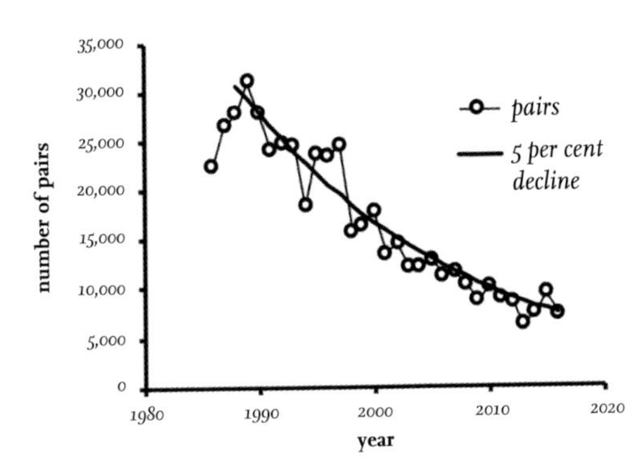

FIG 131. The numbers of pairs of Kittiwakes nesting in the Firth of Forth area, Scotland (excluding Dunbar), between 1986 and 2016. The fitted curve indicates a decline of 5 per cent each year. The point for 1998 includes estimated numbers at two colonies. The decline from 1990 to 2016 was gradual and spread over most of the time. Based on data in the Seabird Colony Register.

(by 2015) (Appendix 3; p. 444). The recoveries from the declines were appreciable, particularly in 2014 when the decline in the previous year was immediately followed by a substantial recovery in all eight colonies. There are several possible causes of the declines. They could have been caused by years of exceptionally high adult mortality (perhaps in the order of 40 per cent in one year), but the study of marked adults on Skomer and the Isle of May at this time did not indicate that there had been a large loss of marked adults in the previous 12 months.

There is no good reason to believe that recruitment suddenly became very much higher and caused the marked recovery in the following two years. The most likely reason is that significant numbers of adults survived but did not breed or even build nests in the years of large declines. If this is a correct interpretation, then the extent of the declines and recoveries in numbers suggest that up to 20 per cent of adult Kittiwakes failed to breed in the years with appreciable declines in nest counts, but they survived and returned within the next two years, built nests and bred. It seems likely that in recent years, the extent of non-breeding among adult Kittiwakes may have has become more frequent.

However, the rate of decrease at individual colonies has differed. Table 55 shows the change between 2000 and

TABLE 55. The percentage changes between 2000 and 2016 in the numbers of Kittiwake pairs nesting at eight colonies in the Firth of Forth, and between 2000 and 2007 at Dunbar.

Colony	*Change in number of pairs*
Inchkeith	–6%
Craigleith	–13%
The Lamb	–23%
Fidra	–24%
Isle of May	–34%
Bass Rock	–51%
Inchcolm	–57%
St Abb's Head	–75%
Dunbar	0%

2016 in the eight colonies in the Firth of Forth, from which it is evident that the extent of the decrease varied markedly between colonies and therefore that these changes could not all be explained by non-breeding. One influence may have been competition for sites by Common Guillemots; at several of the colonies, areas previously used by breeding Kittiwakes had been taken over by the auks – this was particularly obvious at St Abb's Head (p. 270). On the Bass Rock, competition for some nesting sites may have occurred between Kittiwakes and rapidly increasing numbers of Northern Gannets (*Morus bassanus*).

In passing, it is worth noting that a marked decline in numbers of breeding pairs of Kittiwakes also occurred extensively in 2013 (when compared with 2012) at many other colonies in eastern Scotland, eastern England and Wales. This begs the question: was there extensive non-breeding by Kittiwakes in 2013 and has it become much more frequent during this century?

Years with a high proportion of non-breeding adult Kittiwakes, as suspected for 2013, would have produced significant underestimates if a national census had been carried out in that year. While non-breeding behaviour would not account for the large overall declines in Kittiwake numbers, it could appreciably affect the rates of decline between pairs of census years. Between 2000 and 2013, apparently occupied Kittiwake nests in the eight Firth of Forth colonies declined by 65 per cent, while a comparison between 2000 and 2015 numbers resulted in only an overall 48 per cent decline. Both rates of decline are concerning, but there is an indication that non-breeding behaviour by adults may have become more frequent and more variable than in the last century.

Courtship feeding and mating

Some biologists consider that courtship feeding of the female by the male plays an important part in mate selection in gulls and terns. In the case of the

FIG 132. Kittiwake pair. The bird on the left is the female, which is starting to beg from the male. (Mike Osborne)

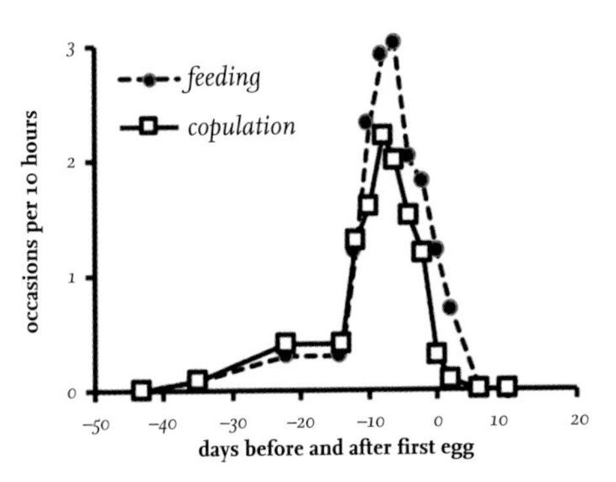

FIG 133. The rates of courtship feeding and copulation in relation to the laying of the first egg in the Black-legged Kittiwake. Based on data collected by John Chardine.

Kittiwake, however, the pair bond is usually firmly developed several weeks before food begging and courtship feeding begin (Fig. 132).

Occasional feeding of the female takes place a month before egg-laying. The frequency of this behaviour increases rapidly and usually leads to copulation about two weeks before laying, reaching a peak at seven to 10 days before the first egg is laid, after which it declines rapidly (Fig. 133). The frequency of courtship feeding and copulation ceases suddenly once the first egg is laid by the female.

Duration of the pair bond

There is a popular, but often incorrect, belief that many species of birds, including Kittiwakes, mate for life. One pair that was studied bred together for 15 years, but this was exceptional, as was their breeding success. During the 15 years, they retained the same nest site, the female laid 29 eggs and the pair fledged 28 chicks. Most Kittiwakes are fortunate if, during their lifetimes, they fledge five chicks. Retention of the same mate usually lasts for two or three years, and on average 34 per cent of females change mates between one breeding season and the next. The chances of retaining the same mate is improved if the pair bred successfully there in the previous year. Young breeders are much more likely to fail in their first breeding attempt, and, as a result, most young birds change their mate when breeding for the second year. About 18 per cent of Kittiwakes must change mates each year because their partner died in the previous 12 months. The remaining 82 per cent of individuals have the opportunity of retaining the same mate, but a quarter of these 'divorced' and paired with a different partner.

Most deaths of adults in study colonies occurred during winter, when Kittiwakes are pelagic, and there is no evidence that pairs attempt to remain together during this period. The use of loggers attached to both members of the pair should produce more detailed evidence of how far apart they are during the winter (p. 364).

Adaptations to cliff nesting in Kittiwakes

Kittiwakes show adaptations to nesting on steep cliffs, as highlighted by Ester Cullen in a classic study published in 1957 and summarised below with a few additions.

1. Fighting with other Kittiwakes occurs only at the nest and in an unusual way. The aggressive individual grasps the beak of the opponent and twists, which often causes the birds to tumble from the nest and spiral downwards, frequently nearly reaching the ground or sea before they release their grip. They then both fly back to the nest site, where the battle often recommences. The advantage usually lies with the bird that manages to gain height and return to the nest site first.
2. The young do not run from the nest as other gulls do when threatened or disturbed, but face inwards and crouch against the cliff face, often hiding their heads. The black neck-band in feathered chicks appears to have an effect of reducing aggression against them.
3. Unpaired males land on a potential nest site and advertise for a female partner using a 'chocking' display. Pairs form only at the nest site.
4. During copulation, females crouch, resting on their tarsi and abdomen (in other gull species, the female stands during copulation). The Kittiwake's habit is presumably an adaptation to reduce the buffeting effects of the strong winds on exposed cliff faces.
5. Nest material is often collected socially from sites such as cliff tops and nearby wet areas that are not visited otherwise (Fig. 134).

FIG 134. Adult Kittiwake collecting nesting material. The legs and bill are discoloured as a result of having recently stood in and collected soft mud. (Mike Osborne)

6. Only the nest or immediate nest site are guarded and defended against other Kittiwakes by the presence of one or both adults during most of the daylight hours. Areas beyond the nest site are not defended, which allows pairs to nest close together and even with their nests touching (Fig. 135).
7. Nest-building is more elaborate than in other gull species, involving the use of mud and algae to attach the base of the nest firmly to the rock surface. This is assisted by prolonged trampling of the nest material to compact it. Grass is used to form the upper part of the nest.
8. Typical nests have a deep cup formed by the female to retain the eggs. In the infrequent event of an egg falling out of the nest, it is not retrieved (in other gull species, an adult will sometimes roll a displaced egg back into the nest).
9. The young take food from the adult's throat rather than the adult regurgitating food onto the nest for the chicks to pick up and swallow, as occurs in many other gull species.

FIG 135. Kittiwakes nesting on the long ledge on the Baltic Flour Mill at Gateshead. Note that their tails all point to the left, as all the birds face into the 'cliff' face. (Becky Coulson)

continued

Adaptations to cliff nesting in Kittiwakes *continued*

10. The young remain on the nest until capable of prolonged flight, resulting in a long fledging period that lasts for an average of 42 days.
11. After each chick hatches, the eggshell is ejected by an adult over the edge of the nest.
12. Parents do not recognise their own chicks and will adopt strange chicks, although this rarely happens on cliff-nesting sites unless the nests are side by side on the same ledge.
13. Chicks (and adults; Fig. 135) standing on the nest face inward or sideways and actively avoid facing out over the cliff. Adults face outwards only immediately before flying away from the nest. This reduces the risk of a chick accidentally being pushed off the nest and falling into the sea or the ground below the nest, which would result in certain death.
14. Both the young and adults have strong toe muscles and markedly curved claws to aid gripping the nest when a returning adult lands on the very restricted nesting area. They also assist the bird in retaining its position in strong winds or when there is an aggressive attempt by another adult to take over the site.
15. The alarm call is given only when predators (including humans) are very close to the nest. Adults usually remain on the nest, covering the eggs or unfledged young, even when a predator is near.
16. Droppings are shed onto or outwards over the edge of the nest, often producing long white marks on the side of the nest and cliff face (Fig. 136).
17. The young are not cryptically marked, either when covered in down or when feathered.
18. The usual clutch size is two eggs, which is fewer than in most gull species. However, three young and two adults on a nest would seem overcrowded.
19. The young are fed only at the nest. When chicks finally leave, they are not accompanied by their parents and they soon go alone out to the open sea.

FIG 136. Kittiwakes nesting on cliffs on Brownsman in the Farne Islands. (Mike Osborne)

Nest site fidelity

Many adult Kittiwakes return to the same nest site in successive years, although the frequency of this varies with age and is appreciably higher in males than females. In the case of females, changing site is often associated with taking a different mate. Most males, including those taking a new mate, show a strong tendency to retain the same nest site as used in the previous year (Table 56).

Effect of mate change

Retaining the same mate in successive years has an associated beneficial effect on the breeding success of Kittiwakes, provided they are breeding for more than the second time. Older females that retain their mate from the previous year lay earlier, produce larger clutches and then fledge more young than comparable pairs that breed together for the first time (Table 57). One effect is that pairs that remain the same as in the previous year shorten the time they spend on preliminary courtship, allowing females to lay earlier.

It is obvious that Kittiwakes can recognise their mates, as the return of a bird to the nest site from a feeding trip frequently produces a reaction from its partner even when it is still about 20 m away. How this recognition is achieved is not clear, but a returning bird usually follows the same approach route to the nest, which may contribute to the early recognition of what is a head-on view. In contrast to mate recognition, parents do not recognise their own chicks.

TABLE 56. The percentage of male and female Kittiwakes of different ages (and hence different breeding experience) returning to breed at the same nesting site as that used in the previous year. Based on 3,565 records at the North Shields colony (data collected by Coulson between 1953 and 1990; note that, unless otherwise stated, this source is used for information relating to the North Shields colony throughout the chapter).

Previous breeding experience (years)	*Percentage of males returning to the same nest site*	*Percentage of females returning to the same nest site*
1	72%	56%
2	74%	61%
3	76%	59%
4 and 5	79%	65%
6 and 7	82%	70%
8 to 10	84%	75%
More than 10	96%	87%

TABLE 57. The average benefit to Kittiwakes of retaining the same mate in the next breeding season compared with those taking a new mate. Keeping the same mate was no benefit for females breeding for the second time, but in older females it increased the numbers of young fledged.

	Benefit for young birds breeding for second time with the same mate	*Benefit to older breeders by retaining the same mate*
Date of laying	No benefit	Three days earlier
Clutch size	No benefit, 2.01 eggs	Increase, from 2.06 to 2.18 eggs
Breeding success	No benefit, 60%	Slight increase, from 62% to 69%
Young fledged per pair	No benefit, 1.21 young	Increase, from 1.28 to 1.50 young

It might be expected that the sexes land at the nest with similar frequency in the pre-breeding period, but this is not so. While both members of the pair make a similar number of feeding trips, males make additional brief departures for a few minutes, before returning to the nest site; the pair then engage in mutual calling on each return, in much the same way as they do when either bird returns to its mate from a feeding trip. In the pre-laying period, males have been recorded arriving 8.1 times per 10 hours, whereas females arrived at the nest only 2.6 times per 10 hours. Thus, males made three times more arrivals than females and two-thirds of these took place after only brief departures. There was also considerable variation in landing and departure among males, with those keeping the same mate as in the previous year making 2.3 times the number of departures and arrivals as the female, while males that changed to a new mate made 4.1 times more arrivals at the nest than the new female partner and so engaged in even more mutual displays (Table 58). These brief departures, which involve flying around the colony, increase the social behaviour of the focus pair and neighbouring pairs, and it is assumed that they play a major role in the behaviour that brings the female into laying condition.

TABLE 58. The number of arrivals per 10 hours at the nest in the two weeks prior to laying the first egg of a clutch of male and female Kittiwakes that either retained or changed their mate from the previous year. Based on nest recordings by John Coulson and direct observations by John Chardine at the North Shields colony.

Category	*Number of arrivals per 10-hour period*
Males, same mate ($n = 58$)	5.9
Males, changed mate ($n = 41$)	10.9
Same/change ratio	0.54
Females, same mate ($n = 58$)	2.6
Females, changed mate ($n = 41$)	2.7
Same/change ratio	0.96

Date of laying and clutch size

The time of breeding in a Kittiwake colony is appreciably influenced by how early it is reoccupied; the start of egg-laying is about one day later for every four days reoccupation is delayed. This effect is also evident in the time of return in different years of individual females. As a result of their delayed arrival, Kittiwakes in Britain are currently laying about a week later than in the 1950–70 period and the average clutch size has declined from 2.0 eggs to about 1.8 eggs, with single-egg clutches now more frequent than three-egg clutches. The clutch size laid by each female in successive years is also influenced by the calendar date of her return, so that in years when a female returns late, her clutch size tends to be smaller.

Two-egg clutches are laid by 70–75 per cent of Kittiwake pairs. Clutches of three eggs are almost invariably laid by early-nesting Kittiwakes, while single-egg clutches are often laid by young birds, which tend to lay later than older adults. However, the first clutches laid at the colony at North Shields in each of 45 successive years was always two eggs, although this was frequently followed at other nests by three-egg clutches. Four-egg clutches are extremely rare, and those that have been recorded were laid by two females forming a pair, with or without a male partner.

Incubation

In a study of marked Kittiwakes at one colony with a high production of young over many years, it was found that egg incubation was less successful than survival of chicks to fledging. This is unusual in other birds and in the Kittiwakes studied was caused by the lack of predation on the chicks and sufficient supplies of food. Incubation lasts about 27 days and usually starts with the laying of the second egg in clutches greater than one egg. Frequently, the second chick hatches at the same time or soon after the first, but the third chick in clutches of three eggs often hatches one to two days later than its siblings, by which time they are distinctly larger. The survival of the third chick is appreciably lower than for the earlier two, and well-grown broods of three chicks are uncommon. If the third chick to hatch survives, its rate of growth is invariably lower than that of its older siblings.

Successful incubation requires careful organisation by members of the pair to keep the eggs covered and incubated. Typically, a successful pair keep the eggs covered for 98 per cent of the time. One study at North Shields used automatic recorders to identify the bird at the nest throughout the incubation period, allowing complete records of incubation by several pairs to be obtained and hence a detailed analysis of the shift system used. Males and females shared incubation equally and individual pairs developed incubation shifts of different lengths. Some pairs had two change-overs each day, others had three and yet more used a system of four shifts per day. Either sex incubated overnight.

The end of an incubating shift was marked by the return of the off-duty partner from feeding; the incubating bird remained on the eggs until the partner arrived and they soon changed over. In one case, the off-duty male was away for more than 48 hours and the female continued to incubate without interruption for the entire period. Similarly, when a male partner died suddenly after being caught in a fishing line while away from the colony, the surviving bird continued incubating without interruption for almost three days, and then left the eggs for less than an hour (presumably long enough for her to drink, but not long enough to feed). She then returned to continue incubating without interruption for a further 24 hours, at which point she finally deserted the eggs and nest.

Each incubation stint lasted several hours and was distinctly longer for the bird present overnight (Fig. 137). The length of the shift decreased progressively for several days at the start of incubation and then stabilised at a low level (Fig. 138). About 30 seconds before it leaves the nest, the departing bird indicates to its mate with a *wak-wak* call that it is about to depart for some time. This call is not used when an adult leaves the nest for only a few minutes. An examination of the trips made by adult males and females (all sexed from mating) showed no meaningful differences (Table 59), although the females made only 42 per cent of the overnight trips compared to 49 per cent of the daytime trips. The time

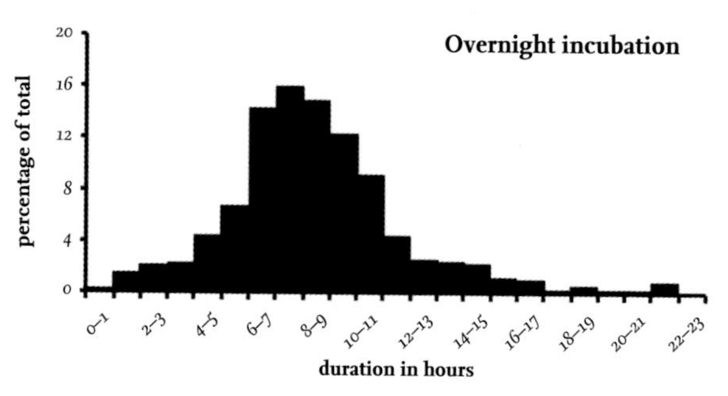

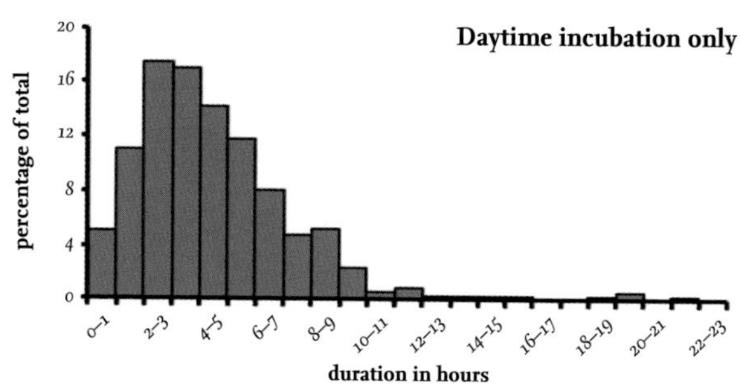

FIG 137. Above: the duration of incubation shifts by Kittiwakes during the day that involved a period in the dark at night. Below: when incubation shift is completed during daylight. Note that those periods that involved overnight incubation lasted about twice as long as the daytime ones. Based on data from the North Shields colony.

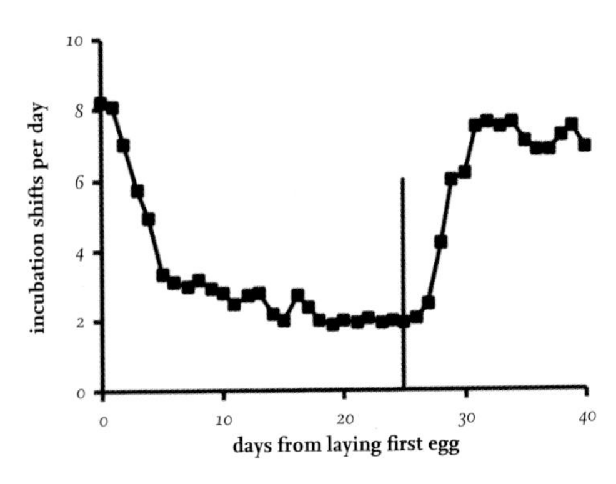

FIG 138. The average number of shifts per day during incubation based on five Kittiwake pairs that were continuously monitored by an automatic recorder at the North Shields colony. The vertical line indicates when the eggs hatched; note that this is followed by a rapid increase in the number of shifts undertaken each day by parents brooding and feeding the chicks.

Kittiwakes depart on feeding trips remained similar between 04.00 and 22.00, with a slight dip in the middle of the day, but very few departures occurred overnight, between 22.00 and 04.00 (Fig. 139).

The duration of feeding trips by Kittiwakes while they have chicks in the nest varies according to the time of day they depart from the colony, being minimal following a morning departure and twice as long when departure occurs during the dark (Fig. 140). For most of the time the chicks are in the nest, one parent remains with them while the other is away feeding, but large chicks are left alone from time to time, with both parents presumably feeding at the same time. However, periods when both parents are away are much more frequent in the evening and overnight, and least during the later morning and afternoon (Fig. 141). This may indicate a pattern whereby the parents tend to remain with the chicks during daylight, when potential predation is more likely.

TABLE 59. A comparison of the duration of feeding trips made by male and female Kittiwakes while feeding young in the nest. Distance travelled was estimated using a flight speed of 42 kph.

	Sex	*Number of trips*	*Average duration (minutes)*	*Most frequent*	*Maximum distance travelled*
Daytime trips	Males	2,216	4 hours 30 minutes	2–4 hours	190 km
	Females	2,132	4 hours 32 minutes	2–4 hours	190 km
Overnight trips	Males	533	9 hours 19 minutes	7–9 hours	390 km
	Females	394	8 hours 50 minutes	7–9 hours	370 km

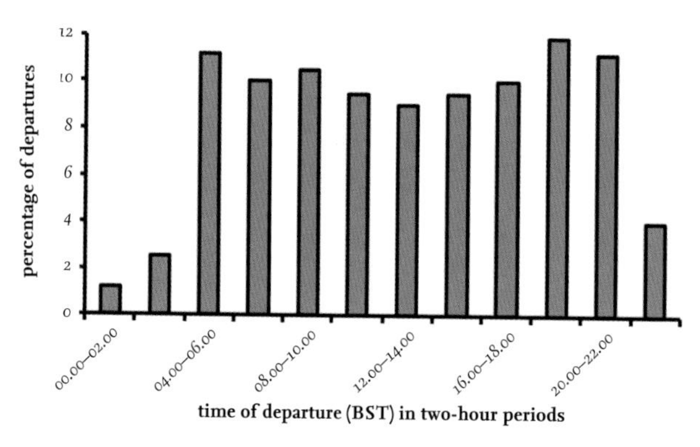

FIG 139. The distribution of the start time of feeding trips by Kittiwakes during incubation. Based on 5,255 departures automatically recorded from the North Shields, Tyne and Wear colony.

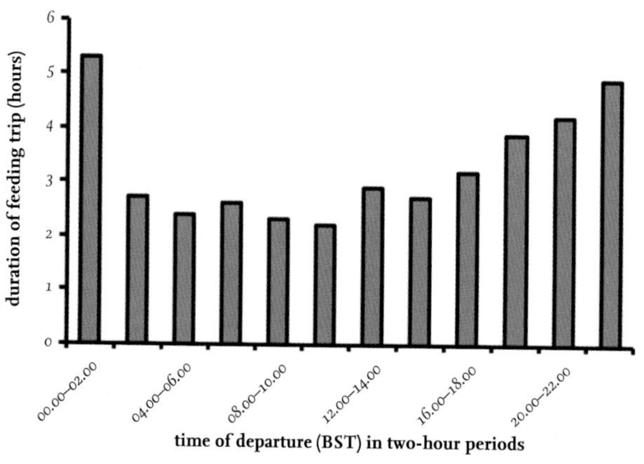

FIG 140. The mean duration of feeding trips by Kittiwakes breeding at North Shields in relation to the time of departure.

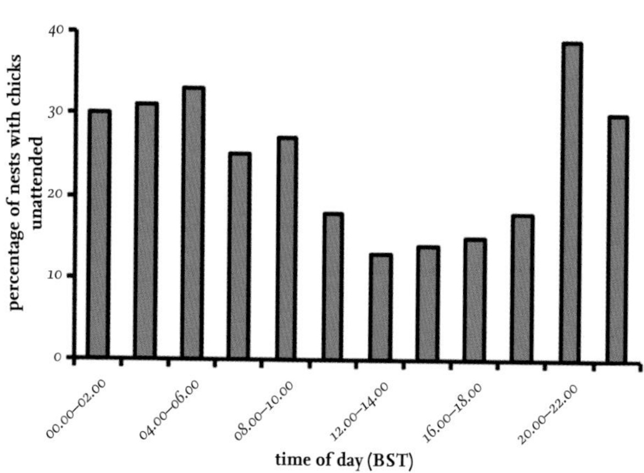

FIG 141. The percentage of Kittiwake nests containing chicks but without an adult present at different times of the day. The sample considered only nests when the chick(s) had already been left alone on a previous occasion. Based on data from the North Shields colony.

Eggs

Black-legged Kittiwake eggs have a paler background colour than those of most other gulls, and are marked with blotches of brown and black. They often become darker as incubation proceeds, picking up mud stains from the base of the nest (Fig. 142). In gulls, egg volume is a good indication of the size of the yolk. While larger eggs also have larger yolks, yolk size increases less rapidly than the volume, and so forms a slightly lower proportion of the total content. While the yolk is the most important factor in the development and size of the chick at hatching, most studies of the importance of egg size have been based on egg volume, calculated from the length and breadth of each egg, because this measurement is not destructive and allows the effect of egg size on the survival of eggs and chicks to be investigated.

As in other gulls, the third egg in a Kittiwake three-egg clutch is smaller and narrower, with a volume about 5 per cent less than either of the other two eggs, which are usually similar in size (Fig. 143). The third egg becomes progressively smaller than the first two eggs when laid as the laying season advances, the difference in size being greater (about 6 per cent smaller) in clutches that are late laid (Fig. 144). The second egg laid in a two-egg clutch is consistently smaller than the first by about 5 per cent whenever laid. Single-egg clutches are frequently laid by young females and on average the egg is 10 per cent smaller than either egg in clutches of two, and even smaller than the third egg in clutches of three (Fig. 144).

In general, the average volume of each egg laid by Kittiwakes becomes gradually larger as the female increases in age. Older females lay eggs that, on average, are broader, but shorter than those laid by younger females and

FIG 142. Kittiwake clutch of two eggs. Note the white droppings around the nest edge. (John Coulson)

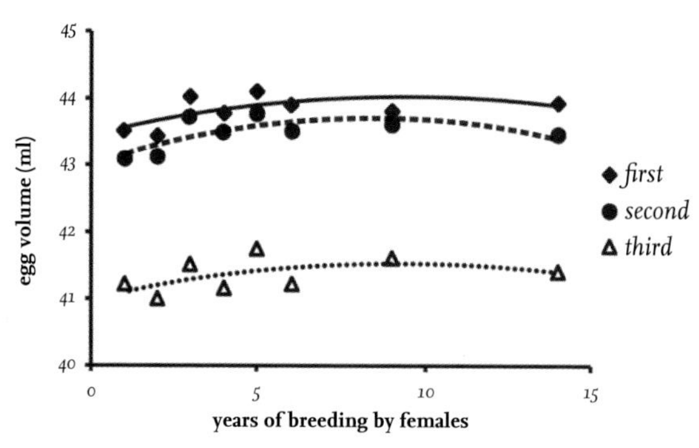

FIG 143. The volume of the first, second and third eggs in clutches of three eggs laid by Kittiwakes of different breeding experience (age). The sizes of the first two eggs in the three-egg clutch are more similar than in clutches of two eggs, while the third egg is consistently smaller. Based on data from the North Shields colony.

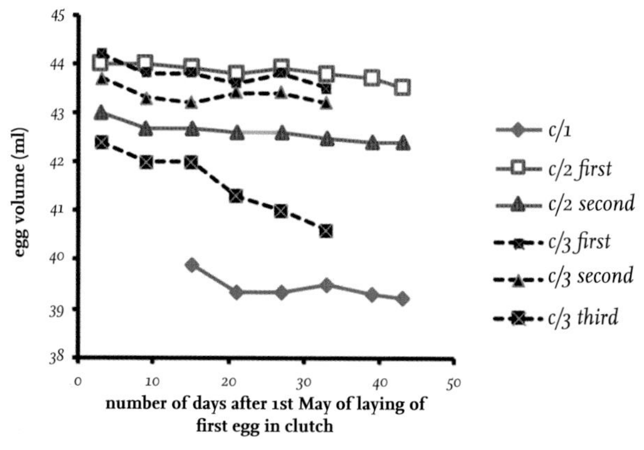

FIG 144. The change in the average volume of a Kittiwake egg in clutches of one, the first and second eggs in clutches of two, and the first, second and third eggs in clutches of three, in relation to the date on which the first egg was laid. Based on data from the North Shields colony.

have a distinctly rounder appearance. The size of eggs laid in two-egg clutches increases for the first three years of breeding by the female, reaching a plateau at about five years of breeding, and declines only slightly in older females (Fig. 145).

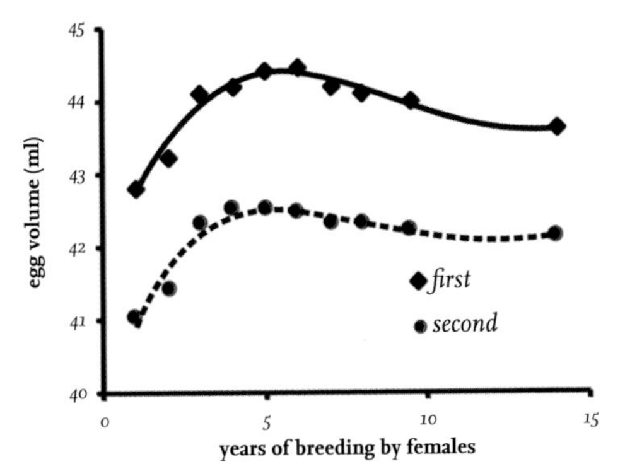

FIG 145. The change in volume of the first and second eggs in Kittiwake clutches of two eggs in relation to the breeding experience (age) of the female. Based on data from the North Shields colony.

Breeding performance in relation to age

A summary of the effect of age of Kittiwakes on their biology and, particularly, their breeding performance is shown in Table 60. Most of the differences relate to young birds making their first and second breeding attempts, and there is very

TABLE 60. The effects of age of female on breeding behaviour in Kittiwakes, based on records from the North Shields colony, 1960–90.

	First-year	*Second-year*	*Third-year*	*First breeding*	*Second breeding*	*10–14-year-old*	*15–20-year-old*
Annual period pelagic (months)	10	9	8	7	6	5	6
Arrival in colony	—	Jun	Apr	Apr	Mar	Feb	Feb
Laying date	—	—	Early Jun	Early Jun	Late May	Early May	Mid-May
Average clutch size	—	—	—	1.6	1.9	2.2	2.0
Proportion re-laying	—	—	—	0%	10%	25%	20%
Success (young fledged per pair)	—	—	—	0.4	1.0	1.4	1.3
Nest site retention	—	—	—	10%	35%	85%	88%
Mate retention	—	—	—	3%	35%	65%	60%
Departure from colony	—	Early Jul	End Jul	Early Aug	Early Aug	Sometimes Sep	Sometimes Sep
Annual mortality estimates	25%	20%	20%	20%	20%	20%	22%

little evidence of any appreciable deterioration in the oldest birds, which have bred for 15–20 years and which could be attributed to senility.

Breeding performance in relation to laying date

By far the commonest clutch size laid by Kittiwakes in Britain is two eggs, and in most colonies about three-quarters of clutches are of this size. However, the average clutch size decreases markedly as the laying date progresses from early May to the middle of June (Fig. 146).

The reason behind this phenomenon is not because there is less food available later in the season, an effect well known in woodland passerines, where the decrease in clutch size appears to be a selective modification to meet rapidly decreasing food supplies. In contrast, availability of fish remains similar during the whole period. Instead, it is possible that the change in clutch size in the Kittiwake is an indication of the condition of the individuals, with later-breeding females having a poorer body condition and therefore being unable to lay more eggs. Because the date of laying is closely correlated with the time of arrival at the colony in spring, it is possible that poorer-quality individuals return later and hence their laying is delayed. It also suggests that because most first-time breeders breed late, they are on average in poorer condition than their elders, a possibility also indicated by their overall poorer hatching and fledging success rates.

More than 91 per cent of all one-egg clutches in north-east England are laid during the second half of the laying season (after 17 May). Many of these clutches are laid by females breeding for the first time. It might be assumed that the smaller size of the egg laid in one-egg clutches could be attributed to inexperience or some physiological constraint imposed when females are developing eggs for the first time. However, this is clearly not the case because experienced females laying single-egg clutches also produce similarly small eggs.

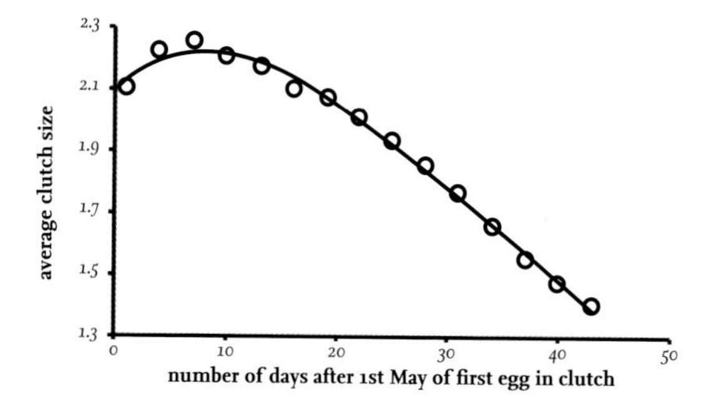

FIG 146. The relationship between the average clutch size laid and date in the North Shields Kittiwake colony. Based on 2,010 clutches laid over 25 years.

The declining size of the final egg in a clutch of three in relation to the date of laying is more difficult to explain; it is not an age response, since females of all ages and breeding experience lay similarly small third eggs, while a smaller size is not evident in the first and second eggs in clutches of three. Note, however, that the last egg of a clutch is reduced in size irrespective of whether it is in a one-, two- or three-egg clutch. The reduced size of these last eggs could be induced by the reduction of material being transferred to the ovaries as the body starts to reabsorb the additional eggs that have been developing there. Studies on several other bird species have also reported a reduced size of the final egg in the clutch, and so it is clearly a widespread effect.

Breeding performance in relation to nest site

The breeding performance of Kittiwakes is not uniform throughout a colony but is more successful at the centre than at the edge. This was first reported in *Nature* in 1968, based on data from the North Shields colony in Tyne and Wear (Coulson, 1968), and has since been found in several other colonial seabirds. Table 61 summarises further data obtained since the original study and confirms the effect. Individual adults breeding at the centre of the colony produced half as many chicks again in their lifetime. Because predation at the nest was negligible in the colony, this was not the cause of these differences. Furthermore, adult survival rates were also higher in birds nesting in the centre and at the edge of the colony, but this is not caused by mortality at the colony, which was negligible. Most adult mortality occurs in the winter months, when Kittiwakes are spread out over the North Atlantic, and is therefore unrelated to their nesting position in the colony. Instead, the higher survival rates at the centre suggest that these birds are better-quality individuals. This is supported by the fact that young potential breeders prefer centre sites to edge sites, and only the fittest individuals are able to cope successfully with the greater competition required to obtain and then retain centre sites. Once they have bred for the first time, extremely few adults move between centre and edge sites in the colony during their lifetimes.

Using the data in Table 61, it is possible to make an approximate estimate of the survival rate of immature Kittiwakes. With about five chicks fledged per pair in the lifetime of an adult and two chicks needed to survive to the average age of first breeding (four years) to replace their parents when they die, the survival to breeding age has to be about 40 per cent in order to maintain numbers. If the survival rate in the second to fourth years of life is the same as in adults (about 83 per cent per year), then the survival during the first year of life following fledging would have to be about 70 per cent to permit two chicks to survive to the

TABLE 61. The average benefit male and female Kittiwakes obtained by nesting at the centre rather than at the edge of the colony at North Shields, Tyne and Wear. Based on data for 1,951 pairs from 1953 to 1990.

	Annual survival rate	*Average number of breeding years*	*Clutch size*	*Breeding success*	*Young fledged per year*	*Young fledged in lifetime*	*Lifetime benefit of breeding in the centre*
Centre males	81%	4.8	2.14	55%	1.18	5.7	58% more fledged young
Edge males	76%	3.7	2.03	48%	0.98	3.6	
Centre females	84%	5.8	2.14	55%	1.18	6.8	45% more fledged young
Edge females	81%	4.8	2.03	48%	0.98	4.7	

age of four, the average age at first breeding. This is a remarkably high survival rate, particularly since the parents do not attend or feed the young birds once they leave the nest.

Breeding success

Just as clutch sizes laid by Kittiwakes decline as the season progresses (p. 266), so does the number of young fledged. In Fig. 147, the loss represented by the difference between the two lines is the result of eggs that failed to hatch and chicks that died. These losses are similar for birds laying during the first 15 days

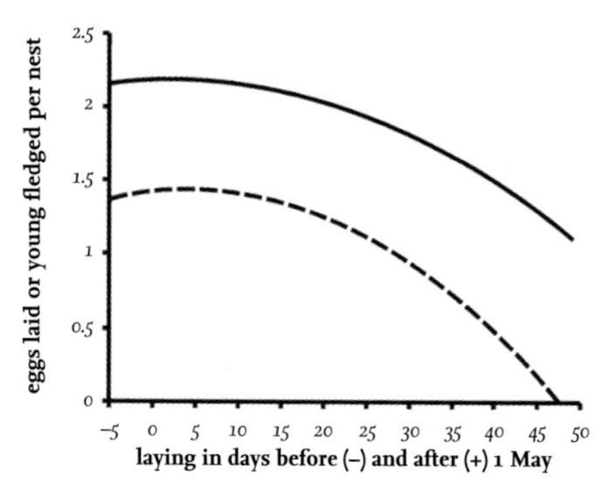

FIG 147. The effect of the date of laying in Kittiwakes on clutch size (continuous line) and numbers fledged per pair (dashed line). All eggs laid after day 43 (12 June) failed to produce young which fledged. Based on 1,881 clutches laid in 1954–90 at the North Shields colony.

of May, and then become proportionately much larger in late breeders. As a result, the few Kittiwakes that delayed laying until the second week of June produced the smallest clutches and almost always failed to fledge any chicks.

The decline in breeding numbers

The reduction in numbers of Kittiwakes breeding in Britain and Ireland is evident in the data in Appendix 4 (p. 444). This overall decline cannot be attributed to increased mortality in winter, when the species is pelagic and individuals from different colonies are mixed, otherwise there would be a more consistent reduction in numbers in colonies throughout the country and the contrasting trends in Shetland and Orkney, and in north-east England, would not occur. The evidence points to failure to breed successfully in certain areas and this, in turn, suggests a reduction in the abundance of sandeels (*Ammodytes* spp.), the birds' main food during the breeding season in Britain and Ireland.

There are clear indications that stocks of sandeels have declined in much of the North Sea. While Kittiwakes in some areas have switched during sandeel shortages to exploiting Sprats (*Sprattus sprattus*) and Herring (*Clupea harengus*) if available, the reduced breeding success in many areas is presumably because of the absence or difficulty in capturing these species. In some restricted areas, avian predation has resulted in the decline of breeding Kittiwakes. For example, in the Scilly Isles the decline is attributed to increased nest predation from Great Black-backed Gulls (*Larus marinus*), while on Foula in Shetland, similar predation by Great Skuas (*Stercorarius skua*) appears to have been a major cause of fewer young being reared, although this in turn may have been driven by food shortage for the predators. In parts of Norway, increased predation by White-tailed Eagles (*Haliaeetus albicilla*) on seabird colonies, including Kittiwakes, is a development causing concern.

I obtained a reliable figure for the productivity required to maintain numbers of adults from a study of 16 colonies in Britain from 1985 to 2015, where both the change in numbers of breeding pairs and data on annual productivity were available (Coulson, 2017). The correlation between change in colony size and average young produced over a 15-year period for those colonies is remarkably good (Fig. 148) and indicates that each pair needs to rear 0.8 chicks each year to maintain the size of the colony. This value is well below the requirement of 1.5 chicks fledged per pair reported in an earlier study (Cook & Robinson, 2010), which is so high that it had been rarely observed in the field, even when Kittiwake numbers were increasing year after year.

The average production of young required to maintain adult numbers in a colony will change if the average adult mortality rate among the breeding birds alters, with more young required if the rate increases. However, there is

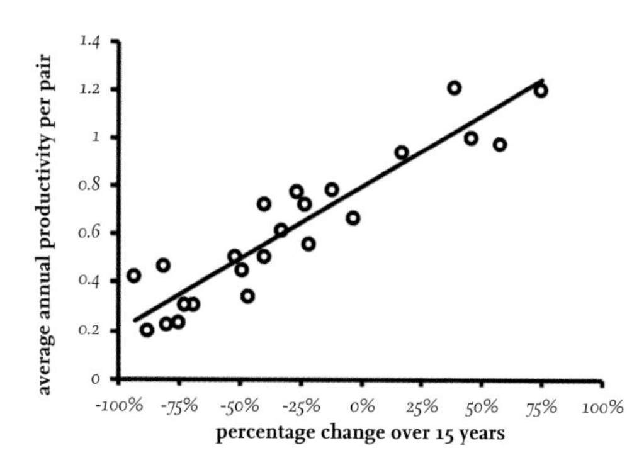

FIG 148. The relationship between change in colony size and the annual production of fledged young at a series of Kittiwake colonies in Britain in the period 1985–99 (or 2000) and 2000–15. The best-fit line crosses the no change value at 0.80 young fledged per pair per year. Based on Coulson (2017).

no evidence to suggest that the adult mortality rate has changed appreciably in Britain in the period 1985–2015. It is informative here to examine the situation between production of fledged young and the change in colony size in several colonies where there is detailed information of these two parameters.

St Abb's Head, Berwickshire, Scotland

There was a sudden change at this colony in 1990, from an increase of 7 per cent per annum to a decline of 6 per cent (Fig. 149). Data collected since 1987 on the productivity of Kittiwakes at St Abb's Head have shown a decline over time and considerable year-to-year variation (Fig. 150). The overall level of productivity is low and declining, with Kittiwakes in two-thirds of the study years fledging fewer than the critical 0.8 young required to maintain colony size. Unfortunately, there is limited historical information on productivity before 1990 (a figure of

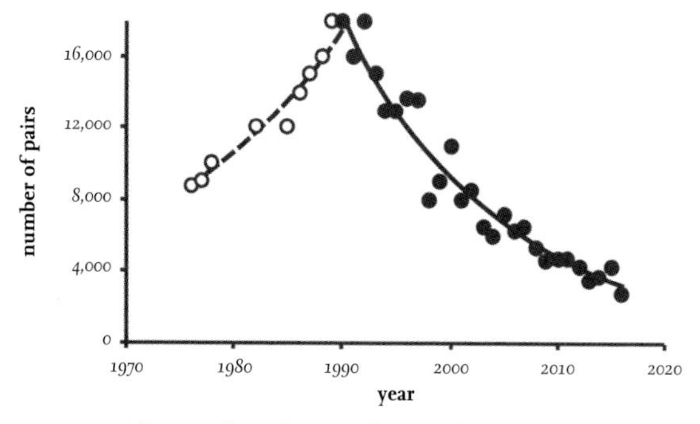

FIG 149. The number of pairs of Kittiwakes nesting at St Abb's Head, Berwickshire, from 1977 to 1989 (open circles), showing an increase of 7 per cent per year, and from 1990 to 2016 (filled circles), showing a decline of 6 per cent per year. The best-fit trend lines are shown. Based on data in the annual reports of the National Trust for Scotland.

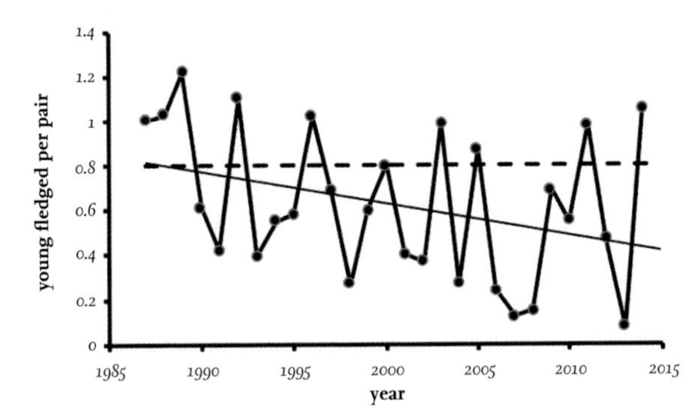

FIG 150. The number of young Kittiwakes fledged per pair at St Abb's Head, Berwickshire, in 1987–2014. The dashed line indicates the number of young required to maintain the colony size, while the continuous line is the best-fit linear trend to the points. Based on data in the National Trust for Scotland annual reports.

1.22 chicks per pair was reported in 1976), but the sudden change from growth to decline has probably involved several factors, including competition for space (following the dramatic increase in numbers of Common Guillemots from 1990 to 2005), which resulted in several areas used by nesting Kittiwakes being completely taken over by these auks. There is, however, no evidence that the changing numbers of Kittiwakes at St Abb's Head have been affected by increased human visitor numbers, despite this conclusion having been reached in a report to the National Trust for Scotland in 2002. The great majority of the nest sites at St Abb's Head were, and still are, on the precipitous cliffs, far from where the public view or approach nesting areas.

Skomer, Pembrokeshire, Wales

Fig. 151 shows productivity data collected between 1986 and 2016 on Skomer Island, south-west Wales, where the pattern is similar to St Abb's Head. Again, production of young fluctuated widely from year to year, and in only a few years was it above the critical 0.8 young per pair. There is only a slight hint of a downward trend. The overall low productivity – fewer than 0.8 young fledged per pair in three-quarters of the years – can be linked to a gradual decline in the number of breeding pairs in the colony since 1995.

Isle of May, Firth of Forth, Scotland

The number of Kittiwakes breeding on the Isle of May increased until 1990 (8,129 pairs), after which there was a long decline of about 5 per cent per year until about 2011 (2,658 pairs), and then numbers began to increase again by 2015 (3,433 pairs)

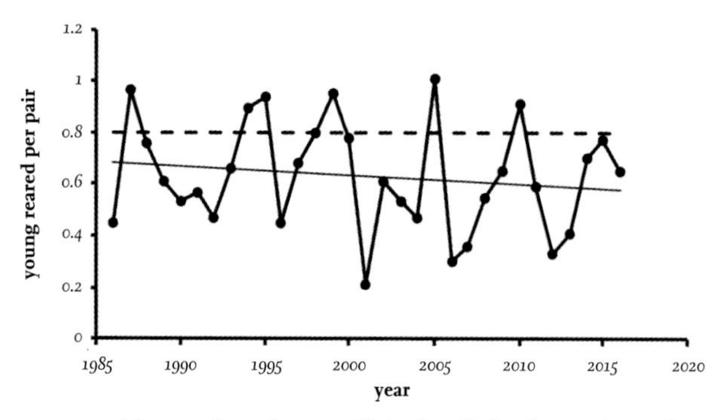

FIG 151. The number of young Kittiwakes fledged per pair on Skomer, south-west Wales, from 1986 to 2016. The horizontal line indicates the number needed to maintain the colony size, while the thin line indicates the long-term trend. The average production over the period declined from about 0.7 to 0.6 chicks fledged per pair. Data from the Wildlife Trust of South and West Wales.

(Fig. 152). The graph of productivity for the Isle of May (Fig. 153) shows that in the great majority of years between 1990 and 2010 (18 out of 21 years), productivity was below the critical level of 0.8 chicks fledged per pair, although there was the suggestion of the start of a recovery in 2011 (Fig. 153). In the period 1986 to 2015, productivity averaged only 0.60 young per pair. The decline in Kittiwake numbers corresponds well with the long period of poor breeding success and the start of improved success in 2011, although young birds do not breed until they are four years old on average and therefore there is a four-year lag before they can contribute to the colony size.

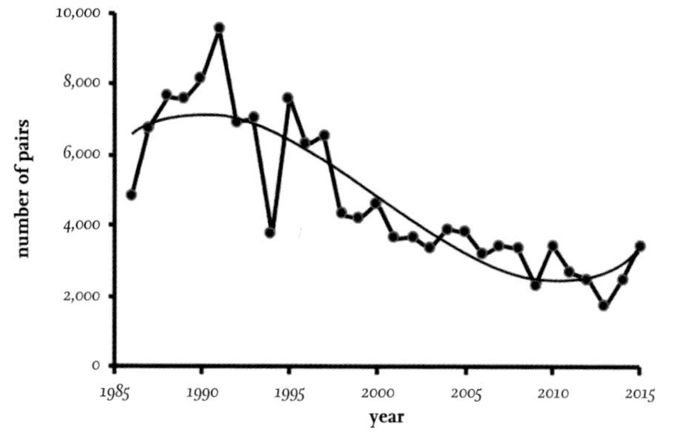

FIG 152. The number of pairs of adult Kittiwakes recorded on the Isle of May in Scotland between 1986 and 2015; the general trend is indicated by the smoothed curve. Data from the Seabird Colony Register.

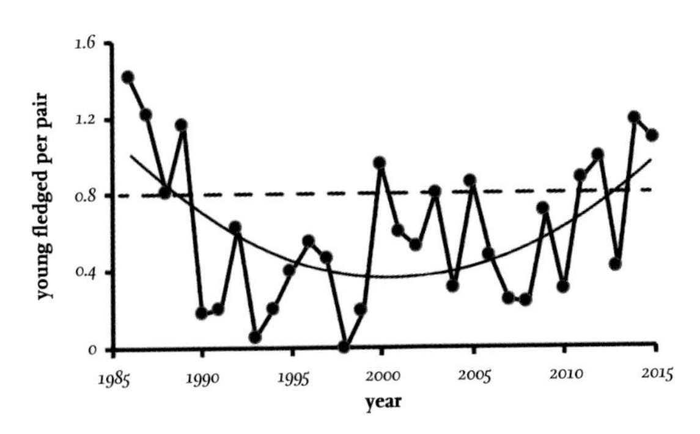

FIG 153. The number of young Kittiwakes fledged per pair annually on the Isle of May in Scotland in 1986–2015, with the general trend shown by the smooth curve. The horizontal dashed line at 0.8 indicates the estimated productivity required to maintain the colony's size. Data from the Seabird Colony Register.

River Tyne area, Tyne and Wear, England

Fig. 154 shows the number of pairs of Kittiwakes breeding in the area in and around the river Tyne in north-east England since breeding first started there in 1932. There has been an almost continuous annual increase in numbers, except in 1998–99, when a sudden, high mortality of adults during the breeding seasons was caused by a lethal natural toxin, which reduced breeding numbers by more than 60 per cent within two years. Following this brief period of high adult mortality, numbers have continued to increase at a similar rate to that which existed for many years previously.

Fig. 155 shows the productivity (young fledged per pair each year) in the river Tyne area from 1954 to 2017, which is the longest continuous series of data available for the Kittiwake. While a slight decline in the annual productivity is

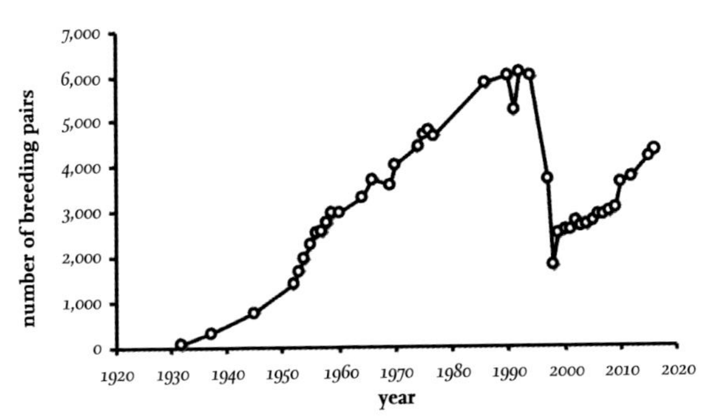

FIG 154. The number of breeding pairs of Kittiwakes nesting in the vicinity of the river Tyne in north-east England since breeding first occurred there in 1931. The long period of annual increase was interrupted in 1998–99 by a dramatic mortality of adults. Since then, numbers have continued to increase at a similar rate to that in the past. Data mainly from Coulson (2017).

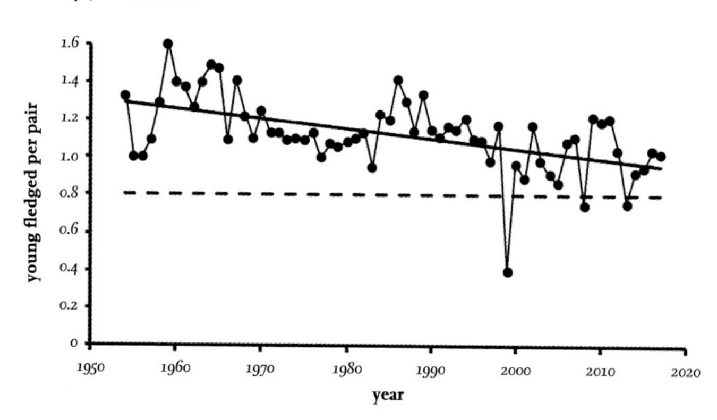

FIG 155. The annual number of young fledged per pair by Kittiwakes breeding in the river Tyne area from 1954 to 2017. The dashed line represents the estimated productivity necessary to maintain the size of the colony. The solid line is the linear trend over the whole period. The low value for 1999 was the result of a high adult death rate, which saw large numbers of chicks dying from the loss of parents rather than from food shortage. Data from Daniel Turner and John Coulson and based on Coulson (2017).

indicated over the 62 years, Kittiwakes there fledged an average of 1.2 chicks per pair from 1954 to 1998 and then 1.1 chicks per pair from 2000 to 2015. There were only three years over this long period when productivity dropped below 0.8 chicks fledged per pair. It is evident that the numbers of young fledged were sufficient both to replace the adult mortality and to permit an annual increase in the size of the colony by about 3 per cent. These figures confirm that the minimum level of production of fledged chicks to prevent numbers of adults declining must be somewhat below one chick per pair each year.

Bempton Cliffs and Flamborough Head, Yorkshire, England

Although numbers of breeding pairs of Kittiwakes are not counted annually at the very large colony at Bempton and Flamborough in Yorkshire, the censuses of nests carried out in 2000 and 2016 suggest that there was only a slight increase (0.03 per cent each year) over this period. The graph of production of young between 1986 and 2015 at Bempton Cliffs (Fig. 156) shows an average figure of 0.92 chicks fledged per pair over this period, which is slightly lower than that of the river Tyne area (see above). It also shows a tendency for the productivity to decline – since about 2002, productivity has been about 0.76, very close to the required level of 0.8 chicks fledged per pair. The breeding success at this large colony would appear to be at a critical stage in determining the trend in numbers of Kittiwakes breeding in England.

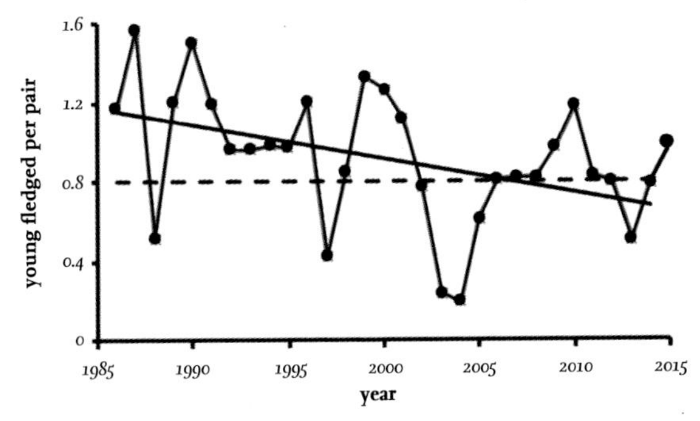

FIG 156. The number of young Kittiwakes fledged per pair annually in 1986–2015 at Bempton Cliffs, with the general trend indicated by the straight solid line. The horizontal dashed line at 0.8 is the estimated productivity required to maintain the colony size. Data from the RSPB and the Seabird Colony Register.

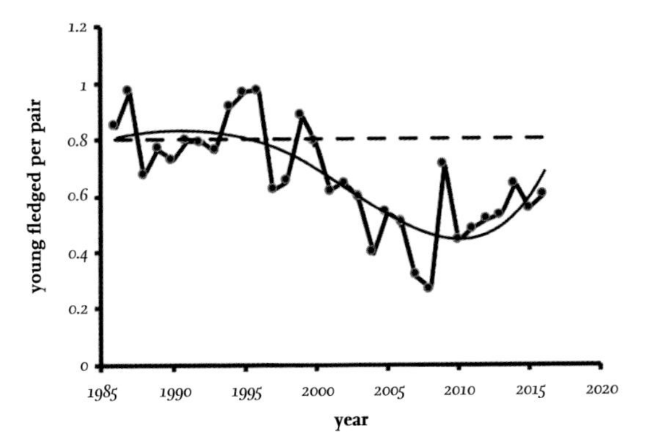

FIG 157. The number of young Kittiwakes fledged per pair in a sample of colonies within Britain but excluding data from the river Tyne area of north-east England. The dashed line indicates the productivity of 0.8 young fledged per pair that is required to maintain the size of the colonies, and the smoothed curve indicates the general trend. Data based on Seabird Colony Register.

Pooled figures on productivity from a sample of colonies around Britain with data available from 1986 to 2016 (Table 62 and Fig. 157) show a changing situation. Until 1999, production averaged about 0.8 fledged young per pair and was probably just enough to prevent these colonies from declining. From 2000 to 2016, the required productivity level was not achieved in any of the 17 years considered, although there is a suggestion that it may be recovering slightly. This pattern of low productivity coincided with widespread decline in numbers breeding in these colonies. Note that Fig. 157 shows much less year-to-year variation than that reported for individual colonies, which suggests that very poor breeding years do not always coincide at all colonies in Britain.

The average values of annual production of young shown in Table 62 suggest that, in the 1985–99 period, 0.73 fledged young per pair was 9 per cent below the required value of 0.80, while between 2000 and 2015 the productivity of 0.57

TABLE 62. Examples of the annual number of young fledged per pair in Kittiwake colonies prior to and since 2000. The values in bold are averages, and those in parentheses are the extreme annual values recorded. The names of colonies still increasing are shown in italics. Data from various published and unpublished sources, including the Seabird Colony Register.

	Years prior to 2000	*Average young fledged per pair (and range)*	*2000 and subsequent years*	*Average young fledged per pair (and range)*
Isles of Scilly	1991–98	**0.47** (0.2–1.2)	2001, 2006–13	**0.22** (0–0.7)
Lundy, Devon	1986–89	**0.58** (0.2–1.1)	2007–15	**0.34** (0.1–0.7)
Berry Head, Devon	1986–99	**0.65** (0.2–1.0)	—	—
St Aldheim's Head, Dorset	1991–99	**0.72** (0.2-1.3)	2000–04, 2007, 2009, 2011–13	**0.50** (0–1.0)
Rinsey Head, Cornwall	—	—	2006–09, 2012	**0.47** (0.3–0.6)
Straight Point, Devon	—	—	2008–10, 2013	**0.56** (0.3–0.9)
Towan Head, Cornwall	—	—	2006–7, 2009, 2013	**0.42** (0.1–0.7)
Bempton and Flamborough, Yorkshire	1986–99	**1.06** (0.5–1.6)	2000–15	**0.78** (0.2–1.3)
River Tyne area, Tyne and Wear	1954–99	**1.21** (0.2–1.6)	2000–15	**1.00** (0.7–1.2)
Coquet Island, Northumberland	1993–99	**0.97** (0.6–1.29)	2000–15	**1.20** (0.4–1.5)
Farne Islands, Northumberland	1987–99	**0.94** (0.3–1.3)	2000–15	**0.72** (0.1–1.2)
Isle of May, Firth of Forth	1985–99	**0.50** (0.2–1.4)	2000–15	**0.77** (0.2–1.2)
Shetland (excluding Foula)	1986–99	**0.43** (0–0.8)	2000–09	**0.30** (0–0.6)
Foula, Shetland	1971–99	**0.98** (0–1.5)	2000–15	**0.20** (0–0.9)
Skomer, Wales	1986–99	**0.67** (0.5–1.0)	2000–15	**0.61** (0.2–1.0)
UK Special Protection Areas	1986–99	**0.70** (0.5–0.9)	2000–14	**0.52** (0.3–0.8)
All	1985–99	**0.73**	2000–15	**0.57**

would suggest a shortfall of almost 29 per cent. Existing data from the national censuses suggest that the decline in breeding numbers of Kittiwakes first started in about 1985 and declined by about 17 per cent between 1985 and 2000. The decline since 2000 has not yet been determined by a national census, but it has been appreciable. The decrease in population index values from 78 to 40 from 2000 to 2014, based on samples of colonies organised by the JNCC, suggests a decline of about 50 per cent in 15 years.

MOVEMENTS

Inland records

There are a few records of small flocks of Kittiwakes flying west in the Firth of Forth in April, gaining height and then flying inland, and it is assumed that these individuals are making flights across Britain. Such cross-country flights probably take about two hours, and it is unlikely that individuals break their journey at inland lakes or reservoirs, while the considerable height of passage mean that such movements probably go unrecorded. Whether such passages are frequent and occur elsewhere is unknown, but some are suspected to occur.

Modest numbers of Kittiwakes are recorded inland each year in Britain. Typically, they involve single individuals, or less commonly small flocks, and are usually reported on large lakes and reservoirs well inland. Records of inland occurrences are scattered, and the online analysis of records in the West Midlands (Fig. 158) by Alan Dean covering 25 years (1986–2010) is informative.

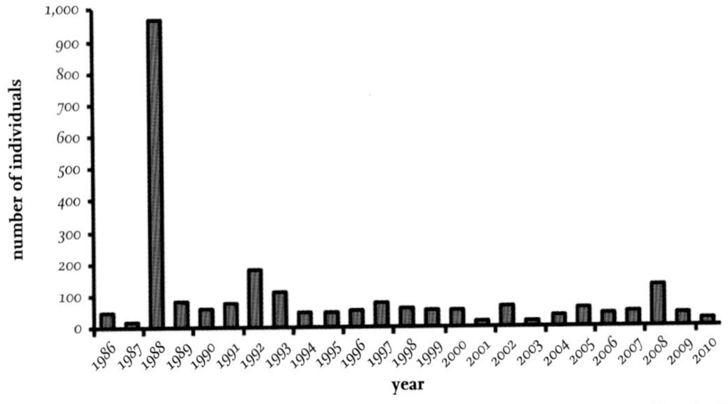

FIG 158. The numbers of Kittiwakes recorded inland in the West Midlands, England, annually between 1986 and 2010. Data based on data collected by Alan Dean.

There was an average of 20 records each year, if double counting and the exceptional year 1988 are disregarded.

Kittiwakes were seen in the West Midlands of England every year, and 1988 was an exceptional year, with almost 1,000 individuals recorded in late February and early March and correctly interpreted as a major wreck of storm-driven birds. On 28 February of that year, there were 200 Kittiwakes inland on Blithfield Reservoir in Staffordshire, associated with severe north-westerly gales affecting Britain. By 6 March, numbers had declined to 113. More were seen at other lakes and reservoirs in the area, and there were additional numbers at the same time in neighbouring counties. This event saw the largest numbers of Kittiwakes recorded inland and was considered to be the first wreck of Kittiwakes in Britain since 1957 (McCartan, 1958); no further major wrecks have been reported (up to 2017). The origin of the birds involved is unknown, nor is it known whether some of the birds were attempting a deliberate cross-country flight rather than driven by the storm accidentally.

Over the 25 years of West Midlands records, it can be seen that there is a major peak in Kittiwake numbers in March (Fig. 159). At this time of year, many adults in Britain are returning to their colonies and those breeding further north are starting their return from an oceanic life. This is confirmed by the large peak of inland sightings reported throughout Britain from 2013 to 2015 (Fig. 160). Examination of records from other years confirms that a major peak of inland Kittiwakes in March is regular and annual. Almost all records are of birds in adult plumage, although this probably means birds over one year old.

In general, inland records of Kittiwakes in their first-year plumage are few. Fig. 161 shows the proportion of the total records for each month that were first-year birds, starting in July, when most are fledging. The proportion was

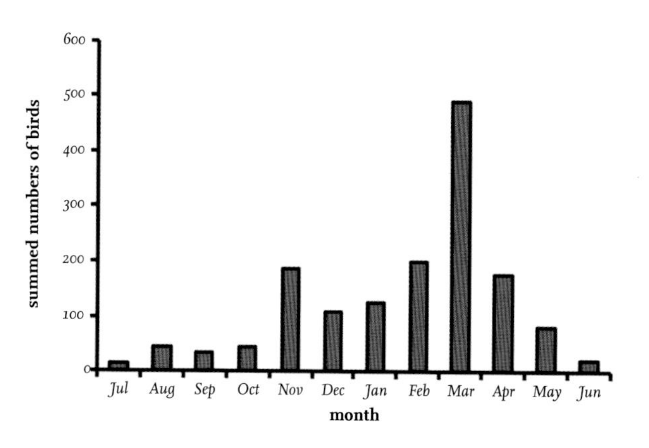

FIG 159. The summed monthly totals of inland records of Kittiwakes in the West Midlands from 1986 to 2010, excluding the large numbers recorded in February and March 1988 during a wreck event. Data based on an analysis of records collated by Alan Dean (n. d.).

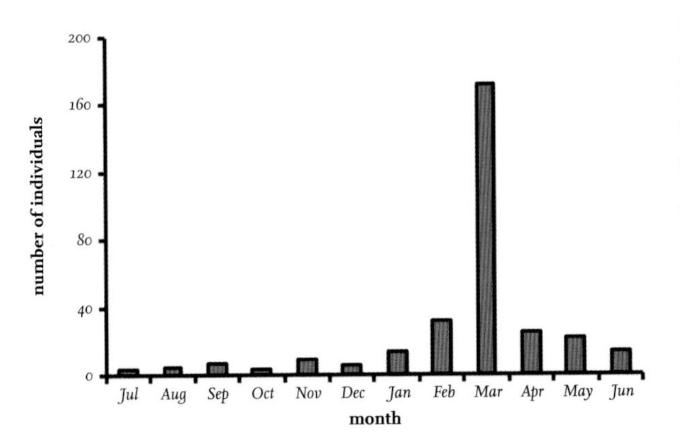

FIG 160. The number of individual Kittiwakes recorded inland in each month in Britain during the period 2013–15.

highest between August and October, soon after fledging, declined during the winter to about 20 per cent of the total number of individuals, and continued at a very low level from March to June. Of particular interest is the small proportion of first-year birds involved in the March peak indicated in Figs 159 and 160. Young, immature Kittiwakes tend to winter further south than adults and most would still be pelagic in March, but their proportions do not increase from April to June.

The large number of records concentrated within March is of considerable interest. Initially, it was thought that most inland records were the result of wrecks caused by adverse stormy conditions, but the restricted period of the majority to one month, year after year, makes this unlikely. While most inland records are of single individuals, the March peak has the highest proportion of sightings, with several birds seen at a time on 24 per cent of occasions between

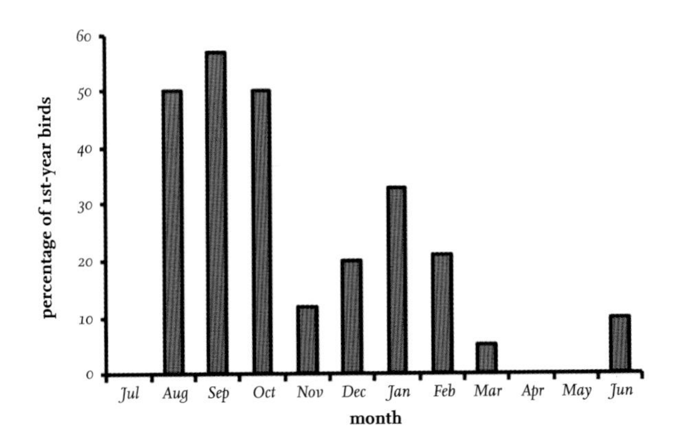

FIG 161. The percentage of first-year Kittiwakes among the aged individuals reported monthly inland in Britain between 2013 and 2015.

TABLE 63. The number of occasions that Kittiwakes recorded inland in England were single individuals, two birds or more than two birds. Based on data from county records since 1980.

Period	*Number*	*Single*	*Two*	*More than two*
Aug–Oct	15	100%	0%	0%
Nov–Jan	27	93%	7%	0%
Feb–Apr	114	76%	15%	9%
May–Jul	34	94%	6%	0%
Total	**190**	**84%**	**12%**	**4%**

February and April, and with the majority of these in March (Table 63). Many sightings report the birds leaving the site (in various directions), which suggests that they are not seriously ill or exhausted. Few of the Kittiwakes reported inland were found dead.

As stated above, most Kittiwakes forming the March peak are aged over one year. It is tempting to suggest that some of the records of two or more birds seen at a time may be paired, but this is unlikely as there is no evidence of Kittiwakes forming pairs other than at the colonies, or of pairs wintering together. It is possible that some of these inland Kittiwakes were making deliberate and successful cross-country passages to colonies, presumably on the east coast of Britain, since birds returning from the Atlantic to colonies on the west side of Britain are unlikely to need to cross the mainland. The attraction of many inland Kittiwakes to lakes and reservoirs is not surprising, as landing and drinking fresh water, rather than seawater, is likely to be a great draw for individuals that have spent several months with only seawater to drink. A final puzzle is the relatively few records of Kittiwakes inland in Scotland. Is this because of a lack of observers, or because of the lack of large freshwater lakes and reservoirs in much of central Scotland?

Inland records of Kittiwakes in winter involving many individuals have also been reported in France, but in different years than the peak years in Britain. I examined a series of skins of Kittiwakes collected in France from such wrecks and preserved in the Musée National d'Histoire Naturelle (National Museum of Natural History) in Paris. Kittiwakes breeding in the Arctic are larger than those breeding in France and Britain and it was evident from the long wings of many of these birds that they came from breeding areas much further north, (p. 369). Interestingly, several of the birds obtained in January were just completing the growth of the last, longest primary, although primary moult is usually thought to be completed two months earlier. This suggests that the birds may have been suffering food shortage for some time and had delayed feather growth, or that the moult of Kittiwakes breeding in the Arctic starts later and may continue well into the winter, although evidence in support of either of these possibilities is not available.

Young Kittiwakes ringed in Britain have died in several inland regions in Europe, including two birds found as far east as the Czech Republic, and an even more remarkable recovery in eastern Ukraine.

Movements in winter

It has been known for some time that Kittiwakes spread out over the North Atlantic in winter – numerous sight records were made from transatlantic ships in the 1930s by Vero Wynne-Edwards, and in the wartime crossings in the 1940s made by Neal Rankin and Eric Duffey. They reported that Kittiwakes had spread south to about 45°N by January, and that this southern limit moved slowly north in the following months (Rankin & Duffey, 1948). By the time of the breeding season, both adults and immature birds had vacated the central ocean area and were only found close to the North American, Greenland and European coasts, with adults moving into their colonies and the immature birds to coastal waters. Ringing recoveries give a distorted picture, because recoveries at sea are very few and are restricted to areas where Kittiwakes are hunted (such as in Greenland) or were captured (in Newfoundland, for use as bait by fishermen on the nearby Grand Banks). Nevertheless, such recoveries have added to our knowledge, particularly of movements across the Atlantic.

Records of Kittiwakes ringed as nestlings in Britain and recovered in the same autumn (Fig. 162) clearly indicate that young birds disperse rapidly. Some cross the Atlantic quickly, with recoveries on the west coast of Greenland within a few weeks of fledging in Europe and even more reported in the following months. These are followed by further transatlantic recoveries from Newfoundland. Other individuals cross the North Sea or move south into the Bay of Biscay. A single young bird was reported from Sicily and two more birds were recovered well inland in Europe. There are fewer recoveries of immature Kittiwakes during the winter, but more transatlantic recoveries have been recorded, while some young birds remain within the North Sea.

In general, young Kittiwakes move further south, with more recoveries in France and Iberia, and proportionately more in Newfoundland compared with those reported in Greenland (Fig. 163). Several Kittiwakes ringed as adults in Britain (Figs 164 and 165) have been recovered in autumn in Greenland and Newfoundland, along with one in Nova Scotia, but none has been reported from the Bay of Biscay

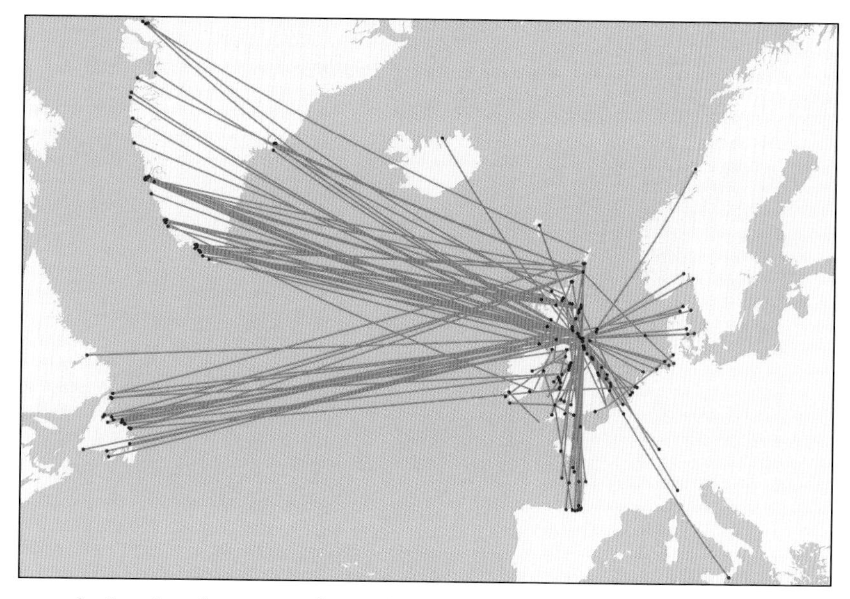

FIG 162. Location of recoveries of Kittiwakes ringed as chicks in Britain and recovered in the same autumn. Reproduced from *The Migration Atlas* (Wernham *et al.* 2002), with permission from the BTO.

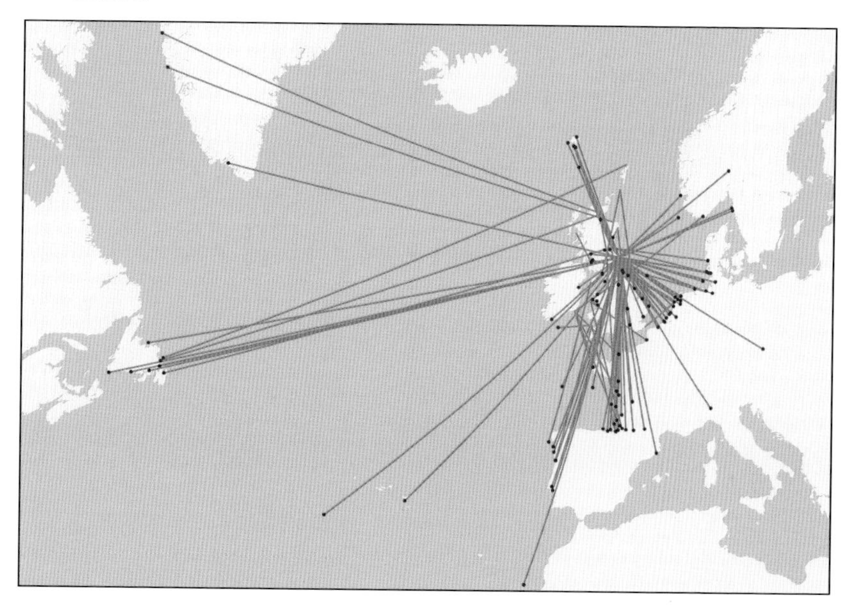

FIG 163. Location of recoveries in winter of immature Kittiwakes ringed in Britain. Reproduced from *The Migration Atlas* (Wernham *et al.* 2002), with permission from the BTO.

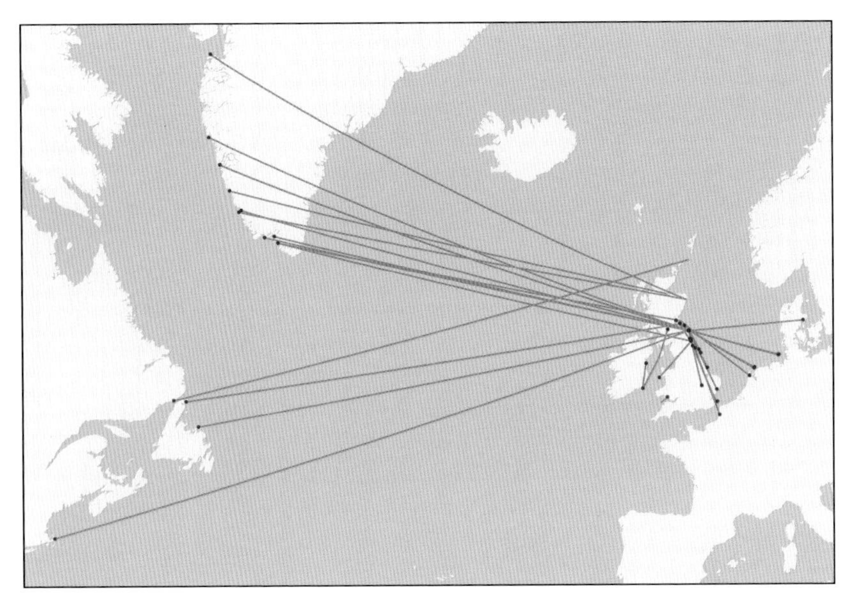

FIG 164. Location of recoveries in autumn of adult Kittiwakes ringed in Britain. Reproduced from *The Migration Atlas* (Wernham *et al.* 2002), with permission from the BTO.

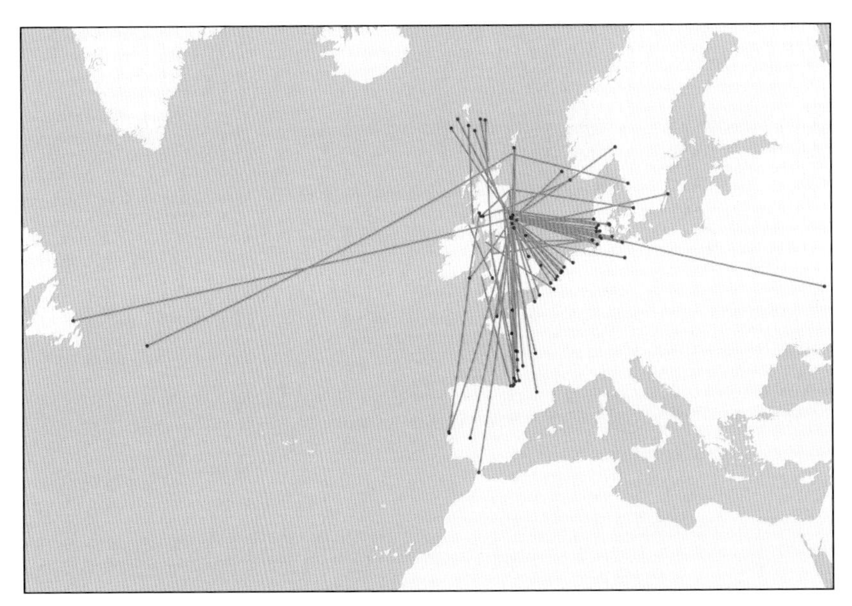

FIG 165. Location of recoveries in winter of adult Kittiwakes ringed in Britain. Reproduced from *The Migration Atlas* (Wernham *et al.* 2002), with permission from the BTO.

region of France or in Iberia before the winter. This suggests that the majority of adults that breed in Britain may remain temporarily in the North Sea after leaving their colonies. Recoveries of Kittiwakes ringed in north-west Russia show movement of these birds into the Atlantic, where they mix with those from Britain, although in general their recoveries in winter tend to be further north.

Tracking by data loggers

The development in the 1990s by the British Antarctic Survey of small data loggers that could be attached to leg rings on gull-sized birds to record their geographical position at regular intervals offered a new insight into their distribution. These data loggers recorded accurately the start and end of daylight each day wherever the individual bird happened to be. The data could only be downloaded from the logger if the bird was captured again in the following breeding season (see Chapter 12 for details of methods used). In general, the data obtained are sound, although cloudy conditions tend to delay the recording of the onset of daylight and advance the end of the day, thereby introducing some errors.

In 2008 and the years following, Kittiwakes breeding in several areas of the North Atlantic were fitted with loggers and a stream of new information rapidly became available. Data were analysed in a scientific paper with no fewer than 31 authors, led by Morten Frederiksen published in 2012.

Transatlantic movements of the Kittiwake, previously proven from only a few ringing recoveries, now became well established and the presence of the birds in winter over the whole of the North Atlantic has been confirmed and quantified. Analysis of the position of individual adult Kittiwakes shows that most winter for long periods well off the coast of Newfoundland and Labrador, or on the eastern side of the Atlantic to the west of Ireland (Fig. 166). Kittiwakes visiting the central ocean areas were probably individuals making cross-ocean movements and, apparently, did not usually remain there for long. The suspected large overlap in wintering Kittiwakes from different colonies and regions was confirmed, and results suggested that up to 80 per cent of Kittiwakes wintered on the western side of the Atlantic, while the birds wintering on the eastern side all bred in mainland Europe. These two main oceanic areas were used by Kittiwakes from the same colonies in the Atlantic Ocean, with the exception of those few birds with loggers attached on Skomer in Wales that did not move further than offshore areas to the west of Ireland.

Wintering records of Kittiwakes marked on the Isle of May in Scotland during the breeding season have led to the suggestion that failed breeders, which tend to leave the colony earlier than successful ones, are more likely to make the

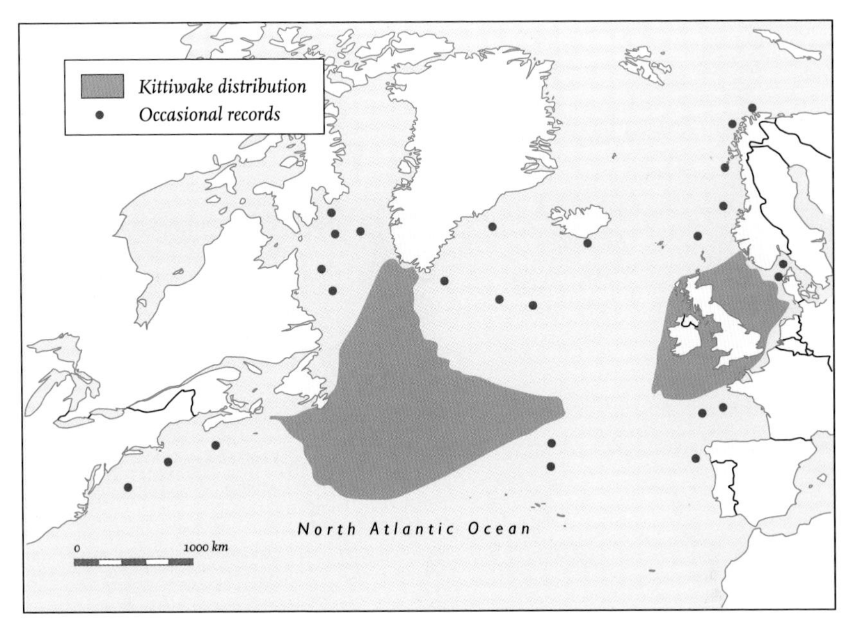

FIG 166. Map of the North Atlantic Ocean, showing the main distribution of Kittiwakes in November and December 2008 (darker blue area), as indicated by data-logger records. Occasional outlying positions are indicated by dots. The mid-Atlantic zone was not occupied in numbers until January, presumably as birds were beginning to return to Europe. Based on data in Frederiksen *et al.* (2012).

movement to Newfoundland waters. However, the difference in the time they leave the colony is so small that this is unlikely to be the cause of differences in their movements. Young adults are much more likely to be failed breeders, and whether the differences found reflect different behaviour between adults of different ages or breeding success remains to be confirmed.

Kittiwakes marked with loggers at Lowestoft in south-east England by the active Kessingland Ringing Group were not included in the meta-analysis. However, their results confirm Fredericksen *et al.*'s (2012) results, with some individuals from that colony moving across the Atlantic towards Greenland and Canada, while another remained all winter on the eastern side of the Atlantic off the west coast of Ireland (Figs 167 and 168). Of particular interest is the first evidence that the same individuals moved to the same wintering areas in successive years (Fig. 168).

Kittiwakes are unlikely to be able to feed in very rough, turbulent seas produced by the deep depressions that typically move from west to east across

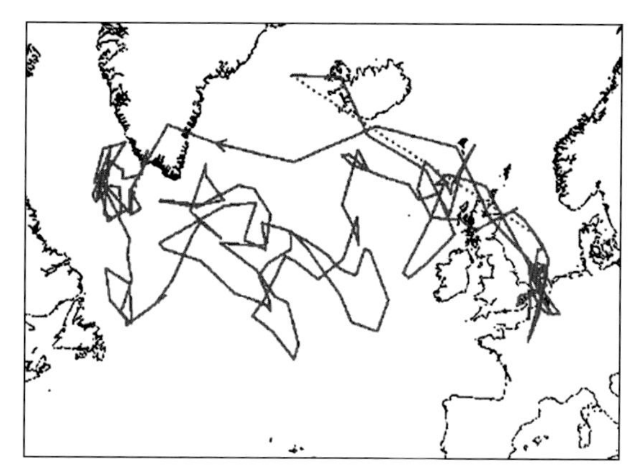

FIG 167. Map of North Atlantic post-breeding season and winter movements of an adult Kittiwake tagged at Lowestoft, Suffolk. Based on maps supplied by Colin Carter and the Kessingland Ringing Group.

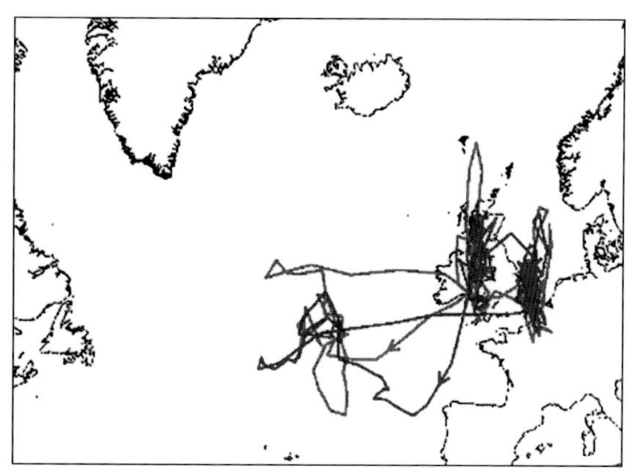

FIG 168. Map of North Atlantic post-breeding season and winter movement of the same individual adult Kittiwake in consecutive years. Based on maps supplied by Colin Carter and the Kessingland Ringing Group.

the North Atlantic. Wind speeds greater than 50 kph make feeding difficult during the breeding season and this would presumably apply in the birds' wintering areas. So how do Kittiwakes react to depressions and high winds while they are oceanic? In October and November, when there are strong easterly winds caused by deep depressions to the south of the North Sea, thousands of Kittiwakes (as well as Gannets, skuas and, occasionally, Little Auks, *Alle alle*) move north along the east coasts of England and Scotland, and then out into the Atlantic via the north of Scotland. Such movements divert the birds away from the worst of the weather. It is possible that similar mass movements take place in mid-ocean to enable Kittiwakes to avoid deep depressions tracking across the Atlantic. Evidence of such oceanic movements do not currently exist, but the employment of loggers (or geolocators) that give day-to-day positions of individuals should soon throw light on how the birds cope with adverse weather while pelagic, and whether they make large-scale movements to avoid atmospheric depressions.

The departure of adult Kittiwakes from colonies and their pattern of movements has changed in recent years. In the 1960s, some adults returned daily to the colonies in autumn and into November, and moved away for less than three months, returning to the colonies in January. Clearly these individuals had much less time in which to make transatlantic crossings than do Kittiwakes today, since colonies now remain unoccupied for more than six months, from mid-August to early March. Have the wintering areas of adults changed along with the present, much later return to the breeding colonies, which in some areas is now two months later than it was 50 years ago?

Movements of young Kittiwakes

Since the early days of ringing, it has been clear that Kittiwakes marked as nestlings migrate and then return to breed very close to where they hatched. The same is true of other gulls, and several scientific papers have stated that seabirds also show this ability, referred to as philopatry. Like other gulls, many Kittiwakes are philopatric, showing a remarkable ability to navigate, find and recognise the colony where they were reared two or more years earlier. However, the tendency toward philopatry has often been exaggerated, mainly because it is easier to locate individuals that have returned 'home' than those that have moved elsewhere to breed. Allowing for this bias, about 50 per cent of surviving young male Kittiwakes return to breed in their natal colony, but only 11 per cent of females do so, resulting in the proportion of philopatric individuals being about 80 per cent males. Further studies on colour-ringed Kittiwake chicks resulted in many being found breeding in other colonies, in some cases several hundred kilometres from where they were reared. Data for Kittiwakes ringed as chicks and recovered in the breeding season when old enough to breed (Fig. 169) give an indication of the extent of these dispersal movements. The few birds reported inland are obviously not at colonies, and the Greenland records could be erroneous, because recoveries from that country are sometimes delayed in being reported and the dates given are sometimes suspect. However, many of those reported elsewhere do indicate genuine movements to breed away from their natal areas, with some of these records supported by the birds being found on nests with eggs or chicks.

This modest level of philopatry in Kittiwakes does not mean that birds reported breeding elsewhere had become lost. Julie Porter, in an intensive study of recruits visiting the North Shields colony in Tyne and Wear, found that, as they reached breeding age, many Kittiwakes marked when chicks visited their natal colony but

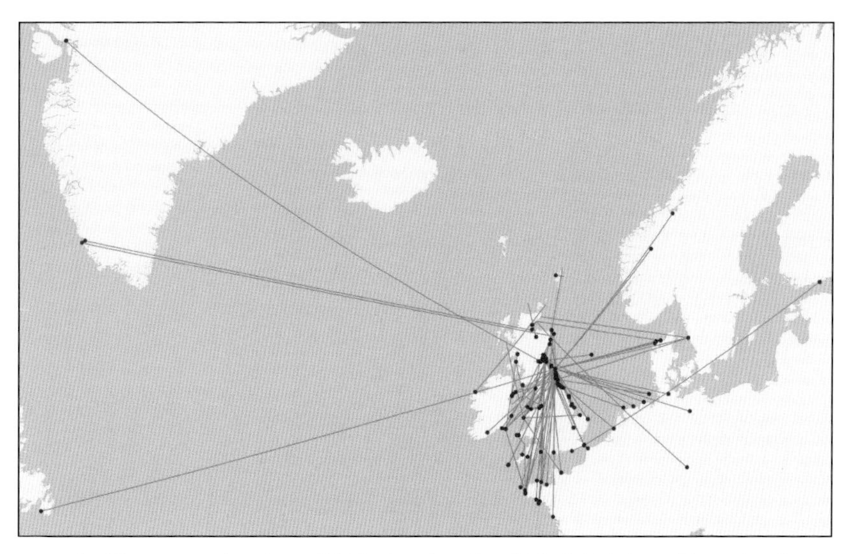

FIG 169. The recovery locations of Kittiwakes ringed as chicks in Britain and Ireland and reported in the breeding season when at least three years old. Most, but not all, of these recoveries were of individuals that had moved to breed away from their natal colony. Reproduced from *The Migration Atlas* (Wernham *et al.* 2002), with permission from the BTO.

then moved away and bred elsewhere (Porter, 1990). Non-philopatric individuals of both sexes are essential for the formation of new colonies, while the sex bias in the intensity of philopatry results in the avoidance of inbreeding. For example, no brother–sister, father–daughter, mother–son or grandfather–granddaughter pairings were found in more than 35 years of detailed study at the North Shields colony, where the status of over 2,000 pairs over many years was recorded.

In contrast, adult Kittiwakes breeding in colonies where young are successfully reared rarely moved to breed in a different colony. At the North Shields colony, only two adults that had bred there moved to breed elsewhere, compared with 2,510 cases of breeders that returned to the colony in the following year. Remarkably, although the two birds that moved were not members of the same pair before they moved, they were found paired together in a colony of 300-plus pairs more than 50 km away. How and when they moved and managed to pair together is unknown, but this was a remarkable event. It suggests that the two individuals might have had a degree of familiarity and recognition before moving, but there are no other records of Kittiwakes forming pairs other than at a colony. Elsewhere, adults have moved to other colonies if breeding is badly disturbed, nest predation is high or if they are subjected to heavy loading of parasites.

FOOD AND FEEDING

Food during the breeding season

Kittiwakes breeding at colonies in Britain and Ireland predominantly feed on fish, especially sandeels, although at a few colonies other fish species form the main diet for at least part of the breeding season (Table 64). Capelin (*Mallotus villosus*), another small shoaling fish, replaces sandeels as the main food of Kittiwakes in Newfoundland and northern Norway, while in the Barents Sea area of Russia, Capelin, sandeels and young Herring all form substantial but varying proportions of the diet. Even further north, in Novaya Zemlya and Spitsbergen, Arctic Cod (*Arctogadus glacialis*) and crustaceans play an important part in the diet in some years.

Data collected on the Isle of May from 1986 to 2015, during the period chicks are present, showed that sandeels formed at least 50 per cent (and usually more than 80 per cent) by weight of the food supplied by the parents (Fig. 170). At some other colonies in Britain, crustaceans are an important food early in the breeding season, and their consumption is evident by the pink colour of droppings on nests and cliff faces. However, the frequency of crustaceans in the diet soon declines and is replaced by Sprats and sandeels,

TABLE 64. Examples of the food consumed by nestling Kittiwakes at different colonies. The data from Pearson (thesis and 1968) and Coulson (unpublished) are based on numbers of individual items, while that of Bull *et al.* (2004) is the percentage of samples containing these taxa.

Location	*Years*	*Sandeels*	*Clupeidae*	*Gadidae*	*Crustacea*	*Polychaete*	*Offal*	*Source*
Farne Islands, Northumberland	1961–62	78%	11%	9%	1%	0%	1%	Pearson (1964)
Farne Islands, Northumberland	1998–2000	99%	15%	3%	1%	0%	0%	Bull *et al.* (2004)
Isle of May, Firth of Forth	1997–2000	95%	16%	4%	2%	0.5%	0%	Bull *et al.* (2004
Inchcolm, Firth of Forth	1997–98, 2000	45%	82%	8%	6%	0%	0%	Bull *et al.* (2004)
Inchkeith, Firth of Forth	1997–99	29%	79%	10%	2%	0%	0%	Bull *et al.* (2004)
River Tyne, Tyne and Wear	1962–70, 1990–95	85%	12%	0%	0%	0%	3%	Coulson (unpublished)

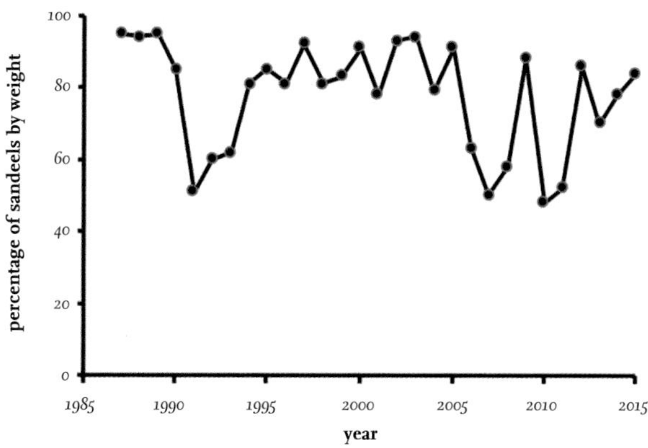

FIG 170. The percentage of sandeels as a proportion of the total food fed to young Kittiwakes on the Isle of May, 1985–2014. Data from Harris & Wanless (1997) and the Seabird Monitoring Programme.

which dominate for most of the breeding season, including the period when chicks are in the nest.

Feeding trips during the breeding season

Direct observations and the recent use of geolocators have shown that Kittiwakes rearing young often fly 70 km or more out to sea to feed – indeed, they travel further to forage than many other seabirds (Table 65), with the exception of the Gannet and Northern Fulmar (*Fulmarus glacialis*).This is puzzling, because terns also feed mainly on sandeels, but they obtain these much closer to the shore. Why do Kittiwakes need to travel so far offshore while terns can feed successfully much closer to the shore and their colonies?

A study made in north-east England by Chris Redfern and Richard Bevan using loggers and published in 2014 found that the duration of Kittiwake feeding trips during the breeding season lasted up to 17.6 hours and adults moved a maximum distance of 111 km from the Farne Islands colony and 122 km from the Newcastle colony, with a maximum distance flown during one journey of 370 km. From the movements recorded, the authors estimated that about half the time on each trip was spent foraging, with the bird frequently changing direction, and the other half was spent in direct flight to and from feeding areas.

Several generalisations can be made about Kittiwake feeding trips from this study. Birds departing from the colony typically fly directly offshore in a fixed direction and at a near-constant speed. A typical path of flight at this stage is shown in Fig. 171a. At some point in time, the flight pattern changes, with the bird frequently switching direction as it presumably searches for food (Fig. 171b) and then from time to time moves to new but nearby areas. When the track within a limited location is examined in greater detail, the repeated turning and searching again over the same area is evident (Fig. 171c). Eventually, the bird begins its return flight to the colony, but wind drift sometimes causes it to reach the coast a few kilometres from the colony, at which point it follows the coastline to reach its nest and chicks.

Records showed that Kittiwakes at sea during the breeding season were active and probably feeding throughout the night, but between 22.00 and 04.00, the hours of darkness, about 20 per cent of birds had periods of inactivity, resting on the sea for short periods and drifting by wind or tide, but not moving far. These periods of inactivity at the feeding grounds were almost five times greater in the dark than at other times of the day, suggesting that the birds' rate of feeding decreased, but was not prevented, in the dark. The extent to which night-time feeding was made in proximity to floodlit fishing vessels actively fishing and hauling nets is not known.

Identification of the feeding areas used by Kittiwakes nesting on the Farne Islands and at Newcastle, 72 km apart, and also at Bempton and Filey

TABLE 65. Estimates of the average and extreme distances from the colony travelled by seabirds while breeding. The values vary from location to location and between years. Data from many sources, including unpublished values.

	Average distance travelled (km)	*Extreme distance travelled (km)*
Little Tern	2	11
Common Tern	5	28
Arctic Tern	7	30
Roseate Tern	12	32
Sandwich Tern	12	57
Black-headed Gull	12	40
Herring Gull	21	93
Common Gull	25	50
Puffin	37	56
Lesser Black-backed Gull	34	181
Common Guillemot	38	134
Black-legged Kittiwake	89	219
Northern Fulmar	122	245
Northern Gannet	232	540

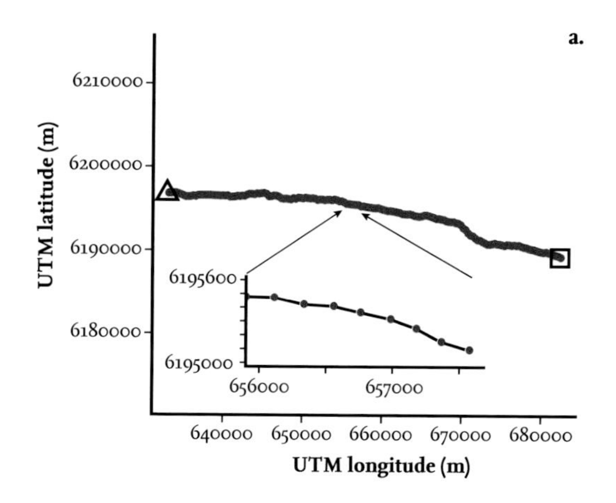

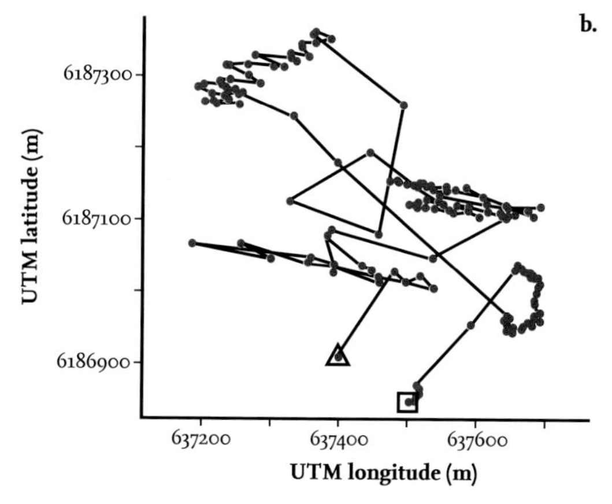

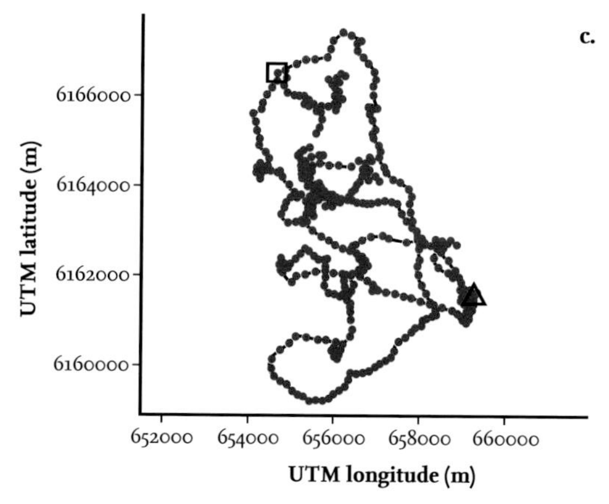

FIG 171. The movements at sea of an adult Kittiwake: (a) the direct flight towards the feeding area; (b) the movement made at a presumed feeding site; and (c) higher resolution of feeding, involving returning to the same area. The triangles and squares show the start and end of the journeys. After Redfern & Bevan (2014).

Brig, Yorkshire, 30 km apart, clearly showed that birds from these pairs of neighbouring colonies overlap in their feeding areas, at least while feeding young. The extent of these overlaps has not been precisely determined, but it raises interesting questions as to why the numbers nesting at neighbouring colonies may be changing at different rates – for example, the declining numbers on the Farne Islands and the increasing numbers nesting on nearby Coquet Island, only 37 km away.

Kittiwakes nesting at Newcastle, some 18 km from the open sea, rarely feed in the river and food given to the young is almost entirely marine organisms and caught at sea. It is not known whether the distance the birds have to travel before they reach the open sea has an effect on the length of time they spend away on each feeding trip. The departure and return journeys along the river add about 45 minutes to a daylight feeding trip, and so could increase the travel time by 20 per cent compared to that for coastal-nesting Kittiwakes. Despite this possible extra flight time, there is no indication that the young at the Newcastle colony are less well fed – the numbers of young per pair fledged there is high, and the annual average is regularly more than one chick per pair.

Feeding methods

Kittiwakes feed in several ways, but the most frequent method is shallow diving into the sea from flight. This type of feeding is rarely seen from the shore, where picking small food objects from the sea surface while in flight or settled on the sea is much more frequent. Kittiwakes also feed behind fishing boats, picking up discards, but are frequently displaced by larger gulls and Northern Gannets. When feeding behind moving fishing boats, they tend to fly much closer to the stern of the vessel than Herring Gulls (*Larus argentatus*), and by so doing they reduce competition from the larger gulls.

It is clear that the Kittiwake is not a food specialist, and takes food items (mainly fish) of a suitable size that it can catch at the sea surface or by diving to shallow depths of about 50 cm. If the usual abundant food source in a particular region temporarily declines, be it sandeels, Herring or Capelin, Kittiwakes will switch to another group of fish or to crustaceans, providing an alternative is available. It is likely that the decline of Kittiwakes in areas like Shetland (Appendix 4, p. 444) stem from the absence of alternative and substantial food sources at times when the availability of sandeels is low, whereas at other localities with stable populations, alternative food has been available, albeit less abundant, less readily captured and of lower energy content than sandeels. Areas lacking alternative food sources should surely be regarded as suboptimal habitats for Kittiwakes, and this includes Shetland and Orkney.

There are still several aspects of the biology of Kittiwakes during the period they are oceanic that remain to be resolved. While breeding, Kittiwakes primarily eat fish, but this is unlikely to be their main source of food in winter. By measuring the carbon and nitrogen isotope concentrations of Kittiwake primary feathers, which are built into them while the feathers are growing, Keith Hobson found that the animal food consumed in winter is from a lower trophic level than that consumed in the summer, and so is unlikely to be fish (Hobson, 1993). More likely, the birds feed on marine molluscs at this time (such as *Clio* spp.), as well as polychaete worms, all of which have flotation devices to keep them in the surface waters of the oceans. Little is known of the distribution of these invertebrates or where and when they are most abundant. An interesting suggestion is that the numbers of the gastropod *Clio* are declining because they are facing increasing difficulty in retaining the calcium carbonate content of their shells. This is because of the slight but consistently increasing acidity in surface waters of the ocean, believed to be caused by higher concentrations of carbon dioxide in the atmosphere being absorbed by the sea, forming carbonic acid and speeding up the erosion of the calcium carbonate in the mollusc shells or slowing down its deposition.

CHAPTER 9

Yellow-legged Gull

APPEARANCE

For many years, the Yellow-legged Gull (*Larus michahellis*) was considered to be a yellow-legged subspecies of the Herring Gull (or what is now known as the European Herring Gull, *Larus argentatus*), but it was given species status by the British Ornithological Union in 2007. It is the same size as a Herring Gull, but the two species display geographical variations. Apart from being distinguished by its yellow legs, the Yellow-legged Gull has slightly darker grey coloration on the mantle and wings than in British Herring Gulls (Fig. 172), the white mirrors

FIG 172. Adult Yellow-legged Gull in summer plumage. (Alan Dean)

FIG 173. A Yellow-legged Gull is remarkably similar to a Herring Gull when the legs are not visible. (Alan Dean)

FIG 174. First-winter Yellow-legged Gull. (Alan Dean)

on the wing-tips of young adults are often reduced in size (as is also the case in four-year-old Herring Gulls), and in the breeding season the eye-ring is red rather than golden yellow.

FIG 175. Adult Yellow-legged Gull with the inner primaries moulted and hardly any white tips on primaries 7–9. (Gzzz, Wikimedia Commons)

First-year birds are paler than Herring Gulls of the same age, except for the darker primaries. Their legs are not yellow as in the adults, but instead are pinkish grey, similar to those of Herring Gulls (Fig. 174). Adults nest at similar sites to those used by Lesser Black-backed Gulls (*Larus fuscus*), including buildings in Gibraltar, Spain, Italy and France.

SUBSPECIES

There are two subspecies of the Yellow-legged Gull. The typical form breeds throughout the Mediterranean and around the south-west coasts of mainland Europe, while the subspecies *Larus michahellis atlantis* breeds in the Azores, Canary Islands and Madeira. Some authors have suggested that *atlantis* individuals should be given species status. However, birds breeding in the Canaries and Madeira are intermediate in both plumage and size between those breeding in the Azores and on the mainland of Europe, so a specific status is probably not justified.

DISTRIBUTION

Numbers of Yellow-legged Gulls have increased in recent years and the birds have spread rapidly northwards in Europe, initially occurring in appreciable numbers in the non-breeding season and then, more recently, some arriving earlier and in time to breed. Most of the sightings in Britain initially occurred in southern England between July and October, and within the past 30 years their numbers have increased.

The increase in Britain is well illustrated by the analysis of records from the West Midlands made by Alan Dean for the period 1973–2000 (Dean, n.d.; Fig. 176). It is interesting that most records in the West Midlands occurred later in the year than those made in Sussex (Fig. 177), suggesting that perhaps birds on the south coast eventually move inland, joining up with the large numbers of other gulls that visit landfill sites between October and January. Numbers decreased in Sussex in winter, while many of the birds in the West Midlands stayed until February. These birds then departed south, leaving at the same time as most adult Herring Gulls at inland landfill sites left for their more northern breeding areas.

BREEDING

Offspring resulting from pairings of Yellow-legged Gulls with Herring Gulls and Lesser Black-backed Gulls have occurred in Belgium since 1996 and in the Netherlands since 1987. In 2002, the first pair of Yellow-legged Gulls was found breeding in Belgium, and breeding has recently spread further north to Denmark and inland to Germany and Poland.

The first attempted breeding in England occurred in 1992, when a single individual paired with a Lesser Black-backed Gull. The first breeding by a pair of Yellow-legged Gulls in Britain took place in Dorset in 1994 and continued at this same site in subsequent years. In 1999, one pair bred, and two mixed pairs were seen at three localities. Breeding attempts occurred in most years thereafter, but mainly as mixed-species pairs. In 2004, the first breeding attempt was recorded in Northern Ireland, in County Fermanagh, again by a mixed pair. By 2010, breeding attempts had been reported in at least six sites in the south of England, with the majority still forming mixed pairs either with Herring or Lesser Black-backed gulls and rearing hybrid offspring. In 2015, two pairs of Yellow-legged Gulls bred in south-west England and reared three young. Numbers of breeding records have increased slowly, although many birds are still forming mixed-species pairs.

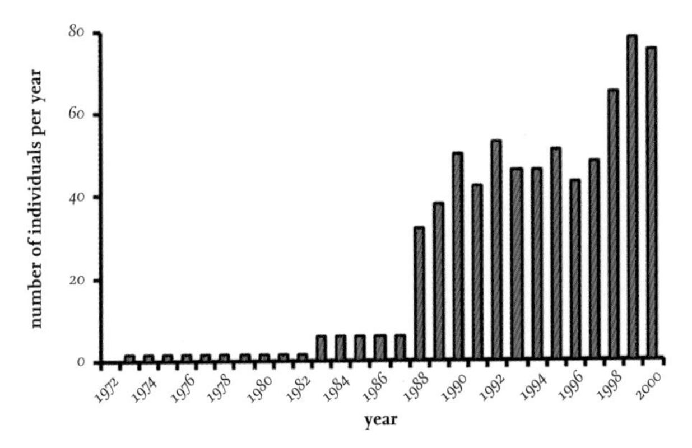

FIG 176. The numbers of Yellow-legged Gulls reported each year in the West Midlands in England from 1973 to 2000. Numbers for 1973–1983 were annual averages over five-year periods. Based on data from Alan Dean (n.d.).

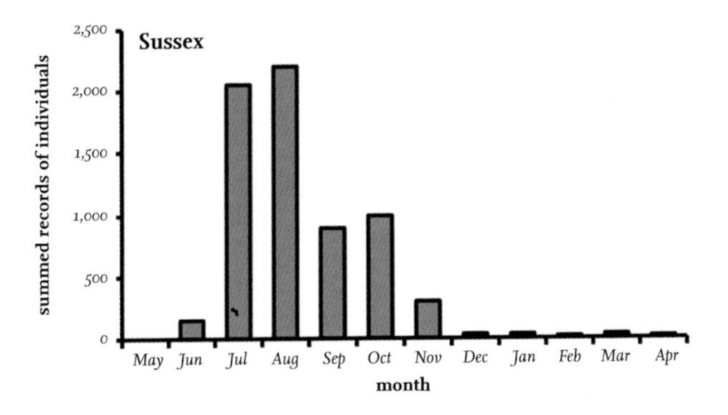

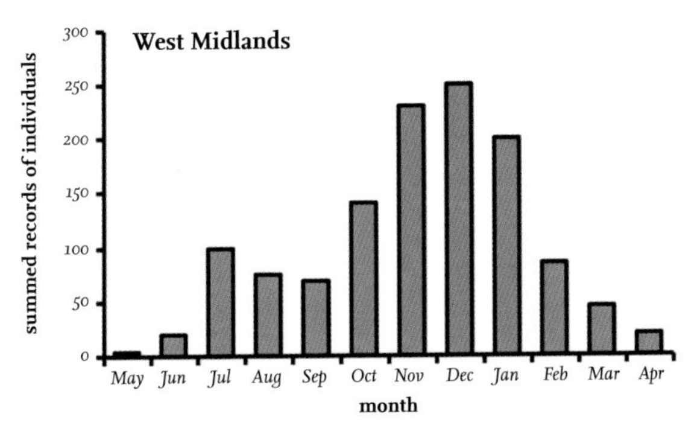

FIG 177. The summed monthly totals of records of Yellow-legged Gulls from 1973 to 2000 for Sussex and the West Midlands, England. The Sussex data have been extracted from the Sussex Bird Reports produced by Richard Fairbank and the West Midlands data are based on an analysis made by Alan Dean (n.d.).

Hybridisation has been frequently recorded between similar-sized gulls when one of the two species occurs in an area in small numbers, as when a species is extending its range. In the case of gulls, the hybrid offspring are viable, but whether they have reduced fertility is unknown. Hybrid young produced by Herring Gull and Yellow-legged Gull mixed pairings are difficult to identify as such in the field while immature, and this would probably also apply to back-crosses produced when the hybrids breed with Herring Gulls.

Following the range expansion and increase of the Yellow-legged Gull in many areas of Europe, it is likely to become a more frequent breeding bird in Britain and Ireland in the next few years, and there will be more same-species pairs breeding at more locations, as has happened in Belgium. In many aspects, this species is likely to compete with Herring and Lesser Black-backed gulls, and the outcome of this will be of considerable interest, although it is difficult to predict in advance. Since the Yellow-legged Gull already breeds in urban areas in southern Europe, it is likely to join the other large gulls breeding in towns and cities in Britain.

CHAPTER 10

Little Gull

APPEARANCE

The Little Gull (*Hydrocoloeus minutus*) is the smallest gull in the world, with a wingspan of just 66 cm, and is appreciably smaller than the Black-headed Gull (*Chroicocephalus ridibundus*; p. 43), a species with which most people are familiar. Adults are readily identified by their small size and dark underwing, which is seen when in flight (Fig. 178). In the breeding season, the black hood of the adult (not dark brown, as in the Black-headed Gull) also aids identification, while the impression of markedly rounded wing-tips is another a useful guide. When flocks are disturbed, they fly in a compact group and in a more erratic manner than other gulls, changing direction in synchrony, more like small waders.

FIG 178. Adult Little Gull in full summer plumage. Note dark underwing. (Ekaterina Chernetsova (Papchinskaya), Wikimedia Commons)

FIG 179. Adult Little Gull in winter plumage. (Alan Dean)

First-year birds have a wing pattern similar to that of Kittiwakes of the same age, but there are minor differences in addition to their smaller size (Fig. 180, right). There is usually a dark cap to the head, which is lacking in the Kittiwake, but the underwing is pale and not dark as in adults.

The Little Gull is a poorly studied species, and several major aspects of its biology remain unresolved.

DISTRIBUTION

The species' world population is uncertain, with estimates of numbers ranging from 97,000 to 270,000 individuals spread across Asia and Europe, plus a few in North America. It is often said to occur as two (or even three) major populations, which are shown separately on most distribution maps of the species in major works of reference: one in Europe and western Siberia; and one or two in central and eastern Siberia. I have searched in vain for evidence to justify these separate groups, and suspect that there is a continuous distribution in northern Europe

FIG 180. Left: first-year Little Gull in flight. Right: first-year Little Gull. (Left, Alan Dean, right, Tony Davison)

and across Siberia, as is the case with the Black-headed Gull (p. 44). This lack of reliable information about the species' distribution emphasises the uncertain knowledge of its abundance, and there is even uncertainty about its numerical trends within Europe.

Separate from the main distribution, a small group of Little Gulls has started breeding in North America (or perhaps they have been overlooked for many years). The first breeding records were near Lake Ontario in Canada in 1962 and 1963, when three nests were found, and several other Canadian localities have been reported more recently. The first record of breeding in the United States was in 1975 in Wisconsin, when four nests were found in a small colony. The breeding range in North America is not known, and confirmed breeding is restricted to parts of the Great Lakes, southern Minnesota and the Hudson Bay Lowlands of Manitoba and Ontario (Ewins & Weseloh, 1999). These North American birds apparently winter on the east coast of the United States, but the numbers counted wintering there greatly exceed those known to breed in the region. Either larger numbers are breeding in North America and are yet to be discovered, or some individuals are regularly crossing the Atlantic from Europe to winter on the east coast of North America. The only support for the latter possibility was a Little Gull ringed in Sweden as a nestling and recovered dead in Pennsylvania, USA.

BRITISH RECORDS

Much of the early history of the Little Gull in Britain was summarised by Clive Hutchinson and Brian Neath (1978), and I have relied on their information here. Up to the end of the 1930s, the species was infrequently recorded in Britain and Ireland, but since then records have become much more common. The increase began in the late 1940s, when flocks of more than 40 Little Gulls occurred annually in the Firth of Tay and Firth of Forth in Scotland. By the 1960s, larger flocks – some numbering more than 500 birds – occurred there. In England, the increase in numbers occurred later, and it was not until 1955 that the first flocks with numbers in double figures were found on the coast and at an inland reservoir in County Durham. By 1966, flocks of at least 200 were seen at the Hurworth Burn Reservoir on several occasions, with individuals apparently moving between there, a coastal roosting site and, presumably, an unknown feeding area, since there were few, if any, records of individuals feeding. By the end of the decade, appreciable numbers were being recorded on the Lancashire coast and smaller numbers in Wales. Between 1970 and 1972, there were large increases in numbers recorded in many English coastal counties, and in 1972 the first flock was reported in Ireland. By the 1970s, numbers reported in Britain totalled thousands each year, comprising mainly birds visiting coastal areas in the autumn, smaller numbers recorded on passage in the spring, and a few immature birds remaining in Britain throughout the summer.

At this time, the first records were made of Little Gulls passing along the English coast in large numbers – for example, 355 flew east at Dungeness on one day in May 1974, and more than 1,000 per day have passed Flamborough Head on several occasions. Sea-watching for birds was becoming increasingly popular, which may have been the reason for the increase in records. In earlier years, such movements were possibly overlooked, and the immature birds (usually much more numerous than adults) quite likely confused with young Kittiwakes when passing at a distance from the shore.

Currently, hundreds of Little Gulls pass through Britain in the spring on their way to their breeding grounds, while in autumn numbers can exceed a thousand individuals at a single location. For example, 1,900 were recorded at Hornsea Mere in Yorkshire in September 1994, and numbers there reached about 3,000 in 1997. In September 2007, up to 21,000 were reported to be roosting on the lake at night.

There have been, and still are, very few Little Gulls breeding in Norway, Sweden, Denmark and the Netherlands. Finland and the Baltic states are the nearest countries to Britain with sizeable colonies of breeding birds, but even these are too few to explain the large numbers of individuals passing through Britain in autumn, many of which must originate from areas even further to the east. The manyfold increase in numbers visiting Britain in the last 50 years does not appear to be mirrored by large and comparable increases in the world population of Little Gulls. The recent increases in records in Britain and Ireland are far in excess of the undoubted greater number of observers, and in any case this increase would not explain the much larger flocks seen in more recent years. The most likely explanation is that many of the birds currently breeding in Finland and beyond have changed their migration route. There is no knowledge of the route taken by these birds prior to 1950, but they seem to have avoided countries along the western coasts of Europe, including Britain. Perhaps they crossed inland Europe to the eastern end of the Mediterranean and the Black Sea. Judging by the numbers of autumnal sightings in recent years, most of the western portion of the Little Gull population after breeding are now moving through countries in western Europe neighbouring the North Sea, and into the eastern part of the Atlantic Ocean. The cause of this change, if indeed it is taking place, is unknown.

BREEDING IN BRITAIN

It is surprising that, despite the large numbers of Little Gulls now passing through Britain, pairs have nested here on only a few occasions. A search of the literature indicates that at least eight attempts of breeding that involved eggs being laid have been recorded in England, including Suffolk (1966), Cambridgeshire (1975), North Yorkshire (1978), Norfolk (1978 and 2007), Hampshire (1982), Nottinghamshire (1987) and Kent (1987). Some of these records may have been rejected or missed by other authors, while the Hampshire record is not listed in *Birds of Hampshire* (1993). Breeding was suspected in Scotland in 1988 and 1991, when recently fledged chicks were seen. All the breeding attempts in England failed to hatch any eggs.

In 2016, the first successful nesting by Little Gulls occurred in Scotland at the Loch of Strathbeg RSPB reserve in Aberdeenshire, where a single pair fledged one, or perhaps two, young from eggs laid near a Common Tern (*Sterna hirundo*) colony. In additional to these records, individuals or pairs have been found in spring in England, Wales and Scotland at potential breeding sites, but without evidence of nesting attempts. While the species often forms small colonies elsewhere, those nesting in England all involved single pairs that attached themselves to Black-headed Gull colonies, and the adults did not return to the same areas in the following year.

The Little Gull usually breeds inland, nesting in freshwater marshes and boggy areas, or on vegetation on or near riverbanks, with the nest built on the ground and usually among vegetation. The usual clutch size is two or three eggs. There is no information on the usual breeding success.

MOVEMENTS

There have been only five ringing recoveries in Britain of Little Gulls marked abroad, with two from Latvia, one from Estonia, one from Sweden and another from Finland. The absence of birds marked further east and found in Britain may reflect the small numbers ringed rather than that those breeding further east do not visit Britain. A hint that Little Gulls from Siberia could visit Britain comes from the recovery in Belgium of a bird marked at Barabinsk, central Siberia (55°21′N, 78°21′E).

The annual start of the post-breeding dispersal is early compared with other gulls, and a few young of the year have already crossed the North Sea to Britain by the end of July. Numbers of Little Gulls visiting Britain and Ireland up to 1973 were clearly dominated by first-year birds (Table 66). In autumn, it might be

TABLE 66. The percentage of first-year Little Gulls recorded in Britain and Ireland in each month of the year from 1948 to 1973. The first-year age category runs from July to June of the following year. Based on data reanalysed from Hutchinson & Neath (1978).

	Jul	*Aug*	*Sep*	*Oct*	*Nov*	*Dec*	*Jan*	*Feb*	*Mar*	*Apr*	*May*	*Jun*
First-year birds as a percentage of total records	54%	59%	66%	64%	64%	57%	48%	41%	65%	66%	89%	85%

expected that less than a third of Little Gulls would be first-years, but 59–66 per cent of those seen between August and November were birds hatched in that year, and so autumn arrivals were heavily biased towards young individuals. This bias persisted throughout the year, and although relatively small numbers of Little Gulls are reported in Britain in winter, about half of those recorded were first-year birds, and again first-year birds dominated sightings in spring and early summer. This suggests that either adults are moving through Britain faster than the younger birds, or that many are using different migration routes. The high proportion of one-year old birds reported in Britain in May and June are presumably individuals spending the summer far from the breeding areas where they were reared.

However, the more recent data from the West Midlands and inland Hampshire do not support the dominance of sightings of first-year birds other than in August to October (Fig. 181). Records from inland areas show a large peak of adults passing inland through England in April and May, which must involve birds returning to their breeding areas in mainland Europe and beyond. It is interesting to note that no marked peak in numbers occurs in the West Midlands in autumn, but it is evident at inland sites in Hampshire, suggesting that adults are indeed following a different route in the autumn, remaining at sea or near the coast. The differences between the West Midlands data in Fig. 181 (1986–2010) and the national data in Table 66 (1948–73) relate to different periods of time and would appear to reflect real changes in the migration, wintering areas and behaviour of Little Gulls that have resulted in the much larger numbers visiting Britain in recent years.

Most Little Gulls passing through Britain may winter further south, frequenting Spain, Portugal and North Africa, but there is only one ringing recovery to support this. A Little Gull ringed on 6 September 1980 in County Durham was recovered in south-west France on 2 January 1982 at Sainte-Marie-de-Ré (46°09′N, 01°17′W), just north of Bordeaux). However, the numbers seen in Iberia and North Africa in winter are relatively few, and the wintering locations of the missing Little Gulls remain unknown.

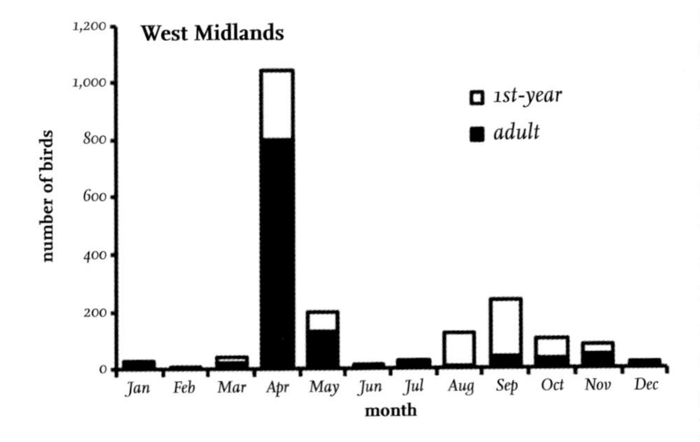

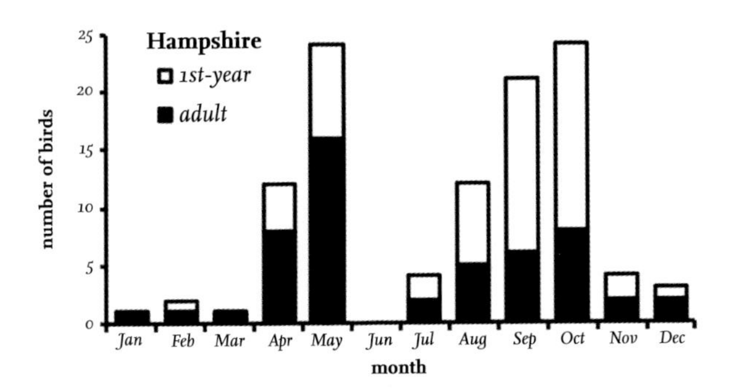

FIG 181. The numbers of adult and first-year Little Gulls reported inland in the West Midlands in 1986–2010 and Hampshire in 1951–92. A few birds that were not aged have been omitted, and data for first-year birds runs from July to June. Based on data collated by Alan Dean.

FOOD AND FEEDING

The Little Gull is essentially an insect feeder, consuming aquatic insects taken on the wing at inland marshes and waterbodies, or collected at or just below the water's surface. The species rarely dives, and small fish are taken only occasionally. On occasions, it has been recorded taking terrestrial insects and worms, and has infrequently been seen following agricultural ploughs, along with Black-headed Gulls.

Questions remain about where individuals feed outside the breeding season. Most autumn records in Britain and Ireland are of birds passing along the coastline, or resting in flocks on the shore or on the edges of reservoirs near the coast, but there are remarkably few records of them feeding. Outside the breeding season the Little Gull becomes a mainly marine species (at least in western Europe), but there is virtually no information on its diet in autumn and winter. Presumably, it consumes small, specialised marine invertebrates that normally float on the sea surface, which also form a key food source for wintering Kittiwakes (p. 293). Small fish are infrequently taken in the breeding areas, but these are unlikely to be consumed regularly at sea.

There is an assumption that Little Gulls outside the breeding season feed offshore, but perhaps within a few kilometres of a coastline to which they return to rest and roost. Some authors have suggested that flocks of Little Gulls seen on the shore have been driven there by strong winds, which prevent them from feeding offshore, but this may be a mistaken interpretation as some of the shore roost sites are visited day after day by resting and preening birds. In 1965, Charles Vaurie suggested that Little Gulls were pelagic, like the Kittiwake and Sabine's Gull, but I am not aware that birds have ever been seen at a great distance from the coast. It is possible that some individuals do cross the Atlantic, producing the many coastal records in North America, but if this species is truly pelagic and occurs far out into the North Atlantic, presumably birds would have been seen and recorded. Records of seabirds seen in transatlantic journeys on ships in the 1930s to 1950s make no mention of Little Gulls, and this includes seabirds seen during the many Atlantic crossings made by Neal Rankin and Eric Duffey in the 1940s, although these authors do not mention seeing Sabine's Gull. The biology of the Little Gull is poorly known and much has yet to be revealed – further investigations are clearly needed.

CHAPTER 11

Rare Gulls in Britain and Ireland

THERE ARE AT LEAST 17 SPECIES of gulls that have not bred in Britain or Ireland but have been recorded here intermittently in varying numbers. Many of these are sufficiently rare for records to be vetted nationally, while others are more frequent and the acceptance of sightings is left to local committees for comment and decision. Some species are at or near the edge of their normal ranges in Britain and Ireland, and occur here occasionally and sometimes in numbers. Other species included in this section are probably only accidental vagrants, occurring far from their normal ranges; they may be individuals displaced by exceptional weather conditions or birds actively pioneering new breeding or wintering areas.

This section includes several species that are difficult to identify in the field and therefore may be under-reported, while others are readily identified but occur infrequently. Many of the British records included have had their identifications critically considered by experts, and while they accept a proportion of the records submitted to them, some are considered insufficiently reliable and are rejected. It is very likely that some genuine records are dismissed because they lack detail, and no doubt a number of possible records are not even submitted for scrutiny. Thus, the total record for any species considered here is an unknown proportion of the actual total occurrences of the gull. Who knows how many rarities reach Britain or Ireland, but are not seen by observers? This number is likely to be large, considering that many accepted records are of birds seen for only one day before they moved away.

The records of several rare species have shown a marked increase in frequency in recent years. These, at least in part, reflect the marked increase of observers in recent years, better knowledge of the key characteristics of some rare species, improved optics and the increased use of photographic evidence in support of sightings. Many published and accepted records from 50 years ago and more were subjected to less vigorous scrutiny, and some have since been rejected. This is evident in the case of the Great Black-headed Gull (*Ichthyaetus ichthyaetus*), which is now recognised as a species that visited Britain on a single occurrence in 1859. The preserved body of the bird still exists in the Royal Albert Memorial Museum in Exeter, and we must trust that it was actually shot in Britain, as was claimed. Several later records of this species have all been rejected – although it is included by David Bannerman in his *Birds of the British Isles* (1953) and also here – presumably owing to a lack of details allowing confirmation of identification or doubts about where the specimens were obtained. Species reported less frequently in recent years are likely to be those that are decreasing as visitors to Britain and Ireland, while those with increased records may have become more frequent in their visits or this may simply be a result of the increased numbers of observers today.

Several gull species rarely recorded in Britain and Ireland breed extensively in North America, and it is interesting and surprising that those such as Franklin's Gull (*Leucophaeus pipixcan*) and the Laughing Gull (*Leucophaeus atricilla*) are reported much more frequently here than are gull species that breed much closer, in Spain and southern France, such as Auduoin's Gull (*Ichthyaetus audouinii*) and the Slender-billed Gull (*Chroicocephalus genei*). These differences would appear to be linked to the normal migration or dispersal directions, which take the European breeding individuals south and away from Britain. The Yellow-legged Gull (*Larus michahellis*; p. 295), American Herring Gull (*Larus smithsonianus*) and Caspian Gull (*Larus cachinnans*) have been added to the British list recently, mainly because of changes in taxonomy, and there is no history of earlier records of these species.

Two factors can contribute to confusion over the identification of rare species. First, hybridisation can produce individuals that have characteristics similar to those of other species. Second, some individuals do not show characteristics that allow them to be separated from related species, and this applies both to the immature stages and adults. Adult American Herring Gulls show wing-tip characteristics that overlap with some European Herring Gulls (*Larus argentatus*); although the possible distinguishing character of a small pale spot on the underside of the tip of the 10th primary has been suggested, its reliability has not been confirmed. A high proportion, but not all, first-year American Herring Gulls show plumage characteristics that differ from European Herring Gulls of similar age reared in Europe. Some Caspian Gulls are identifiable in the field, but other individuals – particularly females – are more difficult to distinguish with certainty from other gull species. Gulls are variable in plumage, and for some species, not every individual is necessarily identifiable in the field. Too little caution is given to the differences in size and structure of males and females of each species, particularly to the size and depth of the bill, which – in males at least – increases in depth for several years. Experience of the observer is important, as is a knowledge of the variability within the commoner species. Hence, numbers of records of Caspian Gulls are reported as 'possible', because the occurrence of hybrids and back-crosses is a major problem, producing a range of variation in individuals, which share some of the characteristics of both parent species.

INCREASED RECORDS OF RARE GULLS

Without doubt, the number of capable field observers has increased during the past 55 years, but little attempt has been made to quantify their effect on the numbers of records. The frequency with which Ross's Gull (*Rhodostethia rosea*) has been recorded in Britain has not changed since about 1980, and records may even have decreased, while many of the other species show varying increases in sightings in recent years. As a preliminary conclusion, it appears that increased numbers of observers may have been responsible for doubling the number of annual records between 1960 and 2017, and only increases greater than this may reflect genuine increases in the occurrences of each species. The change in numbers of observers is an unexplored field that would produce much valuable information.

Approximate numbers of accepted records of rare species of gulls in Britain are shown in Table 67, with records separated into those up to 1949 and those from 1950 to 2017. The obvious lower numbers of records in the earlier years not only reflects the small number of observers, but also the poorer quality of field guides available at the time to aid identification and the lack of awareness of possible species.

In the species presented below, some of their salient features are mentioned but these are not presented for identification purposes. More extensive descriptions and accounts, such as *Collins Bird Guide* by Lars Svensson *et al.* (2010), Peter Grant's *Gulls: A Guide to Identification* (1986) or *Gulls of Europe, Asia and North America* by Klaus Malling Olsen (2004), should be consulted for help with identification. Details of their biology and behaviour can be found in *Birds of the Western Palaearctic* (Cramp & Simmons, 1983).

TABLE 67. The approximate numbers of accepted records of rare gulls in Britain in two periods up to 2017 and in Ireland up to 2014. Note that these are not necessarily all accepted records. The 'Comments' column includes details of years with atypical larger numbers, while dashes indicate that no data are available. Based in part on Reports of Rare Birds published in *British Birds*.

Species or sub-species	*Britain up to 1949*	*Britain 1950–2017*	*Ireland up to 2014*	*Comments*
Ivory Gull	84	66	19	9 in total in 2013
Ross's Gull	1	99	22	8 in total in 2002
Glaucous Gull	Many	Many	Many	Identification confused with Iceland Gull in early years
Iceland Gull	Many	Many	Many	Identification confused with Glaucous Gull in early years
Thayer's Gull	0	First record accepted in 2014	9	Some of these records are considered dubious
Kumlien's Gull	1	Up to 51 possible records	0	None accepted between 1990 and 2013
Sabine's Gull	30–40	100s	—	Many inland. No Irish data
Laughing Gull	1	200	43	Totals of 58 in 2005 and 22 in 2006
Bonaparte's Gull	8	237	76	Totals of 11 in 2006 and 14 in 2013
Franklin's Gull	0	77	17	Maximum of 6 in total in any year
Ring-billed Gull	0	408	Many, total unknown	First recorded in 1973; numerous every year from 1982 to 2017
Glaucous-winged Gull	0	3	0	1 in 2006 and 1 in 2008 in Britain; 1 in Ireland in January 2016; Fair Isle March 2017
American Herring Gull	0	33	96	9 in total in 2002. Difficult to identify
Azores Herring Gull	0	2	13	Difficult to identify
Caspian Gull	0	87+ claimed	0	32 in 1999
Audouin's Gull	0	8	0	First recorded in 2003
Slender-billed Gull	0	10	0	All since 2013
Great Black-headed Gull	1	0	0	1859
Slaty-backed Gull	0	1	1	1 in London and Essex in 2011; 1 in January 2015 in Ireland
Vega Gull	0	0	0	1 record in Ireland, January 2016, remains unconfirmed

Problems of identifying individuals in the Herring Gull and Lesser Black-backed Gull complex

Examination of members of the Herring Gull complex captured for ringing or culled reveals the large amount of variation that occurs among individuals breeding at the same site. Often, such variation is not given sufficient consideration when applied to the identification of live gulls in the field.

1. *Variation in the Herring Gull.* There is considerable variation in the shade of grey of individuals within a given colony and throughout the Herring Gull complex over its wide geographical range. Occasionally, otherwise typical European Herring Gulls with yellow legs are reported, and some breeding birds acquire a yellow tinge to the leg colour at the beginning of each breeding season. Herring Gulls breeding in northern Norway and north-west Russia (*Larus argentatus argentatus*), and wintering in Britain, are typically larger and have darker mantles than those breeding in Britain (*L. a. argenteus*), but the variation within each of these subspecies is considerable.

   While about 95 per cent of the northern *argentatus* birds can be identified as such when examined and measured in the hand, a few (mainly females) that were subsequently found breeding within the Arctic Circle in northern Norway failed to be correctly identified on the basis of biometrics, because of their smaller size and overlapping shades of grey on the mantle and wings. The black wing-tip pattern varies considerably, with the so-called *thayeri*-type pattern occurring in an appreciable minority (Fig. 8), but by no means in all adults breeding in northern Scandinavia and only rarely in birds breeding in Britain.
2. *Age and sex effects.* The black wing-tip pattern varies with age and is best known in the *argenteus* subspecies. The white mirrors tend to increase in size between three-, four-, five- and six-year-old individuals. In European Herring Gulls, and possibly in other species, the depth of the bill (particularly in males) continues to increase with age for several years, although that of females changes little.
3. *Variation within a species.* In Britain, breeding Herring Gulls show considerable variation in the black wing-tip pattern, with differences between individuals in the number of primaries with a black tip and the extent of the white mirror. These differences can occur over quite short distances, such as between the east and west sides of Britain. For example, in one study almost twice the proportion of adults had fewer than six primaries with black pigmentation at or near the tips on the east side of Britain (28 per cent) than those examined on the west side (15 per cent).
4. *Variation due to light and direction.* In the field, the shade of grey on the mantle can appear to vary with the angles of the bird relative to the directions of light and the observer's position, depending on whether the bird is seen in sunlight or under cloudy conditions, and the age (wear) of the plumage.
5. *Hybrids.* These are known to exist between several species of gulls and can lead to incorrect identification. Hybrids often occur more frequently at the edge of the distribution range of a species, where numbers of one of the parent species are low.

RARE VISITORS FROM ARCTIC BREEDING AREAS

Ivory Gull (*Pagophila eburnea*)

This moderate-sized species has a wingspan of about 1 m. It breeds in colonies, often on cliffs, in the High Arctic at Nunavut, on Ellesmere, Devon, Cornwallis and Baffin islands, in Greenland and Spitsbergen, and also on flat ground (Figs 182 & 183) in northern Russia. During the winter, Ivory Gulls are usually found near areas of open water adjacent to sea ice, and they rarely come as far south as Britain and Ireland. Taxonomically, this species is probably related to Sabine's Gull (*Xema sabini*) and Black-legged Kittiwakes (*Rissa tridactyla*), which also have short legs and a similar build. It is often attracted to seal carcasses and other carrion, including the faeces of mammals. In the breeding season it also feeds on fish and crustaceans, and is known to take eggs of other species.

FIG 182. Large colony of Ivory Gulls (*Pagophila eburnea*) on Novaya Zemlya. (Fred van Olphen)

FIG 183. Closer view of Ivory Gulls (*Pagophila eburnea*) on nests on Novaya Zemlya. (Fred van Olphen)

FIG 184. Adult Ivory Gull (*Pagophila eburnea*) with two chicks near fledging. (Fred van Olphen)

The adult plumage is entirely white, the tip of the bill is yellow and the short legs are black. Adults could be confused with albino individuals of other small gulls, and the black legs would not rule out an albino Kittiwake, but the presence of a partially yellow bill might confirm its identity. Immature birds have some black spots on the tips of the primaries and their coverts, and fewer on the secondary feathers, but most of the wing is pure white, while the front of the head is slightly dark.

There are at least 150 British records and 19 Irish records of the Ivory Gull, with 12 individuals recorded in 1973. The great majority are from coastal areas around the whole of the British and Irish coastlines, with the peak of records spread from November to January. About a third of these records are birds in adult plumage. The annual records of Ivory Gulls in Britain have increased by about 50 per cent over the last 50 years, but much of this is due to the influx of individuals recorded in 2013 (Fig. 185). If this one value is excluded on the grounds that it was an exceptional year, there is no convincing evidence of a change in numbers recorded over the past 55-year period. If increased numbers of observers have resulted in increased sightings, then the data suggest that the Ivory Gull is actually declining as a visitor to Britain and Ireland.

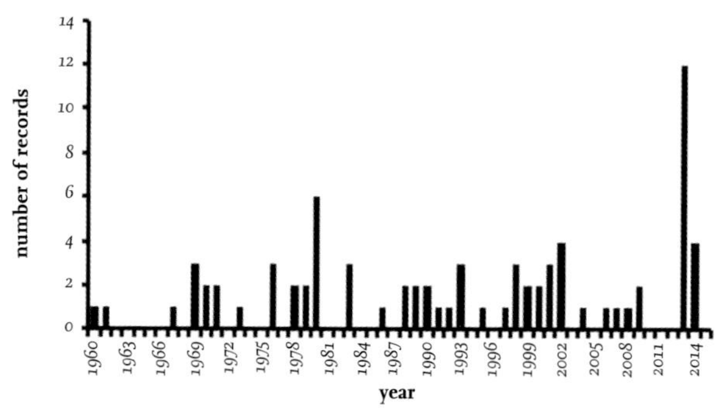

FIG 185. The number of records of Ivory Gulls (*Pagophila eburnea*) in Britain from 1960 to 2014. There is no meaningful trend if the peak numbers in 2013 are disregarded. Based on reports of scarce migrant birds in *British Bird, Rare Bird Alert* and annual county reports.

Ross's Gull (*Rhodostethia rosea*)

A small gull with a wingspan of about 75 cm (only slightly larger than the Little Gull, *Hydrocoloeus minutus*), Ross's Gull breeds in small colonies on the tundra in the High Arctic of North America, Greenland and north-east Siberia. It usually winters at the edge of the Arctic pack ice and is only occasionally recorded further south. In winter, it could be confused with the Little Gull, but it has longer wings and the tail is pointed rather than forked as in the Little Gull (Figs 186 & 187). In summer, the black neck collar and lack of black on the underwing separate adults from the Little Gull.

FIG 186. Adult Ross's Gull (*Rhodostethia rosea*) in winter plumage. (Steven Seal)

FIG 187. First-year Ross's Gull (*Rhodostethia rosea*), Vlissingen, Netherlands. (Michael Southcott)

The first British record was a bird shot in Yorkshire in the winter of 1846/47. There is some uncertainty about the exact date but not about the identity of the species, yet this record is often disregarded. The second British record was 90 years later, in Shetland in April 1936, and the third occurrence was not until 1960. More were recorded in every year between 1991 and 2002, and by 2017 a total of 99 records had been accepted for Britain and 22 from Ireland. The pattern of sightings is very different from those of the Ivory Gull (Fig. 185), with records tending to increase up to 1983, and 16 consecutive years (1974–89) with at least one record each year. While numbers fluctuate from year to year, there is only a hint of long-term changes in the annual numbers of records, in that there have been seven years since 2000 when none was recorded (Fig. 188), although a peak of seven records in

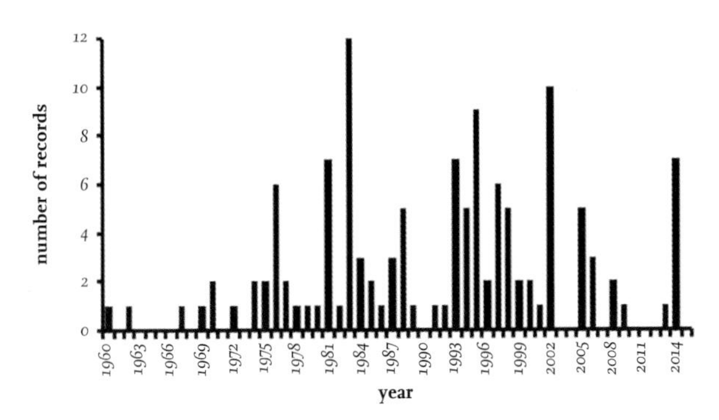

FIG 188. The number of records in Britain of Ross's Gull (*Rhodostethia rosea*) since 1960. There is no obvious trend, but there are fewer records since 2003 and several years during this period in which no individuals were recorded. Based on reports of scarce migrant birds in *British Birds*, *Rare Bird Alert* and annual county reports.

2014 now casts doubt on this possible decline and draws attention to the variability in occurrences. Three-quarters of records are of adults, and most occurred in January and February, with almost all found at localities at or near the coast.

In the breeding season, Ross's Gulls feed mainly on insects, but there is no reliable information on the food consumed in winter, although it is thought to consist of crustaceans and small fish. Occasionally, individuals have been seen walking over mudflats, presumably searching for small invertebrates.

Glaucous Gull (*Larus hyperboreus*)

To the Sea, to the Sea! The white gulls are crying,
The wind is blowing, and the white foam is flying.

– J. R. R. Tolkien, The Return of the King *(1955)*

This large gull, which has a wingspan approaching 1.5 m, is almost the size of the Great Black-backed Gull (*Larus marinus*). It is an Arctic breeding species with circumpolar distribution. In the Atlantic Ocean and neighbouring areas, the Glaucous Gull breeds in Greenland, Iceland, Jan Mayen, Svalbard, Franz Josef Land and Novaya Zemlya. Its large size and the lack of any black pigment on the wing-tips readily separates it from all other gulls (Fig. 189), with the exception of the smaller Iceland Gull (*L. glaucoides*; see below). Immature Glaucous Gulls are also pale but slightly less so than adults (Fig. 190).

Some female Glaucous Gulls overlap in size with male Iceland Gulls; correct field identification of these proved difficult in the past, until more detailed differences between the species were recognised. The most obvious difference is the general overall appearance, often referred to as the jizz, but this requires previous

FIG 189. Near adult Glaucous Gull (*Larus hyperboreus*). (Mark Leitch)

FIG 190. First-winter Glaucous Gull (*Larus hyperboreus*). (Alan Dean)

experience of the two species. Most Glaucous Gulls are robust and have a similar build to the Great Black-backed Gull, while Iceland Gulls tend to be slimmer, with a build similar to that of the Common Gull (*Larus canus*). One valuable difference at rest is the small extension of the wing beyond the tail in the Glaucous Gull, and which is appreciably longer in most Iceland Gulls. However, this is not always a good guide because the outer primaries are moulted in autumn. The growth of the new longest primary is often not completed until late October and in some individuals not until the end of December, making this otherwise useful distinction of less value during several months of the year. Another difference is the size of the bill. While this is a useful distinction year-round, there is considerable variation between the sexes and species in terms of the size and depth of the bill, making the separation between female Glaucous and male Iceland gulls less easy. First-year Glaucous Gulls have black on the tip of the bill, while in Iceland Gulls of the same age, the bill is darker for over half of its length. A further identification problem lies with Glaucous Gulls breeding with European Herring Gulls in Iceland and with Glaucous-winged Gulls (*Larus glaucescens*) in Canada, and producing hybrid individuals with varying amounts of black or grey on the wing-tips.

The only British breeding record is of a Glaucous Gull that paired with a Herring Gull annually between 1975 and 1979 in northern Scotland, and successfully reared young in several of these years. The species is not likely to breed regularly in Britain, as there are few records of adults remaining here during the summer months.

The Glaucous Gull's diet is similar to that of the Great Black-backed Gull (p. 219) and it consumes a wide range of vertebrates and marine invertebrates. It regularly feeds on fish and is a major predator of adult seabirds and their eggs during the breeding season, while also scavenging on carcasses of mammals of all sizes when these are available. The species frequents landfill sites and fish docks in northern areas at other times of year, acting as a scavenger at both locations.

Adults nest in small groups of up to 15 pairs or as single pairs. Nesting areas are often associated with seabird colonies, and the nest is commonly placed on a raised area that gives a good all-round view. The Arctic Fox (*Vulpes lagopus*) is the main terrestrial predator of the species' eggs and chicks.

An appreciable proportion of Glaucous Gulls are resident near the breeding areas, and particularly so if there are refuse dumps or commercial fishing activities nearby throughout the year. Others (particularly immature individuals) move south, and those that reach Britain are at the southern edge of their normal winter range and are usually recorded as solitary individuals, although small groups do occur, particularly in Shetland. Individuals occasionally join flocks of other gulls to feed and roost, although the species has been reported feeding alone at landfill sites and leaving when other gulls come in to feed. Inland sightings are now frequently reported in Britain but are rare among the old records. This increase has developed recently, along with the increase in sightings of other large gulls feeding and roosting in winter at inland locations. The numbers of individuals reported in Britain in winter fluctuate markedly from year to year, the differences possibly reflecting variations in breeding success and the feeding conditions further north.

An analysis of records of Glaucous Gulls in the West Midlands made by Alan Dean clearly shows the fluctuation between years (Fig. 191), with a marked peak of records between December and February, and a total absence between April and September (Fig. 192). Numbers of adults never exceeded numbers of

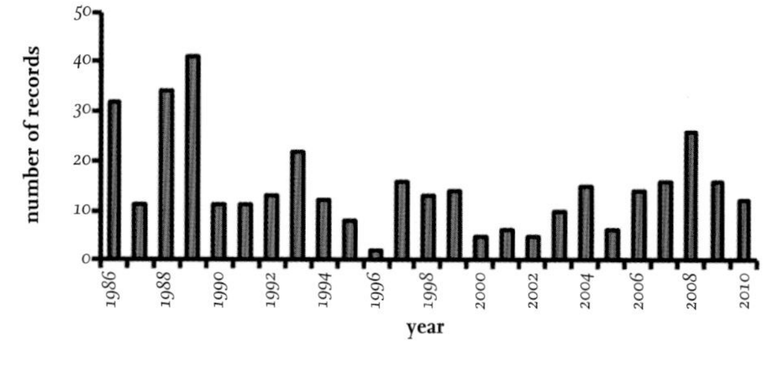

FIG 191. The number of Glaucous Gulls (*Larus hyperboreus*) recorded in the West Midlands, England, each year from 1986 to 2010. Data from Dean (n.d.).

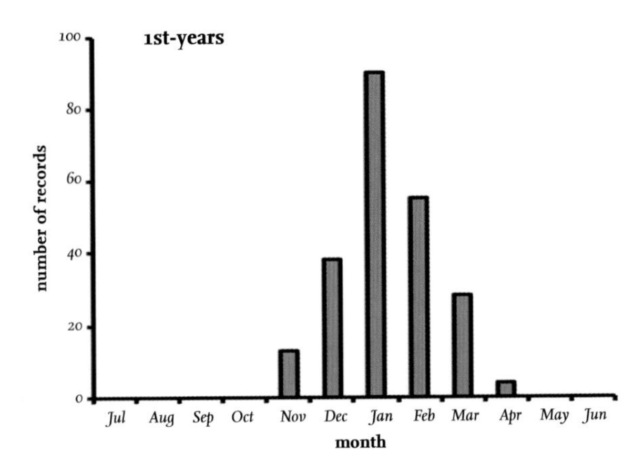

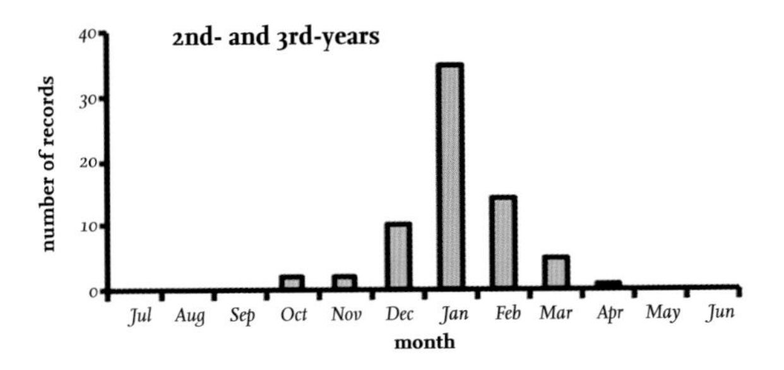

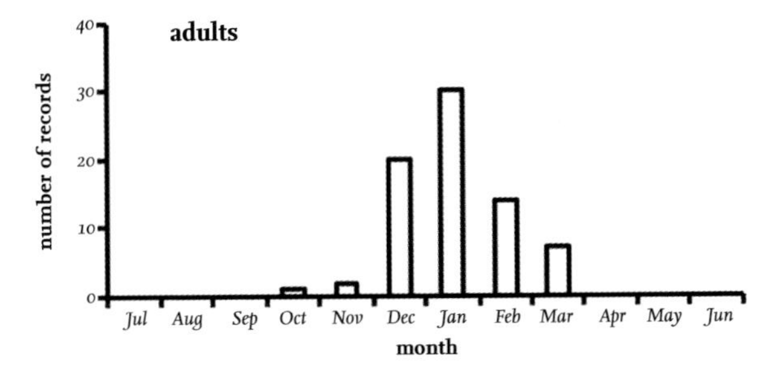

FIG 192. The months in which individual Glaucous Gulls (*Larus hyperboreus*) of different ages were first recorded in the West Midlands, England, in the period 1986–2010. Based on information collected by Alan Dean (n.d.).

immature birds in any year (Fig. 192), and in four years no adults were seen. Most immature birds were in their first year of life, which confirms the general impression that young birds tend to move further south than adults. Most adults did not arrive until December, after completing their primary moult, and first-year birds arrived at the same time despite not moulting. Numbers of Glaucous Gulls reported in the West Midlands vary markedly from year to year, with large numbers in 1986 (32), 1988 (34) and 1989 (41), but only two records in 1996

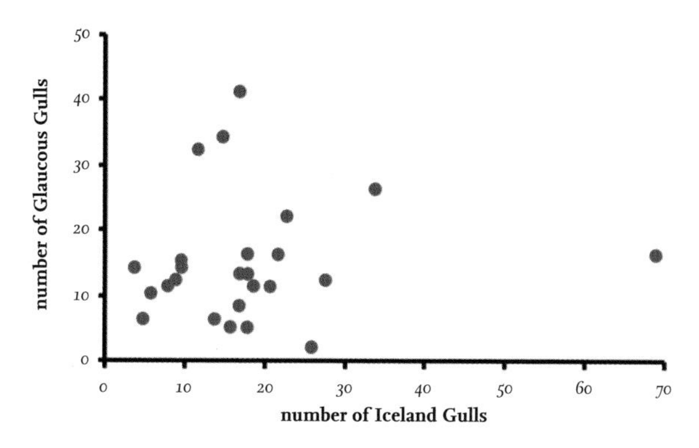

FIG 193. The numbers of Glaucous Gulls (*Larus hyperboreus*) and Iceland Gulls (*L. glaucoides*) recorded each year from 1986 to 2010 in the West Midlands, England. There is no correlation between the numbers of the two species seen each year. Data from Dean (n.d.).

(Fig. 191). Numbers recorded elsewhere in Britain and Ireland fluctuate similarly from year to year, and on average, several hundred individuals probably occur here in most years. There is no correlation between the numbers of Glaucous Gulls and Iceland Gulls recorded each year in the West Midlands (and elsewhere) (Fig. 193), and presumably, different factors cause the annual fluctuations of these gulls visiting Britain.

Both the Glaucous Gull and Iceland Gull are currently given Amber status as species of conservation concern in Britain on the grounds that only small numbers are reported annually here. The global population shows no evidence of a change in abundance, however, and so the only concern in Britain would seem to be that many birdwatchers do not see these species as often as they would wish.

There are eight records of Glaucous Gulls ringed as nestlings and subsequently recovered in Britain or Ireland in winter. Three were marked on Bear Island (74°N, 19°E), one on Svalbard (76°N, 26°E), three in Iceland and one in Norway (61°N, 34°E). Of these eight recoveries, one was in England, one in Ireland and six in Scotland, confirming the greater frequency of Glaucous Gulls visiting Scotland.

Iceland Gull (*Larus glaucoides*)

The Iceland Gull is essentially a smaller and less robust version of the Glaucous Gull (see above), with a wingspan of about 1.3 m. It totally lacks black on the wing-tips (Fig. 194). For many years, there was confusion over the identification of these two arctic breeding gulls in the field, but now that better field characteristics have been identified, most individuals can be reliably distinguished. While Glaucous Gulls are large (most the size of Great Black-backed Gulls), Iceland Gulls are more lightly built and usually slightly smaller than Herring Gulls. They have a less substantial bill than Herring Gulls (Fig. 195), a difference comparable to the size difference of the bills of Herring and Common gulls.

While Glaucous Gulls have a circumpolar distribution, the Iceland Gull is more restricted and, despite its name, breeding is mainly limited to Greenland and north-east Canada. It winters in the North Atlantic as far south as Britain, the east coast of Canada and the extreme north-eastern states of the USA.

There have been two ringing recoveries of Iceland Gulls in our region, both marked as nestlings in Greenland and recovered on the east coast of Britain. A second-year was found recently dead at Teesmouth, north-east England, in July, and the other was a first-year bird found on the east coast of Scotland in January.

FIG 194. Second-winter Iceland Gull (*Larus glaucoides*). (Tony Davison)

FIG 195. Probable second- or third-summer Iceland Gull (*Larus glaucoides*) standing on an ice floe. (Tony Davison)

Iceland Gulls occur in Britain annually as winter visitors, but in markedly variable numbers – this is evident from the data from the West Midlands (Fig. 196), where most arrive in December or later (Fig. 197). As with Glaucous Gulls, Britain is at the southern edge of their normal winter distribution. The winter of 2009/10 was exceptional, with many more records than in any other year. Most of the birds were immature individuals in their first or second winter, and were not associated with the presence of large numbers of Glaucous Gulls. Large numbers of adults were present in Shetland in the winter of 2011/12, but few immature birds were present. Occasionally, an immature bird spends the summer in Britain.

The Iceland Gull, like all gulls of a similar size, has a widely varied diet that includes the eggs of other bird species and fish. In its wintering areas, scavenging appears to be the main method of feeding, particularly discards from fishing boats and refuse at landfill sites.

The numbers of records of Iceland Gulls have increased dramatically in western Europe in the last 70 years. Much of this increase can be attributed to more numerous and better-informed observers, but it is possible that more birds are now wintering towards the south of the species' winter range. There is no evidence that the size of the population of the Iceland Gull has changed in its breeding areas and it cannot reasonably be considered a species of conservation concern in Britain.

Thayer's Gull (*Larus thayeri?*) and Kumlien's Gull (*Larus glaucoides kumlieni*)
Confusion still exists over the taxonomic status of two gulls related to the Iceland Gull, namely Thayer's Gull and Kumlien's Gull, which have varying amounts of grey or black on the wing-tips. Some consider the two to be subspecies of the Iceland Gull, while others believe they represent one or even two separate species.

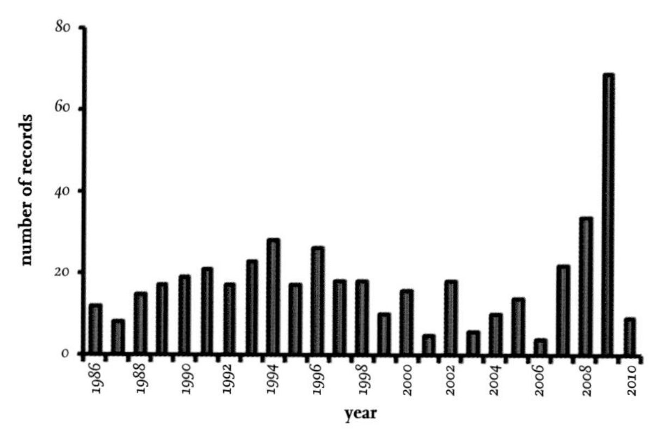

FIG 196. The numbers of individual Iceland Gulls (*Larus glaucoides*) recorded each year in the West Midlands, England, from 1986 to 2010. Data from Dean (n.d.).

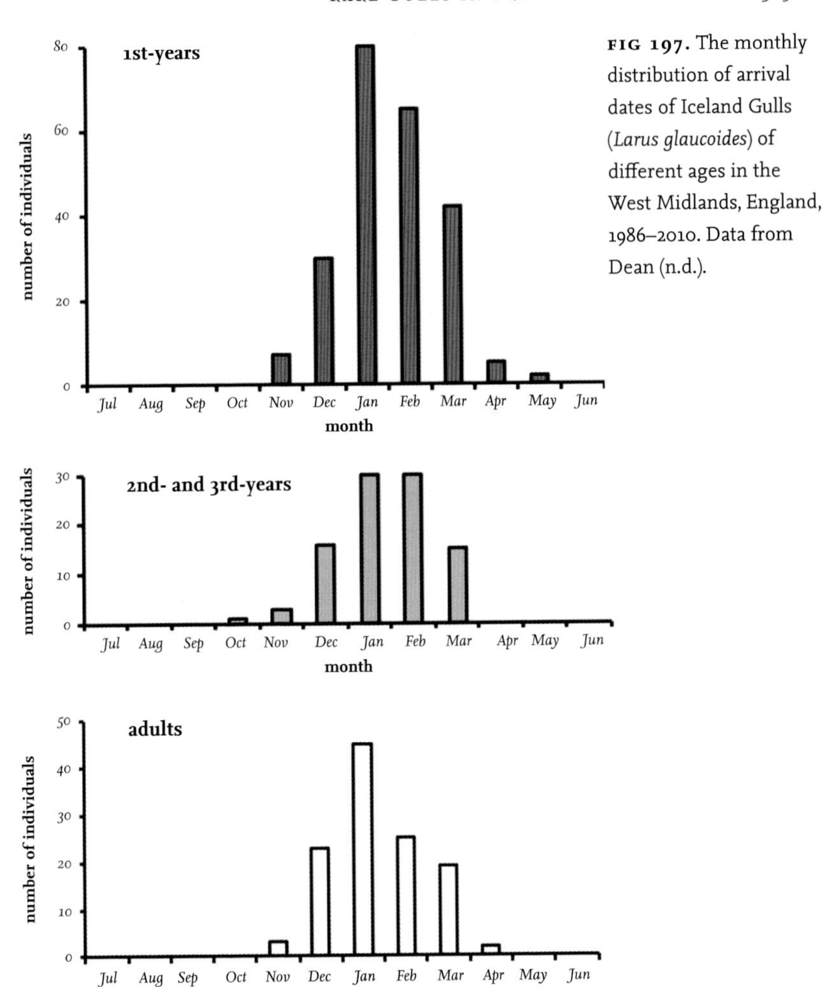

FIG 197. The monthly distribution of arrival dates of Iceland Gulls (*Larus glaucoides*) of different ages in the West Midlands, England, 1986–2010. Data from Dean (n.d.).

Here, the two gulls are considered together, irrespective of their uncertain taxonomic status.

Individuals referred to as Kumlien's Gull are similar in size to the Iceland Gull, but they have a small amount of darker pigment on the wings (Fig. 198) and breed mainly on Baffin Island, to the west of Greenland. Thayer's Gull is of a similar size, but has more extensive dark coloration on the wing-tips and nests in the extreme Arctic of Canada (Banks Island, Southampton Island and Baffin Island), as well as in north-west Greenland. Both forms currently present a complex challenge in avian taxonomy. Individuals regarded as Kumlien's Gull

FIG 198. A probable Kumlien's Gull (*Larus glaucoides kumlieni*). (John Kemp)

are particularly variable in plumage, more so than in any other gull. They also vary considerably in size, with some males reaching a similar size to Herring Gulls, while extreme females are only a little larger than female Common Gulls. The breeding distribution of Kumlien's Gull in arctic Canada overlaps slightly with that of Thayer's Gull, and there is some interbreeding between the two forms. Adding to the confusion, a recent DNA study has suggested that Thayer's Gull may be more closely related to the Glaucous-winged Gull than to either the Iceland Gull or Kumlien's Gull. Because of the degree of variation, they are probably the most difficult gulls to separate and identify in the field, particularly when outside their normal ranges. The plumage of Kumlien's Gull tends to be intermediate between that of the Iceland Gull and Thayer's Gull in the extent of the grey markings on the wing-tips, ranging from white-winged individuals to others that show much dark pigmentation.

The disagreement about the status of these gulls is considerable and goes back more than a century. A few specimens collected in 1901 on Ellesmere Island in Canada were described by W. S. Brooks in 1915 as a new species, and he named it Thayer's Gull (*Larus thayeri*). In 1917, Jonathan Dwight reduced it to a subspecies of the Herring Gull and then later, in 1925, he concluded that Kumlien's Gulls were hybrids between Thayer's Gull and the Iceland Gull. In 1933, Percy Taverner claimed that Kumlien's Gull was a distinct species but in 1937 regarded Thayer's Gull as a subspecies of the Herring Gull. In 1951, Finn Salomonsen classified Thayer's Gull as a High Arctic form of the Iceland Gull and indicated that individuals of Kumlien's Gull were hybrids between the Iceland Gull and Thayer's Gull. In 1961, A. Macpherson reported that the Herring Gull and Thayer's Gull did not interbreed in colonies where they occurred together, and so they were not subspecies, and he concluded that Kumlien's Gull was a subspecies of the Iceland

Gull. In 1966, Neil Smith reported that on Baffin Island, Kumlien's and Thayer's gulls did not interbreed and so should be regarded as separate species. He went on to claim that by changing the colour of the orbital ring, he had induced mixed pairing of Thayer's and Glaucous gulls.

In 1968, G. M. Sutton was the first of several authors to publish sceptical reviews of Smith's 1966 research, but five years later, in 1973 (and perhaps influenced by Smith's work), the American Ornithologists' Union (AOU) concluded that Thayer's Gull was a separate species. However, in 1975 and 1976, B. Knudsen visited the same site used by Smith for his study and found widespread interbreeding between Kumlien's and Thayer's gulls, which was confirmed by Snell when he visited the site a few years later. In 1985, A. J. Gaston and R. Decker also confirmed interbreeding between Thayer's and Kumlien's gulls, this time on Southampton Island in Hudson Bay.

The British Ornithologists' Union currently considers Thayer's Gull and Kumlien's Gull to be subspecies of the Iceland Gull, referring to them as *Larus glaucoides thayeri* and *Larus glaucoides kumlieni*, and suggests that they are part of a gradual change (cline) in characteristics over the geographical range of the Iceland Gull. However, the AOU regarded Thayer's Gull as a distinct species until it changed its opinion in 2017, and several European authorities, including the Irish Rare Birds Committee, continue to give it species status. In recent years, several birds seen in Britain have been claimed as Thayer's Gull, and the first accepted record was a bird seen near Oxford in 2007.

There has been a suggestion that Kumlien's Gull is a series of hybrids between the Iceland Gull and Thayer's Gull, and that these hybrids are fertile, resulting in the production of many back-crosses, each with different genetic make-up and plumage. Variability of this kind is referred to as a hybrid swarm and could possibly explain the large variation in plumage reported in Kumlien's Gull, but this interpretation is not currently generally accepted and confusion over taxonomic status continues.

There is a strong desire among many that Thayer's Gull be regarded as a distinct species, but this is based more on sentiment and a desire for an additional species than on sound scientific evidence. There is still a lack of data, both from the field and from mitochondrial DNA studies. The matter is still not decided, although in 2017 opinion in North America changed and the AOU now considers both Thayer's and Kumlien's gulls as forms of the Iceland Gull. To date, no change in their taxonomic status in Europe has occurred. In this book and for convenience, I have placed Thayer's Gull as a species and, because Kumlien's Gull is so variable, I have, arbitrarily, considered it as a subspecies of the Iceland Gull.

Birds resembling Thayer's Gulls regularly winter on the Pacific coast and around the Great Lakes of North America, while Kumlien's Gulls winter mainly on the east coast of North America. The significance of these differences has yet to be considered.

There are nine accepted Irish records of Thayer's Gull, with the first seen in February 1990 in County Cork, and all the records are restricted to December to March. Seven of these records were birds in their first year and two were adults. The first accepted British record of Thayer's Gull was photographed on 6 November 2010 at Pitsea landfill site, Essex, and was an adult. The second accepted record was a first-year bird reported and photographed at Elsham, Lincolnshire, on 3–18 April 2012, while an immature bird stayed on Islay, Scotland, for about eight weeks. Elsewhere in Europe, there are accepted records from Iceland, Norway, Denmark, the Netherlands and Spain, although the only adult among these was seen in Iceland.

In the field, adult Thayer's Gulls show close similarities with the Herring Gull and males can exceed 1,000 g. As a result, some males are larger than female British Herring Gulls. Most accounts identifying adult Thayer's Gulls using the wing-tip pattern have been written by Americans and are based on comparisons with the American Herring Gull, which has small white wing mirrors. The situation is different in Europe, however, where patterning on the wing-tips of many members of the Herring Gull subspecies *Larus argentatus argentatus*, which breeds in northern Norway, show larger white mirrors and have identical or very similar '*thayeri*' white-and-black wing-tip patterning (see Fig. 8, Chapter 1). Most of the *argentatus* subspecies of Herring Gulls visiting England are darker than British breeding birds (subspecies *argenteus*), but not all can be separated by the shade of grey on the wing. Fig. 9 (Chapter 1) shows the wing-tip patterning of an adult Herring Gull captured in north-east England in winter. Was one of these really a Thayer's Gull that was misidentified? This turned out to be unlikely, as it was later found breeding in Norway.

Adult Thayer's Gulls within their normal breeding range have been reported with both dark and yellow eyes. The eye colour of Herring Gulls changes with age, but this has not been well investigated in Thayer's Gull. Leg colour was recorded as bright pink in the first British record of Thayer's Gull and this may prove to be a good character. Herring Gulls have pinkish-grey legs, but the colour intensity changes as the breeding season approaches. It may simply be the case that some individuals cannot be identified in the field. The identification of Thayer's Gull outside of its usual range is made difficult owing to the presence of similar-looking hybrids, involving Herring Gulls, American Herring Gulls or Glaucous-winged Gulls as one of the parents, mating with Iceland Gulls.

The difficult identification of Kumlien's Gull and its dubious status caused the British Birds Rarities Committee to cease considering records after 1998. As result, the numbers of records of this species in Britain is uncertain but there have been at least 51 possible sightings.

Sabine's Gull (*Xema sabini*)

Sabine's Gull is a small gull, similar in size to the Black-headed Gull (*Chroicocephalus ridibundus*), and has a circumpolar distribution, breeding in the High Arctic tundra of Siberia, northern Greenland, Alaska and Canada. A few authorities still place it in the genus *Larus*, but its closest relation may be the Ivory Gull.

The species has a contrasting patterned wing plumage (Fig. 199). Adults in the breeding season have a dark grey hood, which in some individuals is retained well into the autumn (Fig. 200). Outside the breeding season it is a pelagic species, remaining well offshore and usually out of view of land-based observers unless driven ashore by strong winds. In winter plumage, the adults seem Kittiwake-like, but with more black on the outer leading edge of the wing and a forked tail that, surprisingly, is often not obvious when the bird is in flight.

FIG 199. Adult Sabine's Gull (*Xema sabini*) in flight and in summer plumage. (Phil Jones)

FIG 200. Adult Sabine's Gull (*Xema sabini*) in summer plumage. (Steven Seal)

Immature birds also possess a contrasting wing pattern similar to Kittiwakes of the same age, but there is greater contrast between light and dark shades, and they have a brown-grey mantle, neck and hood (Figs 201 & 202).

Many Sabine's Gulls winter south of the equator in the Atlantic and Pacific oceans. The migration of this gull has been investigated using geolocators on adult birds breeding in north-east Greenland in 2007. Ten of these birds were recovered at the same colony in the following year, which allowed the movement of these individuals to be followed for a full year. Autumn migration took the birds from Greenland towards Britain and the Bay of Biscay, where most had

FIG 201. First-year Sabine's Gull (*Xema sabini*). (Steven Seal)

arrived by the end of August and remained for an average of 45 days. Most left this staging area by mid-October and moved south, close to the west coast of Africa, to winter off the south-west coast of South Africa in a productive area called the Benguela Upwelling. They remained there for about 150 days and did not start their return migration until mid-April, when they moved rapidly to a staging area off the west coasts of Morocco, Mauritania and Senegal, where they remained for about 19 days. They then moved north through the middle of the North Atlantic, well away from the coastline of Europe, to return to their breeding areas in Greenland. The migration covered more than 30,000 km in a year.

The timing of the Sabine's Gull migration agrees closely with the records of birds seen annually but in variable numbers in Britain, where most are recorded between September and November. In contrast, few are recorded in Europe on their springtime return migration to the Arctic, which is rapid and direct, with most individuals remaining far from Britain. This migration pattern probably accounts for many of the records being reported from the western side of Britain and Ireland, and the considerable year-to-year variation in the number of records.

Sabine's Gulls are subject to occasional wrecks, when numbers of individuals are driven onto the European coast as a result of strong winds associated with intense atmospheric depressions. In these circumstances, some are forced inland. Such events have been studied in detail by Norman Elkins and Pierre Yesou

FIG 202. First-year Sabine's Gull (*Xema sabini*). Note the tail does not always look forked. (Dave Turner)

(1998), and the most noteworthy involved more than 300 individuals reported in Britain on 16 October 1987 and successive days, with almost all the reports from the region extending from Dorset to Essex. More frequent autumn wrecks have been reported from France. The largest there occurred in the Vendée region in 1993, when more than 1,600 individuals were reported inland, yet few were seen in England at the same time. Another wreck occurred in 1995, with hundreds recorded in France. In the autumn of 1997, about 1,000 Sabine's Gulls were seen passing south off the coast of Ireland, with 347 recorded from the coast of County Kerry on 29 August and unusually high numbers passing south through the Strait of Dover in October. Surprisingly, most of the individuals reported were full adults, with birds that had hatched that year forming only a few percent of those seen. Such events, but on a smaller scale, have occurred in other years, and the greater frequency in France than in Britain suggests that many birds are aggregating and feeding off the coast south of Brittany in autumn, rather than off Britain and Ireland. Franklin's Gull (p. 337) and Sabine's Gull are the only gull species that make regular and extensive trans-equatorial migrations.

Sabine's Gulls nest on flat ground, usually next to water and sometimes in or adjacent to Arctic Tern (*Sterna paradisaea*) colonies. In the breeding season, the gulls feed on spiders and insects, which are taken by searching plover-like on mud and soil rather than catching them on the wing. Details of the species' diet are lacking, and the suggestion that they feed regularly on tiny springtails (Collembola) seems unlikely, as is a report that they consume small birds. The suggestion that their food away from breeding areas often includes jellyfish has been repeated by several authors, but apparently without new and additional information confirming the original report. As with the Kittiwake (p. 293), the identity of food consumed by Sabine's Gulls while they are pelagic has not been investigated, but it must differ markedly from that consumed during the breeding season.

RARE VISITORS FROM NORTH AMERICA

Laughing Gull (*Leucophaeus atricilla*)

The Laughing Gull is a medium-sized gull with a wingspan of about 1 m. It forms large colonies on coastal marshes and islands along the eastern seaboard of America, from Nova Scotia in the north to Florida, the Caribbean and Venezuela in the south. The northern breeders winter within the southern breeding range of the species, while those birds that breed in more southern areas are mainly sedentary.

ABOVE: **FIG 203.** Adult Laughing Gulls (*Leucophaeus atricilla*) in summer plumage. (Phil Jones)

FIG 204. First-year Laughing Gull (*Leucophaeus atricilla*) in flight. (Steven Seal)

The wings of the Laughing Gull are a darker grey than those of Common and Black-headed gulls, which are of a similar overall size. In addition, adults have only small white mirrors on the black wing-tips, and these are totally absent in immature birds. The legs and bill are long and red. Adults have a black hood in the breeding season (Fig. 203), but this is lost in winter, the season during which most sightings are recorded in Britain. The Laughing Gull can be confused with the smaller Franklin's Gull, which also has dark grey wings, but that species has large white wing-tip mirrors and a shorter bill. First-winter Laughing Gulls have a dusky grey breast and flanks, and all-dark primaries (Fig. 204).

The first record of a Laughing Gull in Britain was in Sussex in 1923 and the second was not until 1950, after which it occurred more frequently. Sightings have been almost annual since 1974, and at least 200 have been recorded in Britain and 43 in Ireland. A large influx of at least 63 individuals occurred late in 2005 and 26 more were reported early in 2006, but there is no evidence that any of these birds stayed to breed. The peak of records occurs in November and about 40 per cent of those recorded have been adults. Numbers of Laughing Gull records have remained low in most years, with no indication of an increase recently apart from the one exceptional influx (Fig. 205). The species is increasing in North America, and the number of European records since 1974 probably reflects both this and the increase in observers.

The Laughing Gull consumes a wide range of foods, including fish and marine invertebrates. It also is attracted to food produced by human activities at beaches, car parks, sports areas and landfill sites, and it frequently engages in kleptoparasitism.

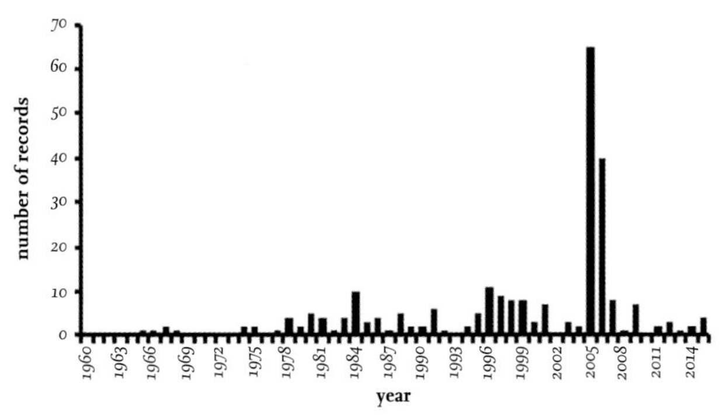

FIG 205. The number of records of Laughing Gulls (*Leucophaeus atricilla*) in Britain and Ireland from 1960 to 2015. Note the exceptional numbers recorded in the winter of 2005/06. Based on *Rare Bird Alert* reports and other sources.

Bonaparte's Gull (*Chroicocephalus philadelphia*)

This is a small gull, just a little larger than the Little Gull, with a wingspan of about 80 cm. Its numbers are believed to be increasing on its breeding grounds, which extend across North America, from Alaska through Canada as far as James Bay. Most pairs nest about 3 m above the ground in trees growing on islands and at the edges of bogs, marshes and ponds, but they avoid densely forested areas. The usual clutch is three eggs. In winter, Bonaparte's Gull moves south to ice-free rivers and lakes along both the Atlantic and Pacific coasts of North America, reaching the Caribbean and Mexico.

This species closely resembles a Black-headed Gull, but adults show a lighter underwing when in flight (Fig. 206). In summer plumage, it has a black head (not chocolate brown, as in the Black-headed Gull). The legs are pink. First-year birds are very similar to Black-headed Gulls of the same age, with minor differences in the distribution of dark feathers on the primary coverts in the wing (Fig. 207). In individuals of all age classes, the small bill is mainly, or totally, black.

FIG 206. Adult Bonaparte's Gull (*Chroicocephalus philadelphia*) in flight. (Tony Davison)

FIG 207. First-year Bonaparte's Gull (*Chroicocephalus philadelphia*). (Nicholas Aebischer)

The first record of a Bonaparte's Gull in Britain was a bird collected on Loch Lomond in Scotland in 1850. Over the next 100 years, records increased to just eight but then began to increase markedly. By the end of 2017, there had been 245 accepted records in Britain and, by 2014, 76 in Ireland. In 38 of the 40 years since 1975, at least one sighting has been recorded (Fig. 208) and an appreciable

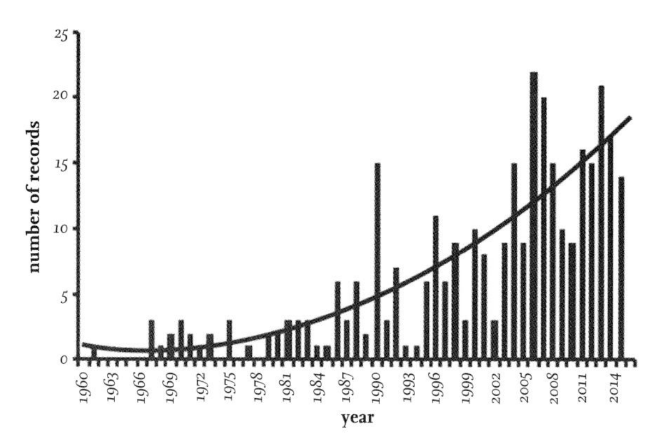

FIG 208. The number of British records of Bonaparte's Gull (*Chroicocephalus philadelphia*), 1960 to 2015, showing a marked increase in records starting about 1986. A smoothed trend line is indicated. Based on reports of scarce migrant birds in *British Birds*, *Rare Bird Alert* and annual county reports.

proportion of these were found inland. The peak of records is from March to May, and more than half of those seen were birds in full adult plumage. The pattern of numbers recorded since 1960 contrasts with that of the Laughing Gull (Fig. 205), another species that breeds in North America. The magnitude of the increase in Bonaparte's Gull records is so large that it cannot be explained by the increase in numbers of observers in recent years, and so must indicate a major increase in the numbers of birds visiting Britain and Ireland. In 2017, a pair bred in Iceland, and with increasing numbers occurring on the east side of the Atlantic, there is a possibility that breeding may occur in Europe, particularly if the present trend continues.

Insects make up the majority of food consumed by Bonaparte's Gulls during the breeding season, while in winter a wide range of invertebrates is consumed. Some feed on the ground on mudflats.

Franklin's Gull (*Leucophaeus pipixcan*)

Franklin's Gull is a North American gull that is similar in size to the Black-headed Gull and has a black hood in the breeding season, when it resembles the Little Gull or a small Laughing Gull (Fig. 209). It has more black on the wing-tips than the Little Gull and larger white wing mirrors when compared with adult Laughing Gulls. All plumage stages are similar to those of the Laughing Gull, but Franklin's Gull has an appreciably shorter bill, which, like the legs, is bright red. Because of the early total moult of the primary feathers (see below), one-year-old birds closely resemble the adults (Fig. 210).

The species commonly breeds inland in colonies that can involve hundreds of pairs, and is found locally throughout the prairie provinces of western Canada and as far east as the Great Lakes and into the USA. It is the most abundant

FIG 209. Adult Franklin's Gull (*Leucophaeus pipixcan*) searching for marine invertebrates on tidal mudflats while wintering in South America. Note the conspicuous wing mirrors and partially black head. (Norman Deans van Swelm)

FIG 210. First-year Franklin's Gull (*Leucophaeus pipixcan*) with a partially darkened head. Note the lack of wing mirrors. (Norman Deans van Swelm)

inland breeding gull in North America, and in this respect is the equivalent of the Black-headed Gull in Europe. It nests on the ground and on floating vegetation in marshes and along inland lakes; the usual clutch is three eggs.

Franklin's Gull winters along the Pacific coast of South America, from Guatemala to Chile. It makes a trans-equatorial migration, mainly avoiding the Atlantic Ocean coast of North America, which probably explains why few reach the eastern side of the Atlantic to Britain and Ireland.

The first record in Britain was as recent as 1970, when two individuals were found in Hampshire. A further 75 individuals were recorded in Britain up to 2017 and 17 more from Ireland by 2014. There was a maximum of 12 recorded in any one year in Britain (Fig. 211). The occurrences are spread widely throughout the year, with the fewest in September, and records peaked in 2005 and 2006. About 40 per cent of sightings are of adults.

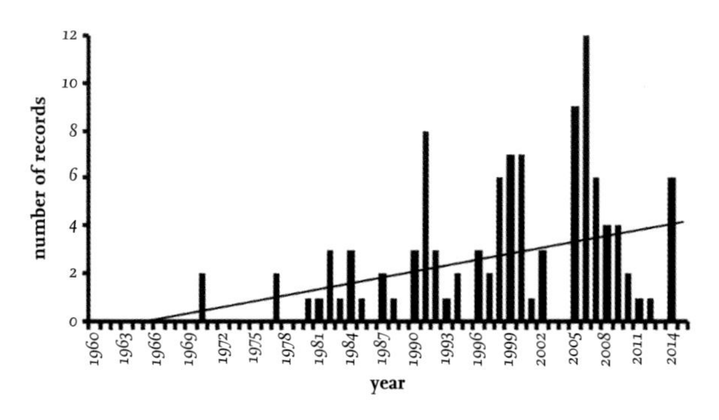

FIG 211. The number of records of Franklin's Gull (*Leucophaeus pipixcan*) in Britain since the species was first reported in 1970. A smoothed trend line is shown. Based on reports of scarce migrant birds in *British Birds*, *Rare Bird Alert* and annual county reports.

Franklin's Gull is unique among gulls in that it moults its primary feathers twice each year and there are records of individuals in primary moult in every month of the year. If the existing data are correctly interpreted, the usual post-breeding moult of the primaries is slow and is spread over about four months, ending in late October, and so includes the period of migration to the species' wintering areas in South America. A further moult of the primaries starts within a month of the completion of the post-breeding primary moult, and again it is in progress during migration as the birds return to the northern hemisphere, apparently being completed before most individuals begin nesting. Some first-time breeders retain the old outer primaries while breeding. There is considerable variation in the state of moult between individuals that have been collected on a similar date, perhaps reflecting an age difference.

It is not currently understood why Franklin's Gulls require two primary moults each year. A second moult of the primaries is a considerable cost in terms of energy, and it is strange that moults are apparently in progress during both annual migrations. Whether the moult is suspended during the actual migration has not been determined. There is a need for more data from additional specimens, particularly from the wintering areas and during migration, and a comparison made with the unusual and complex primary moult in the Common Tern (*Sterna hirundo*), which has a similar migration pattern to Franklin's Gull that also involves crossing the equator. The only suggestion that has been offered so far as to why there are two primary moults each year is that the feathers might deteriorate faster in the warmer weather and high solar radiation experienced by individuals throughout the year. However, this requires supporting evidence, as a double moult does not occur in those tern species whose annual migration also involves moving between the northern and southern hemispheres.

The diet of the Franklin's Gull includes a broad spectrum of food during the breeding season, including many freshwater insects and small fish. It some areas it frequently follows ploughs on farmland to catch the worms, insects and mice that are disturbed, but it rarely attends landfill sites. It is assumed that it takes similar food during its migrations and in its wintering areas, where it searches for marine invertebrates on mudflats and may consume fish more frequently.

Ring-billed Gull (*Larus delawarensis*)

Ring-billed Gulls breed only in North America and have increased markedly in numbers following persecution in the nineteenth century, much as the Herring Gull has done in Britain (p. 128). It is probably the most abundant gull in North America, breeding mainly inland in Canada and the northern part of the United States, particularly around the Great Lakes. The species nests in both small and large colonies, some of which can reach several thousand pairs, on open, bare ground with scattered vegetation beside lakes, along rivers and on the coast. The typical clutch is three eggs. The gulls normally winter along the whole of the Atlantic and Pacific coasts of North America and around the Great Lakes.

The adult Ring-billed Gull looks like a slightly large Common Gull, but differs in having a broad black band across the whole of the bill near the tip, while the Common Gull's bill is entirely yellow in the summer and has an indistinct black band in winter (Fig. 212). The adult Ring-billed Gull has a yellow iris, while the Common Gull has a dark brown eye. Immature individuals are very similar to Common Gulls of the same age, and both have dark-tipped bills and a dark eye, but there are minor differences, such as the shade of grey on the wings and the pale legs. The Ring-billed Gull tends to have a stouter bill than the Common Gull, but the sex difference in bill depth makes this an unreliable character, with male Common Gulls having similar-shaped bills to female Ring-billed Gulls.

The Ring-billed Gull was first recorded in Europe when two individuals marked as nestlings in North America were recovered in Spain, one in 1951 and

FIG 212. Adult Ring-billed Gull (*Larus delawarensis*). (Tony Davison)

the second in 1965. The first confirmed British sighting was in 1973, when two were present in Swansea Bay, Wales, and the first Irish record was reported in 1979. Thereafter, there have been many records in Britain, with at least 11 in 1980, 16 in 1981, 31 in 1982, 27 in 1983 and 65 in 1985, with a peak of 103 in 1992 (Fig. 213). Most birds were seen between November and April. A nestling ringed in June 1980 at Lake Champlain, New York state, USA, was found dead at Doochary, County Donegal, on 28 December 1981.

The large numbers of recent records arise, in part, when observers become alerted to the possibility of the species' presence in Europe. There was a large influx in the 1980s and early 1990s, with some birds staying on eastern side of the Atlantic for several years and raising the possibility that they might remain for many years. One bird returned annually to Gosport in Hampshire for 12 successive winters, and another visited Westcliff-on-Sea in Essex for at least eight winters, but where they spent each summer is unknown. The species continues to be recorded annually in Britain and Ireland, but in lower numbers than in the 1980s. Between 2004 and 2017, the numbers of adults recorded in Britain have declined, with an average of about 15 records each year. In contrast, numbers of first-year birds have declined less markedly, and because of their age, these must be new arrivals from North America each year. Since 2010, first-years have formed the majority of records in Britain (Fig. 214).

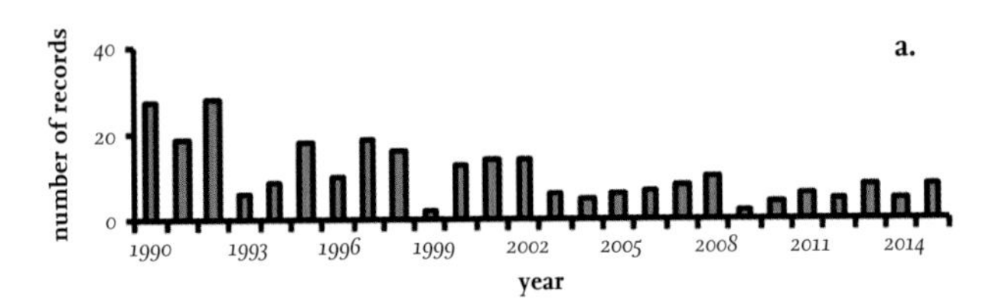

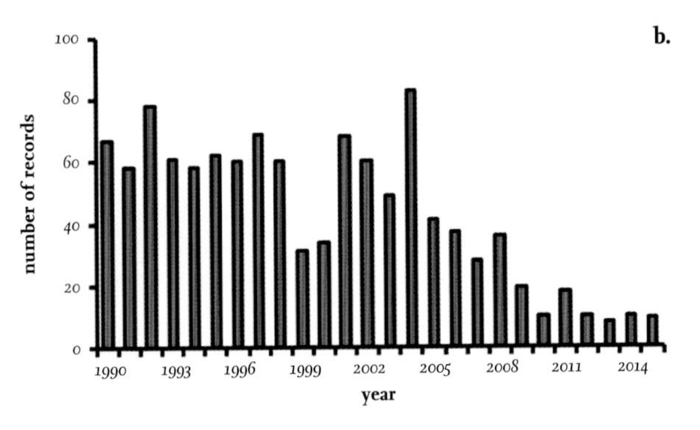

FIG 213. The annual number of sightings of (a) first-year and (b) older Ring-billed Gulls (*Larus delawarensis*) in Britain from 1990 to 2015. Redrawn from White & Kehoe (2017).

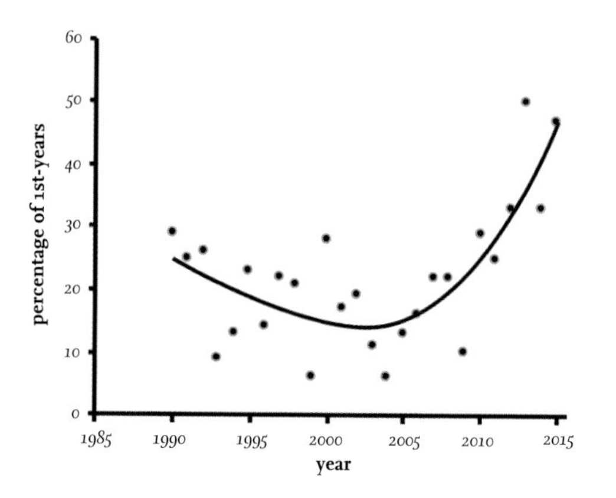

FIG 214. The percentage of first-year Ring-billed Gulls (*Larus delawarensis*) among the annual records in Britain, 1990–2015. A trend line has been fitted to the data points. Redrawn from White & Kehoe (2017).

There are about 15 records each year from Ireland and many records from countries along the west coast of Europe, extending from Spitsbergen to Spain, and the Ring-billed Gull is now by far the most frequently recorded North American gull visiting Europe. It is also the most likely North American species to colonise Europe. So far, no breeding records of pairs of Ring-billed Gulls have been reported in Europe, but in 2004 one adult was seen paired with a Common Gull in a colony in County Down, Northern Ireland, and reared a hybrid chick that year (this was ringed as a Common Gull chick, but when seen as an adult in 2008, its plumage suggested that it was a hybrid). In 2009 and 2014 a Ring-billed Gull was present at a nest in a Common Gull colony in Scotland, but its mate was not seen. In 2016 it was paired with a Common Gull and shared incubation of two eggs at the same colony. Despite this, from 2000–16 there have been a reduced number of sightings of this species in Europe, and the likelihood of this species breeding regularly in Europe has decreased for the time being.

Records of Ring-billed Gulls have become so extensive in Europe that there is a possibility that a change occurred in the migration pattern of part of the population during the 1980s, which resulted in small numbers of adults regularly wintering in Europe. The occurrence of the same individuals wintering at the same locations in different years suggests that either these birds remained in Europe, or that they were making a regular annual migration and deliberately crossing the Atlantic to winter in Britain and other countries along the western coast of Europe. If this is the case, the recent decline of records of adults would suggest that those individuals involved are dying and are not being replaced.

The food of this species is similar to that of Common Gulls, comprising mainly invertebrates and occasional fish caught in intertidal areas or on pastures. In addition, the birds frequently obtain food rejected or dropped by humans at car parks in shopping complexes and near fast-food stores. Individuals in North America have been reported feeding on berries during the autumn and winter. Ring-billed Gulls do not appear to visit landfill sites in Britain, although large numbers frequently do so in North America.

Glaucous-winged Gull (*Larus glaucescens*)

Adult Glaucous-winged Gulls resemble a faded version of the Herring Gull, with the wing-tips reduced to shades of grey rather than being black (Fig. 215). The breeding range of this species extends around the North Pacific Ocean, from the Commander Islands in the east, through the Aleutians, the Pribilofs and the south Bering Sea, to Alaska and south-east to Oregon in the west. It winters from the Bering Sea to northern Japan and Baja California.

Currently, there are only three British and one Irish record of the species, which breeds far from Europe. The first record in Britain was in December 2006, when a third-winter bird was trapped and ringed at Gloucester landfill site. This individual was subsequently seen in March 2007 in Carmarthen, Wales, and then returned to Gloucester, and was last recorded at Beddington, London, in April of that year. A second bird was reported in north-east England in 2008, and a third individual was recorded in 2016 in County Cork, Ireland. In 2017 an adult was found on Fair

FIG 215. Glaucous-winged Gull (*Larus glaucescens*). Note the dark grey primaries. (Dick Daniels, carolinabirds.org)

Isle. There have also been two previous records of Glaucous-winged Gulls in the Western Palaearctic, one from the Canary Islands and the other from Morocco.

Unfortunately, the Glaucous-winged Gull frequently forms mixed breeding pairs with other species of gulls, and hybrids have been reported with the Western Gull (*Larus occidentalis*), Slaty-backed Gull (*L. schistisagus*) and American Herring Gull, while some Glaucous Gull × American Herring Gull and Glaucous Gull × Herring Gull hybrids have occurred in North America and Iceland. All of these hybrids can look very like Glaucous-winged Gulls, and it is often difficult – if not impossible – to be sure that individuals found outside their normal range are not hybrids.

American Herring Gull (*Larus smithsonianus*)

Herring Gulls breeding in North America have long been regarded as a subspecies of the Herring Gulls that breed in Europe (*Larus argentatus smithsonianus*), and the American Ornithological Society still treats them as such. However, in 2007 the British Ornithological Union changed its classification and raised the American Herring Gull to full species status.

The American Herring Gull breeds in Canada, from the west coast to Baffin Island and Newfoundland, and in the USA in Alaska, the Great Lakes and the northern part of the Atlantic coastline. Dispersal in winter is limited, with some individuals moving only to the southern states of the USA or to Mexico, while others remain in their breeding areas.

American Herring Gulls are more readily identified and separated from their European equivalent in their first year of life, and they tend to be darker than European birds of the same age. In addition, they have virtually no white on the tail, while the rump is dark and shows less contrast (Fig. 216). The head tends to

FIG 216. First-year American Herring Gull (*Larus smithsonianus*). (Phil Jones)

be lighter, with a clear demarcation from the darker upper breast. Only 3 per cent of European records of American Herring Gulls to date involve adults and 70 per cent are first-winter birds, suggesting that either older birds are more difficult to identify outside of their normal range or that young birds disperse further.

The first record in Britain was in 1994 on Merseyside, and by 2017 records had increased to 33, while 96 sightings have been reported in Ireland. The records are mainly between December and March, and most are from the Irish coast and the west coast of England. There is only a single record on the east coast of England. Records peaked between 1996 to 2008. To date, there have been no ringing recoveries of an American Herring Gull in Europe, so movements to Europe cannot be confirmed. The nearest was an individual captured in the Atlantic Ocean 400 km off the coast of Spain in November 1937.

A large adult Herring Gull recorded in May 2008 at Chew Reservoir in the Peak District was repeatedly seen by gull experts Andy Davis and Keith Vinicombe, who 'confirmed beyond any reasonable doubt that the bird was an American Herring Gull' (Davis & Vinicombe, 2008). While preening, the bird dropped a feather, which was recovered and sent for DNA analysis to Gareth Jones; the subsequent analysis confirmed that the bird was not an American Herring Gull after all.

The problem with the Chew Reservoir sighting and other similar cases is that the observers and those suggesting key characters have often failed to appreciate the variation that exists within European Herring Gulls, American Herring Gulls and hybrid gulls. Adult American Herring Gulls are unlikely to be safely distinguished in the field in Europe unless new and infallible characters are identified, while immature individuals with characteristics believed to be the American species may be only a small proportion of those moving to Europe from North America. On the other hand, those reportedly identified may not be actually reared in America. As of now, the best that can currently be achieved is to qualify claims of American Herring Gulls recorded in Europe with the word 'probably' unless these sightings involve ringed individuals.

RARE VISITORS FROM SOUTHERN EUROPE AND NORTH AFRICA

Azores Herring Gull or Azorean Yellow-legged Gull (*Larus argentatus atlantis* or *L. michahellis atlantis*)

The taxonomic positions of the large gulls breeding in the Azores, Madeira and the Canary Islands have been confused for many years. The birds were initially considered a subspecies of the Lesser Black-backed Gull and named *Larus fuscus atlantis* by Jonathan Dwight in 1925. Individuals breeding in the Canary Islands and on Madeira were reported to be slightly different from those in the Azores, and so some proposed that subspecies *atlantis* should include only those birds breeding in the Azores. However, by restricting the distribution, it would leave the problem of identifying the birds breeding in Madeira and the Canaries.

Others authors, such as Charles Vaurie, treated *atlantis* individuals as a subspecies of the Herring Gull, *Larus argentatus atlantis*, and more recently some have regarded the birds as a subspecies of the Yellow-legged Gull and so should be named *Larus michahellis atlantis*. To complicate matters further, others suggest that the *atlantis* birds should be given full species status as *Larus atlantis*. A paper on the status and identification of the *atlantis* group appeared in *British Birds* in 2017, discussing its acceptance as a subspecies that has occurred in Britain. This is based on four well-recorded sightings in England in 2005, 2008, 2009 and 2017 all of which were accepted by the BOU Records Committee in 2016 and 2018.

The photographs accompanying the *British Birds* paper show two adults and draw attention to the heavy bill. But these birds are almost certainly males, as females have less substantial bills, although this sex difference is not mentioned. Comments that the birds looked powerfully built and heavy-headed ignored sex differences common to all large gull species. I doubt if individual adult females and males can be identified outside of their breeding range. How much variation exists in the extent and duration of the intense head streaking, and does this overlap with other large, yellow-legged gulls?

The reference to the four-colour bill as a criterion for identifying *atlantis* gulls is not reliable, as this is also seen in some three- and four-year-old Herring Gulls, while the shorter legs and very pale iris are not convincing differences justified by actual measurements. Uncertainty about the 13 Irish records that have been accepted also applies. This subspecies is at the limits of field identification and, if it does indeed occur in Britain and Ireland, a great many individuals (particularly females) will surely go unidentified and fail to be recorded. For certain identification, more extensive ringing is required in the Atlantic islands where the birds breed.

RARE VISITORS FROM EASTERN EUROPE

Caspian Gull (*Larus cachinnans*)

The Caspian Gull is a large gull that was regarded as a subspecies of the Herring Gull until 2007. Its distribution has often been confused with that of the Yellow-legged Gull, although only some adults have yellow legs. It breeds further east in Europe than the Yellow-legged Gull, mainly around the Black and Caspian seas, and as far east as Kazakhstan and China. It has recently spread further north and west in Europe, with small numbers now breeding in Poland, Switzerland and eastern Germany, where the Yellow-legged Gull also occurs. It frequently forms mixed breeding pairs with Yellow-legged Gulls at some of the European breeding sites.

The Caspian Gull (Figs 217 & 218) is difficult to identify in the field and is a species for the specialist with experience of large gulls. In 2014, the BOU retrospectively accepted the description and photographs of a Caspian Gull seen in 1992 on Radipole Lake in Dorset as the first British record. A nestling said to be of this species, ringed in Switzerland in 1997 and captured at a landfill site in Gloucestershire in November 1999, became the second record. This was followed in November 2007 by a first-year bird ringed at Rainham landfill site in Greater London as a Herring Gull, which had its ring number read in a colony of Caspian Gulls breeding near Minsk in Belarus almost seven years later. Two nestlings that were colour-ringed in Poland in 2008 and 2009 as Caspian Gulls have been seen at three different localities in southern England. However, Yellow-legged Gulls also bred in this Polish colony and so possible confusion over identity may occur when ringing takes place.

FIG 217. A possible second-year Caspian Gull (*Larus cachinnans*) (left) with two third-year Lesser Black-backed Gulls (*Larus fuscus*). (Alan Dean)

FIG 218. Immature gull identified as a Caspian Gull (*Larus cachinnans*), but could be confused with an immature Great Black-backed Gull. (Alan Dean)

Many, but not all, Caspian Gulls have long, slender bills and sloping foreheads that are less smoothly curved than in most Herring Gulls. The neck often appears longer than in the Herring Gull and the eyes of adults are often dark (but not always so). Many individuals appear to have longer legs than Herring or Yellow-legged gulls. The shade of grey on the back and wings is within the range of European Herring Gulls, particularly those breeding in northern Scandinavia, and black markings on the outer primaries also overlap with the variation present in Herring Gulls (and hence this is not the reliable character suggested by some). In late autumn, adult Caspian Gulls often have pure white heads, but a few European Herring Gulls are similar (although most have acquired dark streaks by this time of year). First-year Caspian Gulls often have paler heads than European Herring Gulls of the same age. There is still insufficient information about variation within this species, but the head and bill shape of the smaller female appears to be less pronounced and so it is more readily confused with European Herring Gulls. All of the characters mentioned are helpful aids, but not totally reliable in confirming the identification of all individuals seen in the field. A further problem lies in separating adult Caspian Gulls with pink legs from the large, dark subspecies of European Herring Gulls breeding in northern Scandinavia, which regularly winter in Britain.

Records of Caspian Gulls in Britain and western Europe as a whole have increased dramatically since 2000, as they have in the West Midlands (Fig. 219),

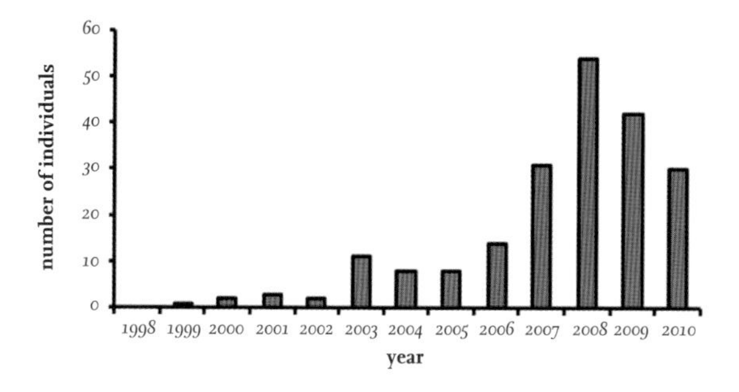

FIG 219. The number of individuals believed to be Caspian Gulls (*Larus cachinnans*) recorded annually in the West Midlands up to 2010. Data from Dean (n.d.).

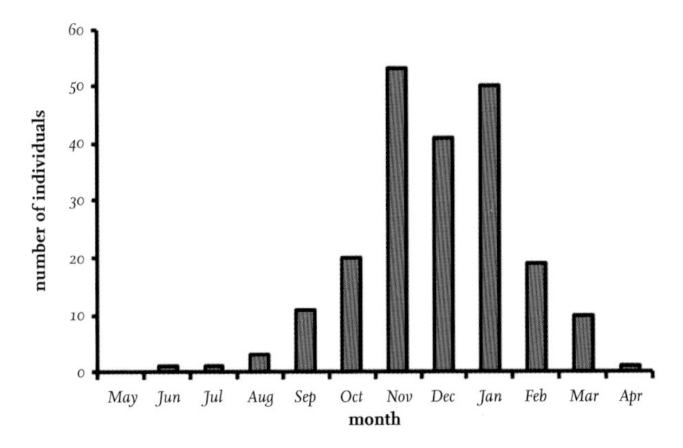

FIG 220. The months in which individuals believed to be Caspian Gulls (*Larus cachinnans*) were first recorded in the West Midlands, based on records from 1999 to 2010. Data from Dean (n.d.).

mainly as a result of observers being alerted to the possible presence of the species. Most British records have been from south-east England and the Midlands, and only a few have been recorded in Scotland. Most records in the West Midlands and elsewhere are in the period October to February (Fig. 220), which is also when large, dark-winged Herring Gulls from northern Scandinavia are present and occur inland in Britain.

Caspian Gulls breed on flat ground alongside lakes and reservoirs, not on cliffs as do many other species belonging to the Herring Gull complex. Their diet is similar to that of Herring Gulls, and many of the recorded individuals have been encountered inland in Britain at landfill sites and roosting on reservoirs.

Audouin's Gull (*Ichthyaetus audouinii*)

This is a medium-sized gull, the adults of which have a distinctive bill that is dark red with a pale tip (Fig. 221). Fifty years ago, Audouin's Gull was considered to be one of the rarest gulls in the world, but its numbers have increased dramatically and are now believed to exceed more than 40,000 adults. It is endemic to the Mediterranean, with small numbers breeding in the eastern Mediterranean in Croatia, Turkey and Cyprus, and many now also breed in the western Mediterranean, particularly in the Chafarinas Islands off the coast of Morocco and in the Ebro Delta in north-east Spain. The species is migratory, with the main wintering areas along the coast of north-west Africa, from Morocco to the Gambia. Despite the large numbers now breeding in the Ebro Delta, there are still few records in France and Britain.

Audouin's Gull is one of the rarest gulls recorded in Britain and has not been reported from Ireland, despite breeding as close as Spain in rapidly increasing numbers. The first British record was of a second-year individual seen at

FIG 221. Adult Audouin's Gull (*Ichthyaetus audouinii*). (Tony Davison)

Dungeness, Kent, on 5–7 May 2003. Up to 2017, only eight more individuals were reported, all from the south and south-east coast of England between May and October.

The gull's diet comprises mainly fish, for which it sometimes forages at night, and fishing waste from commercial boats. The Pilchard (*Sardina pilchardus*) is probably the most frequently taken fish species. In the past, Audouin's Gull was rarely reported inland and scavenged less than other gulls, but there is evidence that it is becoming more frequent inland in Spain, where it is particularly attracted to feeding in rice fields.

RARE VISITORS FROM THE MEDITERRANEAN AND BLACK SEA

Slender-billed Gull (*Chroicocephalus genei*)

The Slender-billed Gull is a medium-sized gull that breeds in Turkey and the Black Sea area, across to Kazakhstan, Afghanistan and Pakistan, and in small numbers around the Mediterranean Sea. Numbers have increased in recent years and breeding has spread to Italy, Spain and the Mediterranean coast of France. In 2010, an inland colony was discovered in Algeria.

Adult Slender-billed Gulls resemble Black-headed Gulls, but they do not have a black head at any time of the year. The elongated head shape is very characteristic, and includes a longer bill and often a long neck (Figs 222 & 223); it has a particularly light-coloured eye. The legs are relatively long, and are pale in immature birds and deep red in adults. The birds often dive to capture fish and also search for food on the ground.

This is indeed a rare gull in Britain, with only 10 records to date. The first confirmed sighting was in Sussex in 1960, and the second was in the same county in 1963. No further birds were reported until 1971, when two were seen. There was then a 16-year absence, until two were seen together at Cley in Norfolk in 1987. A further gap of 12 years without a record followed, until a single bird was recorded

FIG 222. Adult Slender-billed Gull (*Chroicocephalus genei*). (Pep Arcos)

FIG 223. Adult Slender-Billed Gull in flight (*Chroicocephalus genei*). (Pep Arcos)

in Kent in 1999 and two more in the following year, one in Kent and the other in Norfolk. There were then no other accepted records until 2014. The British records are spread from April to August, and are all from the south of England, with none so far recorded in Scotland, Wales or Ireland. The species migrates south to the north-east coast of Africa and into the Indian Ocean, and immature birds often remain within this wintering area. The British records are far less numerous than those for several gull species that breed much further away in North America.

RARE VISITORS FROM ASIA

Great Black-headed Gull or Pallas's Gull (*Ichthyaetus ichthyaetus*)

This species has been commonly called the Great Black-headed Gull for at least 150 years, but the less informative but shorter name of Pallas's Gull has been applied by some since 2000. It breeds in the south of Russia, from the Sea of Azov to the Volga River at Saratov, and in Turkestan and Mongolia. It migrates to the eastern Mediterranean, Saudi Arabia and India, and is rare in western Europe, where the few records are presumably of vagrants. It is surprising that this large gull does not wander more frequently to western Europe, but its main wintering area is to the south and east of its breeding area, and it is even uncommon over most of the Mediterranean.

The large size of the adults, combined with their yellow legs, black head and heavy bill, make them unmistakable in the breeding season. First-year birds have a black tip to the otherwise entirely white tail and are likely to be confused only with immature Great Black-backed Gulls or Yellow-legged Gulls.

There are only a few British records. An adult was shot off Exmouth, Devon, in May or June 1859 and its skin is now in Exeter's Royal Albert Memorial Museum. An individual was reported in Sussex on 4 January 1910 by H. Walpole-Bond, and an adult male was shot in Kent in June 1915 after it had been seen there in May. Another individual was then seen at Bournemouth, Hampshire, in November and December 1924 and the record was accepted at that time by *British Birds*, while a single bird was seen on several days in March 1932 at Cromer, Norfolk. These early records were not subjected to rigorous validation at the time, and without detailed descriptions as evidence to back them up, only the 1859 record is now currently accepted. There have been no Irish records, although there was a single record in Norway in September 2014.

The Great Black-headed Gull feeds on fish, crustaceans, insects and small mammals.

RARE VISITORS FROM THE PACIFIC

Slaty-backed Gull (*Larus schistisagus*)

The Slaty-backed Gull breeds in the northern Pacific on Hokkaido in Japan, in north-east Siberia and south to the Korean Peninsula. Its range has been extended recently to western Alaska, where breeding was first recorded in 1996 within a colony of Glaucous Gulls. There is evidence that its numbers are increasing in some areas and there are currently just over 200,000 adults of this species. Individuals move south in winter to north-east China, the Korean Peninsula and Taiwan, but there are a number of records from North America of vagrants and, most surprisingly, 16 individuals have been recorded in Newfoundland since 2006.

Adults resemble a large Lesser Black-backed Gull, with a slate-grey mantle and pink (not yellow) legs. The most important characteristic of the adult is a string of white tips to the primaries, described as a string of pearls. Immature birds can be confused with those of several other gull species, and claims of sightings in Europe of individuals in their first two years of life are unlikely to be accepted.

The species was added to the British list in 2016, based on one adult seen and photographed at three locations: Rainham in London, and Hanningfield Reservoir and Pitsea, both in Essex, in January and February 2011. A second bird was reported from Ireland in February 2014. These followed the first European record, from Lithuania, in 2008 (and, likely, the same bird was seen in Latvia in 2009). In 2012, individuals were recorded in Iceland, Belarus and Finland. These records are of a gull species that, when in Europe, is at the maximum distance from its main breeding area.

The Slaty-backed Gull breeds in colonies of varying size, usually on rugged coasts and islands, and often nests with or near other seabirds, as do Great Black-backed Gulls (p. 217). The species has a diet similar to that of other large gulls, including fish, refuse, and the eggs and young of other species.

Vega Gull (*Larus vegae*)

The Vega Gull was previously included as a subspecies of the Herring Gull or the American Herring Gull, but is now regarded by some as a separate species. It breeds in eastern Siberia. It is a difficult species to identify outside its normal range. It has not been recorded in Britain, but in January 2016 an adult was reported at Duncannon, County Wexford. If it is accepted by the Irish Rare Bird Committee, this will be the first record in our region. Another bird claimed to be a subadult was seen in France in November 2016.

POTENTIAL RARE VISITORS FROM AFRICA

Grey-headed Gull (*Chroicocephalus cirrocephalus*)

This species is slightly larger than the Black-headed Gull, and has more extensive black on its wing-tips as well as obvious white mirrors. It breeds in sub-Saharan and southern Africa and South America, and a few individuals have been reported in the Mediterranean, including Italy and Spain. Four birds of this species have been reported in southern England, but since about 20 fully winged Grey-headed Gulls are kept at London Zoo in Regent's Park and 18 free-flying individuals are at Birdland near Bourton-on-the-Water in Gloucestershire, the possibility that birds seen in Britain are escapees cannot be ruled out. As a result, all the British sightings so far have been rejected.

The adult has a light grey hood in the breeding season and a bill resembling that of a Slender-billed Gull, but the more extensive black on the wing-tips and a very black-and-white appearance in flight separates it from that species. The first-year birds have black-tipped primary and secondary feathers, making them very different from other immature gulls.

CHAPTER 12

Methods Used to Study Gulls

THE TOOLS USED BY ORNITHOLOGISTS have changed greatly over the years and continue to do so. Shotguns were the first 'tools' used, because the main method of identifying a bird was to shoot it and then study it. Field glasses, which were essentially two small tube-like telescopes coupled together and containing lenses to magnify the image, had a small field of view and were initially made for military use, while smaller opera glasses were used in theatres. The introduction of roof prisms in the place of some lenses at the end of the nineteenth century improved the light quality of binoculars, but these were initially expensive and did not come into general use until after the First World War. They usually gave 7× or 8× magnification, as anything higher than this required a larger and heavier casing, which was difficult to hold steady in the field. Since then, the quality of binoculars has progressively improved, incorporating central focusing, adjustment for eye differences, a larger field of view and coated lenses. While these improved optical aids began to be used in studies of birds, the publication of books to help observers identify birds in the field lagged behind, and it was not until the 1930s that the first modest but useful field guides were produced. A small book written by Norman Joy, *How to Know British Birds*, was published in 1936. This was followed by the five volumes of the *Handbook of British Birds*, published in 1938–41 by Harry Witherby, which were not convenient to carry in a haversack, and it was more than 20 years before the first comprehensive one-volume field guide was produced for the identification of British birds.

For distance viewing and greater magnification, draw-tube telescopes were widely used in ornithology, but they were difficult to keep steady. In the 1950s, smaller and more effective prismatic telescopes were introduced by Bushnell, and these could be attached to a tripod for stability. Such items are now part of the standard equipment for many ornithologists and birders.

Various improvements to photography have taken place over the last century, particularly the move from film requiring exposure meters to modern cameras with automatic focusing and exposure that make digital records of images. Movement- and sound-activated photography can function on small, low-voltage batteries, and video cameras have greatly improved field records of numbers, species and activities of birds. Binoculars, telescopes, tripods, cameras and mobile phones are now the standard field equipment of the keen birdwatcher. At the present time, small pilotless drones fitted with cameras are being trialled to assist with survey work, but how successful they will be has yet to be determined. Concerns about their adverse effects when approaching breeding seabirds and wintering flocks of migrants are already being voiced, and no doubt there will be increasing legislation controlling their use in years to come.

RINGING

Ornithology has developed from asking simply 'What species is it?' to enquiring 'Why, when and how?' A major advance was the ability to obtain information about the same individual bird on different occasions. The first step in this direction was initiated by ringing (known as banding in North America), which involves attaching a small ring to the bird's leg that identifies it and allows the date and location if the ring (and bird) is subsequently found to be reported.

In the nineteenth century, a few birds were marked with rings or bands attached to the leg. This was done on a very local scale and used to satisfy some unknown questions or simply to claim ownership. The first major advance came in 1909, when metal rings with a unique number and an address stamped on them were produced. These were usually attached to the legs of young birds before they fledged. In Britain, Arthur Landsborough-Thomson of Aberdeen University and Harry Witherby, then editor of *British Birds*, independently produced inscribed aluminium rings. The former scheme ceased during the First Word War, while the *British Birds* scheme became the national bird-ringing scheme in 1930. Its rings bore the inscription 'Inform British Museum Nat. Hist., London' and each had a unique number manually stamped on it. The scheme was managed by the British Trust for Ornithology and was entirely run from a small office in the Natural History Museum, London, on a part-time basis and in an honorary capacity by Elsie P. Leach.

Because most gulls are colonial, and the young are often readily accessible, their ringing and recoveries frequently figured in the early days. The scheme soon produced fascinating results, such as that Kittiwakes (*Rissa tridactyla*) crossed the Atlantic and Lesser Black-backed Gulls (*Larus fuscus*) moved between Britain and North Africa. The disadvantage of metal rings was that information was obtained from only a small percentage of the birds that were marked and most of the rings were never seen again.

There are two important reasons why it has been difficult for ornithologists to obtain new information about gulls. The first is that most individuals look just like their neighbours, although a few may have individual characteristics – such as a limp, a missing feather or a damaged bill – that can be used to separate them from others of their own species. And second, while it was known that gulls had immature plumage, it took years of ringing to determine the numbers of years it took them to acquire adult plumage. Adult gulls are of unknown age unless they had been ringed while in the nest. Knowledge of the maximum ages reached by many birds originally depended on records of individuals kept in captivity, and it took more than 70 years of ringing before it was established that a few gulls could survive in the wild for over 30 years. Male gulls tend to be slightly larger than females, but there is no sex difference in terms of the plumage, and so separating the sexes of a gull in the field is difficult and often impossible. Ringing offered a solution to determining answers to many important and previously unknown aspects of avian biology.

The effectiveness of the ringing scheme encountered two major problems. The first occurred in the 1930s and reached a peak in the early 1960s, when sustained attempts were made by national bird protection organisations to stop bird ringing in Britain because of concerns that the rings had an adverse effect on several species, including gulls. Opponents claimed that the rings were too heavy and injured birds, that some birds had been killed because the rings had been caught in vegetation and that ringing caused too much disturbance in colonies. Detailed investigations by the BTO Ringing Committee in the 1960s failed to find evidence in support of these complaints.

Second, the aluminium rings used did not always last well or remain legible for the maximum lifespan of the species being studied. Evidence of this problem was only slowly accumulating in the 1950s, when aluminium rings placed on some seabirds became so worn or corroded within a year or two that the inscription and numbers became illegible. In other cases, rings lost strength, opened at the ends and dropped off long before the bird eventually died. Therefore, information on the survival of individuals and realistic longevity records of gulls and other seabirds was not being obtained, fewer recoveries were being made and questions such as the effect of age on migration could not be answered. For example, only 3 per cent of recoveries of Herring Gulls (*Larus argentatus*) ringed as chicks were reported more than four years after ringing (the age at which they start to breed), and one ringer found that Herring Gulls and Great Black-backed Gulls (*L. marinus*) were able to remove the rings from their legs. Illegible rings were found on Kittiwakes and other seabirds, including Shags (*Phalacrocorax aristotelis*) and Manx Shearwaters (*Puffinus puffinus*), which had been ringed and later recaptured. An examination of a sample of aluminium rings removed from Kittiwakes revealed that they had lost 7.8 per cent of their weight each year, and the inscriptions and numbers had become illegible within three years, and hence they were not reported if found by the public.

In the mid-1950s, Robert Spencer, who had become the secretary of the BTO Ringing Scheme, did much to overcome this problem, introducing more durable rings made of Monel, a copper and nickel alloy unaffected by seawater. Subsequently, several schemes in other countries encountered the same problem and changed to stainless-steel rings on seabirds. I used the first batch of Monel rings on Kittiwakes at my study colony at North Shields, and 20 years later the inscriptions on the rings were still fully legible and the weight loss of the rings had been only 0.5 per cent of their initial weight per year.

Monel rings used on Herring and Lesser Black-backed gulls at first seemed satisfactory, and legible rings have been reported for individuals of both species more than 30 years after they were ringed. However, this confidence in the rings is misplaced. The recovery of 797 Monel rings from Herring Gulls during the huge cull by the Nature Conservancy Council on the Isle of May in the early 1970s allowed a study of weight loss of rings. Their weight on recovery decreased linearly with the numbers of years they had been carried by the gulls, with average loss of 3.8 per cent of their initial weight each year (Fig. 224). Of greater concern, however, was that variation in weight loss increased markedly with time since ringing. Rings on females lost weight less rapidly than those on males, but this was a minor effect. Some Monel rings that had been on Herring Gulls for nine years had lost half of their initial weight and the inscription had become illegible, while others showed little wear after the same time period and had lost only 15 per cent of their weight. Most of the recovered birds had also been colour-ringed and several were found to have retained the coloured ring but had lost the metal one. Some Monel rings had lost 50 per cent of their weight and the inscriptions had become illegible after eight years, but some rings would not have reached a 50 per cent weight loss for 20 years. As a result, there have been fewer recoveries of Herring Gulls that have carried Monel rings for more than 20 years than would be expected had ring wear (or even ring loss) not taken place.

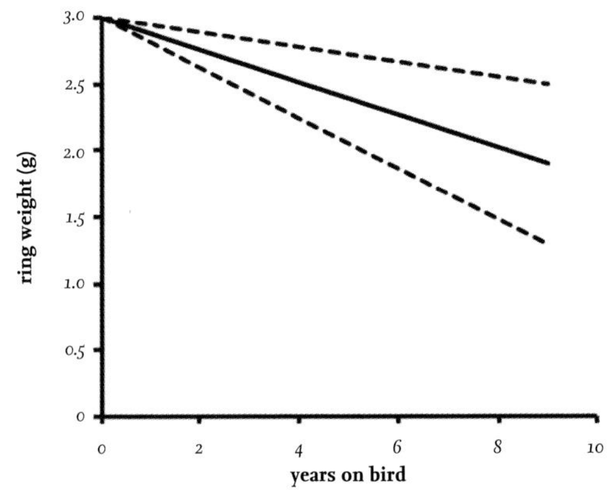

FIG 224. The weight of Monel rings in Herring Gulls (*Larus argentatus*) in relation to time on the bird. The continuous line represents the average weight of the rings, while the dashed lines indicate the upper and lower limits of 99 per cent of individual rings, based on a total of 797 rings. The inscription tended to be illegible once a ring had lost 40–50 per cent of its initial weight. Note that the variation in weights of individual rings increased the longer they were on a gull.

The consequence of this is that while British ringing recoveries using Monel rings can provide a clear picture of movements of gulls of different ages, they must be used with caution when estimating survival rates of large gulls.

It is not obvious why there should be such variation in the wear of rings on Herring Gulls. However, much of the wear appears to occur on the inside of the ring, where it is in contact with the bird's leg. It could be that the variation in wear relates to the feeding areas frequented by the individual Herring Gulls – for example, a bird that is often active on wet sand or mud might have more abrasive material between its leg and the ring, while this would not be found in birds regularly feeding at sea or inland, whose rings would therefore undergo less weight loss.

Wear has also been a problem with aluminium rings on Black-headed Gulls (*Chroicocephalus ridibundus*). Mike Harris measured weight loss on 123 rings and a further 45 rings recovered from our cannon-net captures were examined, some of which had been carried by the birds for up to 20 years. These data have been combined in Fig. 225. The aluminium rings lost 3.5 per cent of their initial weight each year, and as with those used on Herring Gulls, they also showed increasing individual variation in weight loss the longer they were on the bird. A few rings had lost their inscription after six years on a bird, and this probably applied to half of the birds after 15 years, but a small proportion of the rings remained legible for 30 years.

When used on Black-headed Gulls, Monel rings lost about 5 per cent of their initial weight each year (Fig. 226); surprisingly, this is a higher rate than that of

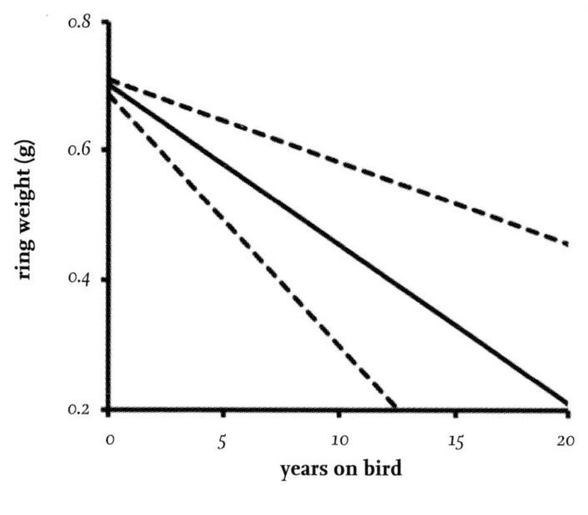

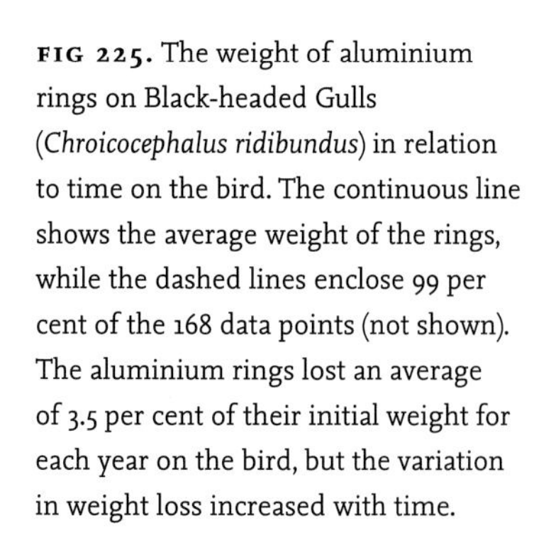

FIG 225. The weight of aluminium rings on Black-headed Gulls (*Chroicocephalus ridibundus*) in relation to time on the bird. The continuous line shows the average weight of the rings, while the dashed lines enclose 99 per cent of the 168 data points (not shown). The aluminium rings lost an average of 3.5 per cent of their initial weight for each year on the bird, but the variation in weight loss increased with time.

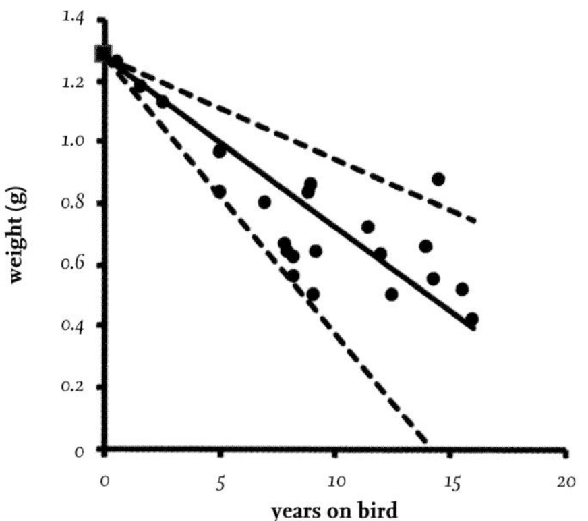

FIG 226. The weight of Monel rings on Black-headed Gulls (*Chroicocephalus ridibundus*) in relation to time on the bird. The continuous line shows the average weight of the rings, while the dashed lines enclose 95 per cent of the data points. The Monel rings lost 5 per cent of their initial weight for each year they were on the bird.

aluminium rings. In addition, the inscriptions did not remain legible any longer than those on aluminium rings. As with Monel rings on Herring Gulls, the illegibility of the inscription over time meant that a proportion of recoveries of Black-headed Gulls were lost. As mentioned above, some countries have moved to using stainless-steel rings on Black-headed Gulls, and while these are more difficult to close and more expensive to produce because the stamping dyes wear out quicker, they lose less of their initial weight each year.

CAPTURING GULLS FOR RINGING

For many years, most birds were ringed as chicks before they could fly. There are no traditional methods of capturing adult gulls comparable to the use of decoys for ducks. Small numbers of gulls were caught in wire-mesh traps placed over eggs, and adult Kittiwakes were caught by a noose or a wire hook placed around a leg while they were standing at the nest. A few adult Black-headed Gulls that became tame enough to take food from people's hands in central London parks were caught in small nets, then ringed and released. Small numbers of gulls have been caught with clap nets while incubating, and the use of baits containing a sedative and anaesthetic drug that induced sleep in selected incubating gulls has allowed their capture, ringing and release. Dazzling birds at night with bright light has also been used, although most roosts are inaccessible.

Large rocket-propelled nets were developed and first used by Sir Peter Scott and the Wildfowl Trust in the late 1950s to capture geese, and the method of deployment was subsequently improved using cannons instead of rockets and applied to capture gulls, both at landfill sites and at locations where they regularly fed or collected in flocks. Cannon netting was a great advance in ringing adults, as several hundred adult gulls could be caught in a single firing, however, a team of experienced people are needed to be on hand to keep the birds passive in darkened holding cages or sacks to await examination and ringing and then to release the birds quickly.

Colour rings

Ringing with metal rings results in only a small number subsequently being reported by the public. In a few cases, ring numbers were read in the field through a telescope by a keen field worker, and one or two people became specialists in this art, but often it was not possible to read the whole number.

Identification of individual gulls is important in detailed studies. When it became obvious that the identification information on metal rings was difficult to read at a distance, coloured celluloid leg rings originally developed to identify domestic hens and geese in the 1930s were used on gulls and a wide range of other species. They were applied successfully to many small birds and terns, and allowed an individual to be identified quickly at a greater distance than for metal rings. One coloured ring could be used to identify birds from a specific locality or reared in a specific year, or each bird could be given a unique combination of two or three coloured rings so that individuals could be identified. I started to use this latter method on adult Kittiwakes nesting at North Shields in 1954 and marked 10 adults in the first year, adding a BTO metal ring to increase the combinations available and so that the organisation would be notified if the bird was found elsewhere. Unfortunately, the celluloid rings changed colour within a year! White and yellow rings both turned to cream, red changed to grey, and blue also faded to a shade of grey. Fortunately, the BTO ring numbers on metal rings allowed the individuals still to be identified.

I searched for another material to use for making colour rings that did not fade and contacted a few firms, including the plastics division of the Imperial Chemical Industry (ICI) in Manchester. I explained the problem and requested small quantities of a suitable material. The lady who took my telephone call said that she would raise the problem with her colleagues and would phone me back later that day. I had encountered similar responses from two other firms and so did not expect a reply! But that same afternoon, she called to say that she thought they had an appropriate material – a thermoplastic called Darvic, which they produced in sheets of bright colours and in a range of thicknesses. It could be cut easily and then moulded into any required shape at temperatures above 80 °C, and it was colourfast and resistant to fading in ultraviolet light. I thanked her and expressed my delight that she had been as good as her word.

Samples arrived by post two days later, and by the end of that day I had made the first Darvic leg rings by cutting a flat strip of the material, heating it in a bath of near boiling water, curling it into a round hole drilled into a block of Perspex and then dipping the whole block into a tray of boiling water to complete the process. The Darvic strip immediately took on the outline of the hole and retained this shape when plunged into cold water. I soon made rings in six bright colours and streamlined the process by using a Perspex block with multiple holes that allowed many rings to be made at a time. A slight modification was to round off sharp corners on the Darvic before shaping, and I used enough material to have two layers in each ring, as a precaution that in case of wear or breakage, a complete ring would remain. This last modification proved unnecessary, but it was a reasonable precaution at the time, as was sealing the ring with an adhesive to prevent a change of shape. I have since examined Darvic colour rings that were on Kittiwakes and Herring Gulls for more than 20 years, and while the surfaces of the rings were scratched, the colour remained almost as bright as when first produced and showed remarkably little wear. Despite having marked more than 3,000 Kittiwakes with Darvic rings, I have never come across a bird that has lost one.

This method was published in *Bird Study* in 1963, and my university technicians took over the production of Darvic rings, selling thousands that were used on birds throughout the world. Currently, several commercial firms produce and sell Darvic rings. Although the rings were obviously a major advance and soon used worldwide in many studies, that publication is by far the least-quoted scientific paper during my entire career.

I soon discovered that ICI also produced sheets of Darvic as laminates of two colours, and found that these could be engraved to reveal the underlying colour. This allowed me to engrave strips of the laminate with unique alpha-numeric inscriptions and then form them into rings in the same way as other colour rings.

FIG 227. Mediterranean Gull (*Ichthyaetus melanocephalus*) with a metal ring on its right leg and an engraved Darvic leg ring on the left. (Norman Deans van Swelm)

I used different colours each year, which allowed me to mark every Kittiwake chick reared in the North Shields colony for more than 30 years with a unique code of colours, letters and numbers, which could be seen and read at a considerable distance without having to catch the bird. Many of the rings were seen by other ornithologists, and these records revealed the movements of young Kittiwakes, when mature, to other breeding colonies, in some cases more than 1,000 km away. Later, ICI came to my aid yet again when they decided to discontinue producing the thickness of laminated Darvic I used for these rings. The company kindly sent me as many sheets of the laminate as I needed for the future, free of charge. The firm probably made little profit in selling the small quantities of Darvic needed to make colour rings for birds, but in doing so they made a huge contribution to ornithology.

Coloured and engraved rings are now widely used on large birds of many species (Fig. 227), and ringing has produced a great deal of new information. For example, the geographical origins of individuals were discovered, such as the large, dark-winged Herring Gulls (*Larus argentatus argentatus*) that breed in northern Norway and Russia, and winter in Britain and other North Sea countries, and the Caspian Gulls (*L. cachinnans*) reared in Poland that visit Britain.

Patagial wing tags

While coloured leg rings are very useful for following the behaviour and distribution of individual gulls, studies on the feeding behaviour of individual Black-headed Gulls on grassland often remained difficult because the rings were at the bottom of the leg and concealed by the vegetation. This problem was overcome by making coloured Darvic wing tags, again with unique engraved numbers, and attaching one to each wing. This method has been widely used, particularly on gulls and birds of prey.

TRACKING SYSTEMS

Radio-telemetry tracking

Radio telemetry was the first tracking method used with gulls, in the late 1960s, but it was limited because of the initial size of the transmitter and because the very high frequency (VHF) radio signals had to be detected by aerials, usually hand-held, which often only received signals over distances less than 1 km. Individual gulls had transmitters attached by a harness and could be followed considerable distances to their feeding or roosting sites, but they were detected only intermittently and the geographical position of the signal (and the bird) had to be obtained via triangulation from multiple locations. Much time was wasted in trying to locate individual gulls in this way because of the large area through which they moved, and the batteries used in the transmitters usually lasted only a few weeks.

Light-level geolocators

The use of light-level recorders for tracking birds was developed and introduced in the 1990s by scientists working for the British Antarctic Survey, where they were used to record the movements of Wandering Albatrosses (*Diomedea exulans*). The system depends on having an accurate clock, a light-sensitive unit that detects daylight, a data-storage system and a power supply. The instrument records when daylight ceases and when it reappears each day, and logs the data in a compact form. The geolocator is light in weight and miniaturised, and is usually attached to a leg ring on the bird. The disadvantage of the method is that it stores the data, so the bird must be recaptured at a later date to allow information to be downloaded. The day length identifies the latitude, while the timing of sunset and sunrise allows the longitude to be calculated. Because identifying the bird's location depends on day length, the accuracy is variable and becomes ineffective near the spring and autumn equinoxes, when day length is the same everywhere. The geographical position of the bird obtained at other times of the year can vary by more than 100 km, and so the method is suitable only for gulls that move considerable distances in the non-breeding season. Data collected can be accumulated and stored for weeks or months according to the components used. Battery size is critical, because its weight is a major factor in determining the size and weight of the geolocator. Other data can be collected and logged at the same time, including temperature and (in the case of seabirds) whether the logger is in the air or in water, indicating whether the bird is flying or sitting on the sea. In general, the data obtained have been outstanding, but spurious locations are not infrequently produced owing to low light levels recorded by the device, such as in conditions of thick cloud cover or even when the bird sits and covers the sensor.

Global Positioning System (GPS) tags

These tags give very accurate locations, usually down to a few metres, but since the data are collected within the device, the bird must be recaptured to recover them. The bird's position can be determined at very short intervals, showing, for example, the route taken by seabirds on individual feeding trips. An example of the results is shown in Fig. 97.

Satellite trackers

Satellite tracking is useful because it is not necessary to recover the tag in order to retrieve data detailing where the birds is or has been. In general, it requires a larger device, which is normally attached to the bird by a harness or glued onto feathers on its back. This is set up in such a way that the transmitter falls off after a given time, such as when the battery has reached the end of its life. Because of their size, satellite trackers tend to increase resistance to movement underwater, so are not suitable for use on species that frequently dive after food, although this is not a problem in most gulls.

The method uses an existing satellite system such as Argos to transmit the bird's position to the data-collecting centre. The Global Positioning System (GPS) of Earth-orbiting satellites is an integral component of the tracking system. The advantage of this system is that frequent locations are reported, and flight speeds can be determined. The locations obtained are usually accurate; while spurious locations are sometimes recorded, these can usually be identified and corrected by the frequent recording of the position.

Satellite transmitters were originally expensive to produce but recent advances have reduced the cost. One experimental study has found that using GPS loggers on Kittiwakes had adverse effects on the behaviour of birds when compared with control birds that did not have loggers attached. Others have claimed no adverse effects, but in some of these, the failure to detect a difference was mainly because only small numbers of individuals made up the treatment category. Other studies on diving seabirds, such as Common Guillemots (*Uria aalge*), have also reported adverse effects. The early transmitters used were probably too large, but sizes have since been reduced using improved technology. That said, studies are still needed to evaluate whether transmitters and recorders currently in use on gulls and other seabirds are having any adverse effects.

SEXING GULLS

Over the last 30 years, the use of DNA analysis has been developed to sex individual birds by taking a small blood sample or even a feather. This method is 99 per cent accurate (the 1 per cent error usually arises from misinterpretation of the results). DNA sexing makes use of a technique called the polymerase chain reaction, which amplifies the male and female chromosomes of birds. Female birds are heterogametic, carrying one copy of the Z sex chromosome and one of the W chromosome (making them ZW), while males are homogametic and carry two copies of the Z chromosome (ZZ). Initially, this method was expensive, and while competition from commercial firms has lowered the cost, it is still expensive if many individuals need to be sexed.

Uniquely colour-ringed gulls can often be sexed by observing courtship feeding or mating. Once the sex of a bird has been determined, this information is recorded and remains known for the lifetime of that individual, as is the sex of the mates with which it pairs in different breeding seasons. In the study of a Kittiwake colony over 30 years, many individuals changed their mates, and so being able to identify the sex of one individual led to the sexing of no fewer than 96 other colour-ringed individuals in that colony. The method was confirmed by dissection of five ringed birds that were found dead during the study, otherwise by head and bill measurements, or when mating with other sexed individuals in past or future years.

Measurements taken of adult gulls captured for ringing also led to the development of methods of sexing individuals using their biometrics. While gulls have a small degree of overlap in size of the sexes, this method allowed the sex of 90–95 per cent of individuals to be determined. Since the difference in size is greater in the larger gulls (p. 16), sexing based on body size works best for these species. Males also tend to have deeper bills, but the discovery that this increases with age in Herring Gulls indicated the need for caution when using this parameter. Many studies have produced discriminant functions to sex gulls, making use of a series of measures. When captures of large numbers of gulls are made, sexing using three or more measurements on each individual (for example, head and bill length, bill length, bill depth, wing length and weight) is unacceptably time consuming and, in several species, there is little to be gained from this. In most cases, the length from the tip of the bill to the back of the head has proved to be the best single measure, and is superior to wing length (which is affected by primary moult and feather wear) and is reliable even when measured by different people. Taking more measurements improves the percentage correctly sexed by only 1–2 per cent, and only a few birds in the overlap zone require further measurements. Weight is an unreliable parameter as it varies during the day and seasonally, and so should be ignored as a means of sexing.

The problem of using biometrics to sex individuals is common to all species of gull and is illustrated here for the Kittiwake. The distribution of wing lengths as

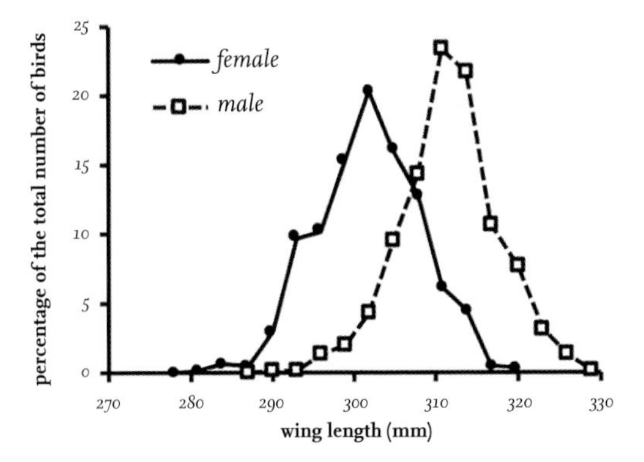

FIG 228. The distribution of wing length of 604 adult male and 592 adult female Kittiwakes (*Rissa tridactyla*) breeding in north-east England, expressed as a percentage of the total of each sex. From Coulson (2009) and personal data.

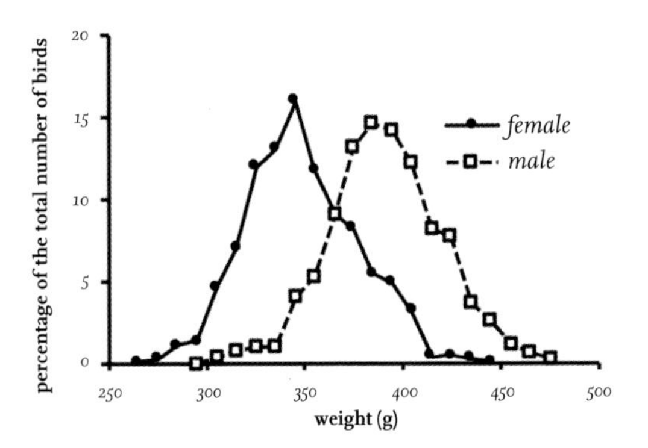

FIG 229. The range of weights of a sample of 759 adult males and 752 adult female Kittiwakes (*Rissa tridactyla*) breeding in north-east England and expressed as a percentage of the total for each sex. From Coulson (2009) and personal data.

seen in Fig. 228 show that males are usually longer-winged than females, but that there is considerable overlap, and so only the shortest-winged females and longest-winged males can be sexed with certainty using this measure. Similarly, while on average male Kittiwakes are heavier than females, there is considerable overlap (Fig. 229) over much of the weight range. Part of this overlap can be attributed to changes in weight that occur in both sexes during the breeding season.

The measurement of the distance from the tip of the bill to the back of the cranium (head and bill length) is both easy to take and more highly repeatable by different people than wing length. It is also the measure that shows the greatest separation between the sexes, although overlap still occurs (Fig. 230). In Kittiwakes, birds with a head and bill length below 89 mm are almost certainly females, while those above 93 mm are certainly males; Fig. 231 shows this separation. The sex is therefore in question only for those individuals with measurements between 89 mm and 93 mm, or about 50 per cent. Those at 89 mm

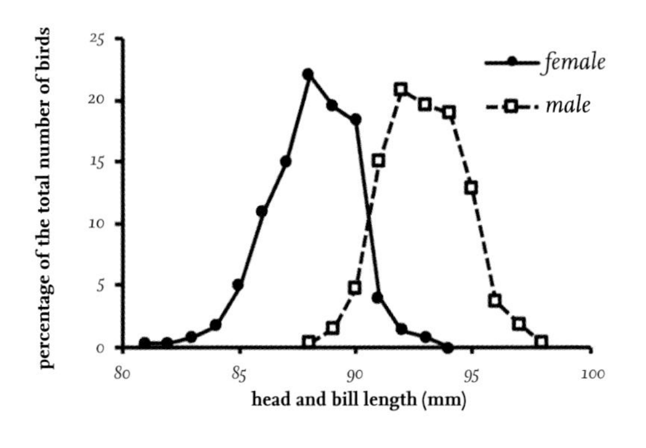

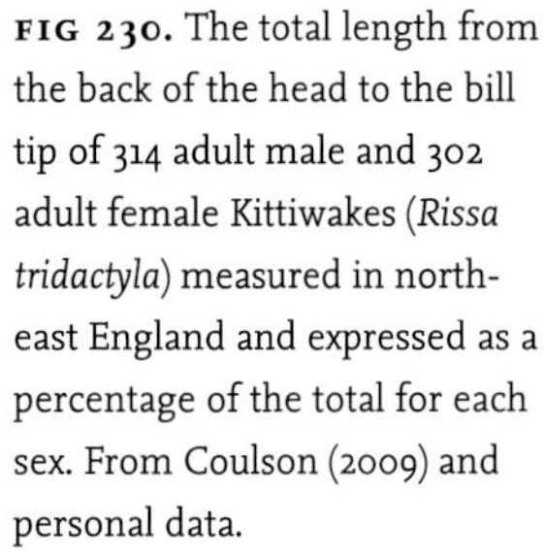
FIG 230. The total length from the back of the head to the bill tip of 314 adult male and 302 adult female Kittiwakes (*Rissa tridactyla*) measured in north-east England and expressed as a percentage of the total for each sex. From Coulson (2009) and personal data.

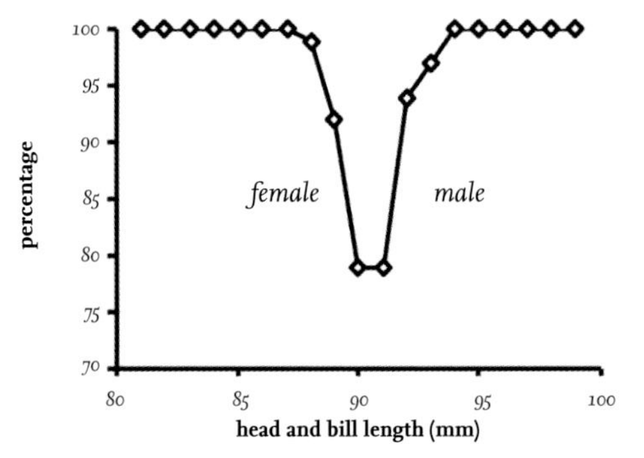

FIG 231. The degree of certainty of sexing Kittiwakes (*Rissa tridactyla*), based on the head and bill measurements of birds in north-east England. From Coulson (2009) and personal data.

and 92–93 mm have a more than 95 per cent chance of being correctly sexed. The dubious sexing is mainly in those birds with 90–91 mm measurements. Note that these values are for Britain, and since Kittiwakes breeding further north in the Atlantic and those in the Pacific are appreciably larger, the zone of overlap in the head and bill measurement must be adjusted accordingly.

PHOTOGRAPHY

Photography has developed rapidly in recent years, and the use of automatic time-lapse, movement and sound-triggered photography can be very effective in determining when individuals are present at the nest site and when incubating pairs change over. In the past, such methods depended on a mains power supply or the use of large batteries, but current methods use compact power supplies and some batteries can be repeatedly recharged by a solar power source. Recording activity at night was also once a problem, but infra-red light sources are now more readily available and economically priced.

Evidence of the identity of rare gulls used to rely on field notes and confirmation from other observers, but most records are now supported by high-definition photographs, which are easily obtained with long-focus lenses. No longer is it necessary to develop film and print paper copies, as detailed photos can be quickly and easily transferred via the Internet. The only drawback I have encountered with digital images is that the colour balance can be altered, which can affect the shade of grey on the wings of Lesser Black-backed Gulls and Herring Gulls.

Studies on the group behaviour of gulls using video recordings has been successful, as they can be replayed over and over again to examine the actions of individuals in situations where large numbers of gulls feed together, and where direct observations miss much of the interactions and the feeding success.

BIRD COUNTS

Prior to the twentieth century, the numbers of gulls in many countries in Europe and elsewhere, including Britain and Ireland, were poorly known and the records that do exist lack quantitative values. Henry (Mick) Southern and James Fisher were probably the first ornithologists to investigate seriously the numbers of seabird species in Britain. Fisher, in his remarkable book *The Fulmar* (1952), also organised a national census of that species' abundance through the BTO, which at that time was a national coordinating and support organisation for research planned, analysed and published by amateur ornithologists, and which had a network of voluntary representatives in each county throughout Britain. This was a very different body from the research organisation it is today, and was run by a single salaried person working in Oxford from a small upstairs room in a side-street building. As described above (p. 356), the ringing scheme was voluntarily organised by Elsie Leach from the entomology section of the Natural History Museum in London.

The first census of a gull species was organised in 1938 by Philip Hollom; it looked at numbers of Black-headed Gulls breeding in England and Wales, and was repeated in 1958. The first census of a gull in Britain that included Scotland was made in 1959 as a joint Northern Fulmar (*Fulmarus glacialis*)–Kittiwake census organised by James Fisher and me, and supported by the BTO. We were greatly helped by the long list of correspondents James had throughout the country and by BTO regional representatives. This was a revealing enquiry, because for the first time the national distribution of Kittiwake colonies was correctly determined, and estimates were obtained of the numbers of nesting pairs throughout Britain. The view held in 1947 by Carl Gibson-Hill was that almost all

British breeding Kittiwakes were in the Hebrides, the western side of Scotland and the west coast of Ireland, with few on the North Sea coastline of Scotland and England. The census discredited this view, with large numbers recorded in Orkney, Shetland and on the eastern coast of Scotland. I was able to combine this enquiry with a search of past literature since 1900, which showed that while numbers of Kittiwakes had recovered between 1900 and 1930, not a single new colony was established in that period and the only change was that existing colonies had grown. It was not until after some 30 years of numerical recovery that the first of many new colonies were formed around Britain.

Soon after its establishment in 1966, the Seabird Group planned Operation Seafarer, with the aim to census all seabirds around Britain. It had one salaried full-time organiser, David Saunders, and was made possible by grants from the Torrey Canyon Appeal Fund (set up after the extensive mortality of seabirds from the oil spill following the wreck of the tanker SS *Torrey Canyon* off Cornwall in 1967), the Royal Society for the Prevention of Cruelty to Animals and the Royal Society for the Protection of Birds. Operation Seafarer resulted in the first national census of seabirds, which was carried out in 1970; the results were published as *The Seabirds of Britain and Ireland* (Cramp *et al.*, 1974). The count was limited to colonies of seabirds on the coast, with gulls nesting inland in Britain and Ireland not included, so national numbers of Black-headed Gulls, Common Gulls (*Larus canus*) and the larger species were underestimated. This restriction did not affect the estimates of the numbers of nesting Kittiwakes. A census was made for a second time, with the remaining species counted for the first time. In 1984, the Seabird Colony Register was established jointly by the Seabird Group and the Nature Conservancy Council, and in the following three years (1985–87) the numbers of coastal seabirds, including gulls, were again estimated and this time published as *The Status of Seabirds in Britain and Ireland* (Lloyd *et al.*, 1991).

In 2000 (although actually covering the years 1998–2002), a further national survey was carried out, and this time inland breeding gulls were included. The data collection was organised mainly by the Joint Nature Conservancy Council and its seabirds team, with financial assistance from several other bodies, including the Seabird Group; results were published as Mitchell *et al.* in 2004.

The spread of gulls nesting in towns and cities was not neglected in censuses. Stanley Cramp published the first survey in 1971, followed by a repeat survey organised by Pat Monaghan and me in 1975, and one in 1995 organised by Susan Raven and me, while the national census in 2000 also included urban nesting gulls. These surveys permitted us to trace the rapid increase of this invasion of urban sites, first in coastal towns and then spreading inland. A detailed study reported that counts of nesting gulls were often grossly underestimated in urban sites, perhaps missing up to 25 per cent of the nests, because they were not visible from a distance using vantage points or from street level without the aid of a mobile access platform. At one time it was thought that small drones might be the answer to this difficulty, but it is becoming evident that legislation may restrict their use in towns and cities, and in any case approval would likely also be required from the owner of each and every property being overflown, if a camera is being used. It is still possible that drones may be useful in counting seabirds at natural coastal sites. The definition of photos taken from satellites is still too poor to facilitate counts of gulls from space, although this method has recently been used on albatrosses.

In addition, annual counts overseen by the JNCC as part of their Seabird Monitoring Programme have been made at a modest number of sites in Britain since 2000, with the aim of obtaining an early warning of any major threat to seabirds. These locations were not selected as representative sites but were those where regular annual counts were possible. However, since these counts were established, they have been used as equivalent to national census data to estimate annual changes in seabird numbers, including Kittiwakes and large gulls. In general, the accuracy in these annual trends, when applied to the national situation, is not clear and there is still a problem as to how representative these sites are of Britain as a whole. In 2017, JNCC withdrew the population trends they had reported annually since 2000 for Herring Gulls and Lesser Black-backed Gulls as they now considered the sample was inadequate. The reliability of these annual samples cannot be evaluated until the next national census is carried out. A new national census was planned for 2015, but it has been repeatedly delayed – it was started in 2017 for some species and areas and will continue until 2020.

Other census work on gulls in Britain includes counts at winter roosts on reservoirs, but these studies suffer from examining only a proportion of roosts and the difficulty of identifying species, particularly of immature gulls, in poor light late on winter afternoons. These surveys are of lesser value in detecting population changes than the national breeding bird counts because they do not include many coastal roosts or those on islands. In recent years, these winter counts have suggested appreciable declines of Black-headed Gulls and Common Gulls, although breeding season surveys have indicated only marginal changes. Winter roost counts include immature gulls and many birds of all ages that have arrived from abroad, while only British and Irish breeding adults are included in the breeding season counts.

These and similar figures play an important part in developing conservation strategies for gulls. However, as those who have participated in a census of breeding gulls know, the work can be difficult and time consuming. For example, gulls nesting in long vegetation are difficult to count, and equally, colonies on some islands can be difficult to access and count.

Readers should note that the numbers in Table 1 are approximate. The main reason for this is that the accuracy of many census totals for gulls is not known. There are preferred methods used to survey and count nesting gulls, and these are noted in the instructions to those making a census. The date of counting is often restricted but is critical, because counts must be made when all pairs have nests. Even then, individual pairs of gulls in a colony can begin egg laying over a 35-day period or more, and therefore there is no single date on which all active nests are occupied. In addition, some pairs build nests late, by which time other nests have been predated, replacement nests have been built and new clutches have been laid. Some pairs of Lesser Black-backed Gulls have been found to build multiple nests but lay in only one. At very difficult sites, the numbers of breeding pairs of gulls must be obtained by estimating the number of individuals flying above the colony when it is disturbed and then applying a correction factor, such as multiplying by 1.6 to convert individual numbers to breeding pairs. In 2015, my wife and I (Coulson & Coulson 2015) showed that up to 25 per cent of gull nests in some towns and cities remained undetected without the use of a cherry picker to view otherwise concealed nests on high roofs.

In some cases, errors probably cancel themselves out when numbers are totalled with other colonies, but similar errors are probable in successive national counts, making the percentage change less likely a result of error. However, the magnitude of error from different causes has yet to be investigated and quantified. Other sources of error arise from the habit of some adult gulls to skip a breeding season and not even attend a colony – studies using marked individuals have shown that the extent of this varies from year to year but can involve up to 15 per cent of adults associated with a particular colony.

Despite these limitations, some analyses of the 2000 seabird census data (Mitchell *et al.*, 2004) have used the estimated national totals (to the last pair!), reporting 116,684 pairs of adult Lesser Black-backed Gulls, of which 10,874 pairs were nesting in towns. Such numbers give a convincing but false impression of the census accuracy. Realistically, the census numbers are probably minimum estimates, but by how much they underestimate the actual numbers remains unknown. The change in numbers between years should therefore not be taken as precise and exact values, although this gives rise to the question: what level of change between years would confidently indicate a population decline? Unfortunately, the confidence ranges on these numbers remain unknown until more detailed studies are made.

CHAPTER 13

Urban Gulls

A pest can be a beautiful rose growing in a cabbage patch.

A SPECTACULAR CHANGE IN THE behaviour of gulls is their rapidly increasing spread to towns and cities to breed. Currently, gulls of at least 10 species are nesting on buildings around the world. Silver Gulls (*Chroicocephalus novaehollandiae*) in Australia and at least three species in North America do so, but the habit is most pronounced in Europe, where at least seven species have been reported regularly nesting on buildings in urban areas; six do so in Britain and four in France. The extent of urban nesting is such that it is now causing problems in many places, and non-lethal methods of management have so far failed to reduce numbers.

Historically, the first record of gulls nesting on buildings in England was in 1909, but there is an even earlier record in mainland Europe, in 1894, of a pair described as 'Herring Gulls' breeding on a building alongside the Black Sea. There was a long gap before the next cases were reported in 1960, when several hundred 'Herring Gulls' (possibly Yellow-legged Gulls, *Larus michahellis*) were nesting on rooftops in Varna, Burgas and Nesebur in Bulgaria.

Urban nesting by gulls is now widespread in Europe, involving countries along the eastern coastline of the Atlantic and the Baltic, from Norway south to Spain and Portugal. In France, urban nesting started in 1970 and currently involves about 30 per cent of all Herring Gulls (*Larus argentatus*) breeding in that country (11,000 pairs) and a minimum of 700 Lesser Black-backed Gulls (*L. fuscus*) nesting in at least 60 towns. The Yellow-legged Gull has also started to nest on buildings in France and Iberia, including Gibraltar. The first urban nesting of this species in Italy was reported in 1971 from Rome, and it has since spread to

Venice, Trieste, Livorno and Naples. Herring Gulls were first reported breeding on roofs in Berlin in 2010.

In the USA, buildings have been used by nesting American Herring Gulls (*Larus smithsonianus*) for several years, mainly on the southern side of the Great Lakes, but also on the east coast at Portland in Maine and in the city of New York. Further north, in Canada, they are now nesting at urban sites in Halifax, Nova Scotia, and at St John, New Brunswick. Glaucous-winged Gulls (*L. glaucescens*) breed on buildings at several locations on the west coast of North America, while Ring-billed Gulls (*L. delawarensis*) nest on buildings around the Great Lakes in USA, and in Toronto and Ontario in Canada.

THE SITUATION IN BRITAIN AND IRELAND

For centuries, gulls in Britain have bred in areas widely separated from those inhabited by humans, using small islands, sea cliffs, bogs, moors and marshlands. While people still managed to exploit them, this was usually achieved by briefly invading their breeding sites in boats, climbing sea cliffs or penetrating wet and boggy areas to obtain eggs, young and, occasionally, adults for human consumption and plumage for household needs.

The Industrial Revolution made access to gulls easier and they were more extensively exploited. Towards the end of the nineteenth century, the numbers of gulls in Britain and Ireland reached an all-time low, as the human population increased rapidly. Small villages were becoming large towns and new urban centres were established. In 1900, the population of Britain and Ireland was approximately 38 million people; by 2017, it was double that at 76 million and still increasing, with an average density of 250 people per square kilometre. While the human population was increasing, we gradually changed from being predators of gulls to active protectors, a change supported by a series of Acts of Parliament.

During the nineteenth and twentieth centuries, an abundance of food for gulls was available both at sea and in harbours from the increasing marine fisheries, and through more intensive farming practices, supported by mechanisation and involving more extensive ploughing and harvesting. From the early 1950s, new and larger landfill sites were created, where household and industrial waste materials were dumped instead of being burnt as was usual in earlier years.

Within a relatively short period of time, both the human and gull populations were expanding, which brought them closer together. For example, gulls were scarce in London in 1900, but in more recent years numbers there have increased, influenced at least in part by people feeding them at lakes, in parks and on the embankments of the Thames. By winter 2000, there were tens of thousands of gulls feeding in London and many were frequently fed by humans, who even threw them bread and other items from windows high up on office and residential blocks. Gulls also competed, often successfully, with ducks and geese on park lakes for food regularly supplied by the public, and fed on items that people rejected, dumped or dropped accidentally in the streets.

Changes in urban colonies over time

The size of most urban colonies of large gulls has increased appreciably over time, in Britain and Ireland (Fig. 232), and also elsewhere. Fig. 233 shows that, as time has passed, the proportion of towns and cities in Britain with fewer than 10 nests has declined appreciably, while the proportion of sites with more than 100 nests has increased from less than 4 per cent to about 20 per cent over the same time period.

Once large gulls start to nest in an urban area, their rate of increase is rapid. Fig. 234 is based on the annual percentage change in numbers at 32 urban colonies of different sizes in Britain and Ireland, and shows that the average rates of increase in urban towns is highest when numbers nesting are initially low. Because of delayed maturity, this early growth occurs at a time when few of

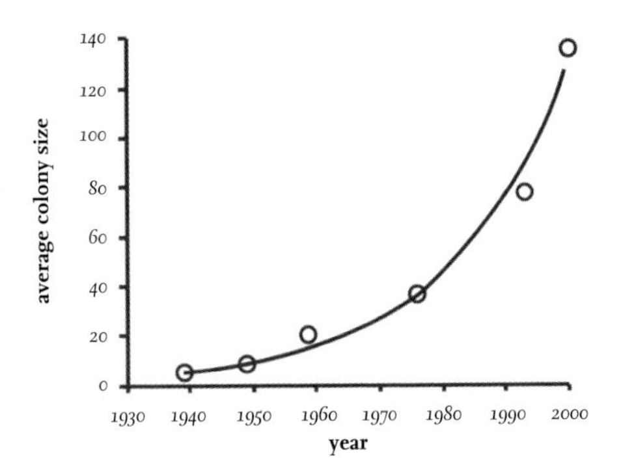

FIG 232. The average number of pairs of large gulls nesting in each town and city where breeding was taking place between 1940 and 2000. The line is the best-fit exponential curve to the data and represents an increase in the average colony size of about 5.6 per cent per year. Sites involving both Herring Gulls (*Larus argentatus*) and Lesser Black-backed Gulls (*L. fuscus*) have been considered as a single colony. Some of the data are taken from Cramp (1971), Monaghan & Coulson (1977), Raven & Coulson (1997) and Mitchell *et al.* (2004) after corrections when additional information became available.

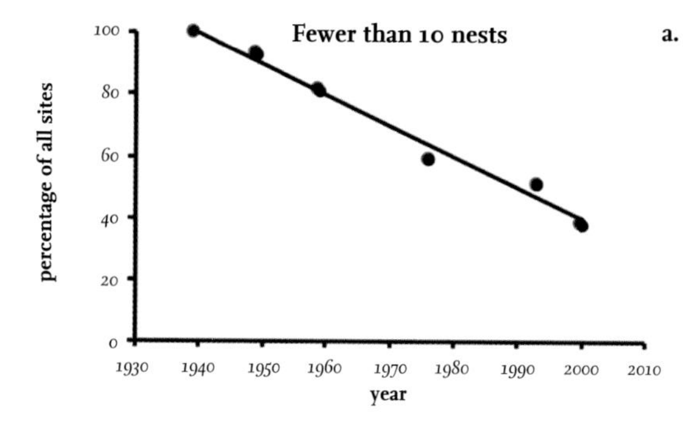

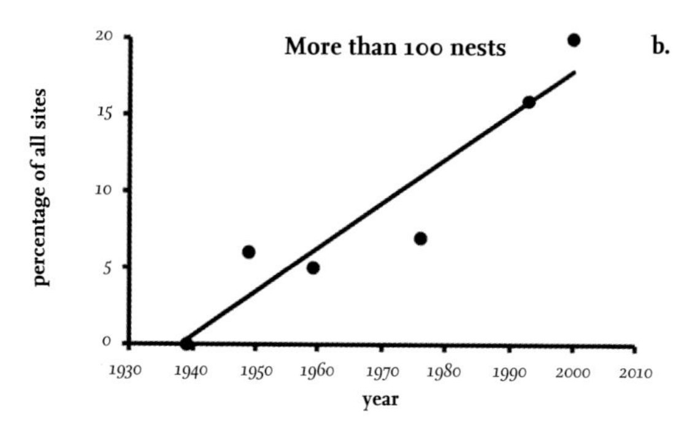

FIG 233. (a) The proportion of British urban towns and cities with nesting gulls that had fewer than 10 pairs has decreased by about 1 per cent each year from 1939 to 2000; (b) the proportion of British urban sites with more than 100 nests of large gulls has increased by about 4 per cent every 10 years from 1939 to 2000.

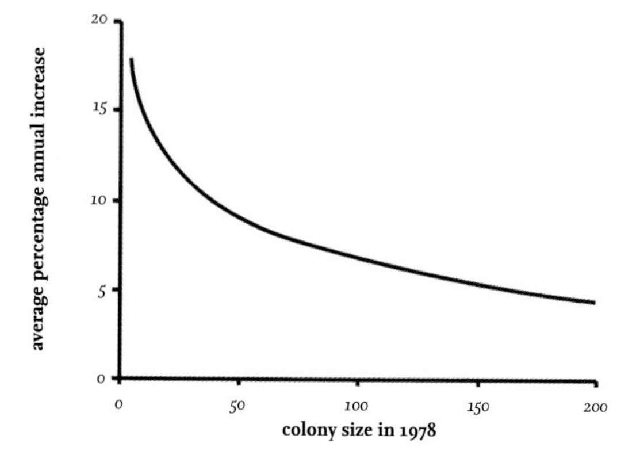

FIG 234. The average annual percentage increase in the size of 32 British urban colonies of large gulls between 1978 and 1994 in relation to their size in 1978. No decreases were reported in any of the colonies during this period. Initially small (usually young) colonies increased much more rapidly than large ones.

TABLE 68. The proportion of urban Herring Gull and Lesser Black-backed Gull colonies of different sizes in Britain and Ireland in 1976 that were subsequently deserted or had been destroyed by 2000. Few urban colonies have disappeared.

Colony size (pairs) in 1976	*Percentage abandoned by 2000*
1–4	10%
5–10	4%
11–20	0%
21–50	0%
51–200	0%
>200	0%

the additional breeding birds each year could have been reared at that site, and therefore it must be driven by appreciable numbers of immigrants.

Table 68 shows the proportion of British urban gull colonies of different sizes that existed in 1976 but had disappeared by 2000. Of the colonies with more than 10 pairs in 1976, none is known to have disappeared in the subsequent 24 years, with or without human intervention. Furthermore, fewer urban colonies have disappeared in recent years (Table 69), possibly because there are now appreciably more moderately sized and large colonies. It is evident that the disappearance of urban colonies is very low and has diminished over time, despite increasing concerns about their presence and management methods used to deter them. As the average size of an urban colony increases rapidly (Fig. 232), the chances that it will disappear after a few years soon declines, even when attempts are made to disrupt breeding. If urban gulls are a concern, new and rigorous measures of management and early intervention are needed.

TABLE 69. The proportion of towns and cities in Britain colonised by nesting large gulls between 1921 and 2016 that were subsequently deserted or otherwise cleared of nesting gulls.

Year of colonisation	*Disappeared by 2016*
1921–40	10%
1941–50	5%
1951–60	4%
1961–70	1%
1971–80	1%
1981–90	0%
1991–2000	0.2%
2001–16	<0.2%

SPECIES NESTING IN URBAN AREAS IN BRITAIN AND IRELAND

The greater proximity of gulls to urban areas during the past century has led to six species starting to nest on buildings in Britain. Currently, numbers of Herring Gulls, Lesser Black-backed Gulls, Kittiwakes (*Rissa tridactyla*), Great Black-backed Gulls (*Larus marinus*), Common Gulls (*L. canus*) and Black-headed Gulls (*Chroicocephalus ridibundus*) now nest on buildings in Britain, and it is likely that the Yellow-legged Gull will join this list within a few years.

Herring Gull and Lesser Black-backed Gull

By far the most numerous and extensive movements of gulls to nest on buildings in urban areas in Britain and Ireland have involved the Herring Gull and, more recently, the Lesser Black-backed Gull. Although the Herring Gull developed the habit first and is still more numerous than other species, numbers of urban-nesting Lesser Black-backed Gulls are increasing more rapidly. Both species nest alongside each other in many urban areas, but the proportions vary markedly from place to place and range from a few per cent to almost total dominance of one over the other. As the numbers nesting in urban areas have increased, so the proportion of towns and cities with both species present has increased. For this reason, it is convenient to consider them together here, particularly since they both contribute to problems and complaints.

The presence of numerous large gulls at harbours and fishing ports around Britain, which forms a large part of our current perception of these birds, has been a reality for only 100 years or so. Numbers at the end of the nineteenth century were so low and persecution near human habitation so high that few gulls were present around towns and villages in coastal areas. As numbers of Herring Gulls recovered and increased through the twentieth century and persecution reduced, the birds were attracted to coastal fishing harbours to waste discarded by the fishing industry. As their numbers increased further, they began to roost during the day on buildings nearby, and eventually a few remained to nest, usually on industrial and commercial building sites and areas in towns that had similarities with their natural nesting areas. In common, these had minimal vegetation and were safe from predatory mammals and disturbance, since humans rarely visited the roofs of houses, dockside buildings and factories. Initially, only a few pairs nested on buildings, but their social behaviour soon attracted other pairs. The numbers nesting at each site typically increased rapidly in the first few years and before any young reared there had reached maturity, so their initial increases had to be driven by immigration of birds reared elsewhere.

This immigration probably continues, while a modest proportion of the young individuals reared there also eventually return to breed.

Initially, urban-nesting gulls were greeted with pleasurable interest by ornithologists and the public alike, but as their numbers increased, concerns developed to the point that many residents started to complain to local councils about the adverse effects caused by the invasion. In part, this was stimulated by frequent and inaccurate coverage by the media (see box, p. 168) and encouraged by wild predictions of future trends. Urban-nesting gulls are now regarded as a genuine problem in many towns and cities, to which conservation bodies, local groups and the government have not yet been able to offer satisfactory solutions.

The first Herring Gulls to breed on a building in Britain are believed to have been a pair that built a nest on an old mill at Gulval, near Penzance in Cornwall, in 1909, but this report remains unconfirmed, as was rooftop nesting by Herring Gulls reported a year later at Port Isaac, also in Cornwall. The first substantiated reports of the habit in England were not obtained until the late 1920s, when nesting occurred at Budleigh Salterton and Torquay, both in Devon. Urban nesting was reported in Dover, Kent, in about 1936, and 10 years later this site had become a colony of some 200 pairs. Further colonisation of buildings occurred in the first half of the 1940s, including several pairs breeding on structures on the West Pier and Palace Pier at Brighton, Sussex, and on the nearby pier at St Leonards. All of these piers were closed to the public during the Second World War, and nesting apparently ceased when access was restored a few years later. In the next few years, there was only one record from the south of England, when a pair nested on a building at Polruan in Cornwall in 1950, although some instances likely went unrecorded.

While it is clear that gulls nesting on man-made structures started on the south coast of England, a second area much further north was soon involved, when roof nesting occurred in 1942 at Whitby and in 1947 at Staithes, both in Yorkshire. By 1961, breeding in South Shields (Fig. 235) and Hartlepool was well established.

In Scotland, Herring Gulls first nested on buildings in Lerwick, Shetland, in the 1940s. Other early records in Scotland were from Cellardyke (Fife) in 1951, and then Peterhead (Aberdeenshire) and Lossiemouth (Morayshire) in 1952.

In Wales, the habit of Herring Gulls nesting on buildings began in about 1945 at Merthyr Tydfil, (Glamorgan, the first inland British site) and it had spread to nearby Hirwaun by 1963. Nesting on coastal buildings in south Wales was first reported in 1951 at Pembrey (Carmarthenshire), Cardiff in 1962 and Newport (Monmouthshire) in 1969, while in north Wales town nesting by Herring Gulls on buildings had started at Llandudno (Caernarfonshire) by 1947 and at Holyhead (Anglesey) in 1958.

Nesting on buildings by Herring Gulls in Ireland started later than in Britain, and was first recorded in the Dublin area and in County Waterford in 1976.

FIG 235. Gulls nesting on a South Shields riverside building in 1966. (John Coulson)

By 1995, the habit had spread to Country Antrim, but no more localities were reported elsewhere in Ireland until 2000.

Lesser Black-backed Gulls first joined colonies of Herring Gulls at British urban sites in 1946. The first mixed colonies in Wales were recorded in 1970 and these were at Cardiff, Merthyr Tydfil, Hirwaun and Newport. Subsequently, the number of sites with mixed species increased rapidly, although the ratio of the two species in individual colonies varies and has a marked geographical pattern. Urban colonies composed of only Lesser Black-backed Gulls are few.

Stanley Cramp (1971) brought together further breeding records in Britain when data were collected nationwide during the first national seabird census in 1970. He reported at least 1,250 pairs of Herring Gulls nesting on buildings at 75 sites in Britain (with 980 pairs in England) and averaging about 17 pairs per site, but at that time Lesser Black-backed Gulls totalled only 61 pairs at five sites, all in south Wales apart from one in nearby Gloucester. Six years later, a survey in 1976 made by Patricia Monaghan and me concluded that the numbers of Herring Gulls and Lesser Black-backed Gulls breeding in urban areas had increased to more than 3,000 pairs in Britain and Ireland consisting of at least 92 urban colonies, with an average of 33 pairs per site. In 1994, a repeat survey I organised with Susan Raven recorded 13,000 pairs, with about 125 pairs per urban colony, even though several key cities and towns in south-west England and south Wales were not covered (Raven & Coulson, 1997). The national census of seabirds made in 2000 (Mitchell *et al.*, 2004) reported 20,200 pairs of urban-nesting Herring Gulls at 225 sites and about 10,800 pairs of Lesser Black-backed Gulls nesting at 64 urban sites, giving a total of 31,000 pairs of large gulls nesting in 230 towns, or an average of 135 pairs per urban area. In contrast, only 83 pairs of Great Black-backed Gulls were nesting in urban sites at that time.

Like all data collected on a countrywide basis, the numbers obtained tend to be underestimates because a few sites are missed. Furthermore, counting gulls nesting in towns is difficult, such that about 25 per cent of nests are missed if the usual census methods are employed. An estimate in 2004 of at least 121,000 and as high as 193,000 breeding pairs (Rock, 2005) in Britain was not based on a national census of data from Britain and Ireland, and so the figures should be treated with caution, as should the prediction made at that time that there would be a million pairs of urban-nesting gulls in Britain by 2014. Nevertheless, there has been a continued increase at many sites since 2000, and a rough estimate based on more recent information (and assuming 650 urban sites, each with an average of 140 pairs) would suggest that there were about 91,000 pairs of large gulls nesting at urban sites in Britain and Ireland in 2014. If accurate, this figure would be an appreciable proportion (perhaps even approaching 40 per cent) of the total number of pairs of Herring and Lesser Black-backed gulls believed to be breeding in Britain and Ireland in 2000.

Proportions of Herring and Lesser Black-backed gulls at urban sites

As mentioned above, Lesser Black-backed Gulls started nesting in urban areas in Britain after Herring Gulls, but they have increased more rapidly. As a result, over time they have formed an increasing proportion of the total (Table 70), such that by 2000 they comprised more than a third of the two species.

The distribution of these two species at urban sites throughout Britain and Ireland is by no means uniform, and the proportions in regional areas show marked differences, as data from 1994 reveal (Table 71). Separating Britain into eastern and western halves highlights a clear contrast, with Lesser Black-backed Gulls being slightly more numerous than Herring Gulls in the western half of the country, but forming only 7 per cent of the total in the eastern half. Breaking the distribution down into smaller areas reveals that the dominance of Lesser Black-backed Gulls is mainly restricted to the Avon and Gloucestershire area of England and to south-west Scotland. The reason for this distinction is not known, but is discussed further on p. 413.

TABLE 70. The percentage of Lesser Black-backed Gulls among the total of Herring and Lesser Black-backed gulls nesting in urban areas in Britain and Ireland at intervals between 1949 to 2000.

	1949	*1959*	*1970*	*1976*	*1994*	*2000*
Percentage of Lesser Black-backed Gulls	2%	3%	5%	10%	19%	35%

TABLE 71. The variation in the percentage of Lesser Black-backed Gulls among large gulls nesting in urban areas in regions of Britain and Ireland in 1994, based mainly on data in Raven & Coulson (1997).

Regions	*Percentage of Lesser Black-backed Gulls*
Eastern half of Britain	**7%**
Western half of Britain	**55%**
South-west Scotland	61%
Avon and Gloucestershire	50%
Wales	18%
Ireland	6%
South-east England	4%
Eastern Scotland	2%
South-west England	<1%

Almost all the urban sites first colonised by Herring Gulls were coastal and close to where other pairs were nesting on natural sites. The use of inland sites started much later (apart from the instance in 1945 at Merthyr Tydfil) and was centred on south-west England and the Forth–Clyde valley in Scotland. More recently, there has been a spread inland using urban sites – for example, breeding was first reported in 2001 in the heart of England at Birmingham, and 10 years

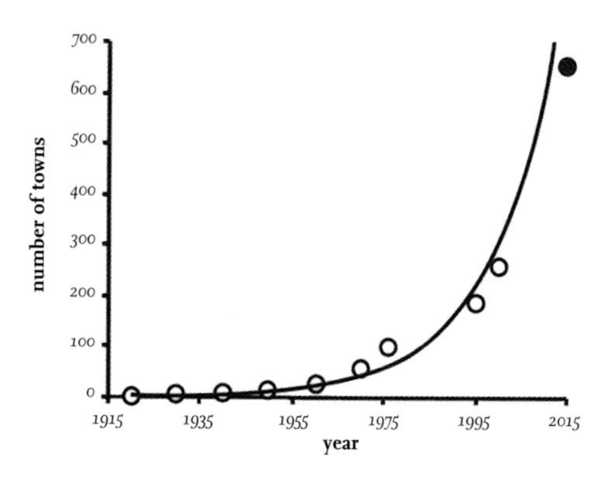

FIG 236. The numbers of villages, towns, cities and other urban areas in Britain with Herring Gulls (*Larus argentatus*) or/and Lesser Black-backed Gulls (*L. fuscus*) nesting on roofs or other man-made structures. The numbers relate to data in publications up to 2004, updated with additional information. The position of the filled circle for 2015 is based on data collected by the author and may be a slight underestimation. The fitted curve indicates an exponential growth, with the number of sites increasing at about 6 per cent each year.

later, Jim Winsper found 550 pairs of Lesser Black-backed Gulls nesting there but only 31 pairs of Herring Gulls.

The increase in the numbers of towns and cities with nesting large gulls in Britain and Ireland is shown in Fig. 236. The trend closely follows an exponential curve, increasing by 6 per cent per year. Eventually, the increase and spread will have to slow down as appreciably fewer towns remain without nesting gulls, but there is only a hint of this so far and more localities will almost certainly be occupied in the next few years. A new national census of urban-nesting gulls is in progress, but census methods used need to be improved.

Kittiwake

Kittiwakes require a different and more restricted type of nesting site to other gulls, and in urban areas they favour narrow ledges on buildings or other structures, close to or immediately adjacent to extensive areas of water, such as the open sea or harbours. In a unique case, they also nest on a large bridge and buildings alongside a wide tidal river up to 18 km from the sea at Newcastle and Gateshead (Fig. 237). Council employees regularly have to clean the paths and ground underneath these sites of droppings and material falling from nests (Fig. 238). In places, spikes have been put up on ledges to prevent Kittiwakes nesting on listed buildings, but these are not always successful – the gulls have sometimes succeeded in building their nest on top of the spikes (Fig. 239).

In 1931, two pairs of Kittiwakes were the first to nest on a man-made structure – the harbour wall at Granton, near Edinburgh – and four pairs nested there in 1933, but all were unsuccessful and no further breeding attempts have occurred there. A more durable colony was established at Dunbar, East Lothian, in 1934 on the window ledges of a harbourside building where grain was dried prior

FIG 237. Kittiwakes (*Rissa tridactyla*) nesting on girders on the Tyne Bridge at Newcastle, which now holds more than 300 nests. (Mike Osborne)

FIG 238. Council workers cleaning the pavement beneath the Tyne Bridge in Newcastle of material and droppings from Kittiwakes (*Rissa tridactyla*). (Mike Osborne)

FIG 239. Kittiwake (*Rissa tridactyla*) nest built on spikes placed to prevent nesting, Newcastle. (Mike Osborne)

to malting (Fig. 118). As numbers increased there in the 1960s, the birds spread across the harbour to nest between weathered sandstone blocks on the ruins of the old castle and on the neighbouring natural cliff (Fig. 121), which formed part of the base of the castle. Later, more bred a little further away on neighbouring rock faces produced where a new entrance had been cut to access the harbour. This is the only colony where Kittiwakes nest side by side on natural and man-made 'cliff' faces. The granary building was demolished, but by that time the gulls were already nesting in numbers across the harbour on the remains of the castle and the displaced birds simply moved there.

In 1949, a small Kittiwake breeding colony was established on a warehouse alongside the Tyne at North Shields, 3 km from the river mouth. This was the first time the species had nested in England away from the coast. Nest sites were mainly on window ledges of the Brewery Store, used by the neighbouring Smith's Dock Ltd (Figs 119 and 120). Numbers of nesting birds increased, and within a few years they had spread to the neighbouring Ferry Offices, to the Clementsen's building (which was run as a ship chandler) and to a neighbouring derelict biscuit factory, until that was demolished. A few then moved to the gable end of a stately building that had been the town residence of the Dukes of Northumberland in the nineteenth century and, in more recent times, part became a public house called the Northumberland Arms but known locally as 'The Jungle'. Extra ledges were built on the Brewery Store to encourage the colony's growth, and eventually it increased to 100 pairs. In 1991, however, the warehouse was converted into luxury riverside residences, and the Kittiwakes were excluded and forced to nest elsewhere. Most moved to the cliffs at nearby Tynemouth and a few across the river to a dockyard building, but one female moved as far as 300 km to breed at Lowestoft in south-east England.

The next colony on the river Tyne was established in 1963 on the windowsill of a sheet-metal workshop, 18 km from the river mouth at Gateshead, but the building was demolished the following year. In the late 1960s, Kittiwakes nested at Newcastle for the first time, using the edges of the roofs of the HM Customs sheds on the quayside, but these sheds were demolished within two years. Between 1963 and 2017, no fewer than 13 different buildings and structures alongside a 22 km stretch from the mouth of the Tyne to Newcastle and Gateshead were used by nesting Kittiwakes, including the roll-on roll-off gantry used for vehicular access to the Newcastle–Norway ferries (Fig. 124). Over this time, all but two of these structures were demolished, causing the birds to move to new sites on other buildings. What grew to be the largest colony on the river was established on the long ledges and windowsills of the Baltic Flour Mill at Gateshead, with 340 pairs at its peak (Fig. 135). The mill was closed and remained empty for several years, and the Kittiwakes continued to breed there on long ledges on the north and south sides that ran almost the whole length of the building; nests touched each other for the whole distance. On the building's east and west sides, pairs nested on window ledges.

In 2002, the mill building was opened as the BALTIC Centre for Contemporary Art, and for a few years Kittiwake nesting on the long ledges continued to be accepted. But then the newly appointed director of the centre decided that the gulls' droppings were a problem, and took steps to make the ledges unusable through the addition of 45-degree concrete slopes along their length. As a compromise, a tower with ledges specifically designed for Kittiwakes was built nearby (Fig. 240), and about a quarter of the gulls transferred there,

FIG 240. Kittiwake (*Rissa tridactyla*) tower at Gateshead, built to accommodate birds displaced from the Baltic Flour Mill. (Mike Osborne)

while others moved across the river to another building. Subsequently, the tower was moved further downstream and many of the Kittiwakes followed it there, but others moved across the river to nest on older buildings, including the Grade I listed Newcastle Guildhall, built in 1655. Steps were taken to exclude Kittiwakes from some of these sites by adding netting or spikes, and many moved onto parts of the Tyne Bridge, where numbers increased annually, reaching 614 pairs out of a total of 1,308 pairs breeding on buildings and structures along the river Tyne in a census carried out in 2017 by Dan Turner (pers. comm.). It is likely that in years to come there will be conflict between the nesting Kittiwakes at this unique inland group of colonies and the maintenance of the bridge and nearby properties.

After an absence of several years, a few Kittiwakes overcame the problem of the slopes added to ledges on the BALTIC Centre for Contemporary Art by carrying in mud and vegetation to level the gradient, and they managed to build new nests there, which persisted from year to year. Numbers nesting increased annually and, fortunately, the centre accepted the birds following discussion with local conservation bodies. A viewing platform now exists at eye level with the nests, and is a major attraction that contributes to the visitor experience, while a video camera has also been installed to provide close-up footage of nesting birds on a large screen.

A colony situated 21 km from the sea became established in the 1970s on another disused flour mill upriver of the Tyne bridges at Gateshead, but that building was soon demolished. Another colony was also established on riverside buildings near Gateshead belonging to Marine Coatings, International Paint Ltd. New ledges have since been put up at this colony and it has increased in size.

In 2015, when a new hotel was constructed close to the Tyne Bridge at Newcastle, the management complained about the noise from the Kittiwakes. In addition, people involved with the Sunday market on the Newcastle quayside have complained about droppings from the birds, and there are several other sources of concern. Spikes were illegally put up on ledges on the bridge to prevent Kittiwakes nesting, which Newcastle City Council removed, but gel applied to deter the gulls remains. As Kittiwakes continue to increase in these unique colonies far from the sea at Newcastle and Gateshead, the question of how many should be allowed to nest and where they should be permitted to do so will have to be addressed. Noise, droppings and damage to property increase as the numbers of birds increase there, yet nationally Kittiwakes are declining in many areas. To date, finding a suitable riverside site and the funds to build artificial cliffs for Kittiwakes at Newcastle and Gateshead have proved unsuccessful.

In 2018, netting that had been put up on some buildings to deter nesting Kittiwakes resulted in 12–15 adults and young birds being entangled and killed and about seven others trapped but eventually released alive. An on-line photograph of a dead Kittiwake and a petition for the netting to be removed achieved over 99,000 signatures and much media attention. The Newcastle Council immediately removed their netting, but much of that put up by others still remains and there is already a conflict between property owners and the general public, and this is likely to escalate as more Kittiwakes attempt to nest on buildings in the area. Eventually, a limit will be necessary if numbers continue to increase, but bird protection organisations have already stated that they would object to any attempt to restrict the size of this unique inland colony, which already includes more than 600 pairs on the Tyne Bridge and quayside at Newcastle and more nearby. A major problem is that it is not known why the Kittiwakes are nesting so far from the sea. They do not feed frequently in the river and obtain most of their food by flying out to sea, which involves a return journey of more than 35 km along the river before they even reach the coast.

Elsewhere, Kittiwakes have established small colonies on ledges of buildings and other structures at Seaham Harbour and Hartlepool in County Durham, Bridlington and Scarborough in Yorkshire, and Liverpool in Lancashire. At Lowestoft in Suffolk, a nesting 'cliff' was built and used by Kittiwakes for many years, but the birds have since spread further and now nest in numbers on a local church (Fig. 241). Kittiwakes are also nesting on offshore marine gas platforms in the Irish Sea, and in the North Sea off the coast of the Netherlands, while a small number nest on a navigation guide in Sweden (Fig. 123).

Young Kittiwakes reared on natural cliffs have frequently joined those breeding on buildings when mature, while many of those reared on buildings have been

FIG 241. Kittiwakes (*Rissa tridactyla*) nesting on a church, Lowestoft, Suffolk. (A. Easton)

found nesting on natural sites. In contrast, only two marked adults out of some 400 birds that have bred at the warehouse colony at North Shields moved away to nest elsewhere while others continued to nest there. In this case, they both moved to a natural cliff site 30 km away (and remarkably, they were paired together for several years, nesting in a colony of 300 other pairs, although they were not paired together at North Shields!). However, where there is poor breeding success, excessive ticks or

FIG 242. Kittiwakes (*Rissa tridactyla*) nesting on the side of a harbour jetty, Lowestoft. (A. Easton)

human disturbance (such as low-flying helicopters), some breeding Kittiwakes in the south of England and France have moved to different colonies.

Great Black-backed Gull

In several areas, single pairs of Great Black-backed Gulls have joined existing urban colonies of other large gulls, just as they have often done at natural coastal nesting areas. In general, they have been later in spreading to urban areas and their numbers are still low, but the habit is becoming commoner and numbers in urban sites are increasing. In 1971, Stanley Cramp could report only a single site in Britain and Ireland, which was on a factory roof at Newlyn, Cornwall. By 1976, there were seven pairs nesting at three urban localities, all in Cornwall. Nineteen years later, in 1995, numbers had increased to at least 11 pairs nesting at nine sites in Britain, with the first records from Scotland (five sites, with six pairs), but only five pairs were recorded in England at four sites (Hampshire, Dorset, Devon and Cumbria) and none was reported in Cornwall. By 2000, breeding at urban sites had spread and increased appreciably to a total of 83 pairs. There were 18 sites recorded in England, involving a total of 55 pairs, with most in south-west England, but one pair bred in Lancashire and three pairs in Cumbria. In addition, six pairs bred in urban sites in the Channel Islands and 21 pairs in Scotland, including nine in Aberdeen, five in Inverness and single pairs in Sutherland and Edinburgh, and the first pair breeding on a building in Wales was found in Dyfed. Since then, breeding by Great Black-backed Gulls has spread to several other towns and cities in Scotland and England, but no recent national census has so far been made to confirm the numbers currently involved. In 2012, about 30 pairs were breeding on coastal buildings in France.

The usual pattern of Great Black-backed Gulls invading towns has been for a single pair to join urban colonies that already have substantial numbers of Herring and/or Lesser Black-backed gulls. As a result, their presence in a number of towns has been readily overlooked for several years.

Common Gull

All records of Common Gulls nesting on buildings in Britain are from Scotland, with the first reported in 1971, when a single pair nested at Inverness Airport at Dalcross. By 1995, 236 pairs were recorded on buildings at 10 locations, mainly nesting in small groups. In 2000, breeding was even more widespread, involving 14 urban areas, and numbers had increased to more than 600 pairs, with at least 280 pairs nesting in Aberdeen, 96 in Inverness, 86 in towns on the north coast of Caithness and a single pair in Shetland. Common Gulls nesting on buildings have also been reported from Denmark, Norway and Sweden.

Black-headed Gull

The Black-headed Gull is the least common gull species nesting in urban areas in Britain, and does so only in small numbers at four locations in Scotland.

Yellow-legged Gull

The Yellow-legged Gull now occurs in large numbers as a visitor to Britain, and a few pairs are nesting at natural sites here. It is soon likely to start nesting on buildings, as occurs in Spain (at least 25 pairs in Barcelona), Gibraltar and also France, where about 100 pairs were recorded breeding on buildings in 2012. The species also nests on buildings in Frankfurt, Germany, where mixed-species pairs with Great Black-backed Gulls have been reported. It is likely that the species will also form mixed-species pairs in urban areas of Britain.

URBAN NESTING SITES

In the early years of urban breeding, large gulls often nested on tall commercial buildings in town centres, where human access to the roofs is difficult, while a few began to nest on flat roofs at industrial sites. In both situations, colonies often became well established and involved many pairs before problems arose from their proximity to people in residential areas.

Today, large gulls nesting in towns and cities use a wide range of sites, including both flat and saw-tooth roofs of industrial and commercial buildings, as well as chimney stacks and roofs of domestic houses. Exceptionally, bridges and piles of rubble left from demolished buildings have been used. One pair even nested in the middle of a traffic island at a busy urban junction at Dumfries, Scotland. Those gulls nesting on residential properties show a very marked preference for sites with chimney stacks that have two, rather than just one, rows of pots. Such stacks tend to be more frequent in older terrace properties, often where two adjacent houses share the same chimney stack. The distribution of these stacks in towns has often determined where gulls nest in residential areas. Another type of site used by gulls is where the chimney stack is not at the roof apex but protrudes through the sloping roof. Nests at these sites are placed against the base of the stack, where it forms an acute angle with the roof, and are particularly difficult to see from ground level or nearby vantage points.

The distribution of nests on flat roofs on industrial and commercial properties by Herring and Lesser Black-backed gulls clearly shows that both species have a strong preference for placing their nest against a raised surface (Fig. 235), such as that provided by a parapet around a roof, a skylight, a ventilation

TABLE 72. The percentage distribution of Herring Gull nests on different site types in towns and cities in England and Scotland. The percentages of all nests on roofs of commercial or industrial properties are also shown in the right-hand column. The Ashington and Annan data relate to counts made a few years after they were colonised. Data for the Lesser Black-backed Gull (LBBG) at Dumfries are also included for comparison with Herring Gulls (HG) nesting in the same town. Data for 1974 and 1975 are from Monaghan (1977).

	Year	*Chimney stacks*	*Sloping roofs*	*Flat roofs*	*Ledges*	*Others*	*Industrial roofs*
South Shields	1974–75	34%	13%	38%	10%	5%	0%
South Shields	1994	36%	21%	30%	13%	0%	34%
South Shields	2001	32%	17%	47%	3%	1%	47%
Sunderland	1974–75	17%	44%	32%	6%	0%	25%
Sunderland	1994	14%	37%	33%	16%	0%	57%
Sunderland	2004	15%	30%	55%	0%	0%	58%
Whitby	1985	60%	25%	15%	0%	0%	10%
Hastings	1976	64%	30%	5%	1%	0%	11%
Staithes	1976	71%	23%	6%	0%	0%	0%
Staithes*	1985	69%	23%	6%	2%	0%	2%
Scarborough	1985	58%	20%	12%	4%	7%	10%
Durham	2014	0%	1%	99%	0%	0%	99%
Ashington*	2000	0%	0%	100%	0%	0%	100%
Dumfries (LBBG)	2001	20%	3%	71%	2%	4%	67%
Dumfries (LBBG only)	2007–08	21%	3%	75%	0%	1%	70%
Dumfries (HG only)	2007–08	24%	4%	72%	0%	0%	64%
Annan*	2016–17	0%	70%	30%	0%	0%	100%
Inverness (part)	2012	3%	27%	66%	0%	3%	95%

* Indicates data based on fewer than 60 nests.

dome or a large item stored on the roof. Differences in the choices of nesting sites between Herring Gulls and Lesser Black-backed Gulls are known at natural sites, with the former more frequently nesting on cliff faces and on bare or sparsely vegetated ground, and the latter favouring longer vegetation and areas with shallow gradients. This separation is by no means complete, however, and in many colonies at natural sites the two species often nest side by side, as they do at urban sites. I have been unable to detect consistent differences in the sites used by Herring Gulls and Lesser Black-backed Gulls within the same urban areas, except for a slight preference of Lesser Black-backed Gulls for large, flat roofs. The sites used by the two species more often depends on the geographical position of the town or city and, to a lesser extent, the types of buildings within the towns, rather than a consistent difference in the choices they make.

Table 72 illustrates the distribution of Herring Gulls on various types of urban nesting sites in England and Scotland, along with a comparison of nest site choice by both Herring Gulls and Lesser Black-backed Gulls in Dumfries in 2007 and 2008. There is little indication of a difference in sites used by the two species in Dumfries, and those used by Lesser Black-backs were virtually the same as in 2001, some six and seven years earlier, although numbers had increased and many of the sites had changed. The frequent use of chimney stacks is characteristic of many towns in Britain, but in warmer areas of Europe where these do not exist, most nesting occurs only on flat roofs.

THE ATTRACTION OF URBAN AREAS TO GULLS

The reason many local councils and the media give for gulls nesting in urban areas is that they were attracted by the abundant food there. In turn, this has led many councils to attempt to reduce food waste in streets, even though there is no sound evidence that carrying out such measures (while desirable for other reasons) has had a discernible effect by reducing numbers of urban-nesting gulls.

I witnessed the first Herring Gulls breeding on urban buildings in north-east England at South Shields, and then Sunderland, in the late 1950s and during the 1960s. Numbers nesting on the only suitable natural sites in the area – two inaccessible coastal stacks – had increased in the late 1940s and early 1950s, and were soon overcrowded. During the early 1960s, surplus adults started to nest on roofs of tall commercial buildings in the centre of South Shields and only subsequently spread to domestic properties. The important point here is that when colonising a new area, the gulls did not attempt to feed in the streets or anywhere within the towns, and availability of food there was not the factor that had attracted them to breed there in the first place. The early urban colonisers fed at sea, often following and gathering food around fishing boats, particularly those where gutting took place and discards were thrown overboard as they return to the North Shields and Sunderland fishing quays. The first reported instances of numbers of Herring Gulls feeding in the towns were not until the early 1970s, 10 years and more after urban breeding had started there.

In the 1950s, food dropped by people in the streets was consumed by Black-headed Gulls and a few Herring Gulls in late winter and early spring, and then in later years they began to exploit food discarded in open-top refuse bins in streets, with this habit in Herring Gulls gradually extending throughout the year (Fig. 243) and exploiting waste dropped by people using the increasing number of fast food shops. At the same time they begin to peck into thin plastic refuse bags left out overnight for morning collection, which resulted in more rubbish being scattered over the streets. In several open urban areas, such as large car parks and on beaches, more people started to feed the birds (Fig. 244), but again this practice did not develop to its present level until long after the towns were first invaded by nesting gulls.

A study of the food consumed by Lesser Black-backed Gulls breeding in Dumfries, south-west Scotland in 2002, 2003 and in 2010 suggested that just 5 per cent was obtained within the town boundaries. Gulls feeding within towns are

FIG 243. Adult and second-year Herring Gulls (*Larus argentatus*) attracted to food left near a litter bin. (Mike Osborne)

FIG 244. Adult and juvenile Herring Gulls anticipating receiving food from a person. (Mike Osborne)

conspicuous, but they represent only a small minority of those nesting there, and many travelled up to 30 km beyond town limits to feed at low densities on extensive agricultural land or to visit landfill sites. Nevertheless, the few individual gulls that do specialise in obtaining food in streets, at shopping centres and on beaches are the cause of many complaints. Some of these gulls have acquired such boldness that they aggressively snatch food from the hands of people.

While large gulls do obtain food in streets, and a few individuals specialise in this behaviour, the amount of food available is clearly insufficient to support the many gulls that now nest in urban areas. If readers are still in doubt about this, then they should ask themselves how the 7,000 adult Herring Gulls that currently breed in Aberdeen could possibly obtain the majority of their food within the city limits.

Another suggestion of the advantage of urban nesting is that urban areas are warmer, and so the gulls were – and are – attracted to towns because roosting and breeding there cuts down on their heat loss. While this is difficult to test, it is unlikely bearing in mind that both Herring Gulls and Lesser Black-backed Gulls breed successfully in colder, more northern natural breeding sites. In the north of England and the south of Scotland, no differences have been detected in the timing of nest-building, egg-laying and breeding success of Herring Gulls and Kittiwakes nesting on natural and at urban sites. The idea that gulls nesting in urban areas benefit from higher temperatures should therefore be treated with caution unless new evidence is forthcoming.

The most convincing explanation is that roofs are safe places to breed. This was first recognised by Patricia Monaghan when she examined urban breeding in South Shields and Sunderland during her doctoral study in 1974 and 1975 (Monaghan, 1977). She found that 494 pairs of Herring Gulls nesting on buildings at that time reared an average of 1.4 chicks per pair each year. This figure is high and has proved to be typical of urban-nesting Herring Gulls elsewhere (Table 73), while studies at natural sites record an average of about 0.74 young fledged per

TABLE 73. Examples of the number of young fledged per pair by Herring Gulls at different nesting sites. The USA data (**) relates to the American Herring Gull.

Location	*Young fledged per pair each year*	*Source*
Rural sites		
Isle of May, Scotland	0.8–0.9	Parsons, 1971b
Priest Island, Scotland	0.9	Fraser Darling, 1938
Kent Island, USA**	0.9	Paynter, 1949
Graesholm, Denmark	0.5	Paludan, 1951
Wilhelmshaven, Germany	0.65	Drost *et al.*, 1961
Eastern USA**	1.1	Kadlec & Drury, 1968
Texel, Netherlands	0.88	Camphuysen, 2013
Terschelling, Netherlands	1.2	Spaans, 1975
Langli, Denmark	0.57	Bregnballe *et al.*, 2015
Treberon Island, Brest, France	1.30	Pons & Migot, 1995
Wales	0.6	Harris, 1964
Skomer, Wales,* 1993–2015	0.62	JNCC
National UK figures, 1989–2014	0.6	JNCC
Average for rural sites	**0.74**	
Urban sites		
Sunderland	1.30	Monaghan, 1979
South Shields	1.58	Monaghan, 1979
Sunderland	1.86	Gibbins, 1991, unpublished
Dumfries	1.31*	Coulson & Coulson, unpublished
Berwick-upon-Tweed	1.23	Coulson, unpublished
Average for urban sites	**1.45**	

* Applies mainly to Lesser Black-backed Gulls.

pair each year – in other words, they are only half as successful as those breeding in urban areas. In a study of the American Herring Gull in Maine, the breeding success of those nesting on buildings in 2011 was 49 per cent higher than in those on a nearby island and in the following year it was 17 per cent greater (Perlut *et al.* 2016).

By using urban areas, Herring Gulls have appreciably increased their breeding success rate – unless residents carry out removal of nests and eggs on an extensive scale (although this did not start until the 1990s in the UK, and then in only a few towns). Urban nesting sites are similar to many natural nesting sites in that they are small islands, where mammalian predators are excluded by the sea and cliffs, or in the case of towns, by roads and buildings with vertical sides.

PROBLEMS CAUSED BY URBAN NESTING

In many towns, the initial colonisation by small numbers of large gulls was of little concern to the human population. The first nests built by gulls in towns were mainly on factories and taller buildings, and people at ground level were usually far enough from unfledged chicks that their presence was not perceived as a threat. The noise the birds produced seldom bothered residents and only occasionally did their droppings cause problems. The adult gulls rarely dived at people to protect their young, but this did occur, such as when gales in early July blew unfledged chicks into the streets and their parents repeatedly attempted to defend their grounded young by swooping at any approaching humans, dogs or cats. Apart from such occasional problems, the early colonisers were generally accepted and even appreciated by the public, particularly at sites where they could be viewed. Most people were delighted to see young gulls being fed, exercising their wings and attempting to take their first flight.

Major concerns developed only after the numbers of gulls nesting in urban areas increased and more moved onto residential properties, building their nests on roofs and chimney stacks, closer to where people lived and slept. Apart from the noise the gulls made, particularly in the early hours of the morning, the remains of their nests and the occasional dead chick blocked roof drains and gutters, leading to water damage to properties. Adults attempting to protect their eggs and chicks by diving at people caused concern, while gulls stealing food from people in the streets made many anxious about the safety of both their children and themselves.

A problem caused by gulls in South Shields was told to me by the headmistress of the local girls' grammar school and is worth repeating here.

During a gale, several large Herring Gull chicks that were still unable to fly were blown off the roofs of neighbouring buildings and landed in the enclosed playing field of the school. As it happened, the school's annual sports day was scheduled for the following day and the member of staff responsible for organising the event went to the field to check that the arrangements were in place. The parent gulls immediately dived at the teacher, and she hurriedly left the field and reported the matter to the headmistress, who in turn telephoned the police for assistance. A police sergeant arrived shortly after, intending to remove the offending young gulls. By coincidence, he had been to the cinema three days earlier and had watched the Alfred Hitchcock film *The Birds*, in which flocks of gulls attack and terrify people. As soon as the adult gulls swooped at the police officer, he turned and, according to the headmistress, ran for the protection of his police car. The sports day was cancelled, and the young gulls remained in the field until they could fly.

People are understandably intimidated by a gull weighing more than 1 kg diving toward them at over 60 kph, usually unexpected and from behind, and often passing less than 30 cm above their head. The draught and shriek made by the gull during each close pass gives the victim the impression that it is coming ever closer. Some attacks are accompanied by the bird defecating on the person as it passes overhead. Victims believe the birds are going to attack them viciously. In fact, I have only ever been hit once by a protective gull and that was, I believe, by a dangling foot, but the incident was sudden and unnerving, and I still react by momentarily crouching. Some people have claimed that diving gulls have pecked at their head, producing a bleeding gash, but it is not to a gull's advantage to endanger itself in a collision. Normally, the birds intend to make a threat rather than an attack, and only very occasionally do they make contact. Carrying an umbrella or wearing a hat with something protruding from the top is enough to ensure safety of passage.

Another problem associated with urban-nesting gulls arose at the workshops of Coles Cranes Ltd, a manufacturer of mobile cranes in Sunderland. Several pairs of Herring Gulls nested on the factory roof and were much admired and appreciated by the office staff, who occupied a building on higher ground overlooking the roof. Two well-grown chicks died on the roof and maggots in the carcasses eventually crawled from the bodies to seek places to pupate. Unfortunately, some crawled through small gaps in the roofing and fell onto the work area below. Not only were the maggots a hazard because they made the workshop floor slippery underfoot, but some landed on red-hot metal and were immediately incinerated, each with a hiss and a puff of smoke. The management decided to cull the adult gulls and the remaining chicks, but the office staff were upset at this, objected and threatened to strike. After a considerable delay and some discussion, all the gulls and those on neighbouring properties were shot early in the morning and the bodies removed while the office staff were absent.

A final concern linked to urban gulls was the higher incidence of food poisoning in humans in coastal towns, and particularly in holiday resorts. It was suggested at the time that the risks were associated with fast-food sales and that the gulls breeding in towns might be carriers of the bacteria responsible for the food poisoning. This led to several studies of *Salmonella* and, later, *Campylobacter* and *Cryptosporidium* in gulls, which showed that up to 10 per cent of adult Herring Gulls were carrying a wide range of these organisms in their guts. While it is obvious that the gulls represented a reservoir of the micro-organisms, it was thought that they were not a major source of infection to humans, but that they were already in the local environment and the birds were becoming infected with them. However, it was found that Herring Gulls wintering in north-east England were carrying *S. montevideo*, and when they returned to eastern Scotland in the spring, this organism was passed to cattle and induced miscarriages. Several other cases of gulls transferring food-poisoning organisms to farm stock have been reported, but the extent to which gulls in towns transfer bacterial infections to people remains unresolved.

The range and number of complaints received by the Dumfries and Galloway Council about gulls nesting in Dumfries over a range of years is given in Table 74. Gulls swooping close to people in the town's streets made up almost half of all complaints received over the 15 years, and most of the remainder related to the mess caused by droppings, gulls opening bin bags, the noise the birds made, food stealing (restricted to a few specific sites) and the presence of chicks in streets that had fallen from buildings. The 'other' category included damage to buildings and concern for pets and postmen.

TABLE 74. The number and nature of complaints received by Dumfries and Galloway Council about urban-nesting gulls in 1999 and 2004–17. Data analysed by Gill Hartley (pers. comm.).

	Swooping	*Mess*	*Noise*	*Food stealing*	*Chicks on ground*	*Other*	*Total*
Number of complaints	462	149	136	90	88	80	1,005
Percentage of total	46.0%	14.8%	13.5%	9.0%	8.8%	8.0%	

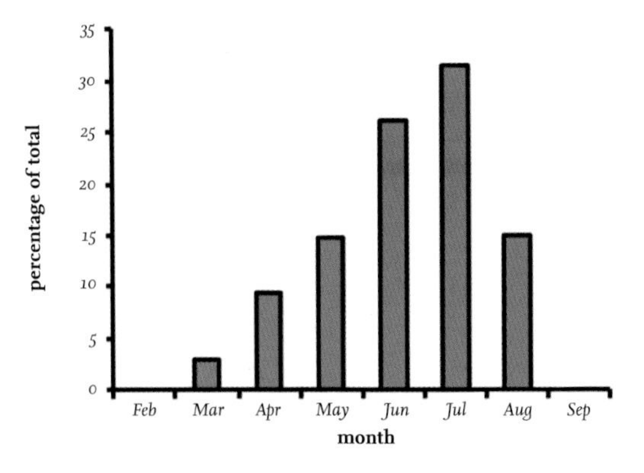

FIG 245. The monthy distribution of complaints about large gulls in Dumfries received by Dumfries and Galloway Council over six years. Note that there were no complaints between September and February.

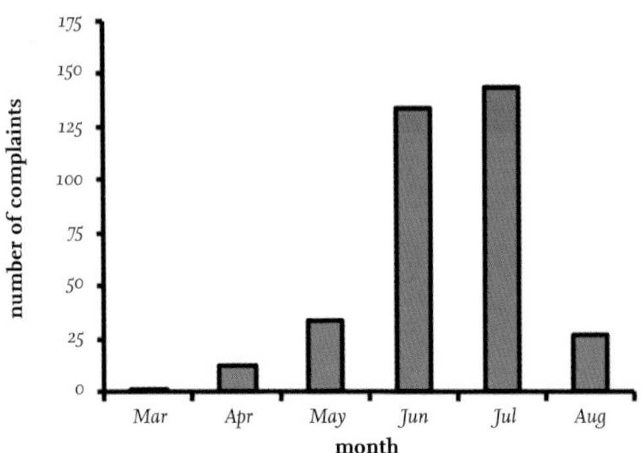

FIG 246. The number of complaints per month about nesting gulls swooping at people in Dumfries received by Dumfries and Galloway Council. Note that there were no complaints from September to February.

Complaints to councils about gulls start in late March, increase month by month, peak in July and then decline rapidly through August once the young have fledged and many adults have left the breeding sites (Fig. 245). Nevertheless, gull-related topics often rank in the top five complaints received each year by councils and they are attractive stories for the media. Cases of large gulls swooping at people are generally restricted to the period when unfledged chicks are present in June and July (Fig. 246).

HISTORY OF URBAN GULL MANAGEMENT IN BRITAIN

Attempts to deter gulls from nesting in towns in Britain have been left to local councils, which often try to address the problem without first seeking expert advice. Having shown an unwillingness to advocate killing the gulls, many local councils decided that removing food sources for gulls within urban areas might be an effective means of reducing their numbers. Some towns have introduced gull-proof bins for rubbish and have requested or legislated to make it an offence for the public to feed gulls. While keeping towns and cities tidy is highly desirable, and it is right that the public should be discouraged from throwing food away in streets, attempts to reduce numbers of nesting gulls by preventing food being made available have failed to drive gulls out or to cause their numbers to decline. Many local authorities have often tried to address public concerns by taking actions that are not controversial, even though these actions may be ineffective. However, such actions may have had the positive effect of reducing the instances of gulls snatching food from people, although this may take a number of years because once acquired, the behaviour may persist in long-lived gulls for their lifetimes.

The use of birds of prey to drive gulls from airfields, airports and landfill has been successful, but the gulls in question were feeding, resting or roosting and unlike nesting birds, they did not have a strong attachment to the sites. Several urban areas have employed falconers to fly birds of prey in an attempt to drive nesting gulls from towns (see below); some of these have claimed success, but this has not been evaluated by unbiased observers, and an independently assessed trial at Dumfries shown that the non-lethal use of birds of prey failed to drive large gulls out of towns.

The first attempt to remove or reduce small numbers of Herring Gulls nesting in towns in Britain took place at North Shields in north-east England in or about 1970, when police gave approval for a council employee to shoot the birds. The marksman involved had worked in pest control for the Ministry of Agriculture and Fisheries, and a few years later he recalled to me that he had shot about 20 pairs of gulls. Once he retired, no further control took place in the town.

A neighbouring town in north-east England also employed a marksman for gull control; he was escorted by a police officer and shot adult birds on properties early in the morning at locations where a specific complaint had been received by the council. This selective culling continued without problems for several years and at least 410 adult gulls were shot, and no complaints about the cull from the public resulted. Only a small proportion of the gulls breeding in the town were removed, however, and numbers nesting there continued to increase. Strangely, the council of a city only a few miles away also decided to shoot gulls as a means of control and while this resulted in several hundred being culled in 1976, the police refused permission in the following years, despite being part of the same authority that continued to allow shooting in the neighbouring town. In several other towns and cities of England and Scotland, requests for police permission to shoot gulls in towns have been rejected.

In Britain and Ireland, the first attempt to remove large numbers of gulls nesting at urban sites was in the late 1970s, when Scarborough Borough Council decided to cull in the towns of Scarborough and Whitby, and Patricia Monaghan and I were able to monitor their efforts. The first method the council considered was to cull the many gulls that visited a landfill site on the northern edge of Scarborough. We expressed caution about this and pointed out that far more gulls visited the landfill than nested in the town. To prove this, we cannon-netted 200 adult Herring Gulls at the landfill site in late spring, and then marked them with dye and released them. Only two of the marked gulls were subsequently found in Scarborough among some 300 pairs nesting in the town, while many of those attracted to the landfill site were found nesting at natural cliff sites up to 35 km away.

A similar result was obtained in a study of Herring Gulls in north-east England, where most of the local adults nesting on buildings fed along the shore and, particularly, offshore in association with fishing boats. The landfill sites available within 50 km of the towns in the area did not attract adult Herring Gulls between April and mid-July, while the many gulls feeding there from August to February were almost all individuals that bred on the eastern side of Scotland or in northern Scandinavia and had moved to winter in north-east England.

When the council accepted that culling at the landfill site would not have a major impact on gull numbers in Scarborough, it decided to kill them at their nests in the town instead. To achieve this, narcotic baits were used – as had been employed during the culls carried out by the Nature Conservancy Council on the Isle of May in the early 1970s (p. 135), and by the RSPB on other Firth of Forth islands and on Coquet Island in Northumberland. Television aerial erectors were employed to place baits in the nests on roofs (owing to health and safety requirements, council employees would have needed scaffolding to reach each nest). As a precaution, council staff patrolled the town in case any treated gulls fell into the streets, but few did so. Each nest was revisited three hours after baiting, when the dead gull on the nest was removed and the site baited again in an attempt to cull the second member of the pair.

The first year of culling was effective, and we estimated that almost half of the breeding gulls in the town were culled. However, in the next breeding season, numbers of breeding pairs were only a third fewer, presumably because new, young breeders or individuals that had missed the previous breeding season had returned to nest in the town.

This was the first and only potentially effective attempt to reduce the numbers of nesting gulls in a town, and while it was successful in reducing numbers, it was less effective than expected. However, further culls in Scarborough did not take place because of threats made to the council. In later years, the gulls breeding in Scarborough and neighbouring towns increased from 150 to 3,000 pairs, and the council is still searching for an effective method to reduce their numbers. The council at Berwick-upon-Tweed also considered culling gulls, but following a demonstration against the proposal it decided to take no further action.

In the UK, laws relating to the use of narcotic drugs to cull birds became more restrictive under the Wildlife and Countryside Act 1981 and the Wildlife (Northern Ireland) Order 1985, and nuisance or minor damage to property were no longer viable legal reasons to kill gulls. Government bodies can issue licences for birds to be killed if there is no suitable non-lethal solution, in order to prevent serious damage to agriculture or the spread of disease, to preserve public health or safety to aircraft, or to protect other animals on nature reserves. While the introduction of these wildlife Acts effectively prevented culling gulls in towns, problem gulls breeding on some nature reserves could be, and were, shot, while thousands nesting on an upland area in the Trough of Bowland were culled because water draining from the colony site was supplied to towns and cities in north-west England and this was deemed to be a hazard to public health.

Part of the current advice on urban-nesting gulls offered by the RSPB on their website under the title 'Urban nesting gulls and the law' states:

> *The birds nesting on roofs of houses are most likely to be herring gulls, whilst lesser black-backed gulls tend to concentrate on the larger expanses of industrial or commercial buildings with flat roofs. Although numbers of roof-nesting gulls, especially lesser black-backed gulls, are still increasing, the overall population of herring gulls is plummeting, making them a red list species… We question the appropriateness of lethal control on a declining, red-listed species and highlight the need to comply with European bird protection law. We believe the best approach to understanding urban gull populations starts with comprehensive research to establish the basics, followed by development of effective deterrent methods for use in situations where gulls are causing problems. These could include rendering nest sites inaccessible, reducing the organic waste taken to landfill sites and, in towns, preventing street littering, and making public waste bins, domestic and business waste containers, and collection arrangements 'gull-proof'. Those best placed to do this include landfill companies, local authorities and statutory bodies with a wildlife management remit, but the behaviour of private individuals is also important.*

Some might question whether the Red-listing of the Herring Gull, which is still numbered in the hundreds of thousands in Britain and Ireland, is justified, particularly since no national census has taken place since 2000 to confirm or contradict that this species is still declining.

The current advice concerning urban-nesting gulls from the RSPCA website is:

> *The most humane way of deterring birds is to remove what attracts them to urban areas – mainly food or shelter. Means of doing this can include reducing food availability or preventing them from accessing roofs or other areas where they could cause disturbance.*

However, as discussed earlier, there is no good evidence that food attracts gulls to nest in urban areas. In many towns, preventing large numbers of gulls accessing roofs has been ineffective and expensive. Applying non-lethal methods have not entirely cleared gulls from nesting in any village, town or city in Britain.

Gull control in Dumfries

Dumfries, a relatively small town in south-west Scotland, was first colonised by gulls sometime in the 1980s. A trial in 2001 to disrupt breeding by Lesser Black-backed Gulls in the town used a Harris's Hawk (*Parabuteo unicinctus*). The raptor caused nearby incubating gulls to leave their nest and eggs immediately, but within 20 minutes of the hawk being removed, all the gulls were back incubating as if nothing had happened. The hawk could not be kept in the area for long periods of time, and in any case it had an effect over only a very limited area for the period it was flown. As a result, it had no lasting impact on the breeding gulls.

A scientifically based study in 2000 using independent observers to assess the effect of repeatedly flying Peregrine Falcons (*Falco peregrinus*) on nesting gulls in Dumfries was carried out by the Central Science Laboratory under contract from the Scottish government. The regular flying of Peregrines through the town in April, before the gulls had started laying, had a short-term effect. The gulls lifted off, circled and gave repeated alarm calls, but within minutes of the falcons being removed they settled down again. Later, when they were incubating eggs, adult gulls did not even rise when a Peregrine flew past within 15 m of them, and they had clearly become habituated to the presence of the birds of prey. Breeding was not disrupted, and the numbers of breeding gulls were not reduced. The only beneficial effect was that immature gulls were disturbed and left the area, which might have affected recruitment in future years had the use of falcons been continued, although this would not have been cost-effective. Deterring gulls from nesting and incubating in towns is much more difficult than moving roosting gulls, and it is clear that breeding gulls rapidly become habituated to the presence of the falcons, which is not surprising since both Peregrines and Herring Gulls often nest close to each other on sea cliffs. When I discussed these results later with a falconer, he suggested that if the falcons had been trained to kill gulls, they would have had a greater impact, but then it would probably be easier and cheaper to cull adult gulls by other means.

The closure of landfill sites within the feeding range of large gulls at Dumfries in Scotland and at Brest in France had no effect on the numbers of gulls nesting in nearby urban areas, although it did lower the breeding success of those nesting in Brest. The current nationwide reduction of open landfill sites and the changes to commercial fishing practices under European Union regulations may already be reducing food available to gulls and having an impact on their numbers, but convincing evidence does not yet exist and will take time to obtain.

The Dumfries and Galloway Council is currently advised by a Scottish government task force on methods to reduce incidences of gulls snatching food unexpectedly from people in the town centre and by the riverside, and of gulls diving at people to protect their young. In addition, the gulls produce numerous droppings, which make the shops, sides of buildings and streets frequently and extensively soiled, while on occasions people walking through the town centre have been hit by 'gifts from on high' and parked cars frequently covered in droppings.

Annual counts of breeding pairs in Dumfries have been made since 1999 and a series of non-lethal measures have been used to manage the gulls, yet numbers have continued to increase and spread outwards from the town centre (Fig. 247). Local residents were (successfully) asked not to feed the gulls, refuse in the town was made unavailable to the gulls and the nearest landfill site closed. Repeated egg and nest removal from buildings is financed and carried out by the council, and they have spent large sums on this programme since it was introduced in 2000. Some roofs have been netted to exclude gulls, and spikes and wires added along roof edges. A few firms broadcast gull distress calls from roofs, repeated at short intervals, while others put dummy owls on roofs, but all these methods failed, with some gulls even nesting against the loudspeakers and the dummy owls within two years! While the actions taken did not stop the annual increase of breeding gulls in the town, they were successful in moving many of the gulls from the town centre. Then, between

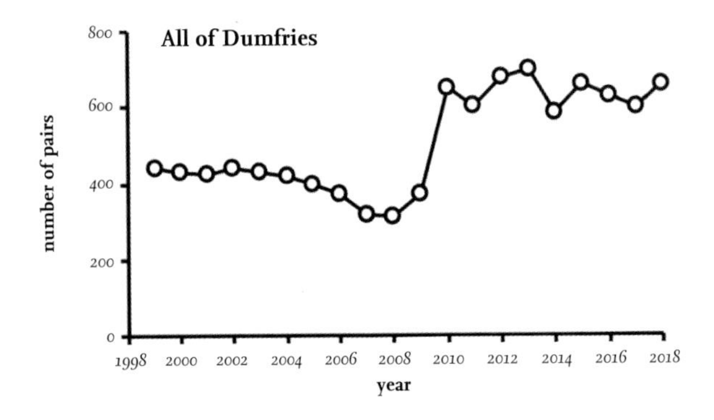

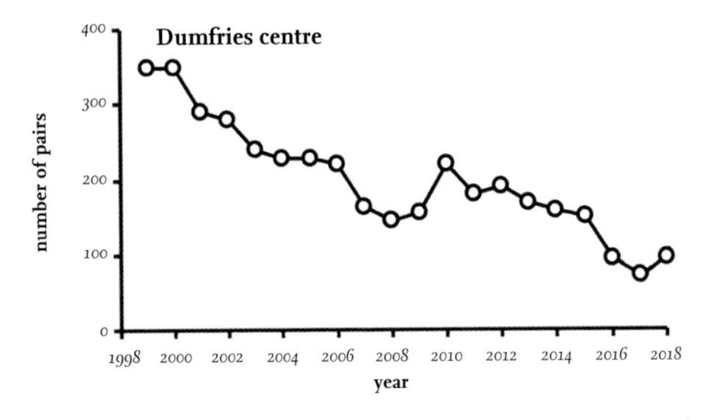

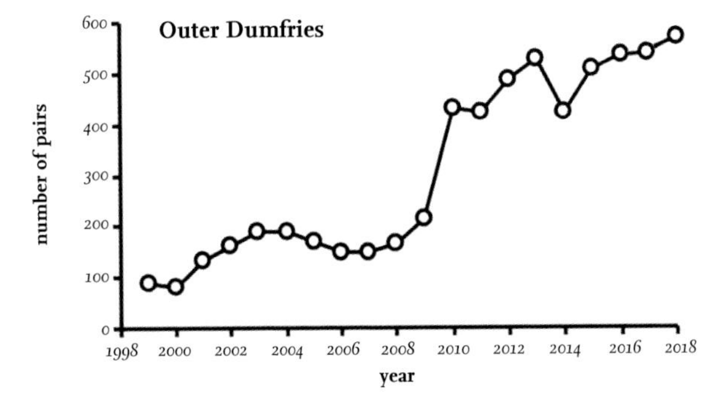

FIG 247. The change in numbers of pairs of large gulls nesting in different parts of Dumfries from 1999 to 2016.

2009 and 2010, there was a sudden and exceptional increase by 76 per cent in numbers of nesting gulls, presumably caused by a major immigration. Arrivals were suspected, but not proven, to have come from a large colony nesting on the nearby Solway marshes, which had been flooded out in several years and had decreased in size.

The successful removal of groups of nesting gulls from flat roofs in Dumfries was achieved by removing their nests and eggs weekly, but similar treatment was less efficient in dealing with isolated pairs on domestic properties. This was because access was more difficult and time-consuming as single nests were usually involved, and visits to most of these sites were soon made only at three-week intervals in an attempt to prevent chicks being hatched. In addition, some commercial property owners refused access to their roofs for security or safety reasons and a few residents refused to have nests removed from their properties. Some roofs were inaccessible, including the top of a Second World War aircraft hangar, but in recent years many such sites have been reached through the use of a cherry picker, albeit at additional cost. This comprehensive removal of eggs and nests in Dumfries has achieved a reduction in the numbers of fledged young produced annually in the town by 80–90 per cent from 2009 to 2017. However, there is no evidence yet that the reduction of fledged young has affected the trends in numbers at the town, and seven years after the practice of egg and nest removal was introduced, numbers of breeding pairs continue to increase, suggesting that an appreciable proportion of birds recruited are probably immigrants reared in other colonies.

The action by the council in Dumfries has had some success in that it has almost certainly prevented the numbers of breeding gulls increasing more rapidly, but the expense has been considerable and the efforts did not prevent or counteract the appreciable immigration in 2010. In part, the council's aims have been achieved, in that the numbers of gulls nesting in the town centre have been reduced and complaints from the public have declined, but more outlying urban areas have been colonised and requests to remove nests and eggs from private properties have increased.

The numbers of complaints made annually to the council about gulls nesting in Dumfries are presented in Fig. 248. Essentially, the level has remained similar over the years, except in the period 2008–11 when there was considerable publicity and comments about the gull problem in the local press and when the council introduced more extensive nest and egg removal. While gull numbers have continued to increase in the town as a whole, the council's actions have reduced the annual number of complaints in relation to gull numbers by about half (Fig. 248).

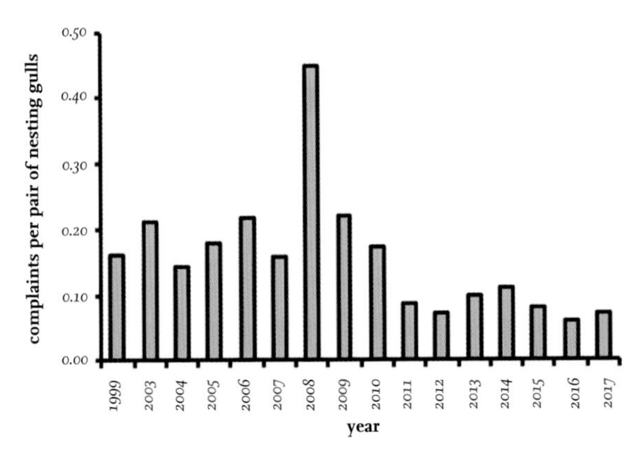

FIG 248. The number of complaints about gulls received by Dumfries and Galloway Council in 1999 and 2003–17 per pair of nesting gulls. Data analysed by Gill Hartley.

It is interesting to note that some newer residential areas of Dumfries have not been colonised. These have houses either without chimney stacks or, if these are present, they have a single vent or the pots are restricted to a single row. Avoiding building chimney stacks with pots in two rows would seem to be a wise precaution in new developments, but altering those already in the town is not feasible. It is not known whether the Dumfries and Galloway Council can continue the management programme in the town in the future as more stringent restrictions on spending are forced upon them. If the programme ceases, then it is most likely that gull numbers will increase dramatically and the current limited success gained over a decade will be lost. Several town councils in other areas have offered nest clearance at a charge to property owners, and not surprisingly, the proportion of nests removed in these areas is invariably low.

URBAN GULL MANAGEMENT

Suggestions for the management of the numbers of urban gulls can be divided into two types: those that alleviate a specific problem at a single locality; and those that aim to clear a town or part of a town of nesting gulls. The problem at a specific site is more readily tackled by causing the breeding gulls to move. However, this does not reduce gull numbers but simply moves the birds to a new site nearby, where they often just become someone else's problem. No satisfactory non-lethal method has yet been developed to clear gulls from all, or even a major part of a town.

Preventing gulls from landing

Preventing gulls from landing at a site by adding netting, spikes or lines can be successful, but a limitation of these methods is that they can make the roof hazardous if it forms part of the escape route for people in the event of fire. Such deterrents can also make it difficult for technicians to reach structures housed on the roof, such as lift shafts, and to maintain waterproofing of the roof surface.

1. Netting the nesting area can be successful as it prevents the gulls from landing, but it is extremely expensive when large roofs are involved. The netting must be of a sufficiently fine mesh and kept taut to prevent the birds from becoming entangled. Netting is ideal for specific sites that cannot easily be reached, but it requires annual checking and maintenance.
2. Using metal spikes to prevent the gulls landing on potential nest sites has been effective at some sites, mainly on chimney stacks, but some have been poorly positioned and failed to stop gulls landing. In some cases, nests have even been built on top of the spikes.
3. Tautly stringing thin nylon or wire lines horizontally about 50–100 cm above the surface of a flat roof and at intervals of about 30 cm apart interferes with gull landings. Attention must be given to the edge of the roof however, because some gulls land there and will walk some distance under the lines to reach their nest. If the wires are placed too far apart, some persistent gulls have developed a technique of landing with partly closed wings to avoid contact with the lines.
4. Repeated removal of nests and eggs at seven- to 14-day intervals from early May and, if necessary, until the end of July often causes failed breeders to move their nesting site. However, this method takes time to be effective and some of the failed breeders move only to another site nearby. There is circumstantial evidence that some gulls that fail to breed successfully move much further and leave the town altogether to breed elsewhere in the following year; however, the proportion doing so has not been determined from marked individuals.

Deterring gulls by disturbance

Preventing gulls from breeding by disturbing them relies on methods that would also usually disturb human residents, and so they are not generally suitable for urban areas.

1. Loudspeakers that broadcast taped distress calls of gulls and gas guns that create periodic explosions (to protect agricultural crops) are both methods

employed to disturb gulls. Where these methods were trialled in a rural gull colony, they worked successfully for a few days, but then habituation set in and the gulls soon ignored the noises. They work for a longer period if used only intermittently or in combination, but this makes them more labour intensive and expensive. The use of laser beams has been developed in the Netherlands to disturb gulls, but it is unlikely that these would be allowed in towns because of the risk to people, and in any case the impact on birds' eyes has yet to be assessed.

2. Extensive areas occupied by nesting gulls have been successfully cleared by the simple but time-consuming policy of preventing the gulls from landing by repeated disturbance as soon as they attempt to return to the colony at the start of each breeding season. This method worked exceptionally well on moorland nesting gulls at Abbeystead, Lancashire, and at Bolton Castle, Yorkshire, where an intensive programme of disturbing the gulls whenever they visited the colony in spring was followed. Areas that had been occupied by several thousand pairs of gulls were completely cleared in one year. It was evident that many of the gulls moved away from the area altogether, but the downside was that a proportion moved onto neighbouring land with different owners who did not appreciate the transfer. This disturbance method is less appropriate at most urban sites, where the relatively small numbers of gulls on individual roofs makes it labour intensive, while regular and easy access to some roofs in the early morning is often not feasible.

Destroying nests and eggs

1. The regular and frequent removal of nests and eggs can be carried out at sites with safe access and in towns where numbers of gulls are nesting on the same roof. Removing the nests as well as the eggs forces the females to build new nests as well as produce new eggs, effectively further delaying the time of re-laying. In addition, studies by Ruedi Nager, Pat Monaghan and David Houston at Glasgow University (2001) found that causing female Lesser Black-backed Gulls to repeatedly lay new clutches induced greater stress in them, shortening their lifespan. For this method to be effective, several visits to nesting sites from May to mid-July are necessary, because many of the gulls whose eggs are removed can lay replacement clutches several times during the same breeding season. Weekly removals are more effective than those at three-week intervals, and the method can be made even more efficient by anchoring at the site a few small but brightly coloured helium-filled balloons or even plastic bags at the end of each clearance. They should be deployed on a cycle of one week on and two weeks off throughout the

breeding season to prevent or reduce the gulls becoming habituated to their presence. In some cases, multiple pairs of gulls on a roof have been totally excluded after a single year of this treatment. Once totally cleared, roofs usually remain so for several years, whereas leaving one or two pairs soon attracts additional pairs.

2. Replacing eggs with wooden or plastic replicas prevents chicks being produced. In many cases, incubation is extended long after the eggs would normally have hatched and eventually the adults desert the nest, often without re-laying. An advantage of this system is that gulls call much less during incubation and it also appreciably reduces incidents of gulls diving at people because there are no chicks to defend. However, the method does not seem to stress the female as much as egg and nest removal, and consequently a much smaller proportion of the adults moved elsewhere in the following breeding season.
3. Spraying eggs with oil blocks the air spaces in the shell and suffocates the developing embryos. Alternatively the embryos can be killed by injecting eggs with formalin. As in the previous method, this prevents chicks hatching, although there are reports that a proportion of the sprayed eggs hatch nonetheless. These methods reduce the noise and instances of aggressive defence of chicks, but they do not have a major effect on numbers of breeding gulls.

Sterilising gulls

It has been suggested that Herring Gulls should be fed a substance that prevents them from breeding – in effect, a contraceptive – and prevents egg production. There are known drugs that could act in this way, but the problem is how to get a specific dose into a gull and how to avoid over-dosing or under-dosing. To place the substance in food would not work because of the way gulls rapidly swallow items, with some individuals consuming large quantities even before others start to feed. Capturing gulls to inject a standard dose individually would not be realistic because of the labour involved.

GULLS NESTING IN TOWNS ARE NOT A CLOSED POPULATION

There is an erroneous belief that the gulls nesting in each town form a closed population and produce all the young required to replace adults that have died. As a result, it has been incorrectly assumed that by preventing large gulls from fledging chicks, this will eventually reduce gull numbers because there will not be enough surviving young birds to replace the adults that die through natural mortality. In fact, many gulls reared elsewhere are recruited into a site as immigrants, and it is therefore quite possible for a colony to continue to grow year on year even if no young were produced.

There are three main reasons why reducing the production of young has little effect on overall numbers. First, only about one in 10 adults dies each year, so preventing recruitment could reduce numbers by a maximum of only 10 per cent each year. Second, the onset of this reduction is delayed by at least four years because of the time it takes gulls to reach maturity. Third, and most importantly, a reduction would occur only if young gulls invariably returned to breed in the town in which they were hatched. While some do so, this applies to only about a quarter of the surviving young, with the majority moving to breed elsewhere (in some cases more than 200 km from the natal site). As a result of high natal dispersion, the majority of new breeding adults settling in a town have been reared elsewhere. Only if the prevention of young being produced is extended over a large area and includes natural nesting sites would this be likely to reduce numbers breeding in towns and cities. As mentioned earlier (p. 404), intensive egg and nest removal has been carried out in Dumfries for seven years, reducing the production of young by about 90 per cent, but there is no indication that the practice has reduced the numbers of breeding gulls there in the past three years, when the first young gulls could return at a breeding age.

THE CURRENT SITUATION

No method currently used to manage large gulls has been entirely successful. New towns are still being colonised by gulls and existing urban colonies are persisting and increasing, despite attempts at non-lethal management. Far too much emphasis is being put on preventing large gulls feeding in towns and, not surprisingly, this has resulted in streets being tidier but has not reduced gull numbers. To say that that the Herring Gull is Red-listed and in danger of extinction in Britain simply confuses the issue. It has declined at natural coastal sites, but the actual numbers breeding in urban areas are increasing. It appears that the difference between urban and natural sites may be a result of different feeding methods, but this needs far more research. What is obvious is that large gulls breeding in inland cities like Birmingham are not feeding at sea during the breeding season because the distance to the coast is too great, and so it is unlikely that reducing food made available by the fishing industry will have a universal effect. Closure of landfills might have an effect as reported in Brest, but feeding by gulls on agricultural land has yet to be investigated in detail throughout the country.

To achieve immediate relief from the problems caused by large numbers of gulls, only one method is completely effective: killing adults, some of which would otherwise live and breed for up to 30 years. However, is it likely the public would approve of this? There were no major objections to the early 1970s cull on the Isle of May in Scotland or to those that took place in a few towns, but there is an appreciable proportion of people in Britain who do object to culling wildlife. On the other hand, large gulls and some other species are currently being killed to protect other breeding bird species, so should this be applied to cases where major public concern exists about the well-being of people in areas where gulls are numerous? The case of Badgers (*Meles meles*) and bovine TB illustrates the stark dichotomy in public opinion.

Currently, there are well over 200,000 adult Herring Gulls in Britain. Virtually all of the methods that protection organisations have recommended to manage gull problems involve disturbance, encouraging the birds to move away from the locations where they nest and cause problems – in other words, they are simply passing the problems on to others. There has been no national study or count of urban gulls since the 2000 census, and there is little indication that bodies like the Department for Environment, Food and Rural Affairs, Natural England and Scottish Natural Heritage intend to become involved in tackling the urban gull problem, despite pressure from local authorities. Councils are spending considerable amounts of money on ineffective management methods that have already been tried and failed elsewhere.

There is little evidence that an 'urban' type of gull is evolving separately from those breeding on natural nesting sites. Movement from natural sites to urban areas has been well established through the extensive invasion of towns in north-east England by young Herring Gulls reared on the Isle of May in Scotland. Herring and Lesser Black-backed gulls regard roofs in urban areas simply as the equivalent of a series of small coastal islands where ground predators are absent. As Herring Gulls are declining on natural sites and have lower breeding success rates there, the young gulls are attracted to places where breeding is successful, and so there is an increased likelihood of town-reared young being attracted to other towns rather than moving to breed on natural sites.

There are only a few records of Herring and Lesser Black-backed gulls reared on urban sites that are known to have moved to breed on natural sites. However, this is to be expected, because much of the data comes from culls, which produce ringing recoveries – most of which have been carried out on natural breeding sites and rarely in towns.

There is also the intriguing difference in the distribution of urban-nesting Herring Gulls and Lesser Black-backed Gulls. Why do few of the latter species nest on buildings in north-east England yet dominate in south-west England and parts of the Forth–Clyde valley in Scotland? One factor that may be involved is that more Lesser Black-backed Gulls tend to breed at inland urban sites than Herring Gulls, but there are many exceptions. Another factor may be that Lesser Black-backs favour areas with a higher rainfall, as on the western side of Britain, but again there are many exceptions. Perhaps Lesser Black-backed Gulls feed more efficiently on earthworms, which are more accessible during the breeding season in areas of high rainfall.

While both species nest on roofs in Dumfries, counts made from 1999 to 2017 showed little indication of a change in the ratio of the two species, with Lesser Black-backed Gulls continuing to account for 90 per cent of nesting gulls even though the total number of all nesting gulls has doubled. This is surprising, since studies elsewhere have suggested that adult Lesser Black-backed Gulls have a higher annual adult survival rate than Herring Gulls. If this is correct, either the survival rates and the recruitment rates of the two species in Dumfries have been very similar for the past 18 years, or Herring Gulls had a lower survival rate that was compensated by a higher recruitment rate to the breeding group to maintain the same proportions. Unfortunately, data to distinguish between these two possibilities do not exist, and there are no adequate data on the proportions of the two species breeding at other urban sites over the past 20 years. All we do know is that in the past, the overall numbers of Lesser Black-backed Gulls breeding in towns have increased more rapidly than Herring Gulls, and therefore the lack of a change in the ratio in Dumfries is surprising.

The abundance of urban-nesting gulls has changed dramatically over the last 50 years, and it is likely to continue to do so during the current millennium, but the future is difficult to predict. Will more limited food supplies eventually cause the urban habit of nesting to decline, or will human intervention be necessary to moderate the current rapid increases in town nesting? Conflict between protection organisations and councils that represent communities will develop further, and effective management will require national rather than local policy decisions. What seems certain is that further sizeable changes in numbers of large gulls in towns in Britain and Ireland are likely, but when, in which direction and why are currently uncertain. What is missing is a current and reliable national census using efficient methods of determining breeding numbers in urban and natural sites. Many current views and policies are based upon numbers obtained at the end of the last century. Adding to the uncertainty is the question over how soon the Yellow-legged Gull will join the regular nesting species in Britain and what impact it will have on numbers of Herring and Lesser Black-backed gulls.

CHAPTER 14

Conservation, Management and Exploitation of Gulls

CONSERVATION INVOLVES THE PROTECTION and encouragement of species and habitats, particularly those considered to be threatened in some way. Less frequently, it may require the reduction of a species that, for sound reasons, is considered too abundant either for its own good or that of other species. An example of the need for management was the problem created by the increase and spread of large gulls over much of the Isle of May, Scotland. The solution to this was the large cull, mainly of Herring Gulls (*Larus argentatus*), carried out by the Scottish section of the Nature Conservancy Council (now Scottish Natural Heritage) in the 1970s (p. 135) to make room again for other seabird species such as terns and Eiders (*Somateria mollissima*) to nest, and to protect the soil and vegetation. This was followed by culls of gulls for similar reasons, made by several protection and conservation organisations, at colonies in south-east Scotland, north-east England and south-west Wales. More recently, attention has turned to mammal pests, with culls to eradicate American Mink (*Neovison vison*) and rats undertaken on several islands to protect colonial seabird species, including gulls. The most recent successful eradication programme was on the Shiant Isles in the Outer Hebrides (starting in 2014) and on some of the Scilly Isles (from 2013), following earlier successful actions to remove rats from Ailsa Craig and Lundy. A programme to remove Stoats (*Mustela erminea*) from Orkney started in 2018. The adverse impacts of rats and other mammals on island seabird colonies is a cosmopolitan problem, and eradication programmes are currently in place in many areas around the world.

THE HISTORICAL RELATIONSHIP BETWEEN GULLS AND PEOPLE

People have been exploiting breeding gulls in Britain and Ireland for thousands of years – there are brief accounts of seabird colonies and of the human exploitation of gulls in the Domesday Book (1086 CE). Historically, people have exploited gulls for food, sport and plumage, but more recently we have become protectors of seabirds. In addition, we have also created extensive new food sources for many gull species in the past century, such as refuse dumped at landfill sites and fish offal discarded from numerous fishing boats. This has allowed some species to increase at such a remarkable rate that they have become too numerous in some places, affecting the abundance of other bird species as well as the lives of the people with whom they now share urban areas.

The historical pattern of exploitation of gulls contrasts with that of their relatives the terns. This group has also suffered from human exploitation, but unlike gulls they have not been able to exploit new food sources created by humans. In many places, terns have been adversely affected by the population explosions of some gull species. In recent times, changes in the abundance of terns and gulls have diverged and followed very different trends, and while numbers of several gull species in some locations are regarded as problematic, the same is not true of any species of tern.

Prior to the Industrial Revolution, gull exploitation was almost entirely localised. There were areas of Britain and Ireland into which people did not frequently venture and where gulls remained undisturbed by the much smaller human populations (about 4.8 million in 1600 and 16.3 million in 1801, compared with 74 million in 2017). However, gulls were probably subjected to greater natural predation than they are now. Comments in historical documents about the abundance of seabirds, including gulls, are probably inaccurate, exaggerated and lack realistic quantitative assessments. They are couched in terms such as 'millions', 'abundant', 'beyond estimate' and 'common', which do not give reliable indications of numbers, and species such as the Great Black-backed Gull (*Larus marinus*) and Lesser Black-backed Gull (*L. fuscus*), and also the Herring Gull and Common Gull (*L. canus*), were confused. Despite the lack of accuracy in reports, it is still evident that, long before the Industrial Revolution, eggs and adult gulls were locally exploited, but at levels low enough to be sustainable.

The situation changed during the Industrial Revolution, which lasted from the mid-eighteenth to the mid-nineteenth centuries. Transport improved both at sea and on land, firearms became cheaper and more efficient, and the human population of Britain increased (although that of Ireland decreased appreciably between 1845–49 due to mortality and emigration associated with the Irish Potato Famine). The distribution of the human population also changed, concentrating where natural resources were abundant and where new industry encouraged employment. There were no laws against taking eggs or killing adult gulls, and people shot birds without restriction. Naturalists who saw a bird that seemed different would shoot it to make an identification, because at the time there were no effective binoculars and no field guides. In some areas, Kittiwakes (*Rissa tridactyla*) were killed for their livers, which were regarded a delicacy. The plumage of seabirds, including gulls, was used to adorn clothing and hats, and for bedding, while eggs were collected by those from all walks of life as a cheap and valued source of protein.

Apart from being taken for food and feathers, many birds – including gulls – were killed for sport. One army officer wagered that he could kill 500 birds in a day, and won the bet by going to coastal cliffs during the breeding season and shooting adult seabirds, including gulls. When engines replaced sails, boats were frequently hired from major ports to take people to visit seabird colonies like those at Bempton in Yorkshire and in the Firth of Forth in Scotland, sometimes to shoot nesting gulls and auks, or simply to produce an impressive visual effect by disturbing them with sirens, causing huge numbers to pour off the cliffs.

Over time, local communities became concerned with the level of killing. Residents living near Bempton and Flamborough, for example, began to notice that numbers of breeding seabirds from which they had collected eggs for generations were decreasing. Eggs of Common Guillemots (*Uria aalge*) were most favoured there, but those of Kittiwakes were also widely collected. The 'climmers' were local people, usually working in teams of four, who collected the eggs. They used ropes to lower one of their group down the high, precipitous cliffs; there he collected whatever eggs he came across, placing them in a basket, and when this was fully loaded both basket and climber were hauled back up to the cliff top.

In 1868, the ornithologist Alfred Newton addressed the British Association for the Advancement of Science 'On the Zoological Aspect of the Game Laws'. In particular, he urged protection of birds of prey and seabirds during the breeding season, and commented on the destruction of seabirds on the Isle of Wight and at Flamborough. A report in the *Manchester Guardian* on 18 November 1868 claimed that more than 100,000 seabirds had been killed at Flamborough in a single season and that one small group of people had killed 1,100 in a single week. A committee called the Association for the Protection of Sea-Birds was appointed, consisting of Frank Buckland, Henry Dresser, William Tegetmeier and Canon Henry Tristram. They proposed a closed season and persuaded the government to support legislation protecting seabirds, which was eventually introduced on behalf of the association by Christopher Sykes, a Member of Parliament. This resulted in the Sea Birds Preservation Act 1869. Sykes' speech to Parliament on 26 February 1869 is recorded in Hansard, and part is repeated below:

> *The sea birds of England were rapidly disappearing from our coasts... A few years ago, the farmers of the East Riding of Yorkshire... were accustomed to see flocks of sea birds following at the heels of the ploughboy and from the newly turned-up earth picking up worms and grubs. [But] he held in his hand a letter from an influential farmer living in the parish of Filey, within a mile of the coast, stating that last summer he did not see a single bird on his farm. He appealed to the House also in the interest of our merchant sailors, for in foggy weather those birds, by their cry, afforded warning of the proximity of a rocky shore, when neither a beacon-light could be seen nor a signal-gun heard. He held in his hand a paper proving that with the decrease of those birds the number of vessels which had gone ashore at Flamborough Head had steadily increased. For the services they rendered to the mariner those birds had earned for themselves the name of the 'Flamborough pilots'. He appealed to the House, likewise, in the interest of the deep-sea fishers, because, by hovering over the shoals of fish, those birds pointed out the places where the fisherman should cast his net. On that ground alone, the Legislature of the Isle of Man had lately passed an Act imposing a penalty of £5 on every man who wilfully killed or destroyed a seagull. Lastly, he made his appeal even in the interest of those thoughtless pleasure seekers themselves who flocked to the coast in the summer months, chiefly from the populous towns of the West Riding of Yorkshire and of Lancashire. Those persons would have themselves to blame if, in a few years, they found that those rocks, which he once remembered as teeming with wild fowl, had become a silent wilderness.*

The 1869 Act protected 'the different species of auk, bonxie, Cornish chough, coulterneb, diver, eider duck, fulmar, gannet, grebe, guillemot, gull, kittiwake, loon, marrot, merganser, murre, oyster catcher, petrel, puffin, razor bill, scout, seamew, sea parrot, sea swallow, shearwater, shelldrake, skua, smew, solan goose, tarrock, tern, tystey, willock'. Apart from specifically mentioning the Kittiwake and its immature young (tarrock), the 'seamew' was presumably the Common Gull, or more likely the Herring Gull, while it and other gull species were also covered by 'gull'. Note that local bird names were used, some of which are no longer common, and no scientific names were given, indicating the confusion at that time over the names of several seabird species, such as shearwaters, petrels and fulmars, as well as bonxie and skuas.

The Act protected seabird species with a closed season running annually from 1 April to 1 August, but eggs were not included in the legislation – indeed, eggs were not effectively protected until the passing of the Protection of Birds Act 1954. The first successful prosecution under the Sea Birds Preservation Act 1869 occurred in the year it was passed into law, when a Mr Tasker of Sheffield was fined £3 19s on 10 July at Bridlington, Yorkshire, for shooting 28 seabirds. However, despite this early prosecution, very few others appear to have been made in the following 10 years.

Further statutory Acts protecting some bird species were passed by Parliament in 1872, 1876, 1881, 1894 and 1896, while in 1891, a group of women formed the Society for the Protection of Birds to campaign against the plumage trade; in 1904, this organisation became the Royal Society for the Protection of Birds.

Because the protection of seabirds was poorly enforced for many years, the numbers of several species in Britain were at an all-time low by the end of the nineteenth century, including the Great Skua (*Stercorarius skua*) and the Kittiwake, Herring Gull and Great Black-backed Gull. However, records of their numbers at that time were rare. Contemporary writers often described abundance of birds by using terms such as 'common' or 'rare', but the meaning of these have changed greatly in their numerical equivalents over time. For example, on finding six Herring Gull nests on the Farne Islands, Northumberland, in 1910, one writer then described the species there as 'common'.

My own literature searches over many years have focused on obtaining a figure for the numbers of pairs of breeding Herring Gulls in England, Wales and the south of Scotland between 1900 and 1910, and have included published records, local accounts and newspaper articles, along with personal diaries and some other unpublished sources. (I failed to find accounts of numbers and distribution for Ireland, northern Scotland, the Hebrides and the Northern Isles.) In the end, I was left with an estimate of well under 200 pairs of Herring Gulls in England, Wales and southern Scotland, with long stretches of coastline completely devoid of gulls during the breeding season – yet at that time this species was repeatedly reported nationally as being a 'common' breeding species.

A consequence of the fact that there were only small numbers of Herring Gulls in the early years of the twentieth century is that the birds were infrequent at many coastal sites, often avoiding harbours and beaches where there was a high risk of being shot. The situation changed dramatically within a hundred years, however, such that by the end of the twentieth century numbers of Herring Gulls in Britain must have been well over 100,000 pairs.

The period from 1900 to 1975 was one when gulls required and received increasing and effective protection from exploitation. Their numbers consequently rose year after year, with some species increasing more rapidly than others. Herring Gulls increased dramatically, by 13 per cent per year from 1900 to 1975 – an exceptional rate of increase, in which numbers doubled every six years. Kittiwakes and Great Black-backed Gulls increased more slowly, at 4 per cent per year or less. Black-headed Gulls (*Chroicocephalus ridibundus*) and Common Gulls also increased, but at unknown rates – although clearly these were much lower than in the Herring Gull.

As Herring Gull numbers increased dramatically and their habits changed to feeding at landfill sites, so the level of concern also increased – for various reasons. In the case of landfills situated near airports, there was an elevated risk of bird strikes; at sites near farms, the gulls passed food-poisoning organisms on to the farm animals; and those near nature reserves produced adverse effects. As a result, attempts were made to exclude gulls from some landfill sites. Shooting occurred at a few sites, but it had little effect, and birds of prey were introduced, but they were used too infrequently to be effective. At a landfill near the military airfield at Lossiemouth, Scotland, the site's bulldozer was equipped with a tape recording of the distress calls of gulls, which was played intermittently through a loudspeaker with the aim of dispersing the gulls. However, within a few months, rather than preventing gulls feeding at the landfill, the call was being interpreted by the gulls as an invitation to feed. I was present on two occasions while the calls were broadcast, and on both a flock of gulls roosting in a nearby field immediately lifted off, flew onto the landfill and started to feed!

The increase in numbers of the Lesser Black-backed Gull appears to have started later than the Herring Gull, but useful estimates for the first half of the twentieth century do not exist, partly because the species was often confused with the Great Black-backed Gull. Since 1960, it has increased at an annual rate similar to that of the Herring Gull, and continued to do so even after Herring Gull numbers started to decline. The Lesser Black-back has been heavily culled in some western areas in recent years, particularly on the Tarnbrook and Abbeystead moors in Lancashire.

MANAGEMENT OF GULL NUMBERS ON MOORLAND: A CASE STUDY

Attempts to manage gull numbers for the benefit of other bird species or for public health concerns have involved several strategies. The culling of Herring

Gulls on the Isle of May, Scotland, is discussed in Chapter 5 and attempts to control large gulls in Dumfries, Scotland, are discussed in Chapter 13. This section describes a case study of the management of gull numbers on moorland in Lancashire.

While Herring Gulls have been culled at some sites, the sizeable inland breeding colony at Tarnbrook Fell (including the Abbeystead, Brennand and Mallowdale estates) in the Trough of Bowland, north-west England, comprised predominantly Lesser Black-backed Gulls. Neil Duncan (1981) reviewed the history of this colony and reported that the area was first colonised in 1938 by a few pairs of Lesser Black-backed Gulls, after which numbers increased rapidly between 1956 and 1974, at about 15 per cent per annum (not dissimilar to the increase of gulls on the Isle of May and the national estimate of the increase in Herring Gulls). By 1978, the colony contained about 50,000 breeding birds, of which 85 per cent were Lesser Black-backed Gulls and 15 per cent Herring Gulls, along with about 40 pairs of Great Black-backed Gulls. At that time, the inland colony covered 1,400 ha of west Pennine moorland managed for Red Grouse (*Lagopus lagopus*) and sheep grazing, at an altitude of 400–520 m. The gulls nested at a low density, averaging just 18 pairs per hectare.

Small culls by shooting were carried out at the colony in the 1960s and early 1970s; while no records of the numbers killed exist, the culls were insufficient to stop the colony increasing in size. In addition, some 640 Herring Gulls were killed for research purposes between 1974 and 1976. In 1978, a licence was obtained from the Ministry of Agriculture, Fisheries and Food by the North West Water Authority (NWWA) to protect public health from unacceptably high levels of potentially dangerous bacteria in a stream that drained from the gull colony and formed part of the water supply to the cities of Lancaster and Preston. Up to 25,000 adult gulls were culled that year using α-chloralose baits, but as was the case on the Isle of May, no detailed records on numbers killed were kept (Fig. 249). Another major concern (although not used to obtain the licence) was that gulls were excluding Red Grouse and sheep from large areas of moorland and that the area dominated by the gulls was increasing in size annually.

Further large-scale culls using α-chloralose baits at nests were carried out in 1979, 1980 and 1982, and the shooting of some 2,000 gulls took place in 1981. In all, over 50,000 gulls were killed between 1978 and 1982, of which 86 per cent were Lesser Black-backed Gulls. By the end of the 1982 breeding season, numbers had been reduced to about a quarter of those present in 1978, before culling started. Further culls organised by United Utilities plc in the 2000s continued to reduce the size of the colony, and a report claims that the colony has now been reduced to about 1,000 pairs of Lesser Black-backed Gulls, although this needs confirmation.

FIG 249. Lesser Black-backed Gull culled by use of a narcotic bait while incubating eggs. (John Coulson)

BELOW: **FIG 250.** Gulls accumulating on a roadway before walking as a group into a netted landfill site, Inverness. (John Coulson)

The problem of contamination of water draining from the gull colony was real, and was first recognised in 1975 by the NWWA. In a 1993 study, water samples were collected on six dates between late March and early July from the intake pipe on the Abbeystead Estate and from a control stream on Hawthornthwaite Fell (10 km away, with similar terrain but no nesting gulls) and then analysed at the NWWA laboratory. Levels of *Escherichia coli* bacteria in the Abbeystead samples were high and increased markedly through the sampling period, reaching a remarkable peak in July, while those at the control site remained low throughout the whole study. By July, the

FIG 251. Large gulls feeding in 1970 at an outfall of untreated sewage causing a change in the seawater colour and is a potential source of food poisoning organisms. (John Coulson)

FIG 252. Herring Gulls (*Larus argentatus*) bathing and drinking at drainage water accumulated at the side of a landfill site and a possible source of botulism. (John Coulson)

concentration of *E. coli* in the run-off water at Abbeystead was 22 times greater than at Hawthornthwaite Fell. *Salmonella* bacteria were also recorded in the samples collected at Abbeystead in May, June and July, but none were found at Hawthornthwaite Fell. Such contamination of water is not a major threat to the public, because of treatment applied before it is supplied to households, but the failure of this treatment has occurred occasionally in the past, requiring large numbers of households to boil water before use.

While culling gulls would, presumably, reduce contamination in the run-off water (although I am not aware of published data showing this), the almost fivefold reduction in the numbers of breeding large gulls from 1978 to 1985 had almost no effect on the sizeable area of moorland occupied by the birds. The culls reduced the density of nesting gulls, but the extent of the colony remained the same and no moorland was reclaimed for grouse and sheep grazing.

In 1992, the Abbeystead Estate manager, on behalf of the owner, asked me to design and execute a programme involving non-lethal methods of reclaiming areas of grouse moorland from the spreading gulls. A series of experiments testing possible methods was applied and evaluated by Mark O'Connell and, subsequently, by Nick Royle. The initial experiments used gull distress calls broadcast by amplifiers, propane gas guns that created small explosions every few minutes, many fluttering plastic flags placed on cane poles, and parallel lines of monofilament nylon stretched about 50 cm above the ground to hinder gulls landing. The techniques were used both singly and in combination, and ran continuously or on repeated cycles of two days with application, and two days without. They failed to totally clear areas of gulls. In all cases, the methods immediately excluded gulls from the study plots, but only for a very limited time, ranging from less than one day to 13 days depending on the methods used. In all cases, the great majority of gulls became habituated to the disturbance methods applied and reoccupied the plots. Some combinations of methods were more successful, yet even the most successful technique of continuously using flags and a gas gun did not succeed in preventing gulls from breeding. The flags were less effective on calm days and the gas gun was less effective in strong winds, while the gulls soon learnt to walk in from the side and under the lines of nylon filaments to reach a nest site.

In other experiments, an early-season start of these management methods beginning in early April proved much more effective than when it was started just before the onset of laying in early May, but still failed to prevent nesting altogether. Based on past knowledge of the edge effects found in colonies of gulls, further experiments examined whether gulls were more readily disturbed and moved from the edges than from the centre of the colony. Results from these showed clearly that the same treatment was twice as successful when applied at the edges of the colony than at its centre. These results led to a new management method that concentrated on rolling back the edges of the colony each year.

By far the most successful method of management proved to be frequent and regular human disturbance, enforced as soon as gulls first attempted to return to the colony in early April. Habituation did not occur with this method, mainly because those applying the disturbance reacted rapidly as soon as gulls were seen to land on potential territories, using firecrackers, rockets and other forms of explosion to scare off the birds. As a result, no gulls landed and remained for more than a few minutes on the edge areas selected for disturbance. Because gulls leave their colony at night before egg-laying commences, disturbance was required only during daylight hours.

Part of the management plan was the creation of a site onto which disturbed gulls could (and did) move. This was placed at the centre of the colony and was called the sanctuary area. This area was clearly marked and access to it was restricted to occasional visits by the researchers. The area was not intended as a permanent sanctuary, but as a means of giving disturbed gulls a nearby alternative area into which they could move from the edge, and which could be cleared in future years.

Frequent human disturbance of the gulls along the edge of the colony was applied from early April to the end of June by patrolling the area selected during most of the period of daylight. It was necessary to increase the numbers of people carrying out the disturbance into two shifts a day in the longer daylight periods of May and June. In the year following the disturbance few gulls even tried to reoccupy the edge areas that had been cleared in the previous seasons, and those that did were easily and quickly deterred, while the disturbance treatment was applied to the new edge areas of the colony. The method was successful, and 500 acres (80 per cent) of the colony on the Abbeystead Estate (except for the sanctuary area) were progressively rolled back over four years and cleared of breeding gulls (Fig. 253) and without killing any individuals. Red Grouse and some waders returned and nested in the cleared areas in the year immediately following the removal of the gulls.

The gulls that were excluded from the area did one of three things. Some moved to the sanctuary area, which fulfilled its planned function – the number of gulls nesting there increased dramatically by 86 per cent in the first year and more thereafter. A proportion of adult gulls appeared to leave the area entirely, possibly moving to the new colony on the Ribble Estuary on the Lancashire coast or to that at Bolton Castle, Yorkshire. Some gulls moved shorter distances, to the adjacent Brennand and Mallowdale estates – for example, the numbers of gulls breeding on the Brennand Estate increased by 57 per cent between 1993 and 1994. The owners and tenants of these estates were unable to finance disturbance teams and expressed their concerns about immigrating gulls. At that point in time, the effective non-lethal management programme at Abbeystead Estate was brought to an end.

There is an important lesson to be learnt from this management programme. While it is possible to cause large numbers of breeding gulls to move from an

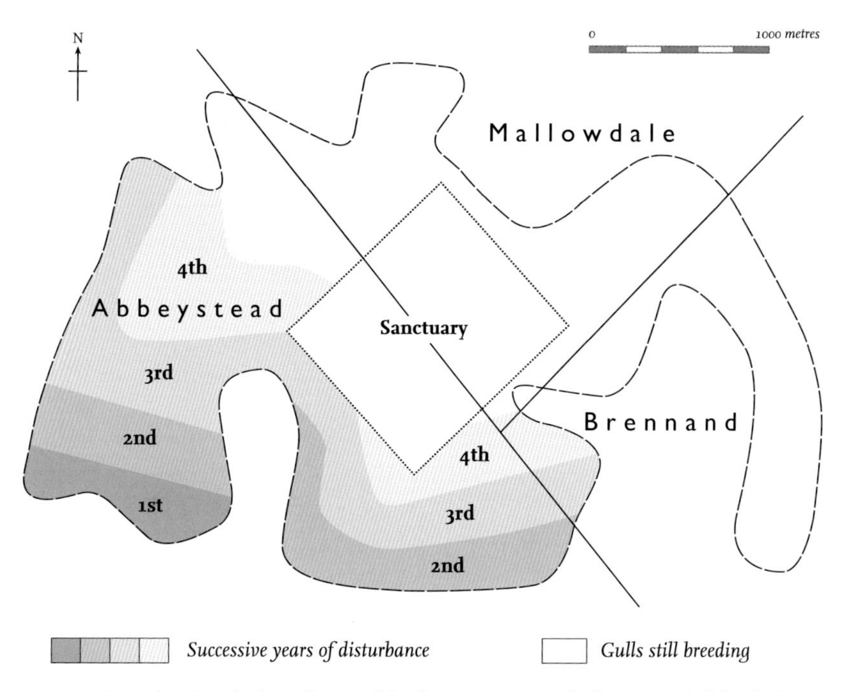

FIG 253. Map showing the boundaries of the three estates on which Lesser Black-backed Gulls were breeding on Tarnbrook Fell (solid lines) and the limits of the gull colony in 1990 (dashed line). The temporary 1 square kilometre sanctuary area is central. The delineated areas were progressively cleared of breeding gulls by disturbance by the pilot study (1st) and then in the following three years (2nd, 3rd and 4th), leaving breeding gulls only to the north of the Abbeystead Estate and in the sanctuary area.

area, this is mainly a case of passing the problem on to someone else and does not reduce the overall numbers of gulls – it just changes their distribution. This comment also applies to non-lethal methods recommended by the RSPB and others for use in towns in attempts to manage gull numbers (p. 402); they are not an effective answer to the management of the urban gull problem.

THE CURRENT SITUATION

No animal species can increase in numbers forever, but it is difficult, if not impossible, to predict in advance when the upper limit will be reached. Apparently, this limit has already been reached for both the Herring Gull and the

Kittiwake in Britain and Ireland, as numbers of both species have been declining for several years. Herring Gull numbers reached a peak by about 1980 and the Kittiwake by 2000.

The increases of the Herring, Lesser Black-backed and Great Black-backed Gull populations during the first three-quarters of the twentieth century were supported and sustained by increased quantities of fish offal dumped overboard from fishing vessels. It was only as recently as 1955 that large gulls discovered that landfill sites offered a new and major source of food, first feeding in large numbers at coastal sites and then, a few years later, also at inland landfill sites. Feeding at landfills did not cause the recovery of the Herring Gull, but it almost certainly permitted numbers to continue increasing for longer than would otherwise have been the case.

In the second half of the twentieth century, increasing numbers of Herring Gulls and (later) Lesser Black-backed Gulls spread to nest in towns. A few were probably encouraged to colonise new nesting sites when some natural breeding grounds became full. Numbers of Kittiwakes also increased and, starting from 1930, new colonies were formed, many of which were on lower sea cliffs. A few Kittiwakes then spread from cliffs to coastal man-made structures and buildings, which fulfilled the same requirements of a nest site, namely that they had vertical faces, were near or over water and had small ledges on which nests could be built. At least two Kittiwake colonies now exist on oil or gas platforms far from the coast, one in the Irish Sea and the other in the North Sea.

So why did the increases in numbers of these gulls come to an end? The most likely cause is the reduction of available food. The management of many landfill sites began to change in about 1970. New regulations meant that newly dumped refuse had to be covered with soil before the end of each working day, and this restricted the time that gulls could feed there, particularly in winter. Later, more refuse was sent for recycling or for incineration to avoid a landfill tax, which was introduced in 1996 and increased annually for the next 20 years. Several major landfill sites have closed within the last 20–30 years, and management of refuse continues to be treated differently, in ways that have made food less available to gulls.

In addition to the changes at landfills, new restrictions and quotas have been placed on commercial fish catches at sea through reforms to the European Union's Common Fisheries Policy. This has not only reduced landings of fish, but also discards at sea. A rolling programme is being introduced from 2014 to 2019, by which time there will be a total ban on discarding unwanted or undersized fish at sea, reducing this food source for gulls, Great Skuas, Northern Fulmars (*Fulmarius glacialis*) and Northern Gannets (*Morus bassanus*). What will happen to the material that was previously discarded is not yet clear, but it is believed that it will be included within the fish quota and will have to be landed. The impact of these changes on gulls will only slowly become apparent, and the extent to which large gulls that have concentrated on feeding on rejected fish will be able to find alternative food sources is unknown. Herring, Lesser Black-backed and Great Black-backed gulls may have to revert to feeding more on free-living marine organisms or on agricultural land, a habit that is already well developed in those birds that nest inland beyond easy reach of the coast.

The overall effect of changes to landfill sites and to commercial fishing is likely to impact on gull numbers and may well already have played a part, along with botulism (p. 133), in the decline of Herring Gulls at the end of the twentieth century. Interestingly, if it is indeed true that botulism in gulls is acquired mainly at landfills, then mortality of gulls from this cause could be reduced – if it has not already done so – as a result of the improved management of sites. Between 2003 and 2013, material sent to landfills was reduced by 20 per cent in Ireland and by 18 per cent in Britain, while between 2006 and 2014, the number of landfill sites in England decreased by about 60 per cent overall or 6 per cent per year.

CONSERVATION CONCERNS

British assessments

Currently, there are only a few problems relating to gulls in Britain and Ireland that require management and protection of the birds, especially in comparison to terns. Both gulls and terns currently benefit from the protection provided at colonies where there had been a risk of predation. One species that is of concern is the Little Gull, which is unlikely ever to be numerous as a breeding bird in Britain and Ireland because suitable predator-free wetland habitats are few. However, my assessments are rather different from those expressed in 'Birds of Conservation Concern 4' (BoCC4) (Eaton *et al.*, 2015).

Gulls in 'Birds of Conservation Concern 4'

Every five or six years, the journal *British Birds* publishes a report entitled 'Birds of Conservation Concern', compiled by a group of conservationists and covering 'native' British birds, including several gull species. The most recent report (BoCC4) was published in December 2015 and classified birds in one of three colour-list categories. The Red List includes those species that the committee consider are of the highest conservation concern and require urgent action. Amber-listed species are the next most critical group, while the third category,

Green-listed species, are currently of no concern in the UK. To be included in either of the first two categories, only one of the stated criteria need apply; the main Red- and Amber-list criteria are detailed below.

Red-list criteria

- Severe (at least 50 per cent) decline in the UK breeding population over the last 25 years, or a longer-term period (the entire period used for assessments since the first BoCC review, starting in 1969).
- Severe (at least 50 per cent) contraction of the UK breeding range over the last 25 years, or a longer-term period.
- Non-breeding population decline by more than 50 per cent.
- Non-breeding range decline by more than 50 per cent.
- Species that are globally threatened.
- Historical decline in breeding populations; species judged to have declined severely between 1800 and 1995.
- Red-listed European status.

Amber-list criteria

- Moderate (25–49 per cent) decline in UK breeding population over last 25 years, or a longer-term period.
- Moderate (25–49 per cent) contraction of UK breeding range over last 25 years, or a longer-term period.
- Moderate (25–49 per cent) decline in UK non-breeding population over last 25 years, or a longer-term period.
- Rare breeder; 1–300 breeding pairs in UK.
- Rare non-breeders; less than 900 individuals.
- Localised; at least 50 per cent of UK breeding or non-breeding population in 10 or fewer sites, but not applied to rare breeders or non-breeders.
- Internationally important; at least 20 per cent of European breeding or non-breeding population in UK.

BoCC4 'considered all naturally occurring *native species* on the British List' (my emphasis), but failed to define these. The report does state that vagrants and species occurring as rare migrants were excluded, yet it automatically includes any globally threatened species. This means, for example, that the Balearic Shearwater (*Puffinus mauretanicus*), which breeds in the Balearic Islands and only visits offshore areas, must be included as a Red-listed species in Britain. This causes bias in the proportion of endangered species, because none of the several numerous North American gull species that reach Britain – even those that occur here annually – is considered or classified, presumably because these gulls are not 'native species'. In addition, the requirement that only one of the Red or Amber criteria needs to be met for a species to be included leads to a 'catch-all' situation and inflates the proportion of Red- and Amber-listed species, yet confusion still remains as to what is a native gull species.

BoCC4 placed no fewer than 27.5 per cent of the British bird species considered on the Red or Amber lists, including 11 gull species: Kittiwake, Herring Gull, Black-headed Gull, Mediterranean Gull (*Ichthyaetus melanocephalus*), Common Gull, Lesser Black-backed Gull, Yellow-legged Gull (*Larus michahellis*), Caspian Gull (*L. cachinnans*), Iceland Gull (*l. glaucoides*), Glaucous Gull (*L. hyperboreus*) and Great Black-backed Gull. However, it is difficult to accept that the status of 92 per cent of the gull species which occur in Britain are currently of conservation concern. BoCC4 recognises only one species, the Little Gull, as being Green-listed and so not of concern, despite the fact that it occasionally attempts to breed in Britain. The Mediterranean Gull, now nesting in appreciable and increasing numbers well beyond the 'rare breeder' limit of 300 pairs, is inexplicably Amber-listed. The Caspian Gull, which is currently better described as a vagrant rather than a native species, has been included for the first time and is Amber-listed, while the Kittiwake, far more numerous than the Caspian Gull in the UK, has been moved from Amber to Red status. It is difficult to believe that the Herring Gull, whose numbers in the UK still total several hundreds of thousands of individuals, should be Red-listed. Is it really in danger of extinction in Britain? Perhaps it is time that the criteria for listing species at Red and Amber levels are revised for gulls and, perhaps, for other species. Are our gull species in such a poor state that most are considered of concern, or has the level of concern been exaggerated?

The 'Birds of conservation concern in Ireland 2014–2019' (BoCCI) (Colhoun & Cummins, 2013) lists eight gull species as a concern. The Herring Gull (which has declined to a greater extent in Ireland than in Britain) and the Black-headed Gull both have Red-list status, while the Little Gull, Mediterranean Gull, Great Black-backed Gull, Lesser Black-backed Gull, Common Gull and Kittiwake all have Amber status. Like BoCC4, the Irish report ignored the North American gull species.

Risk of extinction (Stanbury *et al.* 2017)

In 2017, *British Birds* published a study entitled 'The risk of extinction for birds in Great Britain', using the International Union for Conservation of Nature (IUCN) criteria for the risk of extinction (Stanbury *et al.*, 2017). The authors used the decline in abundance over three generations as the standard for numerical change, which for Herring Gulls is stated as 39 years. (The generation length was calculated by the formula $(1/m) + b$, where m = mean annual mortality rate and b = age at first breeding). While a generation time of 13 years may have been reasonable during the time the Herring Gull was increasing in Britain, the increased mortality rate of adults in the last quarter of the twentieth century has reduced that time span to about 10 years, and so the decline over 30 years or even less, and not 39 years, should have been applied. The use of numerical change over three generations is strange and depends on having a current and accurate measure of the mortality rate of the species. It was introduced in an attempt to have a common basis for all plants and animals, and this is, perhaps, a mistaken goal. Long-lived species tend to have low reproductive rates and so take a long time to recover after a crash in numbers, and so on that basis, generation time would seem appropriate to indicate a potential recovery time. However, mortality from adverse environmental factors such as disease, poisoning and oiling are usually abrupt and often severe, and do not correlate with, or relate to, the generation time of the species considered.

This 2017 report classed 43 per cent of regularly occurring bird species in Britain as 'Threatened'. Of the gulls, the Kittiwake was placed in the 'Critically Endangered' category, while the Yellow-legged Gull, Herring Gull and Great Black-backed Gull were included in the 'Endangered' category, along with wintering numbers of the Black-headed Gull, Caspian Gull, Glaucous Gull and Iceland Gull. However, experts have reservations about the quality of some of the data and methods used to estimate the magnitude of declines and wintering numbers.

Status

The status of each gull species listed in BoCC4 and its classification in the 2017 report are considered below, followed by that given in the risk document.

Black-legged Kittiwake – Red-listed: Critically Endangered

The Kittiwake is the most numerous gull in the world, but it is reported to have declined in Britain by 74 per cent over 25 years, by 62 per cent since 1969 and by 81 per cent over three generations. The decline is genuine, but the extent of it varies geographically – even within Britain, it has ranged from huge declines in Shetland, Orkney and northern Scotland, to much smaller declines elsewhere and even increases on the east coast of England. Large declines have also been recorded around most, but not all, of Ireland, Norway and France, and the poor breeding success in many colonies suggests that shortage of food – particularly sandeels, and Capelin (*Mallotus villosus*) in more northern areas – during the breeding season has been the main contributor. To date, the extent to which the decline and redistribution of sandeels has been directly caused by commercial fisheries has not been fully investigated, but at several locations commercial sandeel fisheries have been banned and there is still a lack of relevant and detailed information on sandeel biology. A seabird species such as the Kittiwake, which depends mainly on a single group of fish such as sandeels for food, is vulnerable when no suitable alternative food source is available, and hence a crash in numbers is to be expected from time to time. It is not possible to predict the future trends in sandeel abundance, nor the future for the Kittiwake. Nevertheless, and despite the declines, breeding Kittiwakes are still much more numerous in most of Britain and Ireland than they were a century ago.

Currently, Kittiwakes need to produce an average of about 0.8 chicks fledged per pair to maintain numbers. In many areas where there has been a marked decline in numbers of breeding Kittiwakes, the average annual number of young fledged per pair lies well below this critical value, while in areas where Kittiwake numbers are maintained or increasing, average productivity is above this value. For example, young production has consistently exceeded the required productivity year after year in the Tyne area of north-east England, where an average of about 1.1 young fledged per pair has been reported, and significantly, numbers of Kittiwakes breeding in that area have increased. Similarly, at the vast colonies at Bempton and Flamborough in Yorkshire, productivity fell below 0.8 young fledged per pair in 10 of the 30 years between 1986 and 2015, but in 16 years it was above the critical level, resulting in a small overall increase in breeding numbers there in recent years.

It appears that feeding conditions in winter in the Atlantic Ocean for Kittiwakes have not deteriorated, and while the adult mortality rates of Kittiwakes breeding on Skomer and the Isle of May have fluctuated, they have not shown an overall increase in recent years. Had the risk of adult Kittiwake mortality increased in winter when there is considerable mixing of individuals from many areas in the North Atlantic, it would have caused a more uniform decline in colony sizes over the whole of Britain and even beyond.

The evidence suggests that availability of food during the breeding season and poor breeding success is causing the decline in numbers at many colonies. The only act of conservation required to assist the Kittiwake would therefore appear to be the difficult task of developing a means of increasing numbers of sandeels, if necessary by introducing greater controls over sandeel fisheries if these are shown to have depleted sandeel stocks or the task of preventing climate change which has been suggested as the cause of the sandeel decline. But how can this be achieved?

Herring Gull – Red-listed; Endangered but also said to be data deficient by Stanbury et al. *(2017)*

The Red-listing and Endangered status of the Herring Gull indicates that the criteria used to identify species of conservation concern need modification. Currently, there are still 200,000 or more Herring Gulls in Britain and Ireland throughout the year, even though breeding numbers have declined appreciably at natural coastal colonies during the last 40 years. The decline in Ireland has been more severe than in Britain, but there is little evidence yet that numbers in Britain continued to decline appreciably or significantly in the period 2001–18. The statement that the Herring Gull has declined by 79 per cent in Britain over the past three generations (39 years), as suggested in 2017 by Stanbury *et al.*, is unacceptable. It seems that this value has been derived from the 60 per cent decline in 25 years reported previously by Eaton *et al.* in 2015, and then simply extrapolated, without justification, for 14 more years at the same annual rate to obtain a figure for the decline over 39 years (three generations). There are no data yet indicating that a decline has indeed continued over the last 39 years, or even since 2001. There has been no national census since 2000 but one is in progress in 2018 and numbers are still increasing in many urban areas. This Red listing has produced a paradox, because the species is now regarded as a pest in many urban areas and a majority of residents affected by the birds want their numbers reduced, while conservationists are claiming that the species in Britain and Ireland is threatened with extinction! Any realistic concern can relate only to the decline in numbers nesting at natural, coastal sites.

A decline in the Herring Gull was predicted more than 30 years ago by those who recognised that a decrease in the available food for this species was imminent as a result of changes to EU policies on fishing discards and the management of refuse. When a species is still abundant and its numbers in Britain are still measured in the hundreds of thousands, there is little chance of it ceasing to breed here and becoming 'extinct' in the near future. It is difficult to identify the need for genuine concern for this species, particularly since a decline in abundance would be expected to follow reductions in available food and numbers might be expected to stabilize at a new, lower level. Perhaps the Red-listed status should not be considered for a very abundant species until its numbers have declined much further. Possibly a more realistic concern for this species in the next few years is the risk that it will lose out to competition from the recent influx and spread of Yellow-legged Gulls.

Lesser Black-backed Gull – Amber-listed; data deficient (Stanbury et al. *2017)*

This species is regarded as of moderate concern because more than 50 per cent of the UK breeding numbers were stated to be at 10 or fewer 'sites' in Britain, and therefore its abundance is considered at risk in the event of an environmental disaster. However, the 'sites' were, for no obvious reason, restricted to Special Protection Areas and Important Bird Areas, and as a result several large colonies were ignored. In fact, more than 50 per cent of all the Lesser Black-backed Gulls breeding in Britain are not restricted to 10 or fewer colonies, and so the species' Amber listing does not appear to be justified. Stanbury *et al.* (2017) did not consider its status because of lack of data, despite the fact that all those used by Eaton *et al.* (2015) two years earlier were available. In any case, appreciable numbers of immature individuals are not at these colonies at any time of the year and their absence moderates the level of risk in the event of a local catastrophe during the breeding season. Furthermore, there is evidence of exchanges of individuals between England, Belgium and the Netherlands. The risk of a catastrophe suddenly threatening the existence of Lesser Black-backed Gulls in western Europe must therefore be extremely small. Is the category of '50% of the UK population… at ten or fewer sites' an arbitrary value or a guess, or does it have a scientific basis I have failed to find? Since at least 200,000 Lesser Black-backed Gulls breed in Britain, and despite extensive culls in north-west England, there is no evidence that its numbers have declined since 2000.

Great Black-backed Gull – Amber-listed; Endangered

Although the Great Black-backed Gull is included on the Amber List because numbers are said to have declined, this has been by only a few percentage points (if at all) as a breeding species in Britain and Ireland. The species is spreading along the east and west coasts of Scotland and England, and along the south coast of England. These changes will have compensated to some extent for possible small decreases in parts of northern Scotland.

There has been an estimated 33–58 per cent decrease in wintering birds, which Stanbury *et al.* (2017) then inappropriately scaled up to a winter decline of 65 per cent over three generations (based on the unsupported assumption that the decline had continued for three generations). This then led to a classification of Endangered based on the weak evidence of a change in winter numbers obtained from a sample of roosts made in only one or two years.

Black-headed Gull – Amber-listed; Vulnerable

Numbers of Black-headed Gulls breeding in Britain (and in several countries in mainland Europe) have changed little over the past 30 years, although a decline has been reported in Denmark). The claimed decline of 47 per cent in wintering numbers in Britain in one of two years is the only justification for the Amber listing. This claimed decline involves the same problems as for wintering Herring and Great Black-backed gulls, when the numbers have been counted at a proportion of winter roosts and then adjusted. A majority of the wintering Black-headed Gulls in Britain are birds that breed in mainland Europe, and there is no evidence that they have declined there. If the numbers of this species have declined in winter in Britain, it is most likely that fewer individuals are crossing the North Sea to winter here and more are staying on the Continent. A similar redistribution in winter, coupled with reduced migration to Britain, has been reported for several wader species and explained by less severe winters on the Continent in recent years. Changes in wintering areas that presumably are of benefit to the species are not a justification for concern. A new analysis of recoveries of Black-headed Gulls ringed on the Continent and recovered in Britain could clarify whether there is indeed justification for the species to be placed on the Amber List and classed as an Endangered wintering species.

Common Gull – Amber-listed; not considered by Stanbury et al. *(2017)*

The numbers of Common Gulls breeding in Scotland and changes in their abundance there are poorly known. It is claimed that there has been a 40–50% per cent decline in wintering Common Gulls in Britain, justifying an Amber listing for the species, but further evidence is needed to confirm this since it is based on only two winter surveys. A reason for the possible decline is unknown, but it may be that, as in Black-headed Gulls, it represents a change in the wintering areas of birds that breed in mainland Europe. It is not apparent why the species was placed on the Amber List in 2015 but was not considered by Stanbury *et al.* (2017), who presumably considered it as not threatened or perhaps that suitable evidence was not available.

Glaucous Gull and Iceland Gull – Amber-listed; Endangered

Both species are arctic breeders, with only a small proportion wintering as far south as Britain, and most of these birds reach only as far south as northern Scotland. It has been estimated by Stanbury *et al.* (2017) that both species average less than 250 individuals in winter, while BoCC4 (by unexplained methods) states that an average of 170 Glaucous Gulls and 240 Iceland Gulls visit Britain each year, although no explanation is given as to why these numbers are the reverse of opinions of the relative abundance of wintering numbers of these two species in Britain. There are occasional years when appreciably more individuals are recorded, presumably because these influxes are driven by poor wintering or feeding conditions further north. It is difficult to understand why these species should be considered as 'endangered and of conservation concern' in the UK context, since they are at the edge of their wintering range in Britain and Ireland, while elsewhere they exist in much greater numbers and where neither is considered as a species of concern or at risk of extinction. It might be a concern to some that these gulls are encountered too infrequently in winter over much of Britain, but that is not a conservation matter.

Mediterranean Gull – Amber-listed by BoCC4; not considered by Stanbury et al. *(2017)*
This species is Amber-listed because breeding numbers as said to be below the arbitrary threshold used for concern, although the numbers of breeding pairs do exceed the 250 pairs required by the IUCN criteria, are increasing rapidly in Britain and have passed 1,000 pairs, which surely must justify a Green status. In the future, the species may be of conservation concern for a different reason, because of their well-known habit of preying on the eggs and chicks of terns and Black-headed Gulls and this has already been reported.

Yellow-legged Gull – Amber-listed; Endangered
The Yellow-legged Gull is spreading northwards from its main breeding areas in the Mediterranean, and in recent years it has occurred regularly in southern England and Wales throughout the year, with several hundred recorded annually. Currently there are only a few breeding records each year in Britain, and about half of these involve birds pairing with Herring or Lesser Black-backed gulls and producing hybrid offspring. The Yellow-legged Gull has rapidly spread and established colonies in France, Belgium and the Netherlands, and so it is likely to become an increasingly frequent breeder in Britain. If so, it is likely to be a competitor of Herring and Lesser Black-backed gulls. Despite only a few individuals currently breeding in Britain, it is difficult to envisage that it is, or will be, a species of conservation concern. The only attraction of this species is that it adds one more species to the number of bird species breeding in Britain and to the tally kept by many keen 'twitchers'.

Caspian Gull – Amber-listed; Endangered
An estimate has been made that there are 90 records of Caspian Gulls in Britain each winter, but this is a species whose identification in the field approaches the limits of what is possible. There are still problems with identification, particularly of females, while the individual variation is increased by the existence of hybrids. In any case, the species is surely as much of a vagrant in Britain as the Ring-billed Gull (*Larus delawarensis*), which has not been considered for listing, which begs the question why the Caspian Gull was added to the BoCC4 Amber List. What steps could or should be taken to meet the concerns of conservationists for

this species? Why would we want it to be more abundant and even breeding in Britain, and is it a 'native' species?

Little Gull – Green-listed; not considered by Stanbury et al. *(2017)*
Despite attempting to breed in Britain on a small number of occasions and in far fewer numbers than the Amber-listed Mediterranean and Yellow-legged gulls, this species is not considered of concern in Britain, although it is included as an Amber-listed species in Ireland (BoCCI). I believe this delightful bird should be included as a species of concern, because of the few attempts it has made to breed in Britain. It is a species that might be attracted to breed more frequently here by the creation of additional suitable wetland breeding sites where predators are excluded. Considered with the Caspian Gull, it surely has more justification to be a concern in Britain, but the criteria currently used do not recognise this.

The Irish assessment: BoCCI
The assessment of the status of gulls in Ireland was made by Kendrew Colhoun and Sinéad Cummins (2013). This is a similar but less complex assessment than that for Britain and place both the Black-headed Gull and the Herring Gull on the Red list of endangered species and six other gulls species (Mediterranean, Little, Common, Lesser Black-backed, Great Black-backed gulls and the Kittiwake) on the Amber list.

THE FUTURE

The criteria currently used to justify the inclusion of a species on either the Red or Amber lists clearly do not work well for gulls. The status of the Black-headed, Common and Great Black-backed gulls are based on a winter census of gulls in Britain made at an incomplete set of winter roosts during the 2003/04 and 2004/05 winters, and the figures obtained by the same means of 740,000 wintering Herring Gulls is clearly too high and obviously incorrect, casting doubt on the estimates of the other three species. Apart from the winter counts, there are no other data suggesting that the numbers of Black-headed, Common and Great Black-backed gulls have decreased dramatically. Similarly, concerns that Lesser Black-backed Gulls are at risk because they are restricted to a few large colonies in the breeding season are unjustified while their numerical abundance both in Britain and in western Europe is ignored.

While BoCC4 places most of the abundant gulls in Britain on either the Red or Amber lists and therefore considers them of conservation concern, it makes little attempt to identify the causes of concern. What could or should be done about the Glaucous and Iceland gulls that visit Britain and Ireland in small numbers simply because our region is at the southern edge of their winter distribution? The Mediterranean Gull is increasing rapidly as a breeding bird in Britain, while it appears that the Yellow-legged Gull is following a few years behind, but why should it be a concern if these species fail to become established as breeding species here? If they do become established in numbers, as they are elsewhere in Europe, they may become competitors of Herring and Black-headed gulls, outcompeting these species to their detriment.

What could be done to change the past and perhaps current decline in gull species that are claimed to be of conservation concern? Opening new and more landfill sites is impossible, and increasing the discard of unwanted fish by the commercial fishing industry will not be encouraged, particularly since it has been regarded as poor practice. It is likely that Herring Gull numbers in Britain will reach a new, lower level, but the species' total loss ('extinction') as a breeding bird in Britain is a very faint possibility.

That almost all gulls breeding in Britain presently are of conservation concern does not seem to be a realistic evaluation. The situation today is vastly better than it was a century ago. A case could readily be made that most gull species are doing well and that concern could be restricted to just one or two species. More than half of the gull species that are claimed to be of concern do not justify this rating.

Identifying realistic conservation needs is not encouraged by the current system of using multiple criteria, some of which are of unproven value, for identifying concerns for gulls and other species. There is a danger that species of concern are being claimed more for political than for scientific aspects of conservation and protection; too little attention is being given to the limitations of the numerical data used and how, in some cases, these data have been poorly or prematurely interpreted. If readers question these conclusions, then they must try to justify why the Caspian, Iceland and Glaucous gulls are Amber-listed in Britain, yet the status of the Ring-billed Gull and other gull species, which breed in North America and frequently visit Britain and Ireland, have not even been considered.

Appendices

APPENDIX 1: THE TAXONOMIC POSITION OF HERRING GULL-LIKE SPECIES

The taxonomy of the Herring Gull group has changed appreciably in recent years, but several aspects are still unclear and need further investigating. Fifty years ago, the species then called the Herring Gull (*Larus argentatus*) was a bird with a large circumpolar distribution in the northern hemisphere, breeding across Europe, North America and Asia, but there have since been many changes in its taxonomic treatment. It is now split into at least five species, restricting the breeding distribution of the Herring Gull in its current sense to north-west Europe and Iceland, and including about 2 million adults.

Herring Gull-like birds breeding in south-east Europe are now called the Yellow-legged Gull (*Larus michahellis*), which is abundant in western France, Iberia, Morocco and much of the Mediterranean, and is spreading northwards. The Caspian Gull (*L. cachinnans*) takes over in the Black and Caspian sea areas, and is spreading northwards as far as Poland. Both species have yellow legs, in contrast with the flesh-grey legs of Herring Gulls breeding in Britain and Ireland.

Some authors believe that the Herring Gull-type birds breeding in the eastern Baltic and parts of north-west Russia, which include individuals with both yellow and flesh-coloured legs, should be named as a distinct species, *Larus omissus*, but currently they are regarded as only a subspecies of *L. argentatus* and some even question whether they even deserve this status. A Russian study suggests that *omissus*-like gulls are most closely related to the Yellow-legged Gull and spread into the eastern Baltic from the Volga basin via inland areas of Russia.

The remaining three species currently forming part of the Herring Gull complex are the Vega Gull (*Larus vegae*) in north-east Siberia, the Armenian Gull (*L. armenicus*) in the Caucasus and Middle East, and the American Herring Gull

(*L. smithsonianus*), which in Europe is now regarded as a distinct species but, confusingly, is still considered a subspecies by the American Ornithological Union. In addition, the subspecies of the Yellow-legged Gull that breeds in Madeira and the Canary Islands is believed by some authors to be a separate species, the Atlantic Gull (*L. atlantis*). Other forms of the Herring Gull, named as *ponticus*, *barabensis* and *mongolicus*, have yet to be studied in detail.

Closer to Britain, those Herring Gulls breeding in Scandinavia, east Germany, Poland and part of north-west Russia are regarded as belonging to the first described (nominate) specimen and as a result they are called *Larus argentatus argentatus*, while those breeding in Britain, Ireland, Iceland, the Faeroes, north-west France, the Netherlands, Belgium and western Germany have paler grey on the wings and mantle, and are named as a different subspecies, *L. a. argenteus*. The zone in which the two subspecies are believed to separate or overlap is uncertain. Mitchell *et al.* (2004) lists both subspecies as occurring together in the breeding season in Belgium, the Netherlands, western Germany and France, while only the subspecies *argentatus* is considered to breed in Denmark, east Germany, Norway, Sweden, Finland, Poland and north-west Russia. However, it is doubtful if gulls appreciate the national boundaries of European countries! It is clear that a new and critical review of the taxonomic status of Herring Gulls throughout northern Europe and Asia is sorely needed.

The separation of the Herring Gull into subspecies within Europe was originally based on a relatively small number of museum specimens collected from a very limited number of localities and before the concept of clines (gradual changes in characteristics over the geographical range) became established. Herring Gulls breeding in the same colony show variation in the shade of grey on the wings and particularly in the pattern of black in the wing-tips, and because of this, the geographical separation, degree of distinction and the exact distributions of the subspecies *argentatus* and *argenteus* within Europe are still uncertain. Even within Britain, adult Herring Gulls on the east side are statistically more likely to have six black-tipped primaries than those on the west side, suggesting that there may be a degree of genetic difference even across Britain.

The Herring Gull was originally described to science by the Danish author Erik Pontoppidan in 1763, and it is presumed that the specimen described was a bird obtained on the Danish island of Christiansø, which may well be within the boundary area where *argentatus* and *argenteus* currently meet and mix. It is not clear which subspecies his original specimen relates to, and in any case it is often impossible to identify with confidence the subspecies to which a single specimen belongs. So, the names given to these two subspecies could be incorrect and may require reversing. Was the original specimen described by Pontoppidan actually a specimen belonging to the subspecies currently called *argenteus*? Resolving the problem using modern genetic methods will be interesting, even if primarily an academic exercise. It may eventually be proven that only a single subspecies of *Larus argentatus* breeds within the countries bordering the North Sea – including southern Norway, Demark, Belgium, northern France, Britain and Ireland – and that within that area individuals show only minor variations in the form of a cline.

If another group of Herring Gulls in Europe justifies species status, it may prove to be the large and darker birds that breed in northern Norway and the Murmansk region of north-east Russia, many of which winter in Britain. These are currently regarded as the same subspecies as the type specimen and those breeding in Denmark, yet many of these northern Herring Gulls differ from more southern breeding birds in several characters detectable in the field, including greater size, wing-tip pattern and migration pattern, from those breeding further south in southern Scandinavia and in the North Sea countries.

That Herring Gull-like birds with yellow legs found in southern Europe belong to a different species is sound, but Herring Gulls with yellow and flesh-coloured legs breed side by side and even together in parts of the Baltic, making this argument a little less convincing. Why do only some Herring Gull-like species have yellow legs but all Lesser Black-backed Gulls (*Larus fuscus*) do? How important is the colour of the eye-ring, which some, but not all, authors believe is an important taxonomic character? Presumably these differences have a genetic basis. Is leg or eye-ring colour biologically important, and if so why? There are many other intriguing questions about the taxonomy of Herring Gulls that remain to be answered. For example, what is the importance of the variations in shades of grey on the wings and mantle, which not only occur between different geographical areas but are also evident in a single colony? And why does the black wing-tip pattern vary, both with age, between individuals and between subspecies and species? The reason why Herring Gulls vary in size geographically is partially answered by Bergmann's rule, which applies to many birds and mammals and states that larger individuals occur towards the poles. This effect is explained as being a response to combating loss of body heat in colder regions. While the rule could explain the size differences of Herring Gulls through north-west Europe, there are several exceptions to this within the Herring Gull complex that require another explanation.

Even with the advantage of DNA technologies, separating species from subspecies that occupy different geographical areas can still involve arbitrary decisions on the part of the investigators. The variation and taxonomy of the Herring Gull complex are puzzling and remain a major academic challenge to the concepts of species, subspecies and varieties.

Finally, there are further problems with the large, dark-winged gulls known as *Larus argentatus heuglini* (breeding in Russia's Kola Peninsula) and *L. a. taimyrensis* (breeding further east in northern Siberia), which are variously called the Siberian Gull, Heuglin's Gull or Tundra Gull. For long, these were regarded as subspecies of the Herring Gull, but currently most authors consider them to be subspecies of the Lesser Black-backed Gull, a species that they closely resemble. Others have suggested that Heuglin's Gull is a separate species, *L. heuglini*, and that *taimyrensis* is a subspecies or even a hybrid of this. The taxonomic situation remains truly confused.

The fluid state of the current taxonomy of what used to be the Herring Gull will continue to develop in the next few years, no doubt producing further disagreements along the way about what is a 'good' species.

APPENDIX 2: A LIFE TABLE FOR THE HERRING GULL

Columns five and six and also seven and eight consider the effect of two different levels (0.8 and 0.9) of numbers of young fledged per pair.

Year	*Alive*	*Breeding birds*	*Immature birds*	*Young fledged/ pair*	*Young fledged by age class*	*Young fledged/ pair*	*Young fledged by age class*
1	1,000	0	1,000	0	0	0	0
2	500	0	500	0	0	0	0
3	435	0	435	0	0	0	0
4	378	0	378	0	0	0	0
5	329	260	69	0.5	65	0.6	78
6	286	276	10	0.7	97	0.8	110
7	249	249		0.8	100	0.9	112
8	217	217		0.8	87	0.9	98
9	189	189		0.8	76	0.9	85
10	164	164		0.8	66	0.9	74
11	143	143		0.8	57	0.9	64
12	124	124		0.8	50	0.9	56
13	108	108		0.8	43	0.9	49
14	94	94		0.8	38	0.9	42
15	82	82		0.8	33	0.9	37
16	71	71		0.8	28	0.9	32

Year	*Alive*	*Breeding birds*	*Immature birds*	*Young fledged/ pair*	*Young fledged by age class*	*Young fledged/ pair*	*Young fledged by age class*
17	62	62		0.8	25	0.9	28
18	54	54		0.8	22	0.9	24
19	47	47		0.8	19	0.9	21
20	41	41		0.8	16	0.9	18
21	36	36		0.8	14	0.9	16
22	31	31		0.8	12	0.9	14
23	27	27		0.8	11	0.9	12
24	23	23		0.8	9	0.9	10
25	20	20		0.8	8	0.9	9
26	18	18		0.8	7	0.9	8
27	15	15		0.8	6	0.9	7
28	13	13		0.8	5	0.9	6
29	9	9		0.8	4	0.9	4
30	6	6		0.8	2	0.9	3
31	4	4		0.8	2	0.9	2
32	3	3		0.8	1	0.9	1
33	2	2		0.8	1	0.9	1
34	1	1		0.8	1	0.9	1
35	1	1		0.8	0	0.9	0
Total	**4,782**	**2,390**	**2,392**		**905**		**1,022**

- At the end of the breeding season, 2,392 immatures (column 4 total) of a total of 4,782 individuals (column 2 total) is equivalent to 50 per cent.
- At the start of the following breeding season before the year's 1,000 new individuals are produced, there is a ratio of about 1,392 immatures for every 3,782 individuals alive, equivalent to 37 per cent.
- The assumed adult survival rate is 0.87.
- The assumed first-year survival rate is 0.50.
- Productivity is assumed to be lower for the first two breeding years. With a productivity of 0.80 chicks fledged per pair, the total number of young fledged per pair is 905, and indicates a decline from the 1,000 at the start of the life table. However, a productivity of 0.90 chicks produces 1,022 young fledged, which indicates that the population would slowly increase. This suggests that the production of fledged young necessary to maintain the population size lies between 0.8 and 0.9 with an adult survival rate of 0.87. While this value varies with the adult survival rate, the model by Cooke & Robinson (2010) indicating that the equilibrium value of productivity lies between 1.3 and 1.5 per pair would seem to be excessively high.

APPENDIX 3: BLACK-LEGGED KITTIWAKE BREEDING NUMBERS IN THE FIRTH OF FORTH

The number of pairs of Kittiwakes (*Rissa tridactyla*) breeding in 2000 and in each year over the period 2012–15 at eight colonies in the Firth of Forth are provided in the table below. The percentage changes between 2012–13, 2013–14 and 2013–15 are also shown. The only colony in the Firth of Forth not included is that in Dunbar harbour.

Colony	*Year*					*Percentage change*	*Percentage change*	*Percentage change*
	2000	*2012*	*2013*	*2014*	*2015*	*2012–13*	*2013–14*	*2013–15*
Isle of May	4,618	2,465	1,712	2,464	3,433	−30.5%	+43.9%	+101%
Craigleith	539	620	293	292	537	−52.7%	−0.3%	+83.3%
Fidra	343	191	128	167	275	−33.0%	+30.5%	+115%
The Lamb	132	95	47	84	99	−50.5%	+78.7%	+111%
Inchkeith	365	325	300	273	260	−7.7%	−9.0%	−13.3%
Inchcolm	111	106	82	65	68	−22.6%	−20.7%	−17.1%
St Abb's Head	11,077	4,314	3,403	3,625	4,209	−21.1%	+6.5%	+23.7%
Bass Rock	670*	395	270	324	441	−31.6%	+20.0%	+63.3%
Total	**17,855**	**8,511**	**6,235**	**7,294**	**9,322**	**−26.7%**	**+17.0%**	**+49.5%**

* Count made in 2001.

APPENDIX 4: CHANGES IN THE STATUS OF BLACK-LEGGED KITTIWAKES

The table below lists the change in status of Kittiwake (*Rissa tridactyla*) colonies at 25 Special Protection Areas (SPAs) as reported by the Joint Nature Conservation Committee, all of which show a decrease since 2000, and at 19 Kittiwake colonies on the east coast of England and Scotland, of which 13 showed an increase over similar periods.

Region	*Name*	*2000 count (pairs)*	*Next count (pairs)*	*Interval (years)*	*Overall percentage change*	*Annual percentage change*
SPAs						
Shetland	Hermaness	643	304	9	–53%	–7.2%
Shetland	Noss	2,395	507	10	–79%	–14.4%
Shetland	Foula	1,934	361	14	–81%	–11.3%
Shetland	Sumburgh Head	877	362	14	–59%	–6.6%
Shetland	Fair Isle	8,204	963	14	–88%	–15.2%
Orkney	West Westray	33,281	12,055	7	–64%	–11.9%
Orkney	Copinsay	4,256	666	12	–84%	–13.3%
Orkney	Marwick Head	5,573	526	13	–91%	–15.5%
Orkney	Hoy	795	397	7	–50%	–8.3%
East coast of Scotland	Troup, Pennan, Lion Head	18,482	14,896	7	–19%	–3.5%
East coast of Scotland	Buchan Ness–Collieston	14,091	12,542	7	–11%	–1.9%
East coast of Scotland	Fowlsheugh	18,800	9,337	12	–50%	–5.2%
East coast of Scotland	Firth of Forth islands	6,632	3,339	14	–50%	–4.8%
East coast of Scotland	St Abb's Head	11,077	3,652	14	–67%	–7.6%
West coast of Scotland	North Rona, Sula Sgeir	4,119	1,253	12	–70%	–8.1%
West coast of Scotland	Handa	7,013	2,715	13	–61%	–6.6%
West coast of Scotland	St Kilda	4,268	957	8	–78%	–15.4%
West coast of Scotland	Shiant Isles	2,006	546	8	–73%	–13.4%
West coast of Scotland	Canna, Sanday	1,274	935	14	–27%	–2.0%
West coast of Scotland	Mingulay, Berneray	5,511	2,878	14	–48%	–4.0%
West coast of Scotland	Ailsa Craig	1,675	228	14	–86%	–14.2%
East coast of England	Farne Islands	5,096	4,175	14	–18%	–1.3%

continued

Appendix 4: Changes in the status of Black-legged Kittiwakes *continued*

Region	*Name*	*2000 count (pairs)*	*Next count (pairs)*	*Interval (years)*	*Overall percentage change*	*Annual percentage change*
East coast of England	Flamborough–Bempton	42,692	37,617	8	–12%	–1.6%
Wales	Skomer	2,257	1,488	14	–34%	–2.9%
Ireland	Rathlin Island	9,917	7,922	11	–20%	–1.9%
Colonies						
East coast of Scotland	North Sutor	447	275	15	–38.5%	–3.2%
East coast of Scotland	Portknockie	104	233	13	+124%	+6.5%
East coast of Scotland	Covesea	384	425	13	+10.7%	+0.7%
East coast of Scotland	Sands of Forvie	601	495	15	–17.6%	–1.3%
East coast of England	Boulby	3,080	1,970	14	–36%	–3.0%
East coast of England	Hartlepool	76	97	10	+28%	+2.5%
East coast of England	Saltburn	3,050	1,620	14	–47%	–4.5%
East coast of England	Howick	530	620	14	+17%	+1.0%
East coast of England	Marsden	2,140	2,568	8	+20%	+2.2%
East coast of England	Tynemouth	92	234	15	+154%	+6.5%
East coast of England	Newcastle–Gateshead	390	755	15	+94%	+4.5%
East coast of England	International Coatings	70	133	16	+90%	+4.0%
East coast of England	Dunstanburgh	499	335	10	–33%	–3.5%
East coast of England	Seaham Harbour	30	70	13	+133%	+6.9%
East coast of England	Coquet Island	80	316	15	+295%	+9.4%
East coast of England	Filey	5,120	6,546	14	+28%	+1.8%
East coast of England	Scarborough	2,004	1,990	15	–0.7%	–0.3%
East coast of England	Sizewell	180	502	14	+179%	+7.5%
East coast of England	Lowestoft	150	196	14	+31%	+1.9%

Select Bibliography and Further Reading

FIELD GUIDES

Svensson, L., Mullarney, K. & Zetterström, D. (2010). *Collins Bird Guide.* 2nd edn. HarperCollins, London.

Grant, P. J. (1986). *Gulls: A Guide to Identification.* 2nd edn. T. & A. D. Poyser, London.

Olsen, K. M. & Larssen, H. (2004). *Gulls of Europe, Asia and North America.* Helm Identification Guides. A & C Black. (Avoid copies without 'Reprinted with corrections 2004' on reverse of title page.)

OTHER REFERENCES

Adriaens, P. & Mactavish, B. (2004). Identification of adult American Herring Gull. *Dutch Birding,* **26**, 151–179.

Aitken, D., Babcock, M., Clarkson, K. & Jeavons, R. (2014). *Flamborough Head and Bempton Cliffs SPA Seabird Monitoring Programme. 2014 Report.* RSPB Bempton Cliffs, Bempton.

Allainé, D. & Lebreton, J.-D. (1990). The influence of age and sex on wing-tip pattern in adult Black-headed Gulls *Larus ridibundus. Ibis,* **132**, 560–567.

Allen, C. (2009). Mediterranean Gulls in Hampshire. *British Birds,* **102**, 635.

Allen, D. & Tickner, M. (1996). Mediterranean Gull: a new breeding bird for Ireland. *Irish Birds,* **5**, 435–436.

Anderson, N. (1990). Investigations into the causes of the decline of the Black-headed Gull (*Larus ridibundus*) colony at Ravenglass, Cumbria. MSc thesis, Durham University.

Baker, R. R. (1980). The significance of the Lesser Black-backed Gull to models of bird migration. *Bird Study,* **27**, 41–50.

Balmer, D. E., Gillings, S., Caffrey, B. J., Swann, R. L., Downie, I. S. & Fuller, R. J. (2013). *Bird Atlas 2007–11: The*

Breeding and Wintering Birds of Britain and Ireland. BTO Books, Thetford.

Banks, A. N., Burton, N. H. K., Calladine, J. R. & Austin, G. E. (2009). Indexing winter gull numbers in Great Britain using data from the 1953 to 2004 Winter Gull Roost Surveys. *Bird Study*, **56**, 103–119.

Bannerman, D. (1962). *The Birds of the British Isles*. Oliver & Boyd, Edinburgh.

Barnes, J. A. G. (1961). The winter status of the Lesser Black-backed Gull, 1959–60. *Bird Study*, **8**, 127–147.

Barth, E. K. (1955). Egg-laying, incubation and hatching of the Common Gull (*Larus canus*). *Ibis*, **97**, 222–239.

Barth, E. K. (1975). Taxonomy of *Larus argentatus* and *Larus fuscus* in north-western Europe. *Ornis Scandinavica*, **6**, 49–63.

Belant, J. L. (1993). Nest-site selection and reproductive biology of roof- and island-nesting Herring Gulls. *Transactions of the North American Wildlife and Natural Resources Conference*, **58**, 78–86.

Belant, J. L. (1997). Gulls in urban environments: landscape-level management to reduce the conflict. *Landscape and Urban Planning*, **38**, 245–258.

Bertolero, A., Martínez-Abraín, A., Molina, B., Oro, D., Tavecchia, G., Mouriño, J. & Genovart, M. (2008). *Gaviotas cabecinegra, picofina, de Audouin y tridáctila, y gavión atlántico en España. Población en 2007 y método de censo* [Mediterranean Gull, Slender-billed Gull, Audouin's Gull, Great Black-backed Gull, Kittiwake in Spain populations in 2007 and method of census]. SEO/BirdLife, Madrid. Accessible online at www.seo.org/media/docs/22gaviesc07.pdf.

Bogdanova, M. I., Daunt F., Newell, M., Phillips, R. A., Harris, M. P. & Wanless, S. (2011). Seasonal interactions in the Black-legged Kittiwake, *Rissa tridactyla*: links between breeding performance and winter distribution. *Proceedings of the Royal Society B*, **278**, 2412–2418. doi: 10.1098/rspb.2010.2601.

Bosman, D. S., Vercruijsse, H. J. P, Stienen, E. W. M., Vincx, M. & Lens, L. (2013). Age of first breeding interacts with pre- and post-recruitment experience in shaping breeding phenology in a long-lived gull. *PLoS One*, **8(12)**, e82093. doi: 10.1371/journal.pone.0082093.

Bourne, W. R. P. (1996). The past status of gulls and terns in Britain. *Sula*, **10**, 156–159.

Bourne, W. R. P. & Patterson, I. J. (1962). The spring departure of the Common Gull from Scotland. *Scottish Birds*, **2**, 1–15.

Brandl, R. & Nelsen, I. (1988). Feeding frequency of Black-headed Gull chicks. *Bird Study*, **35**, 137–141.

Bregnballe, T. *et al.* (2015). Danish national data base.

British Ornithologists' Union (2018). The Simple British List Based on a Checklist of Birds of Britain. 9th ed. *Ibis*, **160**, 190–240.

Brown, A. & Grice, P. (2005). *Birds in England*. T. & A. D. Poyser, London.

Brown, R. G. B. (1967). Breeding success and population growth in a colony of Herring Gulls and Lesser Black-backed

Gulls *Larus argentatus* and *L. fuscus*. *Ibis*, **109**, 502–515.

Buckley, N. J. (1990). Diet and feeding ecology of Great Black-backed Gulls (*Larus marinus*) at a southern Irish breeding colony. *Journal of Zoology*, **222**, 363–373.

Bull, J., Wanless, S. & Harris, M. P. (2004). Local-scale variability in the diet of Black-legged Kittiwakes *Rissa tridactyla*. *Ardea*, **92**, 43–52.

Burton, N. H. K., Banks, A. N., Calladine, J. R. & Austin, G. E. (2013). The importance of the United Kingdom for wintering gulls: population estimates and conservation requirements. *Bird Study*, **60**, 87–101.

Burtt, E. H. (1975). Cliff-facing interaction between parent and chick kittiwakes *Rissa tridactyla* in Newfoundland. *Ibis*, **117**, 241–242.

Butterfield, J., Coulson, J. C., Kearsey, S. V., Monaghan, P., McCoy, J. H. & Spain, G. E. (1983).The Herring Gull *Larus argentatus* as a carrier of salmonella. *Journal of Hygiene (London)*, **91**, 429–436.

Cabot, D. & Nisbet, I. (2013). *Terns*. Collins, London.

Cadiou, B. (2014). *Cinquième Recensement national des Oiseaux marins nicheurs en France métropolitaine: Bilan final 2009–2012*. Agence des aires marines protégées, Brest.

Cadiou, B., Pons J.-M. & Yesou, P. (eds) (2004). *Oiseaux marins nicheurs de France métropolitaine (1960–2000)*. Éditions Biotope, Mèze.

Camphuysen, C. J. (1995). Herring Gull *Larus argentatus* and Lesser Black-backed Gull *L. fuscus* feeding at fishing vessels in the breeding season: competitive scavenging versus efficient flying. *Ardea*, **83**, 365–380.

Camphuysen, C. J. (2013). A historical ecology of two closely related gull species (Laridae); multiple adaptations to a man-made environment. University of Groningen. The Netherlands.

Carboneras, C. I. & Dies, J. I. (2016). A new breeding population of Mediterranean Gull *Larus melanocephalus* in the species' main wintering area maintains independent spatial dynamics. *Ibis*, **158**, 190–194.

Chu, P. C. (1998). A phylogeny of the gulls (Aves: Larinae) inferred from osteological and integumentary characters. *Cladistics*, **14**, 1–43.

Clark, J. M. & Eyre, J. A (eds) (1993). *Birds of Hampshire*. Hampshire Ornithological Society, Hartley Wintney.

Colhoun, K. & Cummins, S. (2013). Birds of conservation concern in Ireland 2014–2019. *Irish Birds*, **9**, 523–544.

Cook, A. S. C. P. & Robinson, R. A. (2010). *How Representative is the Current Monitoring of Breeding Seabirds in the UK?* BTO Research Report **573**. British Trust for Ornithology, Thetford.

Coulson, J. C. (1963). Improved coloured-rings. *Bird Study*, **10**, 109–111.

Coulson, J. C. (1968). Differences in the quality of birds nesting in the centre and on the edges of a colony. *Nature*, **217**, 478–479.

Coulson, J. C. (1976). An evaluation of the reliability of rings used on Herring and Lesser Black-backed gulls. *Bird Study*, **23**, 21–26.

Coulson, J. C. (1983). The changing status of the Kittiwake *Rissa tridactyla* in the British Isles, 1969–1979. *Bird Study*, **30**, 9–16.

Coulson, J. C. (2001). Colonial breeding. In *Biology of Marine Birds* (eds E. A. Schreiber & J. Burger). CRC Press, Boca Raton, FL.

Coulson, J. C. (2009). Sexing Black-legged Kittiwakes by measurement. *Ringing and Migration*, **24**, 233–239.

Coulson, J. C. (2011). *The Kittiwake*. T. & A. D. Poyser, London.

Coulson, J. C. (2017). Productivity of the Black-legged Kittiwake *Rissa tridactyla* required to maintain numbers. *Bird Study*, **64**, 84–89.

Coulson, J. C. & Coulson , B. A. (2008). Lesser Black-backed Gulls *Larus fuscus* nesting in an inland urban colony: the importance of earthworms (Lumbricidae) in their diet. *Bird Study*, **55**, 297–303.

Coulson, J. C. & Coulson, B. A. (2009). Ecology and colonial structure of large gulls in an urban colony: investigations and management at Dumfries, south-west Scotland. *Waterbirds*, **32**, 1–15.

Coulson, J. C. & Coulson, B. A. (2015). The accuracy of urban nesting gull censuses. *Bird Study*, **62**, 170–176.

Coulson, J. C. & Dixon, F. (1979). Colonial breeding in seabirds. In *Biology and Systematics of Colonial Organisms* (eds G. Larwood & B. R. Rosen). Academic Press, London.

Coulson, J. C. & Strowger, J. (1999). The annual mortality rate of Black-legged Kittiwakes in NE England from 1954 to 1998 and a recent exceptionally high mortality. *Waterbirds*, **22**, 3–13.

Coulson, J. C., Duncan, N. Thomas, C. S. & Monaghan, P. (1981). Age-related difference in the bill depth of Herring Gulls *Larus argentatus*. *Ibis*, **123**, 499–502.

Coulson, J. C., Monaghan, P., Butterfield, J. E. L., Duncan, N., Thomas, C. S. & Wright, H. (1982). Variation in the wingtip pattern of the Herring Gull in Britain. *Bird Study*, **29**, 111–120.

Coulson, J. C., Thomas, C. S., Butterfield, J. E. L., Duncan, N., Monaghan, P. & Shedden, C. (1983a). The use of head and bill length to sex live gulls (Laridae). *Ibis*, **125**, 549–557.

Coulson, J. C., Monaghan, P., Butterfield J. E. L., Duncan, N., Shedden, C. & Thomas, C. S. (1983b). Seasonal changes in the Herring Gull in Britain: weight, moult and mortality. *Ardea*, **7**, 235–244.

Coulson, J. C., Butterfield, J. E. L., Duncan, N., Kearsey, S., Monaghan, P. & Thomas, C. S. (1984a). Origin and behaviour of Great Black-backed Gulls wintering in northeast England. *British Birds*, **77**, 1–11.

Coulson, J. C., Monaghan, P., Butterfield, J. E. L. *et al.* (1984b). Scandinavian Herring Gulls wintering in Britain. *Ornis Scandinavica*, **15**, 79–88.

Craik, J. C. A. (1995). Effects of North American Mink *Mustela vison* on the breeding success of terns and smaller gulls in west Scotland. *Seabirds*, **17**, 3–11.

Craik, J. C. A. (1997). Long-term effects of North American Mink *Mustela vison* on seabirds in western Scotland. *Bird Study*, **44**, 303–309.

Craik, J. C. A. (1999). Breeding success of Common Gulls *Larus canus* in west Scotland. I. Observations at a single colony. *Atlantic Seabirds,* **1**, 169–181.

Cramp, S. (1971). Gulls nesting on buildings in Britain and Ireland. *British Birds,* **64**, 476–487.

Cramp, S. & Simmons, K. E. L. (1983). *Handbook of the Birds of Europe, the Middle East and Africa. The Birds of the Western Palearctic. Vol. III: Waders to Gulls.* Oxford University Press, Oxford.

Cramp, S., Bourne, W. R. P. & Saunders, D. (1974). *The Seabirds of Britain and Ireland.* Collins, London.

Cullen, E. (1957). Adaptations in the Kittiwake to cliff nesting. *Ibis,* **99**, 275–302.

Darling, F. F. (1938). *Bird Flocks and the Breeding Cycle: A Contribution to the Study of Avian Sociali.* Cambridge University Press, Cambridge.

Davis, A. & Vinicombe, K. (2008). The 'American Herring Gull' at Chew. In *CVL Birding* [website]. www.cvlbirding.co.uk, accessed 23 August 2018.

Dean, A. R. (n.d.). *Gulls in the West Midlands Region* [website]. http://deanar.org.uk/wmgulls/Kw/kwake.htm, accessed 5 August 2018.

del Hoyo, J., Elliott, A. and Sargatal, J. (1966) *Birds of the World. Vol.* 3. Lynx Edicions, Barcelona.

Drost, R., Focke, E. & Freytag, G. (1961). Entwicklung und Aufbau einer population der Silbermowe *Larus argentatus argentatus. Journal of Ornithology,* **102**, 404–429.

Dubiec, A., Zielinski, P., Zielinska, M. & Iciek, T. (2015). Morphometric sex identification in the Mediterranean Gull (*Ichthyaetus melanocephalus*). *Waterbirds,* **38**, 229–237.

Duncan, N. (1981). The Abbeystead and Mallowdale gull colony before control. *Bird Study,* **28**, 133–138.

Dwyer, C. P., Belant, J. L. & Dolbeer, R. A. (1996). Distribution and abundance of roof-nesting gulls in the Great Lakes Region of the United States. *Ohio Journal of Science,* **96**, 9–12.

Eaton, M. A., Aebischer, N., Brown, A. *et al.* (2015). Birds of conservation concern 4. *British Birds,* **108**, 708–746.

Edgeller, M. L. (1996). First-year Mediterranean Gulls with all-dark wings. *Dutch Birding,* **18**, 241–242.

Elkins, N. & Yesou, P. (1998). Sabine's Gulls in western France and southern Britain. *British Birds,* **91**, 386–397.

Erwin, R. M. (1971). The breeding success of two sympatric gulls, the Herring Gull and the Great Black-backed Gull. *Wilson Bulletin,* **83**, 152–158.

Ewins, P. J. & Weseloh, D. V. (1999). Little Gull (*Hydrocoloeus minutus*), version 2.0. In *The Birds of North America* (eds A. F. Poole & F. B. Gill). Cornell Lab of Ornithology, Ithaca, NY. doi.org/10.2173/bna.428.

Fisher, J. & Lockley, R. M. (1954). *Seabirds. Collins New Naturalist* **28**. Collins, London.

Fitzgerald, G. R. & Coulson, J. C. (1973). The distribution and feeding ecology of gulls on the tidal reaches of the Rivers Tyne and Wear. *Vasculum,* **58**, 29–47

Flegg, J. J. M. & Cox, C. J. (1975). Mortality in the Black-headed Gull. *British Birds,* **68**, 437–449.

Forrester, R. W., Andrews, I .J., McInerny, C. J. *et al.* (2007). *The Birds of Scotland.* Scottish Ornithologists' Club, Aberlady.

Frederiksen, M., Moe, B., Daunt, F. *et al.* (2012). Multi-colony tracking reveals the winter distribution of a pelagic seabird on an ocean basin scale. *Diversity and Distribution*, **18**, 530–542.

Furness, R. W. (2015). *Non-breeding Season Populations of Seabirds in UK Waters: Population sizes for Biologically Defined Minimum Population Scales (BDMPS).* Natural England Commissioned Reports 164. MacArthur Green Ltd, Glasgow.

Furness, R. W., Ensor , K. & Hudson, A. V. (1992). The use of fisheries waste by gull populations around the British Isles. *Ardea*, **80**, 105–113.

Garthe, S. & Hüppop, O. (1996). Nocturnal scavenging by gulls in the southern North Sea. *Colonial Waterbirds*, **19**, 232–241.

Gaston A. J. & Decker, R. (1985). Interbreeding of Thayer's Gull *Larus thayeri* and Kumlien's Gull *Larus glaucoides kumlieni* on Southampton Island, Northwest Territories. *Canadian Field Naturalist*, **99**, 257–259.

Girdwood, R. W., Fricker, C. R., Munro, D., Shedden, C. B. & Monaghan, P. (1985). The incidence and significance of salmonella carriage by gulls (*Larus* spp.) in Scotland. *Journal of Hygiene (London)*, **95**, 229–241.

Goutner, V. (1994). The diet of Mediterranean Gull *Larus melanocephalus* chicks at fledging. *Journal für Ornithologie*, **135**, 193–201.

Grant, P. J. (1978). Field identification of west Palearctic gulls. *British Birds*, **71**, 145–176.

Greig, S. A., Coulson J. C. & Monaghan, P. (1983). Age-related differences in foraging success in the Herring Gull (*Larus argentatus*). *Animal Behaviour*, **31**, 1237–1243.

Gribble, F. C. (1962). Census of Black-headed Gull colonies in England and Wales, 1958. *Bird Study*, **9**, 56–71.

Gribble, F. C. (1976). A census of Black-headed Gull colonies. *Bird Study*, **23**, 135–145.

Gurney, J. (1919). Breeding stations of the Black-headed Gull in the British Isles. *Transactions of the Norfolk and Norwich Naturalists' Society*, **10**, 416–447.

Gyimesi, A., Boudewijn, T. J., Buijs, R.-J. *et al.* (2016). Lesser Black-backed Gulls *Larus fuscus* thriving on a non-marine diet. *Bird Study*, **63**, 241–249.

Hamer, K. C., Monaghan, P., Uttley, J. D., Walton, D. & Burns, M. D. (1993). The influence of food supply on the breeding ecology of Kittiwakes *Rissa tridactyla* in Shetland. *Ibis*, **135**, 255–263.

Hamilton F. D. (1962). Census of Black-headed Gull colonies in Scotland, 1958. *Bird Study*, **9**, 72–80.

Harris, M. P. (1964). Aspects of the breeding biology of the gulls *Larus argentatus, L. fuscus* and *L. marinus. Ibis* **106**, 432–456.

Harris, M. P. (1970). Abnormal migration and hybridization of *Larus argentatus* and *L. fuscus* after inter-species fostering experiments. *Ibis*, **112**, 488–489.

Harris, M. P. (1980). Loss of weight and legibility of bird rings. *Ringing and Migration*, **3**, 41–49.

Harris, M. P. & Wanless, S. (1997). Breeding success, diet and brood neglect in the Kittiwake (*Rissa tridactyla*) over an 11-year period. *ICES Journal of Marine Science*, **54**, 615–623.

Harris, M. P., Morley, C. & Green, G. H. (1978). Hybridization of Herring and Lesser Black-backed gulls in Britain. *Bird Study*, **25**, 161–166.

Hesp, L. S. & Barnard, C. J. (1989). Gulls and plovers: age-related differences in kleptoparasitism among Black-headed Gulls (*Larus ridibundus*). *Behavioral Ecology and Sociobiology*, **24**, 297–304.

Hobson, K. A. (1993). Trophic relationships among arctic seabirds. *Marine Ecology Progress Series*, **95**, 7–18.

Hodges, A. F. (1975). The orientation of adult Kittiwakes *Rissa tridactyla* at the nest site in Northumberland. *Ibis*, **117**, 235–240.

Hollam, P. A. D. (1940). Report on the 1938 survey of Black-headed Gull colonies. *British Birds*, **33**, 202–221, 230–244.

Hosey, G. R. & Goodridge, F. (1980). Establishment of territories in two species of gull on Walney Island, Cumbria. *Bird Study*, **27**, 73–80.

Hudson, A. V. & Furness, R. W. (1988). Utilization of discarded fish by scavenging seabirds behind whitefish trawlers in Shetland. *Journal of Zoology, London*, **215**, 151–166.

Hudson, A. V. & Furness, R. W. (1989). The behaviour of seabirds foraging at fishing boats around Shetland. *Ibis*, **131**, 225–237.

Hume, R. A. (1976). Inland records of Kittiwakes. *British Birds*, **69**, 62–63.

Hume, R. A. (2013). Mediterranean Gulls at a Hampshire pig farm. *British Birds*, **106**, 743–746.

Hutchinson, C. D. & Neath, B. (1978). Little Gulls in Britain and Ireland. *British Birds*, **71**, 563–582.

Ingolfsson, A. (1969). Sexual dimorphism of large gulls (*Larus* spp.). *Auk*, **86**, 732–737.

Ingolfsson, A. (1970). Hybridization of Glaucous Gulls *Larus hyperboreus* and Herring Gulls *L. argentatus* in Iceland. *Ibis*, **112**, 340–362.

Joint Nature Conservation Committee (2018). Herring Gulls *Larus argentatus*. In JNCC [website]. *http://jncc.defra.gov.uk/page-2887*, accessed 27 September 2018.

Kadlec, J. A., & Drury, W. H. (1968). Structure of the New England gull population. *Ecology* **49**, 644–676.

Källander, H. (1977). Piracy by Black-headed Gulls on Lapwings. *Bird Study*, **24**, 186–194.

Kim, S.-Y. & Monaghan, P. (2006a). Sex of the first hatched chick influences survival of the brood in the Herring Gull (*Larus argentatus*). *Journal of Zoology*, **270**, 116–121.

Kim, S.-Y. & Monaghan, P. (2006b). Interspecific differences in foraging preferences, breeding performance and demography in Herring (*Larus argentatus*) and Lesser Black-backed gulls (*Larus fuscus*) at a mixed colony. *Journal of Zoology*, **270**, 664–671.

Kotzerka, J., Garthe, S., Scott, A. & Hatch, S. A. (2009). GPS tracking devices reveal foraging strategies of Black-legged Kittiwakes. *Journal of Ornithology*, **151**, 459–467.

Lewis, S., Wanless, S., Wright, P. J., Harris, M. P., Bull, J. & Elston, D. A. (2001). Diet and breeding performance of Black-legged Kittiwakes *Rissa tridactyla* at a North Sea colony. *Marine Ecology Progress Series*, **221**, 277–284.

Lewis, M., Lye, G., Pendlebury, C. & Walls, R. (2012). *Population sizes of seabirds breeding in Scottish SPAs.* Report to Marine Scotland. Natural Power Consultants, Castle Douglas.

Liebers, D., Helbig, A. J. & de Knijff, P. (2001). Genetic differentiation and phylogeography of gulls in the *Larus cachinnans–fuscus* group (Aves: Charadriiformes). *Molecular Ecology*, **10**, 2447–2462.

Liebers, D., de Knijff, P. & Helbig, A. J. (2004). The herring gull complex is not a ring species. *Proceedings B of the Royal Society*, **271**, 893–901.

Lloyd, C., Tasker, M. I. & Partridge, K. (1991). *The Status of Seabirds in Britain and Ireland.* T. & A. D. Poyser, London.

Lonergan, P. & Mullarn, K. (2004). Identification of American Herring Gull in a western European context. *Dutch Birding*, **26**, 1–35.

Lundberg, C. A. & Vaisainen, R. A. (1979). Selective correlation of egg-size with chick mortality in the Black-headed Gull (*Larus ridibundus*). *Condor*, **81**,146–156.

McCartan, L. (1958). The wreck of Kittiwakes in early 1957. *British Birds*, **5**, 253–266.

MacKinnon, G. E. (1986). Aspects of the ecology of the Black-headed Gull *Larus ridibundus* with comparative data on the Common Gull *L. canus.* Unpublished PhD thesis, University of Durham.

MacKinnon, G. E. & Coulson, J. C. (1987). The temporal and geographical distribution of continental Black-headed Gulls *Larus ridibundus* in the British Isles. *Bird Study*, **34**, 1–9.

MacPherson, H. A. & Duckworth, W. (1886). *The Birds of Cumberland.* Thurnam & Sons, Carlisle.

Marchant, S. (1952). The status of the Black-headed Gull colony at Ravenglass. *British Birds*, **45**, 22–27.

Megyesi, J. (1996). *Restoration of Avian Diversity.* Monomoy National Wildlife Refuge, US Fish and Wildlife Service, Chatham, MA.

Meininger, P. L. & Bekhuis, J. P. (1990). The Mediterranean Gull *Larus melanocephalus* as a breeding bird in the Netherlands and Europe. *Limosa*, **50**, 121–134.

Meininger, P. L. & Flamant, R. (1998). Breeding populations of Mediterranean Gull *Larus melanocephalus* in the Netherlands and Belgium. *Sula*, **12**, 129–138.

Meininger, P. L., Raevel, P. & Hoogendoorn, W. (1993). Occurrence of Mediterranean Gull at le Portel in north-western France. *Dutch Birding*, **15**, 45–54.

Mitchell, P. I., Newton, S. F., Ratcliffe, N. & Dunn, T. E. (eds) (2004). *Seabird Populations of Britain and Ireland: Results of the Seabird 2000 Census (1998–2002).* T. & A. D. Poyser, London.

Monaghan, P. (1977). The utilization of urban resources by the Herring Gull *Larus argentatus.* Unpublished PhD thesis, University of Durham.

Monaghan, P. (1979). Aspects of the breeding biology of Herring Gulls *Larus argentatus* in urban colonies. *Ibis*, **121**, 475–481.

Monaghan, P. & Coulson, J. C. (1977). The status of large gulls nesting on buildings. *Bird Study*, **24**, 89–104.

Montevecchi, W. A., Cairns, D. K., Burger, R. E., Elliot, R. D. & Wells, J. (1987). The status of the common Black-headed Gull in Newfoundland and Labrador. *American Birds*, **42**, 197–204.

Moynihan, M. (1959). A revision of the family *Laridae* (Aves). *American Museum Novitates*, **1928**, 1–42.

Muusse, M., Muusse, T., Buijs, R.-J., Altenburg, R., Gibbins, C. & Luijendijk, B.-J. (2011). Phenotypic characteristics and moult commencement in breeding Dutch Herring Gulls *Larus argentatus* and Lesser Black-backed Gulls *L. fuscus*. *Seabird*, **24**, 42–59.

Nager, R. G., Monaghan, P. & Houston, D. C. (2000).Within-clutch trade-offs between the number and quality of eggs: experimental manipulation in gulls. *Ecology*, **81**, 1339–1350.

Nager, R. G., Monaghan, P. & Houston, D. C. (2001). The cost of egg production: increased egg production reduces future fitness in gulls. *Journal of Avian Biology*, **32**, 159–166.

Paludan, K. (1951). Contribution to the biology of *Larus argentatus* and *Larus fuscus*. *Vedenskabelige Meddelelser Dansk Naturhistorik Forening*, **114**, 1–128.

Parsons, J. (1971a). Cannibalism in Herring Gulls. *British Birds*, **64**, 228–237.

Parsons, J. (1971b). The breeding biology of the Herring Gull *Larus argentatus*. Unpublished PhD thesis, University of Durham.

Parsons, J. (1972). Egg size, laying date and incubation period in the Herring Gull. *Ibis*, **114**, 536–541.

Parsons, J. (1975). Seasonal variation in the breeding success of the Herring Gull: an experimental approach to pre-fledging success. *Journal of Animal Ecology*, **44**, 553–573.

Patterson, I. J. (1965). Timing and spacing of broods in the Black-headed Gull *Larus ridibundus*. *Ibis*, **107**, 433–459.

Paynter, R. A. (1949). Clutch size and the egg and chick mortality of Kent Island Herring Gulls. *Ecology*, **30**, 146–166.

Pearson, T. H. (1968). The feeding biology of sea-bird species breeding on the Farne Islands, Northumberland. *Journal of Animal Ecology*, **37**, 521–552.

Pennington, M. G., Osborn, K., Harvey, P. V. *et al.* (2004). *The Birds of Shetland*. Christopher Helm, London.

Perlut, N. G., Bonter, D. N., Ellis, J. C. & Friar, M. S. (2016). Roof-top nesting in a declining population of Herring Gulls (*Larus argentatus*) in Portland, Maine, USA. *Waterbirds*, **39(sp 1)**, 68–73.

Péron, G., Lebreton, J.-D., & Crochet, P.-A. (2009). Costs and benefits of colony size vary during the breeding cycle in Black-headed Gulls *Chroicocephalus ridibundus*. *Journal of Ornithology*, **151**, 881–888.

Pittaway, R. (1999). Thayer's Gull. Taxonomic History of Thayer's Gull. *Ontario Birds*, **17**, 1–13.

Pons, J.-M. & Migot, P. (1995). Life-history strategy of the Herring Gull: changes in survival and fecundity in a population subjected to various feeding conditions. *Journal of Animal Ecology*, **64**, 592–599.

Pons, J.-M., Hassanin, A. & Crochet, P. A. (2005). Phylogenetic relationships within the Laridae (Charadriiformes: Aves) inferred from mitochondrial markers. *Molecular Phylogenetics and Evolution*, **37**, 686–699.

Porter, J. M. (1990). Patterns of recruitment to the breeding group in the Kittiwake *Rissa tridactyla*. *Animal Behaviour*, **40**, 350–360.

Prévot-Julliard, A.-C., Lebreton, J.-D. & Pradel, R. (1998). Re-evaluation of adult survival of Black-headed Gulls (*Larus ridibundus*) in presence of recapture heterogeneity. *Auk*, **115**, 85–95.

Rankin, M. N. & Duffey, E. A. G. (1948). A study of the bird life of the North Atlantic. *British Birds*, **41 (supplement)**, 1–42.

Rattiste, K. (2004). Reproductive success in pre-senescent Common Gulls (*Larus canus*): the importance of the last year of life. *Proceedings of the Royal Society of London B*, **271**, 2059–2064.

Rattiste, K. (2006). Life history of the Common Gull (*Larus canus*): a long-term individual-based study. PhD thesis, University of Uppsala.

Raven, S. J. & Coulson, J. C. (1997). The distribution and abundance of *Larus* gulls nesting on buildings in Britain and Ireland. *Bird Study*, **44**, 13–34.

Raven, S. J. & Coulson, J. C. (2001). Effects of cleaning a tidal river of sewage on gull numbers: a before and after study of the River Tyne, northeast England. *Bird Study*, **48**, 48–58.

Redfern, C. P. F. & Bevan, R. M. (2014). A comparison of foraging behaviour in the North Sea by Black-legged Kittiwakes *Rissa tridactyla* from an inland and a maritime colony. *Bird Study*, **61**, 17–28.

Reiertsen, T. K., Erikstad, K. E., Anker-Nilssen, T. *et al.* (2014). Prey density in non-breeding areas affects adult survival of Black-legged Kittiwakes *Rissa tridactyla*. *Marine Ecology Progress*, **509**, 289–302.

Robertson, G. S., Bolton, M., Grecian, W. J. & Monaghan, P. (2014). Inter- and intra-year variation in foraging areas of breeding Kittiwakes (*Rissa tridactyla*). *Marine Biology*, **161**, 1973–1986.

Rock, P. (2005). Urban gulls: problems and solutions. *British Birds*, **98**, 338–355.

Ross-Smith, V. H., Grantham, M. J., Robinson, R. A. & Clark, J. A. (2014a). *Analysis of Lesser Black-backed Gull Data to Inform Meta-population Studies*. BTO Research Report **654**. British Trust for Ornithology, Thetford.

Ross-Smith, V. H., Robinson, R. A., Banks, A. N., Frayling, T. D., Gibson, C. C. & Clark, J. A. (2014b). The Lesser Black-backed Gull *Larus fuscus* in England: how to resolve a conservation conundrum. *Seabirds*, **27**, 41–61.

Russell, J. & Montevecchi, W. A. (1996). Predation on adult Puffins *Fratercula arctica* by Great Black-backed Gulls *Larus marinus* at a Newfoundland colony. *Ibis*, **138**, 791–794.

Ryder, J. P. (1980). The influence of age on the breeding biology of colonial nesting seabirds. In *Behavior of Marine Animals. Vol. 4: Marine Birds* (eds J. Burger, B. L. Olla & H. E. Winn). Plenum Press, New York, NY.

Savoca, M. S., Bonter, D. N., Zuckerberg, B., Dickinson, J. L. & Ellis, J. C. (2011). Nesting density is an important factor affecting chick growth and survival in the Herring Gull. *Condor*, **113**, 565–575.

Shrubb, M. (2013). *Feasting, Fowling and Feathers – a History of the Exploitation of Wild Birds*. T. & A. D. Poyser, London.

Sibley, R. M. & McCleery, R. H. (1983). Increase in weight of Herring Gulls while feeding. *Journal of Animal Ecology*, **52**, 35–50.

Smith, N. G. (1966). *Evolution of Some Arctic Gulls* Larus*: An Experimental Study of Isolating Mechanisms*. Ornithological Monographs **4**. American Ornithologists' Union, Washington, DC.

Spaans, A. L. (1971). On the feeding ecology of the Herring Gull *Larus argentatus* Pont. in the northern part of the Netherlands. *Ardea*, **59**, 75–188.

Spaans, M. J. & Spaans, A. L. (1975). Some data on the breeding biology of the Herring Gull *Larus argentatus* on the Dutch Frisian Island of Terschelling. *Limosa* **48**, 1–39.

Stanbury, A., Brown, A., Eaton, M. *et al.* (2017). The risk of extinction for birds in Great Britain. *British Birds*, **110**, 502–517.

Stenhouse, I. J., Robertson, G. J. & Montevecchi, W. A. (2000). Herring Gull *Larus argentatus* predation on Leach's Storm-petrels *Oceanodroma leucorhoa* breeding on Great Island, Newfoundland. *Atlantic Seabirds*, **2**, 35–44.

Stoddart, A. & McInery, C. J. (2017). The 'Azorean Yellow-legged Gull' in Britain. *British Birds*, **110**, 660–674.

Sutcliffe, S. J. (1986). Changes in the gull populations of SW Wales. *Bird Study*, **33**, 91–97.

Tapper, S. (1992). *Game Heritage*. Game Conservancy Ltd, Fordingbridge.

Taylor, C. J., Boyle, D., Perrins, C. M. & Kipling, R. (2012). Seabird monitoring on Skomer Island in 2012. Unpublished report for JNCC, Peterborough.

te Marvelde, L., Meininger, P. L., Flamant, R. & Dingemanse, N. J. (2009). Age-specific density-dependent survival in Mediterranean Gulls *Larus melanocephalus*. *Ardea*, **97**, 305–312.

Thaxter, C. B., Ross-Smith, V. H., Bouten, W. *et al.* (2015). Seabird–wind farm interactions during the breeding season vary within and between years: a case study of Lesser Black-backed Gull *Larus fuscus* in the UK. *Biological Conservation*, **186**, 347–358.

Tinbergen, N. (1953). *The Herring Gull's World*. Collins, London.

Tinbergen, N., Broekhuysen, G. J. F., Feekes, F., Houghton, C. W., Kruuk, H. & Szulc, E. (1962). Egg shell removal by the Black-headed Gull, *Larus ridibundus* L.; a behaviour component of camouflage. *Behaviour*, **19**, 74–117.

Tittensor, R. (2012). The tradition of collecting eggs from the Whitelee Gulls' Hags in the 20th century. *Ayrshire Notes*, **43**, 5–23.

Turner, D. M. (2010). Counts and breeding success of Black-legged Kittiwakes *Rissa tridactyla* nesting on man-made structures along the River Tyne, northeast England, 1994–2009. *Seabird*, **23**, 111–126.

van Dijk, K. & Voesten, R. (2014). Black-headed Gull of 33 years and re-appeal to stop using aluminium rings to mark gulls. *Dutch Birding*, **36**, 249–252.

van Swelm, N. D. (1998). Status of the Yellow-legged Gull *Larus michahellis* as a breeding bird in the Netherlands. *Sula*, **12**, 199–202.

Vaurie, C. (1959) *The Birds of the Palearctic Fauna: A Systematic Reference. Vol. 1 and Vol. 2*. H. F. & G. Witherby, London.

Verbeek, N. A. M. (1977). Comparative feeding ecology of Herring Gulls *Larus argentatus* and Lesser Black-backed Gulls *Larus fuscus*. *Ardea*, **65**, 25–42.

Vermeer, K. (1963). The breeding ecology of the Glaucous-winged Gull on Mandarte Island B.C. *Occasional Papers of the B.C. Museum* **13**, 1–104.

Vernon, J. D. R. (1970a). Food of the Common Gull on grassland in autumn and winter. *Bird Study*, **17**, 36–38.

Vernon, J. D. R. (1970b). Feeding habitats and food of the Black-headed and Common gulls. Part 1 – feeding habits. *Bird Study*, **17**, 287–296.

Vernon, J. D. R. (1972). Feeding habitats and food of the Black-headed and Common gulls. Part 2 – food. *Bird Study*, **19**, 173–186.

Walters, J. (1978). The primary moult in four gull species near Amsterdam. *Ardea*, **66**, 32–47.

Walters, J. (1982). Completion of primary moult in the Black-headed Gull *Larus ridibundus*. *Bird Study*, **29**, 217–220.

Weidmann, U. (1956). Observations and experiments on egg-laying in the Black-headed Gull. *Animal Behaviour*, **4**, 150–161.

Wernham, C., Toms, M., Marchant, J., Clark, J., Siriwardena, G. & Baillie, S. (eds) (2002). *The Migration Atlas: Movements of the Birds of Britain and Ireland*. T. & A. D. Poyser, London.

White, S. & Kehoe, C. (2017). Report on scarce migrant birds in Britain in 2015. *British Birds*, **110**, 518–539.

Winsper, J. (2014). Roof-top nesting gull study: concerning the population of gulls that breed within the Birmingham boundary. *West Midland Bird Report*, **78**, 237–249.

Ytreberg, N.-J. (1956). Contribution to the breeding biology of the Black-headed Gull (*Larus ridibundus* (L.)) in Norway. *Nytt Magasin for Zoologi*, **4**, 6–16.

Ytreberg, N.-J. (1960). Some observations on egg-laying in the Black-headed Gull (*Larus ridibundus* (L.)) and the Common Gull (*Larus canus* (L.)). *Nytt Magasin for Zoologi*, **9**, 5–15.

Index

SPECIES INDEX

Page numbers in **bold** include figures.

Albatross, Wandering (*Diomedea exulans*) 364
Alle alle (Little Auk) 286
Ammodytes spp. (sandeels) 223, 269, **289–90**, 293, 431–2
Anthus petrosus (Rock Pipit) 139–40
Arctogadus glacialis (Arctic Cod) 289
Ardea cinerea (Grey Heron) 76, 223
Auk, Little (*Alle alle*) 286

Badger (*Meles meles*) 25, 76, 185, 412
Branta canadensis (Canada Goose) 52

Campylobacter 398
Capelin (*Mallotus villosus*) 289, 293, 431–2
Chroicocephalus cirrocephalus (Grey-headed Gull) 354
C. genei (Slender-billed Gull) 8, 9, **10**, 310, 312, 354
C. novaehollandiae (Silver Gull) 373
C. philadelphia (Bonaparte's Gull) 8, 9, **10**, 25, 312, **335–7**
C. ridibundus see Gull, Black-headed
Clio (a gastropod) 294
Clostridium botulinum 133–4
Clupea harengus (Herring) 269, 293
Cod, Arctic (*Arctogadus glacialis*) 289
Corvus corax (Raven) 24, 25
Crataegus monogyna (Hawthorn) 38, 77, 80
Creagrus furcatus (Swallow-tailed Gull) 3, 24
Crowberry (*Empetrum nigrum*) 203
Cryptosporidium 398
Cygnus olor (Mute Swan) 52

Diomedea exulans (Wandering Albatross) 364

Eagle, White-tailed (*Haliaeetus albicilla*) 25, 269
Eider (*Somateria mollissima*) 39, 55, 415, 418
Empetrum nigrum (Crowberry) 203
Erinaceus europaeus (Hedgehog) 57, 76
Escherichia coli 422–3

Falcon, Peregrine (*Falco peregrinus*) 25, 110, 403–4
Fox (*Vulpes vulpes*) 25, 48, 52, 53, 55, 56–7, 74, 76, 184, 185
Fox, Arctic (*Vulpes lagopus*) 320
Fratercula arctica (Puffin) 3, 136, **137**, 140, 191, 213, 220, 225, 232, 291, 418
Fulmar, Northern (*Fulmarus glacialis*) 222, 290, 291, 369, 427

Gannet, Northern (*Morus bassanus*) 38, 166, 222, 252, 286, 290, 291, 293, 427
Goose, Canada (*Branta canadensis*) 52
Grouse, Red (*Lagopus lagopus*) 202, 421, 425
Guillemot, Common (*Uria aalge*) 3, 191, 235, 252, 271, 291, 365, 417

Gull, American Herring (*Larus smithsonianus*)
appearance 328, **344–5**
breeding **215**, 217, 395
breeding, inter-species 22, 23, 181, 344
breeding sites 25
breeding success 146–7
British records 312, 345
and European Herring Gull 4, 5, 23, 124, 310, 344–5
status 9
taxonomy 4, 5, **10**, 439–40
urban nesting 374, 396
Gull, Audouin's (*Ichthyaetus audouinii*) 8, 9, **10**, 310, 312, **349–50**
Gull, Azorean Yellow-legged (*Larus michahellis atlantis*) 312, 345–6
Gull, Azores Herring (*Larus argentatus atlantis*) 312, 345–6
Gull, Black-headed (*Chroicocephalus ridibundus*)
appearance **43–4**, 334, 335
breeding, coastal *vs* inland 48
breeding, colonial **49–56**, 73–4 *see also* colonies
breeding, distance travelled from colony 291
breeding, inter-species 22, 23
breeding, rearing young 34, 75–6
breeding sites 25, 49, 50, 51, 71, 74 *see also* nesting; philopatry
breeding, solitary 50
breeding status *see* status
breeding success 46, 50, 52, 57, 76, **86–8**
clines 12
colonies, annual reoccupation of 73–4
colonies, decline of 48, 51–2, 55–6, 56–7
colonies, persistence of 51–2
colonies, size of 49–51
conservation concern 62, 63, 434–5, 437
courtship display **73**
distribution 44–5, **47**, 70
eggs, human exploitation 57–60
eggs/incubation 74–5
feeding, diet 77–8
feeding, food-collecting techniques 80
feeding areas 45–6, 78–9
feeding areas, age structure in different 82–3
feeding areas, use of winter 80–2
fledging period 34
flight, 'panic' 73
incubation period 32, 75
kleptoparasitism 78
longevity 13, 87
movements, winter 12, **61–9**, 71, 121, 435
moulting **37**, 65, 89–90
nesting with Mediterranean Gulls 97, 98
nesting with terns 28, **49**
nesting, urban *see* urban nesting
philopatry **71–3**
population estimates 11, 46–8, 49–51, 62, 420
ringing **359–60**, 361
sexual dimorphism **16**, 17
status 9, 434–5
status, historical 46–8
survival rates **86–8**
survival rates, regional 88–9
taxonomy 3, **10**
urban nesting 390, 393
weight changes **84–6**
winter abundance **61–3**
winter movements *see* movements, winter
winter roosts 62
wintering areas, faithfulness to 62–3, **66–8**
wintering birds, arrival **63–6**, 71
wintering birds, decline 62–3, 434–5
wintering birds, distribution of Continental 70
wintering birds, return to the Continent **68–9**
wintering birds, sex ratio 70–1
Gull, Bonaparte's (*Chroicocephalus philadelphia*) 8, 9, **10**, 25, 312, **335–7**
Gull, Caspian (*Larus cachinnans*) 9, **10**, 12, 22, 310–11, 312, **346–9**, 363, 430, 436–7
Gull, Common (Mew) (*Larus canus*)
appearance **103–5**, 121, 319, 334, 340
breeding, age at first 111
breeding biology/success 110–11
breeding, colonial 26, 28, 109–10
breeding, distance travelled from colony 291
breeding sites 25, 109–10, 389
call 103
conservation concern 108, 430, 435, 437

distribution, Britain and Ireland 106–8
distribution, world 106
food/feeding 115–19
feeding, inland *vs* coastal sites 117–19
feeding, along river Tyne **117–18**
fledging period 34, 111
incubation period 32, 111
kleptoparasitism 116
longevity 13, 105
movements 12, **111–15**
population estimates 11, 106, 108
roosts, night 119–20
sexual dimorphism **16**, 17
status 9, 435
subspecies 12, 108–9
taxonomy **10**
urban nesting 389
weight 119
wintering areas, faithfulness to 115

Gull, Dominican (Kelp) (*Larus dominicanus*) 5, 9, **10**, 22

Gull, European Herring (*Larus argentatus*)
and American Herring Gull 4, 5, 23, 124, 310, 344–5
appearance **36**, **121–4**, 210, 310, 328, 348
breeding, age at first 154
breeding, colonial 27, 28, 51, 129, 141–2 *see also* urban nesting
breeding, distance travelled from colony 291
breeding, inter-species 22, 23, 24–5, 319, 436
breeding, synchronised 177
breeding success 128, 146–51, 396
breeding success and brood size 153
breeding success and nesting density **151–2**
breeding success, urban sites 395–6
calls **123–4**
cannibalism **174–8**
censuses *see* population estimates
conservation concern 403, 411, 430, 431, 433, 437
culling 131–2, **136–40**
drinking **40**
egg collecting 126
eggs/incubation 143–6
eggs, hatching success 146, **152**
distribution and status, Britain and Ireland **126–31**
distribution, compared with Lesser Black-backed Gull **184–9**, 381–3, 413
feeding/food 38, **165–7**, 202
feeding at landfill sites 167–70, 226, 401, 420
feeding at landfill sites and botulism **133–4**
feeding at landfill sites, effect of age **171–3**
feeding at landfill sites, effect of sex 173–4
fledging, post- 153–4
fledging period 34
identification problems 313
incubation period 32, 145
life table 156, 442–3
longevity 13
moulting **36**
movements 12, 121, **156–65**, 189, 199, 401, 412
plumage 121, **122–3**, **124**, 180
plumage, adult variation 18, **19**, **20–1**, 313
population decline, Industrial Revolution 126–8, 374, 378, 419
population decline, late twentieth-century 130–4, 426–7, 431, 433
population estimates 11, 129–31, 371, 412, 419, 437
population recovery, twentieth-century 128–9, 374–7, 378, 419–20
predation by 25, 98
ringing **358–9**, 362
sex ratio, chicks 153
sex ratio, wintering birds 70, 164, 165
sexual dimorphism **16**, 17
status 9, 433
studies 124–6, 162–5
subspecies 348, 353
survival rates, chick 149–51
survival rates, adult 154–6
taxonomy **10**
urban nesting 130, 185, 374, **378–83**, 390–2, **392–6**, 413, 427
weight 121

Gull, Franklin's (*Leucophaeus pipixcan*) 9, **10**, 35, 36–8, 310, 312, 332, 334, **337–9**

Gull, Glaucous (*Larus hyperboreus*) 9, **10**, 12, **16**, 22, 24–5, 312, **318–22**, 324, 344, 431, 435–6, 438

Gull, Glaucous-winged (*Larus glaucescens*) 9, **10**, 13, 22, 129, 312, 319, 326, **343–4**, 374

Gull, Great Black-backed (*Larus marinus*)
appearance 121, 123, **209–10**, 319, 352
biometrics (size/weight) 13–14, 209, 213–14, **216**
breeding **215–17**
breeding, age at first 216
breeding, colonial 26, 28, 211–13, 217 *see also* urban nesting
breeding, inter-species 22, 23
breeding numbers, decline 126, 214
breeding success 216, 217, 224–5, 228–9
cannibalism 217
colony size 217
conservation concern 430, 434, 437
culling, 132, 211, 214, 216–17
distribution **211–13**
distribution, historical 214–15
egg laying/incubation **215–16**
feeding/food 211, 216, **219–25**
feeding at landfill sites 223–4, **225–6**, 269
fledging period 34, 216
incubation period 32, 215–16
kleptoparasitism 223–4
longevity 13, 229
movements **218–19**, 227
population estimates 11, 211, 214
predation by 25, 211, 216, 219, 220, 225, 269
ringing 358
sex ratio, wintering birds 70
sexual dimorphism **16**, 17, 214
site fidelity, winter 229–30
status 9, 434
studies of marked individuals 226–30
survival rates, adult 229
taxonomy **10**
urban nesting 214–15, 389
Gull, Great Black-headed (Pallas's) (*Ichthyaetus ichthyaetus*) 8, 9, **10**, 310, 352
Gull, Grey-headed (*Chroicocephalus cirrocephalus*) 354
Gull, Herring *see* Gull, European Herring
Gull, Iceland (*Larus glaucoides*) 5, 8, 9, **10**, 22, 312, 318, 319, **322–4**, **325**, 326, 327, 430, 435–6, 438
Gull, Ivory (*Pagophila eburnea*) 3, 9, **10**, 25, **30**, 31, 312, **314–16**, 329
Gull, Kelp (Dominican) (*Larus dominicanus*) 5, 9, **10**, 22
Gull, Kumlien's (*Larus glaucoides kumlieni*) 5, 8, 9, 312, **324–8**
Gull, Laughing (*Leucophaeus atricilla*) 9, **10**, 13, 22, 310, 312, **332–4**, 337
Gull, Lesser Black-backed (*Larus fuscus*)
appearance 121, 123, **179–80**, **209**, 353
breeding, colonial 28, 192–4
breeding, distance travelled from colony 291
breeding, inter-species 22, 23, 24, 436
breeding sites 183–9, 202, 297 *see also* philopatry; urban nesting
breeding success 193, 196, 201, 207 *see also* population dynamics
breeding years, missed 207
cannibalism 193
clines 191
conservation concern 430, 433–4, 437
culling 140, 183, 184, 194, 202, 420, **421–2**, 434
distribution, Britain and Ireland **181–3**
distribution, compared with Herring Gull **184–9**, 381–3, 413
distribution, western Europe and North Atlantic 180–1
eggs/incubation 193
feeding/food 38, 196, 200–4
fledging period 34
incubation period 32, 193
longevity 13, 205, 409
migration 192, **196–200**, 357
moulting **37**, 192, 204–5
philopatry **194–6**
plumage variation **18–20**, 190
population dynamics/survival rates 126, **205–8**, 427
population estimates 11, 183, 185, 196, 371, 372, 420
predation by 25, 193, 202
sex ratio 153
sexual dimorphism **16**, 17
status 9, 433–4
subspecies 4, 5, 189–92
taxonomy **10**
urban nesting 183, 185–6, 188–9, 363, **378–83**, 390–2, 393, 413, 427
Gull, Little (*Hydrocoloeus minutus*)
appearance **301–2**, 316
breeding in Britain 305

British records 303–4
conservation concern 428, 430, 437
distribution, world 302–3
fledging period 34
feeding/food 307–8
incubation period 32
longevity 13
movements **305–7**, 308
status 9, 437
taxonomy 3, **10**
Gull, Mediterranean (*Ichthyaetus melanocephalus*)
appearance **91–2**, 105, **363**
breeding, inter-species 22, 24, 93, 97
breeding sites 97
breeding success **97–9**
conservation concern 430, 436, 437
distribution/breeding status **93–6**
egg collecting 60
egg-laying 97
fledging period 34
feeding/food 101–2
immigration/movements 99–101
incubation period 32
longevity 13
nesting with Black-headed Gulls 93, 97, 98
and predation 98
status 9, 436
taxonomy 8, **10**
Gull, Mew *see* Gull, Common
Gull, Pallas's (Great Black-headed) (*Ichthyaetus ichthyaetus*) 8, 9, **10**, 310, 352
Gull, Ring-billed (*Larus delawarensis*) 9, **10**, 13, 22, 103, 105, 106, 312, **339–43**, 374, 436
Gull, Ross's (*Hydrocoloeus rosea*) 3, 8, 9, **10**, 41–2, 311, 312, **316–18**
Gull, Sabine's (*Xema sabini*) 3, 8, 9, **10**, 35, 41–2, 308, 312, **329–32**
Gull, Saunders' (*Saundersilarus saundersi*) 3
Gull, Silver (*Chroicocephalus novaehollandiae*) 373
Gull, Slaty-backed (*Larus schistisagus*) 9, **10**, 312, 344, 353
Gull, Slender-billed (*Chroicocephalus genei*) 8, 9, **10**, 310, 312, 354
Gull, Swallow-tailed (*Creagrus furcatus*) 3, 24
Gull, Thayer's (*Larus thayeri*) 5, 9, **10**, 22, 312, 324–8
Gull, Yellow-legged (*Larus michahellis*)
appearance **295–7**, 352
breeding 22, 23, 24, 298–300
breeding, inter-species 22, 23, 24, 298–300, 436
conservation concern 430, 431, 436, 438
distribution 298
sex ratio 153
status 9, 436
as subspecies 12
taxonomy **10**, 310
urban nesting 373, 390, 414
Gull, Vega (*Larus vegae*) 312, 353
Gull, Western (*Larus occidentalis*) 22, 344

Haliaeetus albicilla (White-tailed Eagle) 25, 269
Halichoerus grypus (Grey Seal) 126, 219, 223
Hawk, Harris's (*Parabuteo unicinctus*) 403
Hawthorn (*Crataegus monogyna*) 38, 77, 80
Hedgehog (*Erinaceus europaeus*) 57, 76
Heron, Grey (*Ardea cinerea*) 76, 223
Herring (*Clupea harengus*) 269, 293
Hirundo rustica (Swallow) 156
Hydrobates pelagicus (Storm Petrel) 6, 7, 31, 35, 219, 418
Hydrocoloeus minutus see Gull, Little
H. rosea (Ross's Gull) 3, 8, 9, **10**, 41–2, 311, 312, **316–18**

Ichthyaetus 3
I. audouinii (Audouin's Gull) 8, 9, **10**, 310, 312, **349–50**
I. ichthyaetus (Great Black-headed (Pallas's) Gull) 8, 9, **10**, 310, 352
I. melanocephalus see Gull, Mediterranean

Kittiwake, Black-legged (*Rissa tridactyla*) 213
appearance **231**, **232–5**
breeding, age at first 249, 267
breeding, colonial 27–8, 51, **235–41**, 246–8
see also colonies
breeding, distance travelled from colony 290, 291
breeding numbers, declines 126, **242–6**, **269–77**
breeding performance and age 265–6
breeding performance and laying date **266–7**
breeding performance and nest site 267–8

Kittiwake, Black-legged *continued*
breeding sites 8, 25, 232, 247 *see also* nesting
breeding success 111, 246, **268–77**, 432
breeding years, missed 249–52
calls 27, 41, 232, **248–9**, 260
chicks 29, 33–4, 41, **231**, **234**
chicks, failure to recognise 41, 256, 257
clines 3, 191, 232
colonies, annual reoccupation 246–8, 287
colonies, changes in departure time 287
conservation concern 242, 430, 431–2, 437
copulation 253
distribution, Britain and Ireland **235–6**, 369–70
distribution, winter **284–5**
distribution, world 231, 235
drinking 40
egg collecting 417
egg laying date and clutch size 259
eggs **30**, 31, **263–4**
feeding, courtship 252–3
feeding, diet 246, 289–90, 293–4, 431–2
feeding methods 293–4
feeding trips, breeding season **260–2**, **290–3**
fledging 29, 33–4
human exploitation 417
incubation **259–61**, 263
incubation period 32
legs/walking 34, 35, 232, 235
longevity 13
mate change, effects of 257–8
moulting 280–1
movements, inland records **277–81**
movements, winter 12, 121, **281–7**, 357
movements, of young 278–9, **287–8**
nesting 29–30, 41, **248**
nesting, adaptation to cliff sites 35, 232, 254–6
nesting, urban *see* urban nesting
pair bond duration 253
'panic flights' 247–8
philopatry 257, 287, 288
population dynamics/estimates 11, **242–6**, 369–70, 420, 426–7, 431–2
ringing 358, 361, 362
roosting, offshore 41–2
sexual dimorphism **16**, 17, **366–8**
size/weight 232, 280
status 9, 431–2
subspecies 12, 232
tracking, data-logger **284–7**, 365
urban nesting **238–41**, 246, **273–4**, 276, 287–8, 289, **383–9**, 427
winter numbers 246
Kittiwake, Red-legged (*Rissa brevirostris*) 232–3

Lagopus lagopus (Red Grouse) 202, 421, 425
Lapwing (*Vanellus vanellus*) 39, 78, 80, 116
Larus argentatus see Gull, European Herring
L. a. argentatus 313, 328, 363
L. a. argenteus 313, 328
L. a. atlantis (Azores Herring Gull) 312, 345–6
L. cachinnans (Caspian Gull) 9, **10**, 12, 22, 310–11, 312, **346–9**, 363, 430, 436–7
L. canus see Gull, Common (Mew)
L. c. brachyrhynchus 109
L. c. canus 12, 108–9
L. c. heinei 12, 108–9
L. c. kamtschatschensis 109
L. delawarensis (Ring-billed Gull) 9, **10**, 13, 22, 103, 105, 106, 312, **339–43**, 374, 436
L. dominicanus (Dominican (Kelp) Gull) 5, 9, **10**, 22
L. fuscus see Gull, Lesser Black-backed
L. f. barabensis 192
L. f. fuscus 18, 189, 191–2
L. f. graellsii 18, 189–92
L. f. heuglini 192
L. f. intermedius 18, 189–91
L. f. taimyrensis 192
L. glaucescens (Glaucous-winged Gull) 9, **10**, 13, 22, 129, 312, 319, 326, **343–4**, 374
L. glaucoides (Iceland Gull) 5, 8, 9, **10**, 22, 312, 318, 319, **322–4**, **325**, 326, 327, 430, 435–6, 438
L. g. kumlieni (Kumlien's Gull) 5, 8, 9, 312, **324–8**
L. g. thayeri see L. thayeri
L. heuglini 192
L. hyperboreus (Glaucous Gull) 9, **10**, 12, **16**, 22, 24–5, 312, **318–22**, 324, 344, 431, 435–6, 438
L. michahellis atlantis (Azorean Yellow-legged Gull) 312, 345–6

L. occidentalis (Western Gull) 22, 344
L. schistisagus (Slaty-backed Gull) 9, **10**, 312, 344, 353
L. smithsonianus see Gull, American Herring
L. thayeri (Thayer's Gull) 5, 9, **10**, 22, 312, 324–8
L. t. kumlieni see L. glaucoides kumlieni
L. vegae (Vega Gull) 312, 353
Leucophaeus 3
L. atricilla (Laughing Gull) 9, **10**, 13, 22, 310, 312, **332–4**, 337
L. pipixcan (Franklin's Gull) 9, **10**, 35, 36–8, 310, 312, 332, 334, **337–9**

Mallotus villosus (Capelin) 289, 293, 431–2
Meles meles (Badger) 25, 76, 185, 412
Mink, American (*Neovison vison*) 25, 48, 52, 56–7, 108, 110, 229, 415
Morus bassanus (Northern Gannet) 38, 166, 222, 252, 286, 290, 291, 293, 427
Mustela erminea (Stoat) 415

Neovison vison (American Mink) 25, 48, 52, 56–7, 108, 110, 229, 415

Oryctolagus cuniculus (Rabbit) 219, 220, **221**

Pagophila eburnea (Ivory Gull) 3, 9, **10**, 25, **30**, 31, 312, **314–16**, 329
Parabuteo unicinctus (Harris's Hawk) 403
Parus major (Great Tit) 151
Petrel, Storm (*Hydrobates pelagicus*) 6, 7, 31, 35, 219, 418
Phalacrocorax aristotelis (Shag) 358
Phoca vitulina (Common Seal) 126, 219
Pilchard (*Sardina pilchardus*) 350
Pipit, Rock (*Anthus petrosus*) 139–40
Plover, Golden (*Pluvialis apricaria*) 39, 78, 80
Puffin (*Fratercula arctica*) 3, 136, **137**, 140, 191, 213, 220, 225, 232, 291, 418
Puffinus mauretanicus (Balearic Shearwater) 429
P. puffinus (Manx Shearwater) 6, 7, 35, 211, **218–21**, 225, 358, 418

Rabbit (*Oryctolagus cuniculus*) 219, 220, **221**
Raven (*Corvus corax*) 24, 25
Rissa brevirostris (Red-legged Kittiwake) 231–2
R. tridactyla see Kittiwake, Black-legged
R. t. pollicaris 232
R. t. tridactyla 12, 232

Salmonella 398, 423
S. montevideo 398
sandeels (*Ammodytes* spp.) 223, 269, **289–90**, 293, 431–2
Sardina pilchardus (Pilchard) 350
Saundersilarus saundersi (Saunders' Gull) 3
Seal, Common (*Phoca vitulina*) 126, 219
Seal, Grey (*Halichoerus grypus*) 126, 219, 223
Shag (*Phalacrocorax aristotelis*) 358
Shearwater, Balearic (*Puffinus mauretanicus*) 429
Shearwater, Manx (*Puffinus puffinus*) 6, 7, 35, 211, **218–21**, 225, 358, 418
Skua, Great (*Stercorarius skua*) 2, 16, 25, 126, 222, 269, 286, 418, 419, 427
Somateria mollissima (Eider) 39, 55, 415, 418
Sprat (*Sprattus sprattus*) 269, 289
springtails (Collembola) 332
Stercorarius skua (Great Skua) 2, 16, 25, 126, 222, 269, 286, 418, 419, 427
Sterna dougallii (Roseate Tern) 291
S. hirundo (Common Tern) 55, 291, 305, 339
S. paradisaea (Arctic Tern) 13, 291, 332
Sternula albifrons (Little Tern) 291
Stoat (*Mustela erminea*) 415
Swallow (*Hirundo rustica*) 156
Swan, Mute (*Cygnus olor*) 52

Tern, Arctic (*Sterna paradisaea*) 13, 291, 332
Tern, Common (*Sterna hirundo*) 55, 291, 305, 339
Tern, Little (*Sternula albifrons*) 291
Tern, Roseate (*Sterna dougallii*) 291
Tern, Sandwich (*Thalasseus sandvicensis*) 28, **49**, 55, 238, 291
Tit, Great (*Parus major*) 151

Uria aalge (Common Guillemot) 3, 191, 235, 252, 271, 291, 365, 417

Vanellus vanellus (Lapwing) 39, 78, 80, 116
Vulpes lagopus (Arctic Fox) 320
V. vulpes (Fox) 25, 48, 52, 53, 55, 56–7, 74, 76, 184, 185

Xema sabini (Sabine's Gull) 3, 8, 9, **10**, 35, 41–2, 308, 312, **329–32**

GENERAL INDEX

Page numbers in **bold** include figures.

Abbeystead Estate, Lancashire 202, 211–13
 gull control 132, 183, 184, 211, 409, 420, **421–6**
 see also Tarnbrook Fell
Aberdeen 110, 154, 168, 389
 University 356
acidification, ocean 247, 294
acorns 77
Aebischer, Nicholas 87
Afghanistan 350
Africa *see* North Africa; South Africa
agriculture 38, 39, 56, 78, 79, 86, 116, **166–7**, 202–3, 307, 339, 420, 428
Ailsa Craig, Scotland 415
airfields/airports 389
 gull control 400, 420
Alaska 106, 109, 238, 329, 335, 344, 353
Aleutian islands, Pacific Ocean 231–2, 343
algae
 and nest-building 29, 255
 toxic 250
Algeria 350
Amber List
 British gulls on 433–7
 criteria 429, 430
American Ornithological Society 344
American Ornithologists' Union 5, 327
Annan, Scotland 391
Anderson, Neil 52
Anglesey, Wales 106
Antarctica 5, 6
Antrim, Country 380
ants 38, 77, 80
Arctic 1, 5, 6, 38, 165, 235, 247, 280, 281
 rare visitors from **314–32**
 see also individual countries
artificial illumination 41, 291
Ashington, Northumberland 391
Asia 4, 5, 44, 45, 106, 192, 302
 rare visitors from 352
Association for the Protection of Sea-Birds 417–18
Atlantic Ocean 3, 5, 211, 219, 222, 232, 235, 247, 267, 304, 323, 432
 transatlantic movements 44, 112, 217, **281–7**, 303, 308, 331, 336, 340–1, 342, 345
 see also depressions, Atlantic
auks 2, 223, 252, 271, 286, 417, 418
Australia 373
Ayrshire Notes 58
Azores 297, 345–6
Azov, Sea of, Russia 352

Baffin Island, Canada 314, 325, 327, 344
Baker, Robin 196
bait *see* narcotic bait
Balearic Islands 429
BALTIC Centre for Contemporary Art 385, 386
Baltic Flour Mill 385, 386
Baltic Sea 63, 65, 108, 191
banding *see* ringing
Banks Island, Canada 325
Bannerman, David 52, 310
Barden Moor, North Yorkshire 59
Barents Sea, Russia 289
barnacles, goose 223
Barrow-in-Furness, Cumbria 132
Barth, Edvard 125
Bass Rock, Scotland 231
Bauer, Kurt 229
Bay of Biscay 217, 330–1
beaches, shingle 106, 110
Bear Island 211, 322
beetles 76, 166, 202
Belarus 347, 353
Belgium
 Black-headed Gulls 66
 Common Gulls 106
 Herring Gulls 23, 129, 159, 160
 inter-species breeding 23, 93, 191, 298
 Lesser Black-backed Gulls 184, 190, 191, **194**, 197–8, 205, 434
 Mediterranean Gulls 93, 96, 98, 99, 100
 Yellow-legged Gulls 298, 436
Bempton Cliffs, Yorkshire 235, **274–5**, 276, 291–3, 417, 432 *see also* Flamborough Head
Benguela Upwelling 331
Bergmann's rule 213
Bering Sea 343
Berlin 374

berries 38, 77, 80, 165, 203, 219, 343
Berry Head, Devon 276
Bevan, Richard 290
bill length, difference between sexes 17
binoculars 355, 356
biometrics *see under individual species*
  sexing using 17, **366–8**
Bird Atlas 2007–11 (BTO) (Balmer *et al.*) **47**, 48, **95**, **107**, **127**, **182**, 211, **212**, **236**
bird counts *see* censuses
Bird Flocks and the Breeding Cycle (Darling) 177–8
bird protection Acts 128, 418–19
Bird Study (journal) 362
The Birds (Hitchcock) 397
Birds of the British Isles (Bannerman) 310
'Birds of Conservation Concern' (*British Birds*) 428
'Birds of Conservation Concern 4' (BoCC4) (Eaton *et al.*) 428–30, 437–8
  status of gulls 431–7
'Birds of conservation concern in Ireland 2014–2019' (BoCCI) (Colhoun & Cummins) 430, 437
Birds of Hampshire 305
birds of prey 25, 363
  as gull control 400, 403–4, 420
The Birds of the Western Palearctic (Cramp & Simmons) 190, 311
Birmingham 382–3, 411 *see also* West Midlands
Black Sea 99, 102, 373
  rare visitors from **350–2**
Blakeney Point, Norfolk 52
Blithfield Reservoir, Staffordshire 278
Bogdanova, Maria 125
Bolton Castle, North Yorkshire 409, 425
Bone, Michael 126
botulism **133–4**, **423**, 428
BOU *see* British Ornithological Union
Bourne, Bill 113
Bournemouth, Hampshire 352
Bourton-on-the-Water, Gloucestershire 354
Bowland, Trough of 402, 421 *see also* Tarnbrook Fell
Brandl, Roland 75
bread 38, 77, 79, 167, 203, 375
breeding
  colonial 26–8
  first, age at 28
  inter-species 22–4
  sites 7–8, 25
  site, fidelity 26 *see also* philopatry
  success 30, 34
  urban *see* urban nesting
  see also eggs; nesting; *and under individual species*
Brehm, Alfred 189
Brennand Estate, Lancashire 421, 425, **426** *see also* Tarnbrook Fell
Brest, France 395, 404, 412
Brighton, Sussex 379
Bristol Channel 125, 200
British Antarctic Survey 284, 364
British Association for the Advancement of Science 417
British Birds (journal) 346, 352, 356, 428
  'The risk of extinction for birds in Great Britain' 430–1
British Birds, Rare Bird Alert **316**, **317**, **334**, **336**, **338**
British Birds Rarities Committee 329
British Ornithological Union (BOU) 5, 108, 327, 344
  Records Committee 346
British Rare Breeding Birds Panel 96
British Trust for Ornithology (BTO)
  Bird Atlas 2007–11 **47**, 48, **95**, **107**, **127**, **182**, 211, **212**, **236**
  Lesser Black-backed Gull survey **203–4**
  Ringing Committee 357
  Ringing Scheme 356, 358
  website 46
  Winter Atlas of Birds 246
  Winter Gull Roost Survey 62
Brittany, France 332
Brooks, W. S. 326
Brough, Trevor 125
Brown, Dick 125
Buckland, Frank 417
Buckley, Neil 223
Budleigh Salterton, Devon 379
Bulgaria 373
Bushnell (telescope maker) 355–6
Butterfield, Jennifer 125

calcium carbonate 294
California 343
Calladine, John 125, 207

calls 40–1
contact 34, 154
distress, loudspeakers to produce 404, 408–9, 420, 424
see also under individual species
Cambridgeshire 305
Camphuysen, Kees 125, 193, 201, 205
Canada 5, 303, 319, 323, 325, 326, 337, 329, 335, 339
American Herring Gulls 129, 147, 155, 344, 374
Common Gulls 106
Great Black-backed Gulls 211, 217, 222, 225
Lesser Black-backed Gulls 4, 181
see also Arctic
Canary Islands 181, 297, 345–6
Canna, Inner Hebrides 134, 222, 229
cannibalism **174–6**, 193
cannon netting, capture by 53, 54, 81, 83, 202, 226, 360–1, 401
Cape Clear Island, Ireland 223
car parks 1, 78–9, 334, 342, 393
carbon dioxide 294
carbonic acid 294
Cardiff 379, 380
Caribbean 332, 335
caterpillars 77
Cellardyke, Scotland 379
censuses
Britain and Ireland 8–12, 129–31, 185, 211, 369–72, 437
count errors 49, 130, 370–1, 372, 381
see also under individual species
Central Science Laboratory (DEFRA) 403
Centre for Hydrology and Ecology 125
Chabrzyk, George 125
Chafarinas Islands, Morocco 349
Champlain, Lake (New York state) 341
Channel Islands 389
Charadriiformes 2
Chardine, John 253, 258
Chew Reservoir, Peak District 345
Chichester Harbour Reserve 96
chicks 33–4
Chile 338
chimney pots/stacks 142–3, 390, 392, 396, 407
China 3, 353
chromosomes 366
Chu, Philip 3
Cley, Norfolk 351
climate change 44, 294 *see also* acidification, ocean
clines 3, 12, 191, 232, 327
Clyde region, Scotland 183, 196, 222, 382, 413
coasts, breeding on 7–8
Coles Cranes Ltd, Sunderland 397–8
Colhoun, Kendrew 437
colonial nesting/breeding 26–8
distance travelled from colony 291
see also under individual species
Commander Islands, Russia 343
Common Fisheries Policy (EU) 222, 404, 427–8, 433
Congo Basin 199
conservation 415
concerns 428–38
see also management *and under individual species*
control *see* management
Cook, Aonghais 146–7
Copinsay, Orkney 217
copulation 41
Coquet Island, Northumberland 28, 56, 157, 183, 276, 401
Cork, County 328, 343
Cornwall 214, 370
Correen Hills, Scotland 110
corvids 24, 25, 146
Cott, Hugh 59
courtship 3, 26, 29, 41
begging/feeding 18, **73**, **252–3**
Cox, C. J. 86, 87
Coxhoe landfill site, Co. Durham 164, 225, **226**
crabs 39, 165, 223
Craik, Clive 86, 110, 229
Cramp, Stanley 370, 389
craneflies 115
Critically Endangered list 431–2
Croatia 349
Crochet, Pierre-Andre 3
Cromer, Norfolk 352
crustaceans 76, 115, 289, 293, 318, 352 *see also* crabs
Cullen, Ester 254
culling
Abbeystead 132, 183, 211, **421–2**
Farne Islands 131, 132, 157, 183, 211
Isle of May 125, 131, **136–40**, 158, 183, 211, 401, 412, 415, 420–1

licences 132, 402, 421
narcotic bait 138, 139–40, 401, 402, **421–2**
Skomer/Skokholm **132**, 183, 211, 220
Sunderland 397–8
urban *see* urban nesting, control
see also individual species
Cumberland County Council 52
Cumbria 52, 70, 106, 215, 389 *see also* Lake District
Cummins, Sinéad 437
Cyprus 349
Czech Republic 281

Darvic rings **362–3**
data loggers 204, 284, **286**, 290, 364–5
Davis, Andy 345
Davis, J. W. F. 125
Dean, Alan 161, 277, 298
Decker, R. 327
Denmark 66, 109, 147, 156, 190, 211, 238, 298, 328, 395, 434
Department for Environment, Food and Rural Affairs (DEFRA) 412
depressions, Atlantic
Kittiwake response to 285–7
and wrecks 278, 279, 280, 331–2
deterrents
distress calls 404, 408–9, 420, 424
disturbance 408–9, 424–5
landing prevention 408, 424
Devon 214, 276, 352, 379, 389
diet *see* feeding/food
diving
at humans 1, 25, 168, 396–7, 404, 410
plunge-diving 38, 79, 223
Dixon, Fiona 27
DNA analysis 2, 3, 109, 326, 345
mitochondrial 3, 4, 190, 327
sexing 365–6
dockyards 132, 385
Domesday Book 416
Doochary, County Donegal 341
Dorset 298, 332
Dover, Strait of 332
Down, County 342
Drent, Rudi 125
Dresser, Henry 417
drinking 39–40
drones 356, 371
droppings 168, **234**, **256**, **263**, 289, 383, **384**, 385, 387, 398, 404
Dublin, Republic of Ireland 126
Duffey, Eric 281, 308
Dumfries, Scotland 200, 201, 202, 203, 215, 391, 413
gull control **403–7**
landfill site closure 201, 404
Dumfries and Galloway Council **398–9**, 404, **406–7**
Dunbar, Scotland **238**, **240**, 383
Duncan, Neil 125
Duncannon, County Wexford 353
Dungeness, Kent 304, 350
Durham, County 59, 60, 82, 108, 306, 391
Common Gull study 114–15
Durham University 81, 125, 135, 139
Dwight, Jonathan 326, 346

earthworms 65, 76, 77, 115, 116, 202, 203
East Riding, Yorkshire 418
Ebro Delta, Spain 190, 349
Edinburgh 215
eggs **30–1**
collecting 46, 57–60, 96, 126, 416, 417, 419
collecting licences 59–60
destroying/replacing/spraying, as control 404, 406, 409–10, 411
turning 33
see also incubation *and under individual species*
Elkins, Norman 331–2
Ellesmere Island, Canada 314, 326
Elsham, Lincolnshire 328
Emmerson, Margaret 125
Essex 328, 332, 341, 353
Estonia, Common Gull research 110–11
estuaries 46
Europe
Eastern, rare visitors from **346–50**
Southern, rare visitors from 345–6
European Bird Census Council 190
European Union, Common Fisheries Policy 222, 404, 427–8, 433
Evans, P. G. H. 217
evolution 2
tree (phylogenetic) 8, **10**

Exmouth, Devon 352
exploitation, human 416–20 *see also* eggs, collecting
extinction, risk of 430–1

Faeroes 44, 106, 181, 190, 191, 199, 211, 217
Fair Isle, Shetland 312, 343
Farne Islands, Northumberland 106, 128, 194, 200, 213, 419
culling 131, 157, 183, 211
Kittiwake colonies 30, 235, **256**, 276, 289, 290, 291–3
Feasting, Fowling and Feathers – A History of the Exploitation of Wild Birds (Shrubb) 46, 57
feathers, primary, moulting **36–8**
feeding/food 38–9
courtship 18, **73**, **252–3**
see also kleptoparasitism, *individual food items and under individual species*
Fermanagh, County 298
Ferns, Peter 125
fidelity
mate 28
site 26 *see also* philopatry
field glasses 355 *see also* binoculars
field guides 190, 311, 355, 447
Filey Brig, Yorkshire 291–3
Finland 66, 112, 304, 305, 353 *see also* Scandinavia
firearms 126, 355, 416
fish docks 168, 223, 246, 320, 378
Fisher, James 235, 369
fishing/fisheries 202, 291, 320
Common Fisheries Policy reforms 222, 427–8, 433
docks 168, 223, 246, 320
effect of gales 226
offal/discards 39, 101, 128, 200, 219, 220–1, 222, 223, 226, 324, 350, 378, 393, 427–8
quotas 222, 427
sandeels 432
Fitzgerald, Gerry 117
Flamborough Head, Yorkshire 274, 276, 304, 417, 418, 432
Flanders Moss, Scotland 184
fledging period 33–4
Flegg, Jim 86, 87
Fletcher, Mark 125
flight 15, 35
'panic' 73, 247–8
Florida, USA 332
Flowers, Kathy 126
formalin 410
Forth, Firth of 251–2, 277, 303, 417 *see also* Isle of May
fossils 2
Foula, Shetland 269, 276
Foulney, Cumbria **55–6**
France 306, 310, 332, 349, 350
Black-headed Gulls 76, 87
Common Gulls 106, 111
Great Black-backed Gulls 211, 389
Herring Gulls 125, 156, 159, 373
Kittiwakes 235, 246, 280–1, 281, 431
Lesser Black-backed Gulls 180–1, 190, 198, 373
Mediterranean Gulls 93, 99
Yellow-legged Gulls 297, 373, 390, 436
Franz Josef Land, Russia 318
Fraser Darling, Frank 26–7, 177–8
Frederiksen, Morten 284
The Fulmar (Fisher) 369
Furness, Bob 228

Galapagos Islands 3, 24
Gambia 349
gas guns, propane 424
Gaston, A. J. 327
Gateshead, Tyne and Wear 385, 386
geographical ranges, current **5–7**
Germany 66, 93, 99, 109, 125, 129, 147, 156, 181, 190, 395
Gibraltar 297, 373, 390
Gibson-Hill, Carl 369–70
Glasgow University 125, 409
Global Positioning System (GPS) tags 365
Goethe, F. 125
Goodbody, Ivan 75
Gosport, Hampshire 341
Grand Banks, Canada 221, 281
Grant, Peter 311
Granton harbour, Scotland 383
Great Lakes 303, 328, 339, 374
Green list 429, 430, 437
Greenland 314, 316, 318, 323, 328, 329, 330–1
Black-headed Gulls 44, 63
Great Black-backed Gulls 211

Kittiwakes 281, 285, 287
Lesser Black-backed Gulls 181, 190, 191, 199
see also Arctic
Greig, Susan 126, 224, 225
grouse moors, gull management *see under* management
Guatemala 338
Gulval, Cornwall 379
Gurney Jr., John 46

habitats 7–8 *see also individual habitats*
Hamford Water, Essex 56
Hampshire 58, 60, 305, 306, **307**
Handbook of British Birds (Witherby) 190, 355
Hanningfield Reservoir, Essex 353
Harris, Mike 24, 125, 207, 359
Hartlepool, Co. Durham 387
Hartley, Clive 71
Hassanin, Alexandre 3
Havergate Island, Suffolk 184, 213
Hawthornthwaite Fell, Lancashire 422, 423
Hebrides 86, 211, 370, 415, 419 *see also* Canna
Helbig, Andreas 190
helicopters 389
Henderson, Eric 126
Herring Gull complex 2, 4, 5, 349
identification problems 313, 345
The Herring Gull's World (Tinbergen) 156
history, gulls in Britain and Ireland
Black-headed Gull 46–8
Common Gull 106–8
Great Black-backed Gull 211
Herring Gull 126–31
Kittiwake 235–8
Lesser Black-backed Gull 181–3
Little Gull 303–4
rare gulls 311–12
Yellow-legged Gull 298
Hobson, Keith 294
Holley, Tony 154
Hollom, Philip 369
Hornsea Mere, Yorkshire 304
Houston, David 409
Hoveton, Norfolk 58
How to Know British Birds (Joy) 355
Hoyt, Del 52
Hudson, Anne 228
Hudson Bay, Canada 303, 327
humans
disturbing gulls, as management method 408–9, 424–5
feeding gulls 39, 77, 375, 393
historical relationship with gulls 416–20
robbed/dived by gulls 1, 25, 168, 396–7, 404, 410
see also culling; eggs, collecting; management
Humber Estuary 82
Hungary 93, 99
Hurworth Burn Reservoir, Co. Durham 303
Hutchinson, Clive 303
Huxley, Julian 3, 191
hybridisation 22–4, 93, 97, 298–300, 310, 311, 313, 319, 326, 327, 344, 436

Iberia 197, 198, 199, 235, 373 *see also* Portugal; Spain
Iceland 44, 106, 190, 191, 199, 217, 318, 322, 328, 353
identification
difficulty with Herring Gull/Lesser Black-backed Gull complex 313, 345
difficulty with rare gulls 309, 310–11
guides 190, 311, 355, 447
immature gulls
plumage variation 14–15, 20
proportion of, Britain and Ireland 10–11
Imperial Chemical Industry (ICI) 362
Important Bird Areas 434
Inchcolm, Firth of Forth 251, 289
Inchkeith, Firth of Forth 251, 289
incubation 31–3
periods 32
see also under individual species
India 352
Indian Ocean 192, 352
industrial estates 110, 186, 378
Industrial Revolution 126, 374, 376, 416–17
insects 38, 77, 78, 79, 80, 101, 115, 116, 166, 167, 307, 332, 337, 339, 352
Institute of Terrestrial Ecology 205
International Union for Conservation of Nature (IUCN) 63, 242, 430–1, 436
Inverness 110, 391, **422**
Airport 389

Ireland, Republic of
Black-headed Gulls 48, 50, 51, 88–9
Common Gulls 106, 108, 111
Great Black-backed Gulls 211, 214, 217, 219, 223
Herring Gulls 126, 133, 156, 379–80, 430, 433
Kittiwakes 235, 242, 243, 245, 370, 431
Lesser Black-backed Gulls 183, 184, 186, 188–9, 196, 382
Mediterranean Gulls 93
rare gulls 312, 317, 322, 331, 332, 334, 336, 338, 342, 345, 353, 437
Irish Potato Famine 417
Irish Rare Bird Committee 327, 353
Irish Sea 221, 222, 427
Islay, Scotland 328
Isle of May, Firth of Forth 213
Herring Gull cannibalism **174–6**
Herring Gull cull 125, 131, **136–40**, 158, 183, 211, 401, 412, 415, 420–1
Herring Gull research 18, 126, **135**, **146**, 147, **149–51**, 153, 155, 157, 158–9
Kittiwakes **271–2**, 276, 284–5, 289, **290**, 432
Lesser Black-backed Gulls 183, 205, 207
Isle of Wight 417
Italy 44, 93, 99, 297, 350, 354, 374

jacanas 2
Jackson, David 87, 125
Jan Mayen 318
Japan 343, 353
jellyfish 332
Joint Nature Conservation Committee (JNCC) 48, 76, 129, 220, 277, 370, 371
Seabird Monitoring Programme 371
Jones, Gareth 345
Joy, Norman 355

Kandalaksha, Russia 203
Kazakhstan 350
Kent 106, 305, 352
Kent Island, Canada 147, 395
Kerry, County 332
Kessingland Ringing Group 285
Keyhaven marsh, Hampshire 59
Killington Reservoir, Lancashire 57
Kim, Sin-Yeon 125
kleptoparasitism 39, 78, 80, 83, 116, 167, 170, 172, 173, 223–4, 225, 334
Knudsen, B. 327
Korean Peninsula 353
Kruuk, Hans 86

Labrador, Canada 44, 284
Lake District 54 *see also* Cumbria
lakes 41, 49, 51, 57, 112, 118, 119, 277, 278, 280, 340, 349
Lancashire 101, 186, 303, 389 *see also* Ribble Estuary; Tarnbrook Fell
Land Girls 59
landfill sites 1, 298, 328, 334, 343, 347, 349, 353, 374
and botulism **133–4**, **423**, 428
closures 39, 128, 198, 201, 404, 412, 427
feeding at 39, 102, 128, 154, **167–74**, 196, 200–1, 223–6, 320, 324, 420, 427
gull control 400, 401, 420, **422**
gull studies 71, 82, 83, 117, 125–6, **162–3**, 164, **170–4**, 202, **223–30**, 361, 401
regulations/tax 427
Landsborough-Thomson, Arthur 356
Langstone Harbour, Hampshire 56, 96, 98
Larne Lough, N. Ireland 56
laser beams, to disturb gulls 409
latitude
and ratio of gull species to tern species **6–7**
variation of gull species with **5–7**
Latvia 305
Leach, Elsie P. 356, 369
Leadenhall Market, London 59
leatherjackets 115, 116, 166, 202
Lebreton, Jean-Dominique 87
Leven, Loch, Scotland 56
Liebers, Dorit 190
light-level geolocators 364
Linnaeus, Carl 3
Lithuania 353
Liverpool 387
Llandudno, Wales 168
Lloyd, Clare 54
Loch of Strathbeg reserve 305
Lockley, Ronald 235
loggers, data 204, 284
Lomond, Loch 336
London 39, 45, 58, 59, 60, 66, 125, 161, 343, 347, 361, 375
Zoo 354
longevity 12–13, 357
Lossiemouth, Scotland 379, 420

Lowestoft, Suffolk 184, 240, 387, **388**
Lundy, Devon 276, 415
Lymington Marsh, Hampshire 59

McCleary, Richard 125
Mackinnon, Gabriella 62, 87, 114
Macpherson, A. 326
Madeira 297, 345–6
Mallowdale Estate, Lancashire 421, 425, **426**
  see also Tarnbrook Fell
mammals, predation by 25, 48, 52, 56–7, 73, 74, 86, 108, 110, 183, 184, 320, 396, 415
management/control
  moorland 409, **420–6**
  urban 407–10
  urban, current situation 411–14
  urban, history of **399–407**
  see also culling; deterrents
Manchester Guardian 417
Maree, Loch 184
Marsden, Tyne and Wear 250
mate fidelity 28
mating, monogamous 24
Mauritania 331
May, Isle of *see* Isle of May
mayflies 38
Mayr, Ernst 4
Mediterranean Sea, rare visitors from **350–2**
Merseyside 345
Merthyr Tydfil, Wales 379
Mexico 335, 344
Middleton Island, Pacific Ocean 246
Midlands, West *see* West Midlands
migration/movements 35
  transatlantic movements 44, 112, 217, **281–7**, 303, 308, 331, 336, 340–1, 342, 345
  trans-equatorial 35, 331, 332, 338
  see also under individual species
Ministry of Agriculture, Fisheries and Food 131–2, 400, 421
Minsk, Belarus 347
Miocene 2
mobile phones 356
molluscs 76, 165, 247, 294
Monaghan, Patricia 125, 126, 157, 225, 370, 380, 393–4, 401, 409
Monel rings 87, 358–60
Mongolia 352
Montrose, Scotland 71
moorland 46, 51, 57, 108, 110, 115, 116, 184, 202
  gull management 409, **420–6**
Morocco 100, 198, 199, 331
Mortlach Hills, Scotland 110
Moscow Zoological Museum 192
moulting **36–8**, 339 *see also under individual species*
movements *see* migration/movements
mudflats 34, 76–7, 318, 337
Muncaster Estate, Cumbria 52, 59
Musée National d'Histoire Naturelle, Paris 280

Nager, Ruedi 125, 409
narcotic bait 138, 139–40, 401, 402, **421–2**
nasal glands 40
National Trust 131
National Trust for Scotland 271
Natural England 412
  culling licences 131, 132
  egg collecting licences 59–60
Natural History Museum, London 356, 369
Nature (journal) 267
Nature Conservancy Council 125, 131, 370
  Isle of May cull 131, **136–40**, 158, 183, 401, 412, 415, 420–1
Neath, Brian 303
Nelsen, Ingrid 75
nesting
  site fidelity 26 *see also* philopatry
  urban *see* urban nesting
  see also under individual species
nests 28–30
  destroying, as control 404, 406, 408, 410–11
Netherlands 409
  Black-headed Gulls 44, 66
  Common Gulls 106, 109, 110, 112
  Herring Gulls 20, 23, 125, 129, 147, 155, 156, 160, 165, 201, 385
  inter-species breeding 23, 93, 191, 298
  Lesser Black-backed Gulls 18, 19, 23, 166, 181, 184, 190, 191, 193, 197, 201, 203, 205, 207, 434
  Mediterranean Gulls 93, 96, 98, 99, 100
  rare gulls 298, 304, 328
  Yellow-legged Gulls 298, 436
netting, as gull prevention 404, 408 *see also* cannon netting
New Brunswick, Canada 374
New Hampshire, USA 181
New York 374

Newcastle, Tyne and Wear 291–3, 383
City Council 387
Guildhall 386
see also Tyne, river
Newfoundland, Canada 217, 281, 284, 285, 289, 344, 353
Newlyn, Cornwall 389
Newport, Wales 379, 380
Newton, Alfred 417
Nicholson, Max 38–9
Norfolk 58, 238, 305, 352
North Africa 99, 100, 196, 198, 306, 331
rare visitors from 345–6, 354
North America
American Herring Gulls 4, 5, 23, 63, 124, 146
Black-headed Gulls 44
Common (Mew) Gulls 109
Great Black-backed Gulls 23, 28, 211, 213, 214, 215, 217
Kittiwakes 281
Lesser Black-backed Gull 199
Little Gulls 303, 308
rare visitors from 310, 316, 328, **332–45**
see also Canada; Greenland; USA
North Rona, Scotland 217, 222
North Sea, movements across
Black-headed Gull 65, 69, 73, 113, 435
Common Gull 112, 113, 114
Great Black-backed Gull 211, 217, 221
Herring Gull 156, 159, 160, 165
Kittiwake 246, 269, 281, 284, 286, 387, 427
Lesser Black-backed Gull 191, 198, 201
Little Gull 305
North Shields, Tyne and Wear
Herring Gulls 393, 400
Kittiwake colonies 238, **239**, **241**, 257, 258, 259, **261**, **262**, 361, 363, 384–5, 388
North Solent nature reserve, Hampshire 59
North West Water Authority (NWWA) 421, 422
Northern Ireland 56, 62, 93, 245, 298, 342
Northern Isles 242, 419 *see also* Orkney; Shetland
Northern Lighthouse Board 135
Northumberland 82, 108
Norway 66, 112, 160, 190, 199, 213, 228, 269, 289, 322, 328, 363, 431 *see also* Scandinavia
Nottinghamshire 305
Nova Scotia 281, 332, 374
Novaya Zemlya 289, **314**, 318
Nunavut, Canada 211, 314

O'Connell, Mark 424
offal, fish *see under* fishing/fisheries
oil, spraying eggs with 410
oil/gas platforms 238, 387, 427
oil installations 110
Olsen, Klaus Malling 311
'On the Zoological Aspect of the Game Laws' (Newton) 417
Onno, Sven 110
Ontario, Canada 374
Ontario, Lake 303
opera glasses 355
Operation Seafarer 370
Orford Ness, Suffolk 183, 184, **204**, 213
Orkney, Scotland 106, 184, 211, 235, 269, 370, 431
Outer Trial Bank, the Wash 184
Oxford University 52

Pacific Ocean, 192, 232, 235, 252, 330, 343, 368
rare visitors from 353
Pakistan 350
Palace Pier, Brighton 379
'panic flights' 73, 247–8
parks/parkland 39, 45, 66, 77, 361, 375
Parsons, Jasper 125, 146, 150, 174, 178
Patterson, Ian 57, 74, 76, 113
Pembrey, Wales 379
Perth, Scotland 71
Peterhead, Scotland 379
petrels 6, 7, 31, 35, 219, 418
philopatry (return to natal area) 26, **71–3**, 156, **194–6**, 287, 288
photography 356, 368–9
phylogenetic tree 8, **10**
pig farms 102
Pitsea landfill site, Essex 328, 353
ploughing 39, 78, 79, 101, 116, **166**, 203, 307, 339
plumage
adult, variation within species 2–3, **18–20**
human exploitation 417
immature, variation 14–15, 20
wing-tip pattern, variation within species **20–1**
see also under individual species
plunge-diving 38, 79, 223

Plymouth 217
poisoning
  botulism **133–4**, **423**, 428
  narcotic bait 138, 139–40, 401, 402, **421–2**
  toxic algae 250
Poland 66, 93, 99, 347, 363
police 397, 400
Polruan, Cornwall 379
polymerase chain reaction 366
Pons, Jean-Marc 3, 8, 125
population estimates, Britain and Ireland 11–12 *see also under individual species*
Port Isaac, Cornwall 379
Port of Tyne Authority 117
Porter, Julie 287–8
Portland, Maine 374
Portugal 10, 159, 181, 190, 196, 217, 235, 306
potato chips 79, 167, 203
Pradel, Roger 87
predation 428
  bird 25, 38, 98, 110, 193, 202, 269, 320
  mammal 25, 48, 52, 56–7, 73, 74, 86, 108, 110, 183, 184, 320, 396, 415
Prévot-Julliard, Anne-Caroline 87
Pribilof Islands, Pacific Ocean 343
Priest Island, Scotland 147
Procellariiformes 31 *see also* petrels; shearwaters
protection 128, 418–20
Protection of Birds Act (1954) 419
Pylewell Marsh, Hampshire 59

radio-telemetry tracking 364
Radipole Lake, Dorset 347
RAF 131
Rainham landfill site, London 347, 353
Rankin, Neal 281, 308
rats 52, 76, 238, 415
Rattiste, Kalev 110
Raven, Susan 117, 125, 370
Ravenglass, Cumbria **52–3**, 57, 59, **74**, 86
Red List
  criteria 429, 430
  gulls on 429, 431–3
Redfern, Chris 290
refuse collection 203, 393, 427 *see also* landfill sites
regurgitation 39, 75, 77, 139–40, 174, 203, 219, 220, 255
reservoirs 41, 49, 51, 57, 112, 118, 119, 277, 278, 280, 307, 349
Ribble Estuary, Lancashire 56, 183, 213, 425
rice fields 350
Richardson, John 126
ringing 13, 87
  aluminium rings 356, 357–8, **359–60**
  attempts to stop 357
  capture for 360–1
  colour rings 361–3
  Darvic rings **362–3**
  Monel rings 87, **358–60**
Rinsey Head, Cornwall 276
'The risk of extinction for birds in Great Britain' (*British Birds*) 430–1
road kill 115, 202
Robinson, Robert 146–7
Rock, Peter 125
Rockcliffe Marsh, Cumbria 183
roosting 41–2, 371
Rosyth, Cumbria 132
Rotterdam, Netherlands 23
Royal Albert Memorial Museum, Exeter 310, 352
Royal Society for the Prevention of Cruelty to Animals (RSPCA) 370
  advice on urban nesting gulls 403
Royal Society for the Protection of Birds (RSPB) 370, 419
  advice on urban nesting gulls 402
  culling 131, 401
  origins 419
Royle, Nick 424
Russia
  Common Gulls 106, 108, 109, 112
  Great Black-backed Gulls 211, 213, 227
  Herring Gulls 12, 121, 130, 161, 164, 313, 363
  Kittiwakes 284, 289
  Lesser Black-backed Gulls 191, 192, 203
  rare visitors from 314, 352, 363
  see also Siberia

St Abbs Head, Scotland 251, **270–1**
St Aldheim's Head, Dorset 276
St Leonards Pier, Sussex 379
Salomonsen, Finn 326
salt excretion 40
Saltholme reserve, Co. Durham **56**
sand dunes 49, 71, 238
sandeels 223, 269, **289–90**, 293, 431–2

Sands of Forvie *see* Ythan Estuary
Sangster, George 190
satellite tracking 365
Saudi Arabia 352
Saunders, David 370
Scandinavia
  Common Gulls 106, 109
  Great Black-backed Gulls 213, 227
  Herring Gulls 12, 18, 121, 130, 134, 161, 163, 164, 170, 213, 313, 348, 349, 401
  Lesser Black-backed Gulls 191, 199
  White-tailed Eagles 25
  see also individual countries
Scarborough, Yorkshire 401
Schiøler, Eiler L. 189
Scilly Isles 214, 276, 415
Scotland
  Black-headed Gulls 48, 50, 51, 55, 58, 70, 89, 390
  Common Gulls 106, 108, 109–10, 113, 389
  Great Black-backed Gulls 211, 214, 215, 222, 225, 229, 389
  Herring Gulls 126, 131, 132, 160, 161, 163–4, 379, 391, 392, 398
  Kittiwakes 235, 238, **243–4**, **250–1**, 252, 280, 286, 370
  Lesser Black-backed Gulls 183, 184, 186, 189, 196, 202, 381, 393
  Mediterranean Gulls 93
  see also Hebrides; Isle of May; Orkney; Shetland
Scott, Sir Peter 361
Scottish Natural Heritage 125, 412, 415 *see also* Nature Conservancy Council
Scoulton Mere, Norfolk 58
Sea Birds Preservation Act (1869) 418–19
Seabird Colony Register, UK 50, 51, 59, 60
Seabird Group 370
Seabird Monitoring Programme (JNCC)
Seabirds (*New Naturalist* 28) 235
The Seabirds of Britain and Ireland (Cramp *et al.*) 370
Seaham Harbour, Co. Durham 387
seals 126, 219, 223, 314
Seamew Crag, Windermere 71
Second World War 59, 379
Sellers, Robin 71
Senegal 331
senility 87
Serbia 99
Severn Estuary Gull Group 204
sewage
  outfalls 79, 83, 116, 117, **423**
  treatment 116, 117
sexes, differences between 15–18
sexing 17–18, **365–8**
shearwaters 211, 219–20, 358, 429
sheep 219
Shetland, Scotland 106, 164, 184, 211, 221, 222, 269, 276, 293, 317, 320, 324, 370, 379, 431
Shiant Isles, Outer Hebrides 415
shopping complexes 1, 45, 342, 394
shotguns 126, 355, 416
Shrubb, Michael 46
Siberia 109, 192, 232, 302–3, 305, 306, 316, 329, 353
silage production 79, 86, 203
size variation
  between species **13–14**, 17
  within species **14–17**
Skokholm, Wales 24, 125, 155, 200, 211, 217, 220, 229
Skomer, Wales
  culling **132**, 183, 211, 220
  Great Black-backed Gulls 211, 217, 219–20, **221**, 228–9
  Herring Gulls 24, **132**, 147, 155, 395
  Kittiwakes 271, **272**, 276, 284, 432
  Lesser Black-backed Gulls 24, 193, **206**, 207
skuas 2, 16, 25, 126, 222, 269, 286, 418, 419, 427
slurry 116
Smith, Neil 327
Smith, Wez 96
Snowdon, Wales 35
soaring 35
Society for the Protection of Birds 419 *see also* Royal Society for the Protection of Birds (RSPB)
Solway marshes, England 406
South Africa 331
South America 338, 339, 354
South Shields, Tyne and Wear 157, 160, 391, 392, 395, 396–7
Southampton Island, Canada 325, 327
Southern, Henry ('Mick') 369
Spaans, Arie 125
Spain
  Black-headed Gulls 44
  Great Black-backed Gulls 217

Herring Gulls 129, 156, 159
Kittiwakes 235
Lesser Black-backed Gulls 181, 190, 196
Mediterranean Gulls 93, 99, 101
rare gulls 306, 310, 328, 340–1, 345, 349, 350, 354, 373
Yellow-legged Gulls 297, 390
Special Protection Areas (SPAs) 129, 243, **244**, 276, 434
speciation 4, 7
species
number, Britain and Ireland 8
number, variation with latitude **5–7**
number, worldwide 4–5
plumage variation within 2–3, **20–1**
size variation between **13–14**, 17
size variation within **14–17**
see also subspecies
The Spectator 58
Spencer, Robert 358
spikes, nest prevention 383, **384**, 408
Spitsbergen 289, 314, 342
sports fields 1, 45, 82, 83, 334
Staithes, Yorkshire 379, 391
Stanford, Norfolk 58
Stanley, Peter 125
status, Britain and Ireland species 9, 431–7
The Status of Seabirds in Britain and Ireland (Lloyd *et al.*) 370
sterilisation 410
Stewart, Peter 204
Straight Point, Devon 276
subspecies 2–3, 4, 5, 12, 108–9, 189–92, 232, 324, 326–7, 348, 353
Suffolk 106, 305
Sunbiggin Tarn, Cumbria **53–4**
Sunderland, Tyne and Wear 157, 160, 391, 395, 397–8
survival rates 12–13
Sussex 298, **299**, 334, 361, 352
Sutcliffe, S. J. 125
Sutton, G. M. 327
Svalbard 211, 318, 322
Svensson, Lars 311
Swansea Bay, Wales 341
Sweden 66, 93, 109, 112, 190, 238, 305 *see also* Scandinavia
Switzerland 106
Sykes, Christopher 418
tagging, wing 364
Taiwan 353
Tapper, Stephen 48
Tarnbrook Fell, Lancashire 193, 194, 202, 421
gull control 132, 183, 184, 211, 409, 420, **421–6**
Taverner, Percy 326
taxonomy 2–3, 8, **10**
Tay, Firth of 303
Teesdale, Upper *see* Upper Teesdale National Nature Reserve
Teesmouth 82, 323
Tegetmeier, William 417
telescopes 355–6
terns 2, 34, 38
conservation concern 428
human exploitation 416
variation of species number with latitude **6–7**
see also under individual species
Thames, river 45
thermals 35
Thomas, Callum 125
Tinbergen, Niko 3, 52, 124
Tittensor, Ruth 58
Toronto, Canada 374
Torquay, Devon 379
Torrey Canyon, SS 370
Torrey Canyon Appeal Fund 370
Towan Head, Cornwall 276
tracking systems
Global Positioning System (GPS) tags 365
light-level geolocators 364
radio-telemetry 364
satellite 365
Tristram, Canon Henry 417
Turkestan 352
Turkey 349, 350
Turner, Daniel 386
'twitchers' 436
Tyne Bridge **383–4**, 386, 387
Tyne, river
Common Gulls feeding along **117–18**
Kittiwake colonies 246, **273–4**, 276, 287–8, 289, **383–9**
see also North Shields; South Shields
Tynemouth, Tyne and Wear 82, 385

UK Seabird Colony Register 50, 51, 59, 60
Ukraine 99, 281

United Utilities plc 421
Upper Teesdale National Nature Reserve, Co. Durham 59, 60
urban nesting
censuses 370–1, 372, 380–1
control 373, 407–14
control, history of **399–407**
control, limitations to 410–11, 426
history of, Britain **374–7**, 379
problems caused by 1, 168, 379, **396–9**
sites 390–2
sites, attraction to gulls **392–6**
species, Britain and Ireland **378–90**
see also under individual species
USA 2, 129, 147, 181, 211, 225, 323, 337, 344, 374, 395

vagrants 8, 9, 309, 352, 353, 429, 430, 436
van Dijk, Klaas 87
Vaurie, Charles 308, 346
Vendée, France 332
Venezuela 332
Verbeek, Nicolaas 125
Vernon, J. D. R. 115
Vinicombe, Keith 345
Voesten, Rob 87
Volga River 352
Vologda, Russia 109
von Blotzheim, Urs Glutz 229

Wadden Sea, North Sea 202
waders 2, 67, 78, 80, 116, 202, 425, 435
Wales
Black-headed Gulls 50, 51, 70, 89
Common Gulls 106
Great Black-backed Gulls 211, 213, 214, 235, 389
Herring Gulls 126, 132, 160, 161, 379, 395
Kittiwakes 235, 243, 252
Lesser Black-backed Gull 183, 186, 188–9, 380, 382
Mediterranean Gull 93, 101
rare gulls 303, 305, 341, 343, 436
see also Skokholm; Skomer
walking 34–5
Walney Island, Cumbria 125, 141–2, 147, 183, 196, 200
Walpole-Bond, H. 352
water contamination 421, **422–4** *see also* botulism
Waterford, County 379
waterfowl 38
Watership Down (Adams) 45
Wear, river 45
weight
variation between sexes **15–16**
variation between species **13–14**
West Midlands
Caspian Gulls **348–9**
Glaucous Gulls **320–2**
Herring Gulls 161, **163**
Kittiwakes **277–8**
Iceland Gulls **324–5**
Little Gulls 306, **307**
Mediterranean Gulls **93–4**
Yellow-legged Gulls 298, **299**
West Pier, Brighton 379
West Riding, Yorkshire 418
Westcliff-on-Sea, Essex 341
Wexford, County 93, 353
Whitby, Yorkshire 168, 379, 391, 401
White, Edward 27
White Sea, Siberia 112, 192, 200
Wildfowl Trust 361
Wildlife and Countryside Act 1981 402
Wildlife (Northern Ireland) Order 1985 402
wind
farms, offshore 203, 204
strength and feeding behaviour **225–6**, 285–7
wing
adaptation to flight over water 35
length, difference between sexes **15–16**
'mirrors' 1, 103, **104**, 105, 111, 121, 295–6, 313, 328, 334, **337**, 354
tags, patagial 364
Winsper, Jim 383
Wisconsin, USA 303
Witherby, Harry 190, 355, 356
wrecks 278, 279, 280, 331–2
Wynne-Edwards, Vero 281

Yesou, Pierre 331–2
Yorkshire 60, 108, 305, 317
Ythan Estuary (Sands of Forvie), Scotland **55**
Ytreberg, Nils 75

Zeebrugge, Belgium **194**